D0845106

Natural Computing Series

Springer
Berlin
Heidelberg
New York
Barcelona
Hong Kong
London
Milan
Paris
Singapore
Tokyo

Hans-Georg Beyer

The Theory of Evolution Strategies

With 52 Figures and 9 Tables

Springer

Author

Dr. Hans-Georg Beyer

Department of Computer Science
University of Dortmund
44221 Dortmund, Germany
beyer@ls11.cs.uni-dortmund.de

Series Editors

G. Rozenberg (Managing Editor)
Th. Bäck, A.E. Eiben, J.N. Kok, H.P. Spaink

Leiden Center for Natural Computing
Leiden University
Niels Bohrweg 1
2333 CA Leiden, The Netherlands
rozenber@cs.leidenuniv.nl

Library of Congress Cataloging-in-Publication Data applied for

Die Deutsche Bibliothek – CIP-Einheitsaufnahme

Beyer, Hans-Georg:
The theory of evolution strategies/Hans-Georg Beyer. –
Berlin; Heidelberg; New York; Barcelona; Hong Kong; London;
Milan; Paris; Singapore; Tokyo: Springer, 2001
(Natural computing series)
ISBN 3-540-67297-4

ACM Computing Classification (1998): G.1.6, F.1.2, G.3–4, I.2.8, J.3

ISBN 3-540-67297-4 Springer-Verlag Berlin Heidelberg New York

Springer-Verlag Berlin Heidelberg New York,
a member of BertelsmannSpringer Science+Business Media GmbH
http://www.springer.de

Cover Design: KünkelLopka, Heidelberg
Typesetting: Camera ready by the author
Printed on acid-free paper SPIN 10761860 45/3142SR – 5 4 3 2 1 0

In Memory of my Parents

Preface

> *Most of us – we should be aware of this –* love *our hypotheses, and it is, as I said once, a painful, though a rejuvenating and healthy, early-morning exercise, to throw daily a favorite hypothesis overboard.*
>
> KONRAD LORENZ

Evolutionary Algorithms (EA), such as Evolution Strategies (ES), Genetic Algorithms (GA), and Evolutionary Programming (EP), have found a broad acceptance as robust optimization algorithms in the last ten years.

The idea of optimizing systems through imitating nature and applying the "genetic operators" such as selection, mutation, and recombination has a certain appeal. First of all, the fundamental algorithms are attractively simple: In principle, every secondary school pupil can understand and program them. Secondly, these algorithms are quite universal; that is, they can be applied with relatively minor changes and without problem-specific *a priori* knowledge to the different optimization problems. To put it another way: EAs give – just like the biological evolution itself – successes in the short term.

In contrast to the wide propagation and the resulting practical prosperity, the theoretical analysis of EAs has not progressed that much. The aim of this monograph is to provide a theoretical framework for the ES research field.

The analysis applied in this book is based on viewing the evolutionary algorithm as a *dynamical system*, i.e. a primary aim of the analysis is to make *predictions* on the temporal behavior of the evolutionary system

ES & FITNESS FUNCTION.

This presupposes the formulation of the equations of motion in the search space, which describe the change in the system state over time and necessitates the computation of the driving microscopic "forces."

The central quantity of such an analysis is the progress rate φ, introduced by Rechenberg (1973). This is a measure for the state change of the system toward the optimum. Most of the investigations in this book aim at the approximate calculation of φ. If this calculation succeeds, then the dynamics of the system can be described in general. Knowing the system dynamics, the questions concerning the time complexity of the system or the (local) convergence order can be answered.

This monograph contains seven chapters.[1] The first two chapters have an introductory character. Particularly, the first chapter serves – among other aims – to give the reader the author's personal "EA philosophy," without

[1] A detailed overview of the contents can be found at the end of the first chapter (Sect. 1.5).

giving the elaborate mathematical substantiation. These proofs – as far as possible for today – are derived in the following chapters. The second chapter serves as the starting point: the essential notions, concepts, and also the general framework of the analysis are defined there. The third chapter is dedicated to the progress rate of the $(1 \overset{+}{,} \lambda)$-Evolution Strategies for the sphere model. Both the N-dependent (N is the search space dimension) and asymptotic progress rate formulae are derived; moreover, the effect of disturbances of the fitness measurement process is investigated. The fourth chapter follows with the analysis of the quality gain of the same strategies. The fifth chapter is devoted to the technically difficult theoretical problem of population formation in (μ, λ)-strategies. In the sixth chapter, the recombination operator is investigated. The question of the actual use of "sex" is followed up there. The statistical error correction ("genetic repair," or GR for short) occurring in these strategies will prove to be the reason for the performance gain. This remarkable finding suggests that the performance-increasing effect of recombination is due to the extraction of similarities from the parents. This is in contrast to the commonly believed building block hypothesis, which offers the combination of "above average traits" as the performance-increasing source of recombination. The seventh chapter is devoted to the analysis of σ-self-adaptation using the $(1, \lambda)$-ES as an example.

No special knowledge is required for reading and understanding this book. However, a certain familiarity with the Evolutionary Algorithms in general and with the Evolution Strategies in particular is recommended.[2] Since ES is a probabilistic algorithm, the utilization of probability theory is inevitable, but familiarity with the knowledge offered in standard textbooks suffices. Special mathematical techniques beyond these are handled in the appendices. The same holds for some complicated integrals and approximation techniques. One intention of writing this book was to free the reader from the burdensome searching and ploughing through the special literature.

The contents of this monograph mirrors the state of the research in ES theory. However, establishing crossreferences to the "relatives," i.e. to Genetic Algorithms, is also intended. Future developments will show to what extent the common EA working principles postulated in this work will prove universal.

The research area "theoretical analysis of Evolutionary Algorithms" is still almost uncharted; there are more open problems than solved ones. Therefore, the book contains references to open problems, to new problem formulations, and to future research directions at the relevant places (as well as in the index). By following this road, the reader will be able to get directly to the frontiers of research. Actually, the first version of this monograph (in German) served as a textbook for my Ph.D. students.

[2] The books of Rechenberg (1994) and Schwefel (1995) are recommended for this purpose.

This book is the result of a two-phase process. The first phase started in the early 1990s and was sponsored after a certain initial time by a DFG[3]-Habilitandenstipendium.[4] It finally culminated in the book: "Zur Analyse der Evolutionsstrategien" (in German) and was successfully submitted as a "Habilitationsschrift" to the Computer Science Department, University of Dortmund. For that time period I owe Prof. H.-P. Schwefel, Dortmund, a debt of gratitude. Furthermore, my thanks go to Dipl.-Ing. M. Wolle, Rödermark-Urberach, as well as Prof. H. W. Meuer, Director of the Computing Center at the University of Mannheim. I also thank the DFG, especially Dr. A. Engelke for his support and also the anonymous referees for their cheering and supporting remarks.

The second phase began when my former Ph.D. student, Dr. A. I. Oyman, started working on ES theory. He was the first to study ES theory from my book. As an exercise, he translated the German version into English. While his translation has gone through many changes, it was the basis for the final version. This final version reflects the comments, suggestions, and influence of many colleagues who have read former versions of this book. I am especially grateful to Dr. D. B. Fogel, San Diego, California, and Dipl.-Inform. D. V. Arnold, Dortmund. I am also grateful to Prof. D. E. Goldberg, Urbana-Champaign, Illinois, for his critical comments. Last, but not least, I want to express my thanks to Prof. G. Rozenberg, Leiden, the Managing Editor of the "Natural Computing Series" and especially to Dr. T. Bäck, Dortmund and Leiden, Series Editor, for his personal support in getting my book published in this series.

Bochum, Spring 2001 Hans-Georg Beyer

---

[3] DFG – Deutsche Forschungsgemeinschaft (German Research Foundation)

[4] Scholarship to qualify for a professorship, DFG grant No. Be1578/1.

On the Symbols Used

A list of all symbols and notation used is avoided here. Instead, the most important conventions will be introduced. The rest can be understood from the corresponding context. Moreover, references to important symbols can be found in the index.

Vectors are displayed as boldface, lower case Latin letters (e.g. $\mathbf{y}$), and their components in normal italic type with subscripts (e.g. y_1). The same holds for matrices, which are presented as boldface, capital Latin letters. $\mathbf{E}$ is the unit matrix. The unit vectors are indicated using[5] $\mathbf{e}$ and have a subscript, where appropriate, to characterize the direction. The transposition of vectors and matrices is marked with the superscript "^{T}". The gradient is usually symbolized by the nabla operator ∇

$$\mathrm{grad}_x F(\mathbf{x}) = \nabla_x F := \sum_{i=1}^{N} \mathbf{e}_i \frac{\partial}{\partial x_i} F(\mathbf{x}).$$

The generation count in the formulae is indicated by superscripts with parentheses "$^{(g)}$". This form of using g denotes a description of a system in the *time discrete* domain. Time-continuous descriptions are used as approximations. In this case, the notation "$\cdot(g)$" is used instead of the superscript "$^{\cdot(g)}$" (e.g. $R^{(g)} \to R(g)$). The generation counter is symbolized almost exclusively by g and a time interval of generations by G.

The superscript with parentheses is used in addition to the generation counters and also by the progress coefficients and progress rates in order to specify the kth order. The "E" symbol with curly brackets denotes the expected value of a random variate f in the probability theory. Alternatively, the random variate f will also be expressed with an overline

$$\overline{f} \equiv \mathrm{E}\{f\}.$$

The variance of a random variate is expressed as $\mathrm{D}^2\{f\}$.

The probability density functions are symbolized only with $p(\cdot)$, and the cumulative distribution functions with $P(\cdot)$. The inverse function of $P(\cdot)$, the quantile function, is denoted as $P^{-1}(\cdot)$.

[5] Exceptions are the Euler number ($\mathrm{e} \approx 2.71828\ldots$) and the generalized progress coefficients $e^{\alpha,\beta}_{\mu,\lambda}$.

Dirac's delta function $\delta(x)$ can be regarded as a very special probability density function. In the one-dimensional case one has

$$\int_{-\infty}^{\infty} \delta(x - x_0)\,\mathrm{d}x = 1 \qquad \text{and} \qquad \int_{-\infty}^{\infty} f(x)\,\delta(x - x_0)\,\mathrm{d}x = f(x_0).$$

Occasionally, the symbol δ is used in this book also to characterize (small) differences. In contrast to the delta function, the respective variable is written after the δ as usual, and not in parentheses. Alternatively, the capital delta Δ is used in some places to indicate differences. A confusion with the N-dimensional Laplace operator Δ_y

$$\Delta_y := \sum_{i=1}^{N} \frac{\partial^2}{\partial y^2}$$

is avoided, since this will be used in Sect. 2.3.2.3 only.

The (Gaussian) normal distribution plays an important role in the Evolution Strategy. The normally distributed random variables are symbolized as $\mathcal{N}(\overline{x}, \sigma^2)$,[6] where $\overline{x}$ is the expected value and σ^2 the variance. In the multidimensional case, the notation $\mathcal{N}(\overline{\mathbf{x}}, \mathbf{C})$ is used, where $\mathbf{C} = \mathbf{C}^{\mathrm{T}}$ stands for the correlation matrix and $\overline{\mathbf{x}}$ for the expectation vector.

The mean values of the populations and the vectors are marked with angular braces. A subscript characterizes the sample size where appropriate

$$\langle x \rangle := \frac{1}{\mu} \sum_{m=1}^{\mu} a_m, \qquad \langle \mathbf{y} \rangle_\rho := \frac{1}{\rho} \sum_{r=1}^{\rho} \mathbf{y}_r.$$

The equality of asymptotic calculations is symbolized by "$\sim$", e.g.

$$\sqrt{N - \sqrt{N}} \sim \sqrt{N}.$$

It is used when the relative error vanishes for $N \to \infty$. If also the absolute error goes to zero, we will use "$\simeq$", e.g.

$$\sqrt{N - 1} \simeq \sqrt{N}.$$

The error terms in series approximations are denoted mostly by $\mathcal{R}$ and with a subscript where appropriate.

References to equations occur principally in the parenthesized form "(X.Y)", where "X" stands for the chapter and "Y" for the continuous number in the chapter.

To simplify the text flow, the abbreviations (such as ES, EA, GA, etc.) are normally not declined.

[6] $\mathcal{N}$ should not be mixed up with N; the latter always stands for the dimension of the parameter space.

Contents

1. **Introduction** 1
1.1 A Short Characterization of the EA 1
1.2 The Evolution Strategy 4
1.2.1 The $(\mu/\rho \overset{+}{,} \lambda)$-ES Algorithm 4
1.2.2 The Genetic Operators of the ES 7
1.2.2.1 The Selection Operator 8
1.2.2.2 The Mutation Operator 9
1.2.2.3 The Reproduction Operator 11
1.2.2.4 The Recombination Operator 12
1.3 The Convergence of the Evolution Strategy 14
1.3.1 ES Convergence – Global Aspects 15
1.3.2 ES Convergence – Local Aspects 17
1.4 Basic Principles of Evolutionary Algorithms 18
1.4.1 Evolvability 18
1.4.2 EPP, GR, and MISR 20
1.4.2.1 EPP – the Evolutionary Progress Principle 20
1.4.2.2 GR – the Genetic Repair Hypothesis 21
1.4.2.3 The MISR Principle 21
1.5 The Analysis of the ES – an Overview 22

2. **Concepts for the Analysis of the ES** 25
2.1 Local Progress Measures 25
2.1.1 The Quality Gain $\overline{Q}$ 26
2.1.2 The Progress Rate φ 28
2.1.3 The Normal Progress φ_R 30
2.2 Models of Fitness Landscapes 31
2.2.1 The (Hyper-)Sphere Model 32
2.2.2 The Hyperplane 33
2.2.3 Corridor and Discus – Hyperplanes with Restrictions 34
2.2.4 Quadratic Functions and Landscapes of Higher-Order 35
2.2.5 The Bit Counting Function `OneMax` 36
2.2.6 Noisy Fitness Landscapes 36
2.3 The Differential-Geometrical Model for Non-Spherical Fitness Landscapes 37

2.3.1 Fundamentals of the Differential-Geometrical Model .. 38
2.3.1.1 The Local Hyperplane $\partial Q_{\mathbf{y}}$ ... 38
2.3.1.2 The Metric Tensor $g_{\alpha\beta}$ ... 39
2.3.1.3 The Second Fundamental Form ... 40
2.3.1.4 The Mean Curvature $\langle\varkappa\rangle$ ... 41
2.3.2 The Calculation of the Mean Radius $R_{\langle\varkappa\rangle}$ ... 43
2.3.2.1 The Metric Tensor ... 43
2.3.2.2 The $b_{\alpha\beta}$-Tensor ... 44
2.3.2.3 The Computation of the Mean Radius $R_{\langle\varkappa\rangle}$ .. 45
2.4 The ES Dynamics ... 47
2.4.1 The R-Dynamics ... 48
2.4.2 The Special Case $\sigma^*(g) = \text{const}$ ... 49

3. The Progress Rate of the $(1 \overset{+}{,} \lambda)$-ES on the Sphere Model 51
3.1 The Exact $(1+1)$-ES Theory ... 51
3.1.1 The Progress Rate without Noise in Fitness Measurements ... 54
3.1.2 The Progress Rate at Disturbed Fitness Measurements 57
3.2 Asymptotic Formulae for the $(1 \overset{+}{,} \lambda)$-ES ... 61
3.2.1 A Geometrical Analysis of the $(1+1)$-ES ... 61
3.2.1.1 The Asymptote of the Mutation Vector $\mathbf{z}$ ... 62
3.2.1.2 The Progress Rate of the $(1+1)$-ES on the Sphere Model ... 64
3.2.1.3 The Success Probability P_{s1+1} and the EPP . 68
3.2.2 The Asymptotic $\varphi^*_{1 \overset{+}{,} \lambda}$ Integral ... 69
3.2.3 The Analysis of the $(1,\lambda)$-ES ... 71
3.2.3.1 The Progress Rate $\varphi^*_{1,\lambda}$ ... 71
3.2.3.2 The Progress Coefficient $c_{1,\lambda}$ ... 74
3.2.4 The Analysis of the $(1+\lambda)$-ES ... 77
3.2.4.1 The Progress Rate ... 77
3.2.4.2 The Success Probability $P_{s1+\lambda}$ ... 79
3.2.4.3 The Progress Function $d^{(1)}_{1+\lambda}(x)$ ... 79
3.3 The Asymptotic Analysis of the $(\tilde{1} \overset{+}{,} \tilde{\lambda})$-ES ... 80
3.3.1 The Theory of the $(\tilde{1} \overset{+}{,} \tilde{\lambda})$ Progress Rate ... 80
3.3.1.1 Progress Integrals and Acceptance Probabilities ... 80
3.3.1.2 The Asymptotic Fitness Model and the $p(\tilde{Q}_{|x}|x)$ Density ... 83
3.3.1.3 The Calculation of $P_1(\tilde{Q})$ ... 85
3.3.2 On the Analysis of the $(\tilde{1},\tilde{\lambda})$-ES ... 86
3.3.2.1 The Asymptotic $\varphi^*_{\tilde{1},\tilde{\lambda}}$ Formula ... 86
3.3.2.2 The Dynamics of the $(\tilde{1},\tilde{\lambda})$-ES ... 89
3.3.3 On the Analysis of the $(\tilde{1}+\tilde{\lambda})$-ES ... 93

3.3.3.1 The Asymptotic $\varphi^*_{\tilde{1}+\tilde{\lambda}}$ Integral and $P_{s\tilde{1}+\tilde{\lambda}}$ 93
3.3.3.2 An (Almost) Necessary Evolution Criterion for $(\tilde{1}+\tilde{\lambda})$-ES 95
3.3.3.3 Some Aspects of the $(\tilde{1}+\tilde{1})$-ES 97
3.3.4 Convergence Improvement by Inheriting Scaled Mutations 102
3.3.4.1 Theoretical Fundamentals 102
3.3.4.2 Discussion of the $(\tilde{1},\tilde{\lambda})$-ES 103
3.4 The N-Dependent $(1,\lambda)$ Progress Rate Formula 104
3.4.1 Motivation 104
3.4.2 The $p(r)$ Density 105
3.4.3 The Derivation of the φ^* Progress Rate Formula 107
3.4.4 Comparison with Experiments 109
3.4.5 A Normal Approximation for $p(r)$ 111

4. The $(1 \overset{+}{,} \lambda)$ Quality Gain 113
4.1 The Theory of the $(1 \overset{+}{,} \lambda)$ Quality Gain 113
4.1.1 The $\overline{Q}_{1\overset{+}{,}\lambda}$ Integral 113
4.1.2 On the Approximation of $P_z(z)$ 115
4.1.3 On Approximating the Quantile Function $P_z^{-1}(f)$ 118
4.1.4 The $\overline{Q}_{1,\lambda}$ Formula 119
4.1.5 The $\overline{Q}_{1+\lambda}$ Formula 120
4.2 Fitness Models and Mutation Operators 122
4.2.1 The General Quadratic Model and Correlated Mutations 122
4.2.2 The Special Case of Isotropic Gaussian Mutations 125
4.2.3 Examples of Non-Quadratic Fitness Functions 126
4.2.3.1 Biquadratic Fitness with Isotropic Gaussian Mutations 126
4.2.3.2 The Counting Ones Function `OneMax` 127
4.3 Experiments and Interpretations of the Results 131
4.3.1 Normalization 131
4.3.1.1 Quadratic Fitness, Isotropic Gaussian Mutations and the Differential-Geometric Model 131
4.3.1.2 The Normalization for the Biquadratic Case 134
4.3.2 Experiments and Approximation Quality 135
4.3.3 Quality Gain or Progress Rate? 137

5. The Analysis of the (μ,λ)-ES 143
5.1 Fundamentals and Theoretical Framework 143
5.1.1 Preliminaries 143
5.1.2 The (μ,λ) Algorithm and the $\varphi_{\mu,\lambda}$ Definition 144
5.1.3 The $\varphi_{\mu,\lambda}$ Integral 146

5.1.4 Formal Approximation of the Offspring Distribution $p(r)$ 147
5.1.5 Estimation of $\overline{r}$, s, and γ, and of $\overline{r}_{|R_m}$, $M_{2|R_m}$, and $M_{3|R_m}$ 150
5.1.6 The Simplification of $\overline{r}_{|R_m}$, $M_{2|R_m}$, and $M_{3|R_m}$ 153
5.1.7 The Statistical Approximation of $\overline{r}$, s, and γ 156
5.1.8 The Integral Expression of $\overline{\langle(\Delta R)^2\rangle}$ 159
5.1.9 The Integral Expression of $\overline{\langle(\Delta R)^3\rangle}$ 162
5.1.10 Approximation of the Stationary State – the Self-Consistent Method 166
5.2 On the Analysis in the Linear γ Approximation 168
5.2.1 The Linear γ Approximation 168
5.2.2 The Approximation of the $\varphi_{\mu,\lambda}$ Integral 169
5.2.3 The Approximation of $s^{(g+1)}$ 172
5.2.3.1 The I_A Integral 173
5.2.3.2 The I_B Integral 175
5.2.3.3 Composing the $s^{(g+1)}$ Formula 176
5.2.4 The Approximation of $\gamma^{(g+1)}$ 176
5.2.4.1 The I_C Integral 177
5.2.4.2 The I_D Integral 178
5.2.4.3 The I_E Integral 181
5.2.4.4 Composing the $\gamma^{(g+1)}$ Formula 182
5.2.5 The Self-Consistent Method and the $\varphi^*_{\mu,\lambda}$ Formulae 183
5.3 The Discussion of the (μ, λ)-ES 185
5.3.1 The Comparison with Experiments 185
5.3.1.1 Data Extraction from ES Runs 185
5.3.1.2 The ES Simulations for $\sigma = \text{const}$ *and* $\sigma^* = \text{const}$ 186
5.3.2 Simplified $\varphi^*_{\mu,\lambda}$ Formulae and the Progress Coefficient $c_{\mu,\lambda}$ 188
5.3.2.1 The Derivation of the $\varphi^*_{\mu,\lambda}$ Formulae 188
5.3.2.2 EPP, the Properties of $c_{\mu,\lambda}$, and the Fitness Efficiency 189
5.3.3 The (μ, λ)-ES on the Hyperplane 192
5.3.4 The Exploration Behavior of the (μ, λ)-ES 194
5.3.4.1 Evolution in the (r, γ_i) Picture 194
5.3.4.2 The Random Walk in the Angular Space 195
5.3.4.3 Experimental Verification and Discussion 199
5.3.4.4 Final Remarks on the Search Behavior of the (μ, λ)-ES 199

6. The $(\mu/\mu, \lambda)$ Strategies – or Why "Sex" May be Good 203
6.1 The Intermediate $(\mu/\mu_I, \lambda)$-ES 203
6.1.1 Foundations of the $(\mu/\mu, \lambda)$ Theory 204
6.1.1.1 The $(\mu/\mu_I, \lambda)$ Algorithm 204

6.1.1.2 The Definition of the Progress Rate $\varphi_{\mu/\mu,\lambda}$ ... 205
6.1.1.3 The Statistical Approximation of $\varphi_{\mu/\mu,\lambda}$ 206
6.1.2 The Calculation of $\varphi_{\mu/\mu_I,\lambda}$ 210
6.1.2.1 The Derivation of the Expected Value $\overline{\langle x \rangle}$.... 210
6.1.2.2 The Derivation of the Expected Value $\overline{\langle \mathbf{h}^2 \rangle}$... 213
6.1.2.3 The $(\mu/\mu_I, \lambda)$ Progress Rate 215
6.1.3 The Discussion of the $(\mu/\mu_I, \lambda)$ Theory..... 216
6.1.3.1 The Comparison with Experiments and the Case $N \to \infty$..... 216
6.1.3.2 On the Benefit of Recombination – or Why "Sex" May be Good 218
6.1.3.3 System Conditions of Recombination 222
6.1.3.4 The $(\mu/\rho_I, \lambda)$-ES and the Optimal μ Choice .. 224
6.1.3.5 The Exploration Behavior of the $(\mu/\mu_I, \lambda)$-ES 229
6.2 The Dominant $(\mu/\mu_D, \lambda)$-ES 232
6.2.1 A First Approach to the Analysis of the $(\mu/\mu_D, \lambda)$-ES . 232
6.2.1.1 The $(\mu/\mu_D, \lambda)$-ES Algorithm and the Definition of $\varphi_{\mu/\mu_D,\lambda}$ 232
6.2.1.2 A Simple Model for the Analysis of the $(\mu/\mu_D, \lambda)$-ES 233
6.2.1.3 The Isotropic Surrogate Mutations 235
6.2.1.4 The Progress Rate $\varphi^*_{\mu/\mu_D,\lambda}$ 237
6.2.2 The Discussion of the $(\mu/\mu_D, \lambda)$-ES 239
6.2.2.1 The Comparison with Experiments and the Case $N \to \infty$..... 239
6.2.2.2 The MISR Principle, the Genetic Drift, and the GR Hypothesis..... 240
6.3 The Asymptotic Properties of the $(\mu/\mu, \lambda)$-ES..... 246
6.3.1 The $c_{\mu/\mu,\lambda}$ Coefficient 247
6.3.1.1 The Asymptotic Expansion of the $c_{\mu/\mu,\lambda}$ Coefficient 247
6.3.1.2 The Asymptotic Order of the $c_{\mu/\mu,\lambda}$ Coefficients 250
6.3.2 The Asymptotic Progress Law 250
6.3.3 Fitness Efficiency and the Optimal ϑ Choice 252
6.3.3.1 Asymptotic Fitness Efficiency $\eta_{\mu/\mu,\lambda}$ of the $(\mu/\mu, \lambda)$-ES 252
6.3.3.2 The Relation to the $(1+1)$-ES 252
6.3.4 The Dynamics and the Time Complexity of the $(\mu/\mu, \lambda)$-ES..... 255

7. The $(1, \lambda)$-σ-Self-Adaptation..... 257
7.1 Introduction 257
7.1.1 Concepts of σ-Control 257

7.1.2 The σ-Self-Adaptation ... 259
7.1.3 The $(1, \lambda)$-σSA Algorithm ... 261
7.1.4 Operators for the Mutation of the Mutation Strength ... 261
7.2 Theoretical Framework for the Analysis of the σSA ... 263
7.2.1 The Evolutionary Dynamics of the $(1, \lambda)$-σSA-ES ... 264
7.2.1.1 The r Evolution ... 265
7.2.1.2 The ς Evolution ... 266
7.2.2 The Microscopic Aspects ... 268
7.2.2.1 The Density $p_{1;1}(r)$ of a Single Descendant ... 269
7.2.2.2 The Transition Density $p_{1;\lambda}(r)$ ... 270
7.2.2.3 The Transition Density $p_{1;\lambda}(\varsigma)$ ... 271
7.2.2.4 The $\varphi^{(k)}$ and $\psi^{(k)}$-SAR Functions ... 272
7.3 Determination of the Progress Rate and the SAR ... 275
7.3.1 Progress Integrals $\varphi^{*(k)}$ ... 275
7.3.1.1 Numerical Examples for the Progress Rate φ^* 275
7.3.1.2 An Analytic φ^* Formula for $p_\sigma = p_{\mathrm{II}}$, $\lambda = 2$ ... 276
7.3.1.3 The $\tau \to 0$ and $\beta \to 0$ Approximation for D_φ^* ... 279
7.3.2 The SAR Functions $\psi^{(k)}$ ... 280
7.3.2.1 Numerical Examples for $\psi^{(1)}$, Discussion, and Comparison with Experiments ... 280
7.3.2.2 An Analytic ψ Formula for $p_\sigma = p_{\mathrm{II}}$, $\lambda = 2$ ... 283
7.3.2.3 Bounds for ψ ... 286
7.3.2.4 Analytic Approximation of $\psi^{(k)}$ for Small ς^* and τ – General Aspects ... 289
7.3.2.5 Approximations for ψ, $\psi^{(2)}$, and D_ψ ... 294
7.4 The $(1, \lambda)$-σSA Evolution (I) – Dynamics in the Deterministic Approximation ... 299
7.4.1 The Evolution Equations of the $(1, \lambda)$-σSA-ES ... 299
7.4.2 The ES in the Stationary State ... 300
7.4.2.1 Determining the Stationary State ... 300
7.4.2.2 Optimal ES Performance and the $1/\sqrt{N}$ Rule 302
7.4.2.3 The Differential Equation of the σ Evolution, the Transient Behavior for Small $\varsigma^{*(0)} < \varsigma_{ss}^*$, and the Stationary r Dynamics ... 303
7.4.2.4 Approaching the Steady-State from $\varsigma^* \gg \varsigma_{ss}^*$ ... 308
7.5 The $(1, \lambda)$-σSA Evolution (II) – Dynamics with Fluctuations ... 309
7.5.1 Motivation ... 309
7.5.2 Chapman-Kolmogorov Equation and Transition Densities ... 309
7.5.3 Mean Value Dynamics of the r Evolution ... 313
7.5.4 Mean Value Dynamics of the ς^* Evolution ... 315
7.5.5 Approximate Equations for the Stationary σ_∞^* State ... 317
7.5.5.1 The Integral Equation of the Stationary $p(\varsigma^*)$ Density ... 317

7.5.5.2 An Approach for Solving the Momentum Equations 319
7.5.6 Discussion of the Stationary State and the $1/\sqrt{N}$ Rule 320
7.5.6.1 Comparison with ES Experiments 320
7.5.6.2 Special Analytical Cases 321
7.5.6.3 The τ-Scaling Rule 323
7.6 Final Remarks on the σ Self-Adaptation 324

Appendices 327

A. Integrals 329
A.1 Definite Integrals of the Normal Distribution 329
A.2 Indefinite Integrals of the Normal Distribution 331
A.2.1 Integrals of the Form $I^{\alpha,\beta}(x) = \int_{-\infty}^{t=x} t^{\beta} \, \mathrm{e}^{-\frac{1+\alpha}{2}t^2} \mathrm{d}t$ 331
A.2.2 Integrals of the Form $I_{\phi}^{\beta}(x) = \int_{-\infty}^{t=x} t^{\beta} \mathrm{e}^{-\frac{1}{2}t^2} \Phi(t) \, \mathrm{d}t$ 332
A.3 Some Integral Identities 334

B. Approximations 337
B.1 Frequently Used Taylor Expansions 337
B.2 The Hermite Polynomials $\mathrm{He}_k(x)$ 339
B.3 Cumulants, Moments, and Approximations 341
B.3.1 Fundamental Relations 341
B.3.2 The Weight Coefficients for the Density Approximation of a Standardized Random Variable 344
B.4 Approximation of the Quantile Function 349

C. The Normal Distribution 351
C.1 $\mathcal{N}(0,1)$ Distribution Function, Gaussian Integral, and Error Function 351
C.2 Asymptotic Order of the Moments of $\frac{x}{R}$ 354
C.3 Product Moments of Correlated Gaussian Mutations 355
C.3.1 Fundamental Relations 355
C.3.2 Derivation of the Product Moments 356

D. $(1,\lambda)$-Progress Coefficients 359
D.1 Asymptotics of the Progress Coefficients $d_{1,\lambda}^{(k)}$ 359
D.1.1 An Asymptotic Expansion for the $d_{1,\lambda}^{(k)}$ Coefficients 359
D.1.2 The Asymptotic $c_{1,\lambda}$ and $d_{1,\lambda}^{(2)}$ Formulae 360
D.1.3 An Alternative Derivation for $c_{1,\lambda}$ 361
D.2 Table of Progress Coefficients of the $(1,\lambda)$-ES 363

References 365

Index 373

1. Introduction

This chapter gives an introduction to the Evolutionary Algorithms (EA) in general and particularly to the Evolution Strategies (ES). Different EA variants are handled under a unified approach and way of thinking. This will be sometimes only possible in an informal as opposed to a theoretically formal manner, since the theoretical analysis of Genetic Algorithms (GA) is still at the stage of finding suitable models of explanation. Therefore, the views presented here result essentially from the analysis of ES models. These are carried out in the following more technical chapters. An overview of the underlying work as a whole is given at the end of this chapter.

1.1 A Short Characterization of the EA

Evolutionary Algorithms are methods which are suggested by the Darwinian paradigm of evolution.[1] The *principle of variation and selection* can be considered as the fundamental principle of the Darwinian evolution. This principle, combined with the change of the generation (reproduction), builds up the fundamental components of the *evolutionary loop* (see Algorithm 1).

At a very abstract level, the evolution can be regarded as a process of selecting structures or states (Beyer, 1989a), where the selection is determined by the fitness properties of the structures. Which structures will be able to "survive," i.e. to reproduce themselves, whether a stable structure will come up (stationary state), or whether periodic/chaotic features will occur, is left open in the framework of this characterization. Nevertheless, Darwinian evolution is often viewed analogous to optimization, and the algorithms inspired from it are applied almost exclusively for the optimization of fitness functions. However, this is not a necessity. The evolutionary algorithm developed by Beyer (1989a, 1990b) for dissipative systems does not perform any optimization; rather, it simulates stationary states in dissipative systems in an evolutionary manner.

---

[1] It is assumed that the reader is familiar with terms such as standard EA, ES, GA, and EP (Evolutionary Programming). The aim here is to present the author's principal "EA Philosophy." A more formal definition of the standard EA used for optimization purposes can be found, e.g., in various survey papers of Schwefel et al. (see, e.g., Bäck & Schwefel, 1993; Bäck et al., 1997).

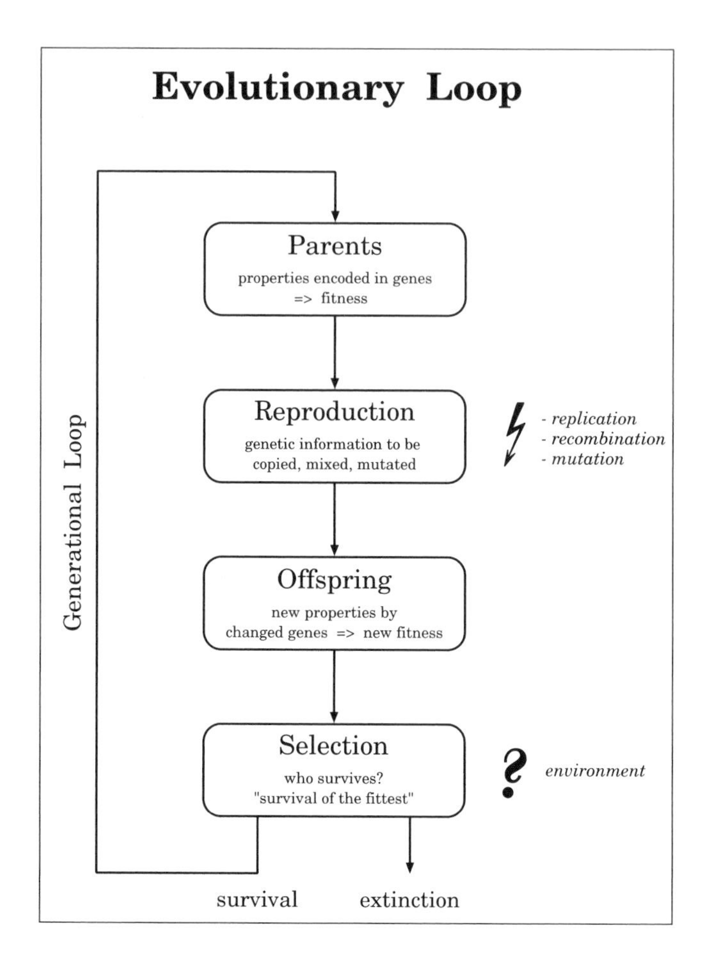

Algorithm 1. The general evolutionary algorithm

Even when applied as an optimization method, the Darwinian principle does not guarantee successful optimization at all. An especially good example is the standard genetic algorithm (sGA) of Holland (1975), also referred to as "simple" GA, which uses fitness proportional selection. Conditioned through the stochastic process of selection and the necessity of mutations, the sGA principally *cannot* be a function optimizer (DeJong, 1992), i.e. for $g \to \infty$ (g: generation counter) there is no guarantee that the system is at the optimum with probability $P = 1$. This leads to one of the central questions of the EA theory, the question of the convergence of the EA, which is tackled in different parts of this work.

As already mentioned, there are different EA classes or "schools." If only the optimization methods are considered, there exist three established main classes of EA:

- Evolution Strategy (ES) (Rechenberg, 1973; Schwefel, 1995),
- Genetic Algorithms (GA) (Holland, 1975; Goldberg, 1989),
- Evolutionary Programming (EP) (Fogel et al.,1966; Fogel, 1992).

Although these classes were easy to distinguish from each other in the beginning phase, it is difficult to build disjunct sets today.[2] Only one differentiation is possible even today – the boundary to the EP: usually, recombination is not used at all in contemporary EP algorithms (see, e.g., Fogel & Stayton, 1994); the mutation is consequently the decisive search operator. On the other hand, some GA proponents regard the mutation operator only as a background operator; however, the number of these adherents seems to be of decreasing tendency.

The ES has an intermediate position. Usually, mutation and recombination have equal importance, as far as real-valued parameter optimization is considered, and they are applied to all individuals by default. The most important differences between the ES and the sGA variants (as described in Goldberg, 1989) are the method of selection, and *whether* the sensible strategy parameters are adjusted (during the evolution) or not (i.e. the constant mutation rate in GAs versus self-adaptation of the mutation strength in the ES).[3] The emphasis of the usage of bitstrings in genetic algorithms is comparatively unimportant.[4] Also, the differentiation with respect to the selection is not valid anymore for many modern GA variants: meanwhile, (μ, λ) and $(\mu + \lambda)$ selection also exist in the class of GA in a similar form, though under the name "truncation selection" or "elitist selection." And self-adaptation is experimented with also in Genetic Algorithms.

Besides the three main classes, *simulated annealing* should also be mentioned at this point (Kirkpatrick et al., 1983), since it fits very well to the concept of Algorithm 1, and since it can be principally interpreted as a (1+1)-ES with time-dependent selection pressure, but with a constant mutation rate.

The existence of different EA classes with different emphasis of the role of the genetic operators (essentially: selection, reproduction, mutation, and recombination) is definitely not only a matter of marking or separation of the protagonists. The main reason may be the lack of a unifying theory. Such a

[2] The author does not consider this as a real problem. The much more important fact is that EAs are clearly based on the same common principles. The extraction of these principles should be one intention of the EA theory.

[3] In the early period of GAs, self-adaptation was considered in a work of Rosenberg (1967).

[4] This assertion is related to the principal difference between ES and GA: the fitness-proportional selection is analogous to the natural selection model of Fisher (although *tournament selection* would definitely be a more adequate description of nature), whereas selection in the ES corresponds to *breeding*. The optimization in the search space of bitstrings is, in comparison, not a distinguishing feature. The ES works on bitstrings as well (Rechenberg, 1973, pp. 83–86). In addition to the selection method used, another difference is the conversion to the binary coding in the GA for real-valued optimization problems, whereas, the ES works in the "natural problem representation," i.e. directly in the parameter space.

theory should characterize the effect of the operators and would allow for an objective comparison between the different combinations of the operators as to their *optimization power*. The deficiency is especially evident in the GA class. Only in recent (last six) years, has a change been observed from the *schema theorem* as the dominating theory to alternative approaches (see also footnote 5).

The history of the GA was governed more through pure pragmatism, and their theory persisted in the schema theorem over a long time. In contrast, the history of ES contained from the beginning the endeavor to understand the algorithms theoretically, i.e. to estimate the performance of a genetic operator regarding its *optimization performance*.[5] Therefore, in the earlier phase of the ES, the *progress rate* measure φ (in German: "Fortschrittsgeschwindigkeit"; also "Fortschrittsrate" is used synonymously) was introduced (Rechenberg, 1973), which measures the local advancement toward the optimum. The underlying work is mainly on this φ measure.

1.2 The Evolution Strategy

1.2.1 The $(\mu/\rho \overset{+}{,} \lambda)$-ES Algorithm

Consider an optimization problem for the fitness function $F(\mathbf{y})$, $F \in \mathcal{F}$,

$$F(\mathbf{y}) \to \text{Opt.} \tag{1.1}$$

where $\mathbf{y}$ is an N-dimensional *object parameter vector* in the object parameter space $\mathcal{Y}$, $\mathbf{y} \in \mathcal{Y}$, and the location of the optimum is labeled as $\hat{\mathbf{y}}$

$$\mathbf{y} := (y_1, \dots, y_N)^{\mathrm{T}} \qquad \text{and} \qquad \hat{\mathbf{y}} := (\hat{y}_1, \dots, \hat{y}_N)^{\mathrm{T}}. \tag{1.2}$$

The type of the components y_i of $\mathbf{y}$, and therefore the search space $\mathcal{Y}$ spanned by them, depends on the optimization problem. There are no restrictions to the applicability of the ES algorithm, i.e. all of the alternatives $y_i \in \mathbb{R}$, or $y_i \in \mathbb{N}$, or $y_i \in \mathbb{B}$ are allowed; moreover, mixed variants of these are realizable, as well as more complex data structures. A concretization is necessary first in the theoretical investigations, especially for the calculation of the progress rates.

The ES, invented by Rechenberg, Schwefel, and Bienert in the mid-1960s (Rechenberg, 1965, 1973; Schwefel, 1974), operates with populations $\mathfrak{P}$ of size $(\mu \overset{+}{,} \lambda)$. In this notation, μ stands for the number of parent individuals and λ for the number of offspring. An individual consists of an object parameter

[5] Theoretical investigations of the optimization performance (in the sense considered here) in GAs exist; however, they are relatively new. The reader is referred to the works of Goldberg et al. (Thierens & Goldberg, 1994; Miller & Goldberg, 1997) and especially to Prügel-Bennett et al. (Prügel-Bennett & Shapiro, 1994; Shapiro et al., 1994; Prügel-Bennett & Shapiro, 1994, 1997; Rattray & Shapiro, 1997).

set $\mathbf{y}$, the *endogenous* (i.e. evolvable) *strategy parameter set* $\mathbf{s}$, and its fitness value $F(\mathbf{y})$

$$\mathfrak{a} := (\mathbf{y}, \mathbf{s}, F(\mathbf{y})). \tag{1.3}$$

The endogenous strategy parameter set $\mathbf{s}$, $\mathbf{s} \in \mathcal{S}$, serves for the *self-adaptation* of the ES algorithm – a "specialty" of ES and EP. It does *not* take part in the calculation of the fitness of the individual; however, it is passed to the offspring depending on the fitness value of the individual.

The state of an individual $\mathfrak{a}$ is defined by the three-tuple $(\mathbf{y}, \mathbf{s}, F(\mathbf{y}))$ completely. The total of these elements composes the state space $\mathcal{A}$,

$$\mathcal{A} = \mathcal{Y} \times \mathcal{S} \times \mathcal{F}, \tag{1.4}$$

where evolution takes place.

The individuals $\mathfrak{a}$ build up a population, consisting of μ parents $\mathfrak{a}_m$, $m = 1, \ldots, \mu$, and λ descendants $\tilde{\mathfrak{a}}_l$, $l = 1, \ldots, \lambda$. The parameters μ and λ are *exogenous* strategy parameters, i.e. they are not changed by the ES, however, they can be evolved using a Meta-ES (see footnote 6). The populations of the parents and descendants at time g are symbolized as $\mathfrak{P}_\mu^{(g)}$ and $\tilde{\mathfrak{P}}_\lambda^{(g)}$, respectively

$$\begin{aligned} \mathfrak{P}_\mu^{(g)} &:= \left\{\mathfrak{a}_m^{(g)}\right\} = \left(\mathfrak{a}_1^{(g)}, \ldots, \mathfrak{a}_\mu^{(g)}\right) \\ \tilde{\mathfrak{P}}_\lambda^{(g)} &:= \left\{\tilde{\mathfrak{a}}_l^{(g)}\right\} = \left(\tilde{\mathfrak{a}}_1^{(g)}, \ldots, \tilde{\mathfrak{a}}_\lambda^{(g)}\right). \end{aligned} \tag{1.5}$$

Using the notations given above, one can express the $(\mu/\rho \overset{+}{,} \lambda)$-ES as the conceptual Algorithm 2.

The algorithm formalized in this way contains in principle all the conceivable ES variants,[6] including the $(\mu, \kappa, \lambda, \rho)$ variant with parameter κ for the lifespan suggested by Schwefel (Schwefel & Rudolph, 1995), since the lifespan of an individual can be coded in the strategy parameter set $\mathbf{s}$.

In addition to these exogenous strategy parameters μ and λ, the exogenous parameter ρ appears in the $(\mu/\rho \overset{+}{,} \lambda)$ notation. It determines the number of parents which take part in the procreation of a *single* individual (group mating number, mixing number), where $1 \le \rho \le \mu$. The parents are "married" by the reproduction operator via random selection and make alltogether the set of the parent family, which is a ρ-tuple $\mathfrak{E}$, and which consists of ρ members each

$$\mathfrak{E} := \left(\mathfrak{a}_{m_1}, \mathfrak{a}_{m_2}, \ldots, \mathfrak{a}_{m_r}, \ldots, \mathfrak{a}_{m_\rho}\right). \tag{1.6}$$

For the case $\rho = 1$ we have asexual reproduction, i.e. the recombination operators described are **1**-operators, consequently, which do not change the

[6] Hierarchically organized ES, also referred to as Meta-ES or nested ES, suggested by Rechenberg (1978) and realized, e.g., by Herdy (1992), fit into this schema indirectly, if one interprets the ES at the lower level of the hierarchy as the fitness function of the higher level. The Meta-ES optimizes appropriately defined properties of the ES at the lower level of the hierarchy.

```
Procedure (μ/ρ +, λ)-ES;                                              line
Begin                                                                    1
  g := 0;                                                                2
  initialize(𝔓_μ^(0) := {(y_m^(0), s_m^(0), F(y_m^(0)))});               3
  Repeat                                                                 4
    For l := 1 To λ Do Begin                                             5
      𝔈_l := reproduction(𝔓_μ^(g), ρ);                                   6
      s_l := s_recombination(𝔈_l, ρ);                                    7
      s̃_l := s_mutation(s_l);                                            8
      y_l := y_recombination(𝔈_l, ρ);                                    9
      ỹ_l := y_mutation(y_l, s̃_l);                                      10
      F̃_l := F(ỹ_l)                                                     11
    End;                                                                12
    𝔓̃_λ^(g) := {(ỹ_l, s̃_l, F̃_l)};                                      13
    Case selection_type Of                                              14
      (μ, λ) :   𝔓_μ^(g+1) := selection_{F_μ}(𝔓̃_λ^(g));                  15
      (μ + λ) :  𝔓_μ^(g+1) := selection_{F_μ}(𝔓̃_λ^(g), 𝔓_μ^(g))          16
    End;                                                                17
    g := g + 1;                                                         18
  Until stop_criterion                                                  19
End                                                                     20
```

Algorithm 2. The $(\mu/\rho \overset{+}{,} \lambda)$-ES

genetic information. The case $\rho = 2$ is the standard case in biology. For $\rho > 2$, *multirecombination* takes place. It is observed in nature only in the family of unicellulars (protozoans), especially in bacteria, but also in ciliates (Margulis & Sagan, 1988).[7]

Every EA contains a selection mechanism as the fundamental component. In the ES, the selection is symbolized by $(\overset{+}{,})$, where *either* $(+)$-selection ("plus"-selection) *or* $(,)$-selection ("comma"-selection) is used (see Sect. 1.2.2, p. 8).

The ES Algorithm 2 can be understood in the sense of the general Algorithm 1, using the explanations given. The evolutionary loop is realized by the successive application of the genetic operators on the population $\mathfrak{P}$ of individuals. The genetic operators used in the ES are (see Sect. 1.2.2) reproduction, recombination, mutation, and selection. The generation cycle takes place in the Repeat-Until-Loop (lines 4–19, Algorithm 2). Every gen-

[7] To be more precise, the conjugation behavior of these individuals, which is mainly observed under high environmental pressure (selection pressure, "survival of the fittest"), can be interpreted as a multirecombination.

eration g consists of a parent population $\mathfrak{P}_{\mu}^{(g)}$ from which the λ offspring are generated, which form together the population of descendants $\tilde{\mathfrak{P}}_{\lambda}^{(g)}$. This generation procedure occurs in the lines 5–13. Every descendant is produced step by step. First, its parents are selected in line 6. If the self-adaptation facility is also implemented, the set of strategy parameters is treated next. For $\rho > 1$, recombination takes place, $\rho = 1$ corresponds to the simple reproduction, i.e. asexual reproduction. Mutation is applied in line 8. The set of strategy parameters obtained in this way controls the treatment of the object parameters. For the case of $\rho > 1$, recombination takes place in line 9; the responsible operator does not depend on $\tilde{\mathbf{s}}_l$.[8] In contrast, the mutations are controlled by the set of the strategy parameters. The new set of object parameters is obtained after mutation and then it is evaluated according to its fitness (line 11).

The selection (lines 14–17) follows the procreation of the offspring population $\tilde{\mathfrak{P}}_{\lambda}^{(g)}$. Lastly, the new parent population $\mathfrak{P}_{\mu}^{(g+1)}$ is built according to the type of the selection used. Thereafter, the "game of evolution" starts anew, until the predefined stopping criterion is fulfilled.

1.2.2 The Genetic Operators of the ES

The genetic operators are the "flesh" of every EA. Besides the choice of the suitable problem-specific data structure, the operators determine essentially the *performance* of the EA. The development of such operators is certainly a kind of "art," and it seems to be hard to give general design rules. Heuristics dominates the research arena; a satisfactory background is missing in most of cases. Especially in the GA domain, this led to a wild growth of genetic operators. Fortunately, this inflation is limited in the domain of the ES. However, generally there exists a phase shift between practice and theory: the fast practical successes in implementation hinder very often a thorough theoretical analysis.

Occasionally, the opinion is offered in the EA community that there are *no* general EA principles. Instead, every problem class and EA class ought to have its own specific principles; moreover, ES and GA have almost no similarities. As will be described in Sect. 1.4, the author will offer some other ideas leading to basic EA principles that might have a broader validity. However, proofs of these basic principles succeed only to some extent in the ES field. The verification or falsification of these principles and the exact determination of the domain of validity is an ambitious research program for the future.

The theoretical analysis of complex EA processes depends always on the model considered. Therefore, it is a good idea to concentrate on some genetic operators which have proven themselves to be successful in practice. As a

[8] Extensions to the strategy are of course imaginable; however, their practical benefit is questionable from the current theoretical viewpoint.

consequence, not all possible genetic operators are defined here. Only those will be mentioned which are analyzed – at least partially – in the following chapters.

1.2.2.1 The Selection Operator. The selection operator of type ($\dagger$) produces the parent population $\mathfrak{P}_\mu$ of the next generation $g+1$ through a *deterministic* procedure, which chooses the best individuals from the set of γ individuals $(\mathfrak{a}_1, \ldots, \mathfrak{a}_\gamma)$ according to their fitness value $F(\mathbf{y})$ (*truncation selection*)

$$(\mathfrak{a}_{1;\gamma}, \mathfrak{a}_{2;\gamma}, \ldots, \mathfrak{a}_{\mu;\gamma}) := \mathtt{Selection}_{\mathtt{F}_\mu}(\mathfrak{a}_1, \ldots, \mathfrak{a}_\gamma), \qquad \gamma \geq \mu. \tag{1.7}$$

The symbol $(.)_{m;\gamma}$ introduced here stands for the individual with the mth best fitness value or the respective components of the data structure of this individual. $(.)_{1;\gamma}$ corresponds to the best individual and $(.)_{\gamma;\gamma}$ to the worst individual, respectively. The advantage of this notation is the independence of the optimization type (maximization or minimization). It presents a generalization of the $(.)_{m:\gamma}$ symbolism used in the theory of order statistics (David, 1970), which expresses the order relation

$$F_{1:\gamma} \leq F_{2:\gamma} \leq \ldots \leq F_{m:\gamma} \ldots \leq F_{\gamma:\gamma}. \tag{1.8}$$

The difference between the (μ, λ)- and $(\mu + \lambda)$-selection is defined by the set of individuals participating in the selection. In the (μ, λ)-selection (cf. Algorithm 2, line 15), only the offspring population $\mathfrak{P}_\lambda$ is considered in the selection. In other words, the parents die out *per definitionem*. On the other hand, the selection pool contains both the offspring and the parent individuals in the $(\mu + \lambda)$ variant (cf. Algorithm 2, line 16).[9] Therefore, the parents can survive for many generations.

Both selection variants may have their advantages and disadvantages; therefore, they have specific application areas. The usage of (μ, λ)-selection is recommended in the field of real-valued parameter optimization, based on simulation results (Schwefel, 1987). If we consider the combinatorial optimization problems, the $(\mu+\lambda)$-selection is rather to be recommended (Beyer, 1992a). The reason for that is the principal failure of the self-adaptation of the mutation strength of the permutation operator: the strength of the real-valued mutations (their variance) can be decreased as necessary, whereas "half" or "quarter" permutations do not exist. Therefore, the evolutive progress becomes a seldom event, i.e. the success probability[10] P_s is very low; one observes a $P_\mathrm{s} \propto g^{-\beta}$ rule, with $\beta \approx 1$. The evolutive progress can be attained only through conservation of good states, which is exactly performed by the $(\mu + \lambda)$-selection with simultaneous maintenance of a certain genetic variability.

[9] This kind of selection is called *elitist selection*, especially in the GA community (DeJong, 1975).

[10] P_s is the probability to generate at least one descendant in the next generation which is better than the parents.

The theoretical studies of ES with $(\mu \overset{+}{,} \lambda)$-selection are concentrated mainly on the $(1+1)$-variant (Rechenberg, 1973; Beyer, 1993) and the $(1, \lambda)$- or $(1+\lambda)$-selection (Schwefel, 1981; Beyer, 1993, 1994a). Since we have $\mu = 1$ in all these cases, a population as such actually does not exist. The main reason for the concentration in the case $\mu = 1$ can be found in the enormous mathematical difficulties which emerge in the self-consistent determination of the population distributions. The analysis of the (μ, λ)-ES in Chap. 5 will bear witness to this fact.

1.2.2.2 The Mutation Operator. Besides the selection operator, the mutation operator is the most important "ingredient" for an ES. It is the source of the genetic variations. Mutation operators are problem-dependent, and the correct design of them has an essential influence on the performance of the ES. The following rules may help for the design:

1. *Requirement of reachability*: Since the operator performs search moves in the state space $\mathcal{A}$ of the individual, it should be realized so that it can transfer the individual from state $(\mathbf{y}, \mathbf{s})$ to any other state $(\tilde{\mathbf{y}}, \tilde{\mathbf{s}})$ in finite time g, $g < \infty$. The fulfillment of the reachability requirement is a *necessary* condition for the functioning of the ES (which is also valid for the other EA).
2. *Requirement of scalability*: The length of the search steps (mutation strength) should be tuneable for the calibration of the locally optimal mutation strength. This requirement is easy to fulfill for real-valued parameter spaces (see below); however, for integer-valued or combinatorial problems, it can be carried out only to a certain extent or it is not possible at all.
   Instead of scalability, there is the notion of "minimal move step" for the combinatorial problems, i.e. the mutations should be built up using unit permutations, which change the individuals in the smallest possible way and which have the smallest cardinality of neighborhood.[11] For instance, for the traveling salesman problem (TSP), this is the so-called *Lin-2-Opt* step (see Lin & Kernighan, 1973), which is called occasionally *inversion* in the ES literature (Herdy, 1991).
3. *Absence of biases*: The mutation distributions should be chosen according to the maximum entropy principle (for the maximum entropy principle, see Jaynes, 1979). For real-valued object parameter spaces without constraints, this requirement leads naturally to Gaussian normal distribution. In integer search spaces, we have the geometrical distribution, which is a special case of the negative binomial distribution (Rudolph, 1994).
   No theoretical studies exist for the combinatorial problems to date.[12]

[11] The number of neighbors which can be reached by a single move step.

[12] Only recently, a first attempt has been made in understanding the optimization dynamics of TSPs (Nürnberg & Beyer, 1997).

4. *Symmetry*: This is strongly connected to Requirement 3, but is not equivalent to it. The symmetry expresses the assumed isotropy of the search space. As a result, the expected value of the state changes induced by the mutations should be zero.
   In the case of combinatorial optimization or operating on bitstrings, the microscopic *reversibility* substitutes the "expected value = zero" requirement. It states that the transition from a state to another should have the same probability as the reverse process (Aarts & Korst, 1989).

The requirements list given here describes guiding principles for the design of mutation operators. Not all of these requirements are absolutely important, nor are they definitely necessary. The robustness of the ES is guaranteed mainly through deterministic selection. Merely, Requirement 1 expresses a certain minimal condition which should be fulfilled.

The theoretical analysis of the ES concerned up until now almost exclusively real-valued parameter optimization. According to Requirement 3, the object parameters are mutated using a normal distribution with an expected value of zero (Requirement 4). That is, the descendant $\tilde{\mathbf{y}}_l$ is produced around the parental recombination result $\mathbf{y}_l$ (cf. Algorithm 2, line 10) through the addition of a Gaussian random vector $\mathbf{z}$

$$\tilde{\mathbf{y}}_l := \mathbf{y}_l + \mathbf{z}, \qquad \text{with} \qquad \mathbf{z} := (z_1, z_2, \ldots, z_N)^{\mathrm{T}}. \tag{1.9}$$

The mutations are isotropic in the simplest case. Every z_i component fluctuates $\mathcal{N}(0, \sigma_i^2)$, and is statistically independent of other components with the same standard deviation $\sigma_i = \sigma$

$$\mathbf{z} := \sigma\, (\mathcal{N}(0,1), \mathcal{N}(0,1), \ldots, \mathcal{N}(0,1))^{\mathrm{T}}. \tag{1.10}$$

The density function is therefore

$$p(\mathbf{z}) = \frac{1}{\left(\sqrt{2\pi}\right)^N} \frac{1}{\sigma^N} \exp\left(-\frac{1}{2}\frac{\mathbf{z}^{\mathrm{T}}\mathbf{z}}{\sigma^2}\right), \tag{1.11}$$

where the standard deviation σ – the mutation strength – is the only endogenous strategy parameter $\tilde{\mathbf{s}} = \sigma$.

In the most general case of Gaussian mutations, correlations between the z_i-components are considered, so that the density function is given by

$$p(\mathbf{z}) = \frac{1}{\left(\sqrt{2\pi}\right)^N} \frac{1}{\sqrt{\det[\mathbf{C}]}} \exp\left(-\frac{1}{2}\mathbf{z}^{\mathrm{T}}\mathbf{C}^{-1}\mathbf{z}\right), \tag{1.12}$$

where $\mathbf{C}$ is the correlation matrix and $\mathbf{C}^{-1}$ is its inverse. The mutation is not parameterized using a scalar mutation strength in this case, but using a set of $N(N+1)/2$ strategy parameters, which are the elements of the symmetric correlation matrix $\tilde{\mathbf{s}} = \mathbf{C}$.

Principally, ES can also work on bitstrings – the domain mainly addressed by the GA.[13] The components y_i of the object parameter vector $\mathbf{y}$ are then elements of the set $\mathbb{B} = \{0,1\}$, $\mathbf{y} \in \mathbb{B}^N$, with $N = \ell$, the length of the bitstring. Interestingly, some of the analysis methods developed for the real-valued case can be applied to the functions on bitstrings (e.g. the counting ones function "`OneMax`").

Mutations on bitstrings are characterized by the mutation rate p_m. This is the probability of the negation of a single bit position i, $y_i \to \neg y_i$. The probability density of the descendant $\tilde{\mathbf{y}}$ can be expressed by *Dirac*'s delta functions

$$p(\tilde{\mathbf{y}}) = p_m^N \prod_{i=1}^{N} \delta(\tilde{y}_i - 1 + y_i) \; + \; (1-p_m)^N \prod_{i=1}^{N} \delta(\tilde{y}_i - y_i). \tag{1.13}$$

Mutation operators for the strategy parameters $\mathbf{s}$ are not described here. They will be introduced in Chap. 7, Sect. 7.1.4, in relation to the analysis of the $(1,\lambda)$-σ-self-adaptation.

1.2.2.3 The Reproduction Operator. The reproduction operator selects the parents $\mathfrak{E}$ which take part in the procreation of *one* offspring individual

$$\mathfrak{E} := \begin{cases} (\mathfrak{a}_{i_1}, \ldots, \mathfrak{a}_{i_r}, \ldots, \mathfrak{a}_{i_\rho}), & \text{for } \rho < \mu \\ (\mathfrak{a}_1, \ldots, \mathfrak{a}_m, \ldots, \mathfrak{a}_\mu), & \text{for } \rho = \mu. \end{cases} \tag{1.14}$$

with

$$\begin{aligned} &\forall r \in \{1,\ldots,\rho\}: \quad i_r := \texttt{Random}\{1,\ldots,\mu\}, && \text{for } \rho < \mu \\ &m = 1,\ldots,\mu, && \text{for } \rho = \mu. \end{aligned} \tag{1.15}$$

In the case of $\rho = \mu$ one has $\mathfrak{E} := \mathfrak{P}_\mu$; i.e. all individuals of the parent population take part in the production of one offspring. Whereas for $\rho < \mu$ a random choice is performed among the parents, and every individual of $\mathfrak{P}_\mu$ is selected with the same probability $1/\mu$. In order to make the algorithm simpler, every selected individual is returned to the pool, thereby having the effect of an infinitely large "mating pool" (mating selection with replacement). Such a procedure includes "inbreeding." It can be clearly avoided using a respective selection algorithm; however, here arises the question whether the extra computational effort is worth doing. Anyway, it is recommended to use "$\rho = \mu$"-recombination to attain the maximum progress rate, which will be substantiated in Chap. 6.

---

[13] It is often wrongly claimed that early ES applications did not address the optimization of binary problems. However, the *mimicry experiment* by Rechenberg (1973, pp. 83–86) actually used mutation and k-point crossover (!) to optimize a binary pattern matching problem.

1.2.2.4 The Recombination Operator. It is still not explained satisfactorily why the recombination operators may provide a performance gain. There are empirical studies, especially in the GA field, which aim to explain the usefulness of "sex." However, there are also empirical studies and theoretical considerations showing that recombination does not provide any conclusive performance increase (Fogel & Stayton, 1994).[14]

The GA theoreticians substantiate their crossing-over operator with the *building block hypothesis* (BBH): the good properties of the parents should flow via crossover to the offspring (Goldberg, 1989). This explanation model has some difficulties in explaining the performance of the *uniform crossover* (Syswerda, 1989). What can be concluded, favorably assuming that the recombination brings performance increase? Basically, just that the BBH does not give a suitable explanation for the working of such operators.

Are there any other explanations for the working of the recombination? There is another plausible model: the *genetic repair* (GR) hypothesis. It is founded theoretically for the ES in Chap. 6. At this point, we will only mention those aspects which might be important for the construction of recombination operators. The explanation model is in contraposition to the BBH: it is not the different good properties of the parents – their apparent differences – flow into the offspring,[15] but their *common features*. The recombination produces *similarities*; the children are "similar" to their parents. But why is the recombination in EA benefical at all, if it mainly drives "egalitarianism"? Recombination itself has usually *no* benefit; it is useful mainly in combination with a relatively large mutation strength and a sufficient environmental pressure (selection): mutations are absolutely necessary for the production of new variants and for the evolutionary progress.[16] However, most of them are more harmful than useful. The duty remaining for the selection is choosing the suitable mutants. The recombination extracts then the common features – the similarities – of these selected individuals and may reduce the statistically uncorrelated parts of the mutations. This point of view suggests that the similarities in the selected individuals (high fitness) are those components which are statistically the benefical ones. Put it another way, by this process of conserving the similarities and rearranging the remaining parts, recombination performs something like rudimentary *self-adaptation* using the parental

[14] Recently, Jansen and Wegener (1999) have shown for a special steady-state GA, optimizing a specially chosen fitness function from the unitation class, that uniform crossover can reduce the time complexity.

[15] Of course, there is also a certain probability of that, but usually it is small.

[16] This statement might be a bit misleading, since there are GA implementations which do *not* use "physical" mutations at all. These algorithms need a large initial population variance. The random initialization of a population, however, can be regarded as the result of a physical mutation process. In other words, population variance can be interpreted as a result of a virtual surrogate mutation process (for the application of the surrogate mutation concept, see Sect. 6.2.1.2, pp. 233ff).

population distribution. (A detailed discussion follows under Sects. 6.1.3.2 and 6.2.2.2.)

Provided that this hypothesis is correct, we can list some criteria for the design and use of recombination operators:

1. *The problem-specificness*: Recombination operators should be constructed in a problem-specific manner. They should work directly in the respective parameter space of the optimization problem and have a certain "capability to abstract," which makes the extraction of the statistical similarities possible (see also Point 3).
2. *The absence of biases*: Every individual taking part in the recombination should contribute to the recombination result with equal probability.
3. *GR Quality*: The effect of the recombination can be seen as a statistical estimation method, or as a failure correction mechanism, genetic repair (GR). The quality of estimation increases with the number ρ of the samples which take part in the recombination, though it conflicts with the selection. That is, the choice $\rho = \lambda$ implies $\mu = \lambda$ and therefore no selection pressure. From the viewpoint of GR, $\rho = \mu$ is recommended. However, the optimal value of μ is problem-dependent (see Sect. 6.1.3.4).
4. *Permanence*: Recombine always; there is no reason (in the scope of GR hypothesis) to operate with a probability $p_c < 1$. The empirical results with smaller p_c values in some studies in the GA field point rather to an unlucky choice of the mutation rate p_m. The adjustment of the optimal mutation strength is of crucial importance for the effective working of the recombination.

The criteria proposed here do not identify the recombination operator completely. Particularly, Point 1 does not give any practical construction rules. This catalog of criteria, however, can be used to single out already existing operators.

For the ES in real-valued parameter spaces, there are two kinds of recombination – the intermediate recombination and the dominant/discrete recombination. The definition of the operators is offered for a ρ-tuple of vectors $(\mathbf{x}_1, \dots, \mathbf{x}_\rho)$. The symbol $\mathbf{x}$ stands for the set of the object parameters $\mathbf{y}$, consisting of $D = N$ components, or for the set of strategy parameters $\mathbf{s}$, with $D = N_s$ components. The symbol `y_recombination`$(\mathfrak{E}_l, \rho)$ in Algorithm 2, line 7, and `s_recombination`$(\mathfrak{E}_l, \rho)$ in line 9 intimate the respective information from the set of ρ parents $\mathfrak{E}$ selected for the reproduction. For the case of $\rho > 2$, we have *multirecombination*.

Intermediate ρ-recombination: The recombined descendant $\mathbf{r}$ is generated by vectorial formation of the center of gravity from ρ randomly selected parents

$$\mathbf{r} := \langle \mathbf{x} \rangle_\rho := \frac{1}{\rho} \sum_{\nu=1}^{\rho} \mathbf{x}_\nu. \tag{1.16}$$

Dominant ρ-recombination: Every component i of the recombined descendant $\mathbf{r}$ is obtained by random selection from the ρ i-components of the selected parents

$$\mathbf{r} := \sum_{i=1}^{D} (\mathbf{e}_i^{\mathrm{T}} \mathbf{x}_{m_i}) \, \mathbf{e}_i, \qquad m_i := \texttt{Random}\{1, \ldots, \rho\}. \tag{1.17}$$

The symbol $\mathbf{e}_i$ stands here for the unit vector in the ith direction; the scalar product gives the ith component of the respective vector $\mathbf{x}_\rho$. In other words, the ith component of the recombined descendant is specified exclusively by the ith component of the (randomly) selected parent $\mathbf{x}_m$, which one can also interpret as "dominance," if one contrasts the mean value formation in (1.16).

The most widespread variant of the dominant recombination is the $\rho = 2$ version, which is called in the field of GA *uniform crossover*[17,18] (Syswerda, 1989) and in the ES domain *discrete recombination* (Schwefel, 1975). The case $\rho = \mu$ trades in the ES under the name *global discrete recombination.*

1.3 The Convergence of the Evolution Strategy

In the scope of this work, convergence in the broader sense means everything which can be subsumed under the notion "approaching the optimum." At first glance, this definition of the term seems to be trivial; however, e.g., in the field of GA, there exists a definition of gene convergence which originates from DeJong (1975) where the gene position is considered converged if at least 95% of the individuals of the population have the same allele. This definition does not present a sufficient criterion for convergence to the optimum. In the most inconvenient case, the gene convergence can occur for any state (genetic drift). Gene convergence is therefore at most a necessary but not sufficient criterion for convergence to the optimum.

The problem formulations which are interesting for optimization are global convergence and its convergence order (time complexity) as qualitative evaluation aspects of the EA, as well as the performance of the algorithm. The global convergence and convergence order or time complexity, respectively, can be regarded as *global* convergence aspects on the *macroscopic level*, whereas the term *performance* applies to the *microscopic level*, i.e. to the local behavior in the state space $\mathcal{A}$ (see p. 5). Therefore, the subject of convergence is divided into global and local aspects.

[17] In his original work Syswerda produced two offspring per recombination. Discarding one of them randomly provides equivalence to the $\rho = 2$ dominant recombination considered here.

[18] Uniform crossover was re-invented by Syswerda (1989). However, its history goes back to the very beginning of EA in the 1950s and 1960s (Fraser, 1957; Bremermann et al., 1966; Reed et al., 1967).

1.3.1 ES Convergence – Global Aspects

The question which every EA-user is usually interested in first is the global convergence of the algorithm. Since EA are probabilistic algorithms, there exists always a certain probability that the same state $\mathbf{y} \neq \hat{\mathbf{y}}$ is generated continuously. Therefore, a rigorous convergence proof can only be given in the sense of a "Pr = 1"-probability for infinitely many steps (generations) g. In discrete optimization problems, this is achieved by proving[19]

$$\lim_{g\to\infty} \Pr\left(\mathbf{y}^{(g)} = \hat{\mathbf{y}}\right) = 1 \qquad \text{and} \qquad \sum_{\gamma=1}^{\Gamma} \lim_{g\to\infty} \Pr\left(\mathbf{y}^{(g)} = \hat{\mathbf{y}}_\gamma\right) = 1, \tag{1.18}$$

respectively, and in the "fitness picture"

$$\lim_{g\to\infty} \Pr\left(F(\mathbf{y}^{(g)}) = \hat{F}\right) = 1. \tag{1.19}$$

In the case of real-valued optimization problems, Condition (1.19) should be changed so that the location of the optimum $\hat{F} = F(\hat{\mathbf{y}})$ has to be reached within an arbitrarily small ε-vicinity of $\hat{F}$

$$\lim_{g\to\infty} \Pr\left(\left|F(\mathbf{y}^{(g)}) - F(\hat{\mathbf{y}})\right| \le \varepsilon\right) = 1, \qquad \varepsilon > 0. \tag{1.20}$$

Proofs for the Condition (1.19) have been performed particularly in the field of *simulated annealing*. The survey by Laarhoven and Aarts (1987) is recommended here. Because of the similarity to the (1+1)-ES, the methodology of proof has a certain relevance for the (+)-strategies. The same is valid for the results of Fogel (1992) in the field of EP.

A proof of the global convergence for $(\mu + \lambda)$-strategies has not been published up until now. However, it should be relatively easy to carry out, at least for the case of constant mutation strength. For the (μ, λ)-ES with constant mutation strength, however, such a proof is principally not possible, since a reached optimum can be left anytime with a certain probability. A more precise analysis is necessary to see whether it is possible to decrease this escape probability to zero through an appropriate adiabatic decrease of the mutation strength. The condition of reachability (cf. Condition 1, p. 9) must be fulfilled simultaneously.

In the real-valued case, a sketch of the proof of (1.20) was provided for the (1+1)-ES with constant mutation strength by Rechenberg (1973). A detailed proof can be found in Born (1978). The extension to the $(\mu + \lambda)$-strategies should be easy to accomplish. The (μ, λ)-strategies, however, are generally *not* convergent. They principally require a suitable control of the mutation strength, in order to be able to converge at least locally. We will come back to this issue at different places.

[19] The second formula (1.18) is for the case with a *degenerated* global optimum (a ground state with a degeneration level Γ).

Although it is good to know that the EA can be globally convergent, this information is of minor practical value, since nobody is willing to wait $g \to \infty$ generations. Evolutionary algorithms are expected to give good approximate solutions in the shortest possible time. This is only possible at the expense of convergence security. If one regards convergence security as the only goal then one must enumerate the search space completely[20], which is only possible in exceptional cases or for small problems. After acknowledging this fact, another important aspect comes to the foreground: the question of the convergence rate of the EA, or of its expected time complexity. The interesting point is the *dynamics* of an evolutionary system in the state space $\mathcal{A}$.

The investigation of the evolutionary dynamics in the state space $\mathcal{A}$ is by no means simpler than the previously mentioned problem of convergence. On the contrary: the problem of convergence is a special case of the EA dynamics which can be reduced to the question of whether $\hat{\mathbf{y}}$ is the only stable attractor of the EA system ($\hat{=}$ EA & fitness function). After considering the matter from this point, it becomes clear that the statements on convergence rates or on the global dynamics of the system are *necessarily* of example character, since the EA system does not only contain the optimization algorithm, but also the fitness function. If one wants to learn more on the dynamics of an EA, one must specify the fitness function (at least to a certain extent). In other words, suitable fitness models should be developed which are on the one hand simple (to be able to succeed the mathematical derivations); on the other hand, they should be so general that they can reflect the typical behavior of a large class of fitness functions. In that way it is possible to make statements on the convergence order of an EA or even to determine its EA dynamics.

Rappl (1984, 1989) obtained results on the convergence order for the class of convex functions. He studied random search methods which are very similar to the $(1+1)$-ES, e.g., a $(1+1)$-strategy which works with spherical probability distribution for mutations and a step length control for the radius of the sphere distribution. Such an algorithm shows a *linear* convergence order (Rappl, 1989), i.e. the expected value of the deviation from the optimum is given by

$$\mathrm{E}\left\{F\left(\mathbf{y}^{(g)}\right) - \hat{F}\right\} = \mathcal{O}\left(\beta^{g}\right), \qquad \beta \in (0,1). \tag{1.21}$$

In the case of constant mutation strength without step length control, one obtains a *sublinear* convergence order (Rappl, 1984)[21]

$$\mathrm{E}\left\{F\left(\mathbf{y}^{(g)}\right) - \hat{F}\right\} = \mathcal{O}\left(g^{-\alpha}\right), \qquad \alpha > 0. \tag{1.22}$$

[20] Of course, it is implicit at this stage that no efficient algorithm is known which solves the optimization problem.

[21] It is interesting to note that sublinear convergence order is also observed in ES for combinatorial problems, e.g., the traveling salesman problem, firstly observed in (Beyer, 1992a); for an explanation model, see Nürnberg and Beyer (1997).

The information obtained from the knowledge of the convergence order is only of restricted practical interest. Actually, all ES variants exhibit a linear convergence order locally at smooth fitness functions in real-valued parameter spaces – as long as the mutation strength is controlled appropriately. Therefore, the performance of different ES variants cannot be compared using only the respective convergence orders.

More precise statements on the performance must be derived either from the evolution dynamics or through a *local performance analysis*. The former is again the harder problem, requiring a priori knowledge of the local performance. Thus, we are guided to the *microscopic* aspects of the convergence.

1.3.2 ES Convergence – Local Aspects

Evolutionary algorithms operate locally; they generate offspring, which in turn lie in the state space $\mathcal{A}$ in the neighborhood of their parents. The structure and size[22] of the neighborhood are determined by the genetic operators applied (mutation and recombination). The restriction to the sufficiently small neighborhood is crucial for the effective working of the EA. A too large neighborhood lets the EA degenerate into a Monte Carlo method. For instance, if the mutation rate of a real-valued ES is set too large, the probability to generate a descendant better than its parent decreases drastically. The mutations "get lost" in the parameter space. The neighborhood is too large in this case and the local convergence of the algorithm too slow; accordingly one may even have divergence in the case of (μ, λ)-selection.

The evolutionary progress is obtained only in a small, restricted neighborhood. This most important discovery had already been formulated by Rechenberg (1973). The concept of the *evolution window* is shaped thereafter. An essential part of this book is devoted to the computation of the evolution window, i.e. the determination of the strategy parameter region where the ES locally operates optimally.

Appropriate measures for the local performance will be introduced in order to quantify the local behavior of an EA in dependence on the size of the neighborhood. An important quantity is the progress rate φ, which measures the expected change of the population with respect to a reference point in the parameter space $\mathcal{Y}$ from generation g to generation $g+1$. In contrast, the quality gain $\overline{Q}$ measures the expected change in the fitness values. Both measures describe *microscopic* aspects of the local evolution.

Under specific circumstances, the local performance measures can be used to investigate the macroscopic dynamics. For example, the performance of an algorithm over time can be predicted from the progress rate, e.g. for the ES

[22] The notions "neighborhood," "structure," and "size" are used here intuitively. They are frequently used in the field of *simulated annealing*. The reader is referred to the related works and to the introduction of Laarhoven and Aarts (1987).

algorithm on the sphere model. Moreover, the mathematical analysis gives a deeper insight into the mechanisms of the EA. It shall even be claimed that an EA is only understood if one is able to analyze the important aspects of it analytically, i.e. if one can derive (approximate) formulae for the progress rate analytically. Of course, this is possible only for some very simple fitness models, and the statement may sound exaggerated. In fact, the essential discoveries on the basic mechanisms (see Sect. 1.4) of the recombinant ES presented in this book are obtained exactly in this way. It is important to emphasize that the point of interest is the derivation of *analytical approximations*. On the way to such approximations, we are permanently forced to identify the *relevant* terms and to neglect the *irrelevant* parts. Exactly this selection process is in the end the source for a more in-depth understanding of the functioning of the ES.

1.4 Basic Principles of Evolutionary Algorithms

There exist bizarre ideas on the working mechanisms and performance of EA. The spectrum of these ideas is very wide. Some opponents of the EA compare these methods often with purely tossing coins (Monte Carlo methods). Some GA specialists emphasize the binary coding of the variables of the fitness functions, i.e. the usage of a minimal alphabet, as the fundamental requirement for an "efficient" global optimization (implicit parallelism; Holland, 1975; Goldberg, 1989).

However, in this book a rather local point of view will be taken. In other words, the working mechanisms of an EA are essentially determined through the neighborhood properties of the parameter space, or more generally, of the state space. Depending on the high dimensionality of these spaces, the evolutive progress can only be attained by small steps (small neighborhood size). Escaping from local attractors by "jumps" or the "tunneling" of high potential walls are very unlikely events as long as the algorithm does not "know" in which direction a better attractor is to be found. This leads us to the notion of *evolvability*, i.e. the ability to evolve by (small) meliorization steps, to be discussed briefly in the next section.

Besides the evolvability, the following principles are suggested in this book as prerequisites for the working of the EA (Beyer, 1995a): the *evolutionary progress principle* (EPP), the *genetic repair hypothesis* (GR), and the principle of *mutation-induced speciation by recombination* (MISR), to be discussed in the following sections.

1.4.1 Evolvability

Global optimization in multimodal fitness landscapes can function only if there is a certain ordering in this landscape, i.e., if a certain "tendency" to the global optimum exists. This obviously intuitive prerequisite[23], which should be fulfilled by the fitness landscapes, has been designated by Rechenberg (1994) as *causality.* This notion is used in this context to characterize the smoothness property of the fitness landscape with respect to a given set of genetic operators, and not the necessary and impelling relation between causes and effects (which is also called "causality," see Lenk & Gellert, 1972, pp. 728f), which always exists, if the fitness function is deterministic.

Rechenberg distinguishes between *strongly* and *weakly causal worlds* (fitness landscapes). In the former one, the ES always finds the global optimum by just following the local gradient by a respective diffusion process. The weakly causal landscapes do not have this property, and the ES is expected to work there not better – or even worse – than Monte Carlo methods. Between the two extremes "strong" and "weak," Rechenberg expects a certain domain where the ES can work conditionally. Born calls these worlds *semi-causal* and tries to give a mathematical definition for them(see, Voigt et al., 1991), which postulates a monotonic order of the local optima in $\mathbb{R}^N$: Assuming that $\hat{\mathbf{y}}$ is the global minumum and $\{\mathbf{y}_\alpha\}$ the set of the remaining local minimal positions, $\hat{\mathbf{y}}$ is designated as the *monotonic minimum attractor*, if the condition

$$\forall \alpha, \beta: \quad ||\hat{\mathbf{y}} - \mathbf{y}_\alpha|| \le ||\hat{\mathbf{y}} - \mathbf{y}_\beta|| \quad \Longleftrightarrow \quad F(\mathbf{y}_\alpha) \le F(\mathbf{y}_\beta) \tag{1.23}$$

is fulfilled. In empirical studies, the ES exhibits very often (conditional) global convergence properties in landscapes having the order properties defined by Condition (1.23). Unfortunately, a satisfactory theoretical explanation for this observation still remains to be offered. It is also not clear whether or not (1.23) is a necessary criterion for the convergence of the ES in a stronger sense. It is very probable that the "ES hardness" of a problem is not satisfactorily defined by (1.23), which is also not claimed by the authors. Condition (1.23) gives a qualitative criterion for the intuitive notion of semi-causality; however, "quantitative nuances" of the semi-causality are also easy to imagine. The existence of a measure for the degree of order is assumed thereby. Possibly the correlation functions from statistics can be used as a rudimentary approach to describe the order properties of the fitness landscape.

It is interesting to note that similar questions are also being discussed in the area of *simulated annealing.* Some authors believe that the fitness landscapes which are suitable for the simulated annealing also have certain order properties. For example in Sorkin (1991), it is assumed that the fitness landscape of the traveling salesman problem and also of similar problems should possess certain properties of self-similarity to make efficient simulated annealing possible. He defines the fitness function $F(\mathbf{y})$, i.e. the tour length, as a function of the city permutation $\mathbf{y}$, as a random fractal, where the conditional probability distribution of $F(\mathbf{y}')$ is $\mathcal{N}(0, \sigma^2)$-distributed under the

[23] This prerequisite is – of course – only obvious if we accept that the EA operate *locally.*

condition $F(\mathbf{y})$, with $\sigma^2 \propto \text{dist}(\mathbf{y}', \mathbf{y})^{2H}$. The "dist" stands for the distance between the permutations and $H \in (0, 1)$ is a parameter related to the fractal dimension. After computing the expected value of $(F(\mathbf{y}') - F(\mathbf{y}))^2$, the relation

$$\mathrm{E}\left\{(F(\mathbf{y}') - F(\mathbf{y}))^2\right\} \propto \text{dist}(\mathbf{y}', \mathbf{y})^{2H} \tag{1.24}$$

is obtained, which can be empirically verified for distances "dist" which are not too large. The fundamental message of (1.24) is: small changes of the state cause small fitness changes and larger changes in the state cause larger fitness changes. This interpretation of (1.24) corresponds exactly to the semi-causality and represents a statistical pendant to (1.23).

At the beginning of this section, it was intuitively argued which prerequisites should be fulfilled by the fitness landscape so that the EA can exhibit an approximately global optimization behavior. The attempt to obtain a mathematical definition of the problem led to the definitions (1.23) and (1.24). Thereby we reached a starting point for theoretical and empirical investigations. As already mentioned, the argumentation was intuitive; a more satisfactory theoretical analysis of the problem of evolvability remains as a task for future research.

1.4.2 EPP, GR, and MISR

1.4.2.1 EPP – the Evolutionary Progress Principle. The EPP stands for the statement – which is almost trivial in the ES domain – that each change of the gene set, i.e. the change of the individuals in the state space $\mathcal{A}$, can result in fitness gain as well as fitness loss. The sum of these two tendencies makes up finally the evolutionary progress:[24]

Evolutionary Progress Principle (EPP)

EVOLUTIONARY PROGRESS	=	PROGRESS GAIN	−	PROGRESS LOSS

This principle is confirmed in all (analytical) formulae of the ES, both for progress rate φ and for the quality gain $\overline{Q}$. As an extension to the EPP, one can formulate that the evolutionary progress is a function of the mutation strength σ.

An important part of the EA analysis concerns the extraction of the gain and loss parts, caused by the genetic operators (depending on the mutation

[24] As a contrast, considerations in the GA domain are mostly concentrated on the frequency (or occupation numbers) of schemata. Unfortunately, the correlation between the schemata fitness and the fitness of the individual is mathematically ambiguous. Therefore, the estimates on the optimization performance of the GA based on the schema theorem are not so reliable. A more detailed discussion of this problem can be found in Beyer (1995a).

strength). The EA designer should construct algorithms and data structures in such a way that the gain parts are as large as possible and the loss parts are as small as possible. This leads us to a central problem of the EA theory: the question of the benefit of the recombination.

1.4.2.2 GR – the Genetic Repair Hypothesis. Why are the recombinant ES, either intermediate or dominant, substantially better than the nonrecombinant strategies, i.e. have larger (local) progress rates?

The standard answer, which is almost already a dogma in the GA domain, is the *building block hypothesis* (BBH). It seems to be obvious: different building blocks (genes) with "good properties" (high fitness value) of different parents are (re)combined and build the offspring which are better than their parents. The most feared phenomenon in the GA is therefore *disruption*, i.e., the tearing apart of the building blocks. Disruption appears with a negative sign in the schema theorem (Holland, 1975; Goldberg, 1989). The negative effect of disruption is reduced by decreasing the recombination probability.[25] Even if we ignore the fact that the schema theorem is not appropriate to explain the functioning of the GA quantitatively (see, e.g., Mühlenbein, 1992), it cannot be used to explain the benefit of the recombination qualitatively. Strictly speaking, one could even conclude from the schema theorem that it is the best to avoid recombination at all because of its disruptiveness.

In order to avoid misinterpretations, it is to be emphasized that neither the schema theorem nor the BBH are totally wrong. However, the schema theorem describes aspects of the evolutionary process which are not directly (or in a mathematically unambiguous manner) related to fitness optimization, whereas the BBH *describes* an *effect* and not the cause of an obvious observation made: building blocks must inevitably come together in order to constitute the final solution, i.e. this process can be observed in *any* iterative optimization procedure. Cause and effect are mixed up at this point. The combination of building blocks is not *the* specific property of the sGA or EA.

The *genetic repair hypothesis* (GR) is suggested in this work as an alternative working principle of recombination. The GR principle is obtained almost inevitably in the analysis of the $(\mu/\mu, \lambda)$-ES (for the definition, see (1.16), p. 13). The principle of the statistical error correction (GR) was already mentioned on p. 12 and will be studied in Chap. 6, Sect. 6.1.3.2, in detail.

1.4.2.3 The MISR Principle. Dominant recombination, i.e. the exclusive and random transfer of genes from the parents to the descendants, does not seem to fit into the GR concept, since it does not contain an averaging in the algorithm. However, one obtains from the analysis that "genetic repair" indeed occurs, though implicitly and "statistically blurred," and therefore these strategies attain lower progress rates than the intermediate counterparts. Another interesting issue is observed at this point – the *genetic variability* and

[25] Also other tricks are used to avoid the disruption of the building blocks; however, there is no general recipe for this.

the concept of the *species*. In short, the following happens: (small) physical mutations[26] are transformed into genetic diversity by the dominant recombination. This genetic diversity can be interpreted as supermutations generated from an imaginary parent state (at least from the viewpoint of probability theory). It is worth noting that an "evolution loop" which is carried out iteratively, and which consists of mutation and reproduction/recombination, but does *not* have a fitness related selection operator (flat fitness landscape), does not lead to a dispersal of genetic information. Instead, the population distribution stays concentrated around a "mean individual" which in biology is called the *wild type*. It is really important to note that this effect *does not* require (fitness related) selection and is totally independent of the type of mutation distribution used as long as a finite dispersion is guaranteed. Dominant recombination thereby causes the cohesion of the population. What can be stated at last are the typical properties of a *species*: variability and cohesion. For the functioning of this game, mutations are as necessary as the (dominant) recombination. The species consists of recombined mutants. This is the content of the third suggested principle "MISR" (*mutation-induced speciation by recombination*).

It is important to underline the fundamental difference between the common explanation model and the concept represented here: Recombination is often considered as a search operator which performs global information processing by combining the partial solutions from different regions of the state space and performs global optimization in this manner. In this book a contrasting explanation is given by proposing that the dominant recombination is a locally working operator, which transforms the mutations by a random sampling process in genetic variability. Similar phenomena are also observed in nature (by way of analogy). The recombination between different species does not seem to work. An overlap of the genetic material to some degree, i.e. a sufficient similarity of the genomes, is apparently necessary in order to be able to generate viable descendants. We will return to this discussion again in Chap. 6, Sect. 6.2.2.2.

1.5 The Analysis of the ES – an Overview

The primary objective of this book is to develop analysis methods to predict the local performance of the ES. A certain theoretical framework is necessary for this purpose, where the essential terms like progress rate and quality gain are defined. This framework is established in Chap. 2. As already mentioned, concrete statements on the performance can only be obtained for appropriate fitness functions. A very essential fitness function is the sphere model. Under certain circumstances, other fitness landscapes can be approximated

[26] Random bit flips in the GA and normally distributed mutations in ES and also in EP.

locally by the sphere model. This formal approximation occurs using differential geometry techniques.[27] The sphere model also serves as the basis for an elementary treatment of the evolution dynamics.

Chapter 3 is devoted to the progress rate φ for the $(1 \overset{+}{,} \lambda)$-strategies. An exact model of the (1+1)-ES under noise is also developed in this chapter. The $N \to \infty$ asymptotics[28] of the $(1 \overset{+}{,} \lambda)$-ES is investigated for fitness functions with and without noise. Interestingly, the asymptotic results for the $(1 \overset{+}{,} \lambda)$-ES under noise were obtained earlier (Beyer, 1992b, 1993). At that time, it was believed that the complication of the N-independent quadratic progress law[29] of the $(1, \lambda)$-ES using N-dependent terms did not have a practical relevance. The usefulness of the N-dependency was recognized later for certain problem formulations. The access to the N-dependent theory of the $(1, \lambda)$-ES is found in the analysis of the σ-self-adaptation (Chap. 7). In chronological terms, these results already existed at the time of the analysis of the (μ, λ)- and the $(\mu/\mu, \lambda)$-ES. This gives an explanation for the fact that in the following chapters the N-dependent results are obtained, whereas for the noisy ES (in the third chapter) only asymptotic results will be considered.

Chapter 4 is dedicated to the quality gain $\overline{Q}$ on $(1 \overset{+}{,} \lambda)$-strategies. The approximations obtained there are valid for any arbitrary fitness functions and mutation distributions, since the quality gain can be expressed as a function of certain statistical parameters of the mutation-induced fitness fluctuations. The concrete application is offered for isotropic and correlated Gaussian mutations, and it is shown that this approach is also appropriate for the treatment of the `OneMax` bit counting function. Moreover, a connection to the normal progress, i.e. the progress rate in the normal direction, is established and it is shown that the results are comparable with those that use the model obtained from the differential geometry approach under moderate conditions.

The treatment of the (μ, λ)-ES turns out to be the technically most complicated chapter. It succeeded here for the first time to calculate a distribution of a *finite* population in a self-consistent and approximately analytical manner. The results obtained after very long derivations can be expressed as a simple formula for the asymptotic limit case $N \to \infty$: In principle, the application of the $(1, \lambda)$-φ^*-formula (3.101) from Chap. 3 suffices (see also footnote 29); the progress coefficient $c_{1,\lambda}$ should just be substituted using $c_{\mu,\lambda}$. Besides the investigation of the (μ, λ)-performance, the *explorative behavior* of (μ, λ)-ES is discussed for the sphere model. In contrast to the view presented sometimes that the population follows the local gradient ("gradient diffusion"), it will be shown that the ES performs a *random walk* on the surface of the hypersphere (this topic will be also discussed in the sixth chapter).

[27] The justification of this model as a local approximation to the more complicated fitness landscapes takes place in Chap. 4, in which the quality gain is analyzed.

[28] N is the search space dimension of the optimization problem.

[29] For the normalized progress rate φ^* the quadratic progress law can be expressed as $\varphi^* = c_{1,\lambda}\sigma^* - {\sigma^*}^2/2$. It holds exactly for $N \to \infty$.

Chapter 6 deals with the multirecombinative $(\mu/\mu, \lambda)$-ES. The intermediate as well as dominant recombination is analyzed. It is shown that the benefit of the recombination is to be explained by a statistical error correction mechanism, which is called "*genetic repair*" being similar to the DNA repair hypothesis of Bernstein et al. (1987). Thereafter, the effect of the dominant recombination is investigated using a simple mutation-recombination model. This model explains the emergence of populations, which corresponds very closely to the notion of the species in biology. The underlying principle is termed mutation-induced speciation by recombination (MISR) (Beyer, 1995a). For practical purposes, it is important to note that the progress rate formulae should be formulated in the N-dependent form for the multirecombinative ES. In contrast to the (μ, λ)-ES, the effect of N is remarkable, e.g. at the optimum number of parents μ for a given number of offspring λ. Table 6.2 will give the detailed information. Furthermore, the benefit of recombination in the case of GAs will be treated by a speculative discussion. Finally, the asymptotic behavior of the multirecombinant ES will be analyzed. An essential result of the asymptotic analysis will be that an optimally working $(\mu/\mu, \lambda)$-ES can exactly attain a progress rate λ times larger than that of the optimally tuned $(1+1)$-ES.

The study of the self-adaptation forms the seventh chapter. For the previous chapters, the theoretical framework established in Chap. 2 suffices completely. However, the analysis of the self-adaptation – even for the case of the $(1, \lambda)$-ES investigated here – requires an extension of the framework. Firstly, the whole state space $\mathcal{A}$ must be considered instead of $\mathcal{Y}$, since also the evolution of the mutation strength σ must be considered in the theoretical analysis. Some adequate quantity, similar to the progress rate φ, is to be introduced – the so-called self-adaptation response ψ. Secondly, the stochastics of the σ-evolution proves itself to be nonlinear. As a result, the problem can only be conditionally reduced to the dynamics of the mean value. Instead, the corresponding Chapman-Kolmogorov equations must be investigated. Finally, some suggestions for the setting of the learning parameter τ and a deeper understanding of the self-adaptation process are provided.

2. Concepts for the Analysis of the ES

In this chapter the theoretical framework will be defined in which the analysis of the ES will be conducted in the following chapters. In particular, the progress measures will be defined here, which can be used in the evaluation of the optimization performance of an EA (i.e. not only ES). These measures describe the local and microscopic behavior of an EA. The macroscopic behavior (i.e. the dynamics of ES) is handled in Sect. 2.4. The calculation of the progress measures requires the selection of appropriate fitness landscapes, which can be used as the models of the real landscapes. Sect. 2.2 is devoted to this topic. The differential-geometric model – which can be used for the approximation of the local (differentiable) fitness landscapes – is introduced in Sect. 2.3.

2.1 Local Progress Measures

From the mathematical perspective, every EA can be described as an inhomogeneous Markovian process, which transforms a parental population $\mathfrak{P}_\mu^{(g)}$ at time g (time or generation are used synonymously) to the parental population $\mathfrak{P}_\mu^{(g+1)}$ at time $g+1$ using a stochastic mapping (for the definition of $\mathfrak{P}$ see (1.3) and (1.5) on p. 5)

$$\mathfrak{P}_\mu^{(g)} \quad \longmapsto \quad \mathfrak{P}_\mu^{(g+1)}. \tag{2.1}$$

That is, the evolution of the system is described completely by the state of the parental generation $\mathfrak{P}_\mu^{(g)}$. The stochastic evolution of the system can be expressed by so-called Chapman-Kolmogorov equations (see, e.g., Fisz, 1971), which give integral equations in the case of real-valued ES. They transform the population density $p^{(g)}(\mathfrak{P}_\mu)$ of the parental state at generation g according to

$$p^{(g+1)}(\mathfrak{P}_\mu) = \int \cdots \int p(\mathfrak{P}_\mu | \mathfrak{P}'_\mu)\, p^{(g)}(\mathfrak{P}'_\mu)\, \mathrm{d}\mathfrak{P}'_\mu \tag{2.2}$$

into the density at time $g+1$; $\mathrm{d}\,\mathfrak{P}'_\mu$ denotes here the volume element of the Cartesian product of all μ state spaces $\mathcal{A}$ defined by (1.4).

Unfortunately, the direct analytical solution of this equation is almost always excluded. Moreover, it is mostly not possible to specify the integral kernel $p(\mathfrak{P}_\mu | \mathfrak{P}'_\mu)$ that describes the transition density of the stochastic process analytically. We will come back to this point in Chap. 7.

It is quite clear that the direct treatment of (2.2) does not appear very fruitful. Furthermore, the density $p^{(g)}(\mathfrak{P}_\mu)$ does not have a great practical importance, since the information contained in it is difficult to interpret: the actual interesting quantities are certain expected values which should be derived from this density. The *local performance measures* enter the stage at this point. They are defined as expected values of the change of certain functions of the population $\mathfrak{P}_\mu$ from the gth to the $(g+1)$th generation. The adjective *local* refers to the Markovian character of these quantities, i.e. the expected values are defined completely by the state of the system at the gth generation. This process does not have a memory. Therefore, the evolutionary dynamics of the expected values are governed by first-order difference equations, and these in turn can be approximated under certain conditions by differential equations (see Sect. 2.4).

Local performance measures are expected values of certain functions of the population. Since there are infinitely many such functions, the choice of the correct function depends on the question one wants to ask of the EA. If one is interested in the evolution of binary schemata in a GA (Holland, 1975), the schemata frequencies are adequate quantities. Whereas, if one wants to measure the optimization performance of an EA or the effect of the local "improvement of the population state," the *quality gain* and the *progress rate* are the natural quantities. Both progress measures are introduced in the following sections. Furthermore, if self-adaptation processes are considered, one has to deal with an additional quantity: the *self-adaptation response* (SAR). It measures the expected change in the endogenous strategy parameters. For the case of a single common mutation strength σ, the self-adaptation response ψ will be introduced in Chap. 7.

2.1.1 The Quality Gain $\overline{Q}$

The quality gain $\overline{Q}$ measures the expected fitness change of the parent population $\mathfrak{P}_\mu$ from generation g to $g+1$. There are many different ways to define this fitness change. If a population of μ parents with fitness values $F_m^{(g)} = F(\mathbf{y}_m^{(g)})$ is considered, the following definition

$$\overline{Q} := \mathrm{E}\left\{ \langle F^{(g+1)} \rangle_\mu - \langle F^{(g)} \rangle_\mu \right\} \quad \text{with} \quad \langle F^{(g)} \rangle_\mu := \frac{1}{\mu} \sum_{m=1}^{\mu} F_m^{(g)} \tag{2.3}$$

is reasonable. The expected value E{.} is to be calculated over the conditional density function $p(F_1, \ldots, F_\mu \mid \mathfrak{P}_\mu^{(g)})$. Consequently, one obtains

$$\overline{Q} = \frac{1}{\mu} \sum_{m=1}^{\mu} \int \cdots \int F_m \, p\Big(F_1, \ldots, F_\mu \mid \mathfrak{P}_\mu^{(g)}\Big) \, \mathrm{d}F_1 \ldots \mathrm{d}F_\mu - \langle F^{(g)} \rangle_\mu . \tag{2.4}$$

The problems with this approach concentrate on the calculation of the conditional probability density $p(F_1, \ldots, F_\mu \mid \mathfrak{P}_\mu^{(g)})$.

Until now, only the theory for the $(1 \overset{+}{,} \lambda)$ case has been developed, i.e. without a real population and therefore also without recombination. Therefore, the definition (2.4) simplifies to

$$\overline{Q}_{1 \overset{+}{,} \lambda} := \int F\, p_{1;\lambda}(F \mid \mathbf{y}^{(g)}, \mathbf{s}^{(g)}, F^{(g)})\, \mathrm{d}F \; - \; F^{(g)}, \tag{2.5}$$

where $\mathbf{y}^{(g)}$ is the set of the parameters of the parent, $\mathbf{s}^{(g)}$ is the corresponding set of strategy parameters, $F^{(g)}$ is the fitness $F^{(g)} = F(\mathbf{y}^{(g)})$ of the parent and $p_{1;\lambda}(F \mid \mathbf{y}^{(g)}, \mathbf{s}^{(g)}, F^{(g)})$ is the density of the best descendant. In this case, it is convenient to introduce the *local quality function* $Q_{\mathbf{y}}(\mathbf{x})$, which relocates the zero point of the quality value to the parental state $\mathbf{y} = \mathbf{y}^{(g)}$

$$Q_{\mathbf{y}}(\mathbf{x}) := F(\mathbf{y} + \mathbf{x}) - F(\mathbf{y}) . \tag{2.6}$$

State changes in the parameter space, i.e. the mutations, are described by $\mathbf{x} \neq \mathbf{0}$. Therefor, using (2.5) one obtains

$$\overline{Q}_{1 \overset{+}{,} \lambda} = \int Q(\mathbf{x})\, p_{1;\lambda}(Q \mid \mathbf{y})\, \mathrm{d}Q(\mathbf{x}). \tag{2.7}$$

The density of the best descendant generated by the mutation $\mathbf{x}$ is denoted by $p_{1;\lambda}(Q)$.[1]

The quality gain $\overline{Q}$ is a performance measure for populations with a finite number of individuals $\mu, \lambda < \infty$. In the asymptotic limit, infinite populations can be considered. If the *truncation threshold* (selection strength) ϑ of the (μ, λ)-selection is defined as

$$\vartheta := \lim_{\mu, \lambda \to \infty} \frac{\mu}{\lambda}, \qquad 0 < \vartheta < 1, \tag{2.8}$$

$\vartheta \cdot 100\%$ of the offspring are reproduced. If one assumes that the probability densities $p_N^{(g)}$ of the fitness distribution of the offspring population are known, the quality gain (in the case of maximization) can be defined as

$$\overline{Q} := \frac{1}{\vartheta} \int_{F = P^{-1}(1-\vartheta)}^{\infty} F\, p_N^{(g)}(F)\, \mathrm{d}F \; - \; \int_{-\infty}^{\infty} F\, p_E^{(g)}(F)\, \mathrm{d}F. \tag{2.9}$$

The lower limit of the first integral is the $(1 - \vartheta)$th order quantile, i.e. one integrates over the values of F which belong to the ϑ·100% best offspring. The definition (2.9) is used mostly in quantitative genetics and is called *selection response* R (Blumer, 1980).

The selection response model was applied by Mühlenbein and Schlierkamp-Voosen (1993) to the analysis of GAs. Since this model uses an infinite population size, the results obtained have principally an approximative character.

---

[1] In order to simplify the notation, the dependency on the strategy parameters of the density $p_{1;\lambda}(Q)$ is not indicated in (2.7).

However, the main problem lies – as in the case of models with finite population size – in the self-consistent determination of the offspring distribution $p_N^{(g)}(F)$. Rudimentary approaches made to estimate this distribution contain phenomenological coefficients. A complete deduction of these coefficients from the stochastic GA process has not succeeded until now, not even for the case of the `OneMax` function.[2]

2.1.2 The Progress Rate φ

While the quality gain $\overline{Q}$ measures the change of the scalar fitness in the $\mathcal{F}$ space, the progress rate φ is a distance measure in the parameter space $\mathcal{Y}$. It measures the expected change in the distance of certain individuals of the parent population $\mathfrak{P}_\mu$ to a fixed reference point $\hat{\mathbf{y}}$ from generation g to generation $g+1$.

As far as fitness optimization is concerned, it is reasonable to identify $\hat{\mathbf{y}}$ with the coordinates of the optimum. This measure is only uniquely defined with respect to $\hat{\mathbf{y}}$, if the optimum is not degenerate, i.e. if we do not have several optima with the same fitness value, which is the case for all of the fitness landscapes investigated in this work.

The identification of $\hat{\mathbf{y}}$ with the place of the optimum, however, is not mandatory. The "progress" in the parameter space $\mathcal{Y}$ can also be defined in other ways, e.g., as the "fictitious distance gain" in the direction of the local gradient $\nabla_x Q_{\mathbf{y}}(\mathbf{x})$ (cf. Sect. 2.1.3). However, since the gradient coincides only in special cases (e.g., the sphere model) with the direction toward the optimum, the information obtained from this definition is not equivalent to the progress rate with respect to $\hat{\mathbf{y}}$.

In general, the progress rates can be defined in arbitrary directions in space, even perpendicular to the optimum direction, e.g., if one is interested in the distribution of the population at the parabolic ridge test function. The application of such measures is still at the initial stages and is therefore not included in this book.

The investigation of the ES in real-valued parameter spaces $\mathcal{Y} \subseteq \mathbb{R}^N$ will be based on the following definitions of the progress rate. The symbol $\tilde{\mathbf{y}}_l$ denotes the object parameter set of the lth descendant at generation g, and $\mathbf{y}_m$ the mth parent. We compare the parents of generations g and $g+1$. The reference point is the location of the optimum.

Principally, there are two possibilities to specify the change in the parent population: The "averaged evaluation of the parent individuals" and "collective evaluation" of the population. In the first case, the Euclidean distances of the parents to the optimum is considered. The definition of the progress rate φ is then

---

[2] There is an approach by Prügel-Bennett and Shapiro (1994, 1997) that is able to determine the dynamics of the `OneMax` function (and also for other functions). However, it uses a special form of proportionate selection, called Boltzmann selection, which is usually not used in practical GA applications.

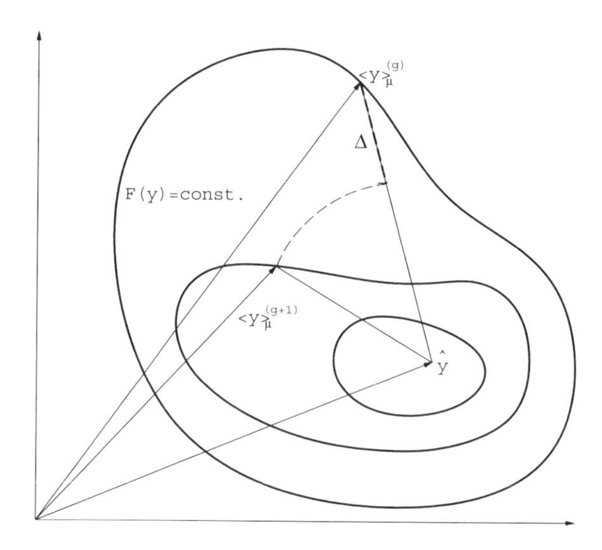

Fig. 2.1. On the geometrical interpretation of the progress rate $\varphi_{\mu/\rho \overset{+}{,} \lambda}$ with "collective evaluation," i.e. computation of the centroids of the parent populations in the parameter space $\mathcal{Y}$: φ is the expected value of the change in the distance $\Delta := \|\hat{\mathbf{y}} - \langle \mathbf{y} \rangle_\mu^{(g)}\| - \|\hat{\mathbf{y}} - \langle \mathbf{y} \rangle_\mu^{(g+1)}\|$, see Eq. (2.13).

$$\varphi_{\mu \overset{+}{,} \lambda} := \mathrm{E}\left\{ \frac{1}{\mu} \sum_{m=1}^{\mu} \left\| \hat{\mathbf{y}} - \mathbf{y}_m^{(g)} \right\| - \frac{1}{\mu} \sum_{m=1}^{\mu} \left\| \hat{\mathbf{y}} - \mathbf{y}_m^{(g+1)} \right\| \right\}. \tag{2.10}$$

Here $\frac{1}{\mu} \sum_{m=1}^{\mu} \|\hat{\mathbf{y}} - \mathbf{y}_m^{(g)}\|$ and $\frac{1}{\mu} \sum_{m=1}^{\mu} \|\hat{\mathbf{y}} - \mathbf{y}_m^{(g+1)}\|$ stand for the averaged distances of the parents to the optimum. For comma strategies $\mathbf{y}_m^{(g+1)} = \tilde{\mathbf{y}}_{m;\lambda}^{(g)}$. Thus, one obtains with (2.10)

$$\varphi_{\mu,\lambda} = \frac{1}{\mu} \sum_{m=1}^{\mu} \mathrm{E}\left\{ \left\| \hat{\mathbf{y}} - \mathbf{y}_m^{(g)} \right\| - \left\| \hat{\mathbf{y}} - \tilde{\mathbf{y}}_{m;\lambda}^{(g)} \right\| \right\}. \tag{2.11}$$

This equation will be the starting point for the calculation of the progress rate on the (μ, λ)-sphere model.

An alternative to the progress rate definition (2.10) is to consider a "collective evaluation" of the population. For this purpose, the centroids $\langle \mathbf{y} \rangle_\mu^{(.)}$ of the parent generations g and $g+1$ are calculated

$$\langle \mathbf{y} \rangle_\mu^{(.)} := \frac{1}{\mu} \sum_{m=1}^{\mu} \mathbf{y}_m^{(.)}. \tag{2.12}$$

The expected value of the change in the distance with respect to $\hat{\mathbf{y}}$ is the progress rate. It will be used for the analysis of the recombinative ES

$$\varphi_{\mu/\rho \overset{+}{,} \lambda} := \mathrm{E}\left\{ \left\| \hat{\mathbf{y}} - \langle \mathbf{y} \rangle_\mu^{(g)} \right\| - \left\| \hat{\mathbf{y}} - \langle \mathbf{y} \rangle_\mu^{(g+1)} \right\| \right\}. \tag{2.13}$$

The geometrical interpretation of this quantity is shown in Fig. 2.1. The centroid of the $(g+1)$th parent generation, $\langle \mathbf{y} \rangle_\mu^{(g+1)}$, can be represented in the case of comma strategies by the descendants $\tilde{\mathbf{y}}_l$ of the gth generation

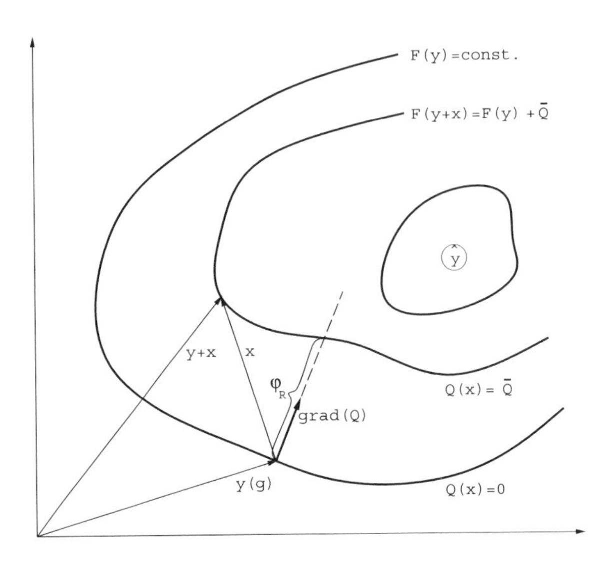

Fig. 2.2. On the geometrical interpretation of the normal progress φ_R defined by Eq. (2.19).

$$\langle \mathbf{y} \rangle_\mu^{(g+1)} = \frac{1}{\mu} \sum_{m=1}^{\mu} \tilde{\mathbf{y}}_{m;\lambda}^{(g)}. \tag{2.14}$$

The definitions (2.11) and (2.13) are not equivalent in general, since the operations for averaging and taking the absolute value are not exchangeable. However, in the case of $(1 \overset{+}{,} \lambda)$ strategies, the results are equivalent, and the definitions reduce to

$$\varphi_{1 \overset{+}{,} \lambda} = \mathrm{E}_{\overset{+}{,}} \left\{ \left\| \hat{\mathbf{y}} - \mathbf{y}^{(g)} \right\| - \left\| \hat{\mathbf{y}} - \tilde{\mathbf{y}}_{1;\lambda}^{(g)} \right\| \right\}. \tag{2.15}$$

The progress rates using (2.11) and (2.13) are not comparable in general. However, if populations are considered which show the structure of a "species," i.e. the population is concentrated around the center of mass of the parents $\langle \mathbf{y} \rangle_\mu$, and the distances of the parents to the reference point $\hat{\mathbf{y}}$ are large compared to the extension of the population, the $\varphi_{\mu/\rho \overset{+}{,} \lambda}$ behaves asymptotically like $\varphi_{\mu \overset{+}{,} \lambda}$

$$\varphi_{\mu/\rho \overset{+}{,} \lambda} \xrightarrow{N \to \infty} \varphi_{\mu \overset{+}{,} \lambda}. \tag{2.16}$$

The relation (2.16) is particularly valid, because recombinative strategies generate population distributions which are distributed almost normally around the centroid of the population and which have a ratio of the averaged population extension to the residual distance to the optimum of the order of $\sqrt{\mu/N}$ (sphere model). This will be shown in the relevant chapter.

2.1.3 The Normal Progress φ_R

The derivation of analytical results on the progress rate φ is tied to considerable mathematical/numerical problems. Even if the parent population size is reduced down to $\mu = 1$, the treatment of correlated mutations or of fitness models, which are more complicated than the sphere model, is very difficult. On the other hand, the quality gain for $(1 \overset{+}{,} \lambda)$-strategies can be calculated relatively easily for a wide range of fitness functions. Probably due to this reason, Rechenberg (1994) has suggested defining the progress in the direction of the parental gradient. Figure 2.2 visualizes this idea: the local parental gradient is perpendicular to the isometric line of equal fitness $Q(\mathbf{x}) = 0$ of the local fitness function $Q_{\mathbf{y}}(\mathbf{x})$ and intersects the neighboring isometric line $Q(\mathbf{x}) = \overline{Q}$. The distance defined thereby was denoted in Rechenberg (1994, p. 56) as progress rate φ'. We will call this progress rate *normal progress* φ_R; the subscript "$_R$" is used to denote Rechenberg and "normal" refers to the normal direction progress, since this progress measure gives the (fictitious) distance gain in the gradient direction.

Assuming that $\mathbf{x}$ is sufficiently small, the local quality function can be expanded at $\mathbf{x} = \mathbf{0}$ using the Taylor series

$$Q_{\mathbf{y}}(\mathbf{x}) = \nabla_x Q_{\mathbf{y}}(\mathbf{x})\big|_{\mathbf{x}=\mathbf{o}} \delta\mathbf{x} + \dots . \tag{2.17}$$

If the quality gain $\overline{Q}$ is calculated as the $\delta\mathbf{x}$ in the direction of the local gradient according to Fig. 2.2

$$\delta\mathbf{x} := \frac{\nabla_x Q_{\mathbf{y}}(\mathbf{x})\big|_{\mathbf{x}=\mathbf{o}}}{\left\| \nabla_x Q_{\mathbf{y}}(\mathbf{x})\big|_{\mathbf{x}=\mathbf{o}} \right\|} \varphi_R, \tag{2.18}$$

one obtains, by neglecting the higher-order terms in (2.17), the definition equation of the normal progress φ_R

$$\varphi_R := \frac{\overline{Q}}{\left\| \nabla_x Q_{\mathbf{y}}(\mathbf{x})\big|_{\mathbf{x}=\mathbf{o}} \right\|}. \tag{2.19}$$

The normal progress φ_R can serve only as an "approximate value" to the rigorous φ-definition, which was introduced in the previous subsection. Indeed, one should not expect from definition (2.19) any information beyond the quality gain $\overline{Q}$. Actually, (2.19) is a *local normalization* of the quality gain $\overline{Q}$, which is invariant with respect to "gauge transformations" $Q(\mathbf{x}) \mapsto aQ(\mathbf{x})$, $a = \text{const}$.

For fitness landscapes $Q_{\mathbf{y}}(\mathbf{x}) = 0$ which are almost symmetrical with respect to $\mathbf{x} = \mathbf{0}$ and which have the optimum $\hat{\mathbf{y}}$ in the neighborhood of the symmetry axis, the normal progress gives results in accordance with the rigorous φ-definition. We will come back to this point in Chap. 4.

2.2 Models of Fitness Landscapes

The analytical calculation of the progress measures requires suitable fitness models which can be handled using approximate-analytical methods. The most prominent model is definitely the (hyper-)*sphere model*. In the case of very small mutation strengths, it degrades to the *hyperplane*. One can apply constraints to the hyperplane, yielding the *corridor* and the *discus models*. Besides these very simple fitness models, one can analyze more complicated fitness landscapes using the quality gain $\overline{Q}$ measure. The general cases of quadratic fitness functions and functions with higher-order Taylor expansions among such landscapes should be mentioned. The `OneMax`-function – a very popular model in GA research – can also be investigated in the framework of the $\overline{Q}$-analysis.

All the functions mentioned up to now are deterministic. However, the optimization of technical systems is usually affected by noise. Particularly, the measurement of the fitness value is subject to stochastic perturbations. Therefore, noisy fitness landscapes form an important problem class which can be described by appropriate models.

The fitness functions considered in this subsection – except the `OneMax`-function – are mappings of type $\mathbb{R}^N \mapsto \mathbb{R}$. For the `OneMax` we have $\mathbb{B}^\ell \mapsto \{0, \mathbb{N}\}$.

2.2.1 The (Hyper-)Sphere Model

This model describes sphere-symmetrical fitness landscapes $F(\mathbf{y})$ with $\hat{\mathbf{y}}$ as the center of symmetry, i.e. "isometric surfaces" $F(\mathbf{y}) = \text{const}$ draw concentric (hyper-)sphere shells around $\hat{\mathbf{y}}$. For notational convenience, the origin is used as the center of symmetry, and the radius vector $\mathbf{R}$ is introduced

$$\mathbf{y} = \hat{\mathbf{y}} + \mathbf{R}. \tag{2.20}$$

Therefor, $F(\mathbf{y})$ becomes a function of type $Q_{\hat{y}}(\mathbf{R})$, which only depends on the length $R := \|\mathbf{R}\|$, $R \geq 0$

$$F(\mathbf{y}) = F(\hat{\mathbf{y}} + \mathbf{R}) =: Q_{\hat{y}}(\mathbf{R}) = Q(\|\mathbf{R}\|) = Q(R), \quad Q(0) = F(\hat{\mathbf{y}}). \tag{2.21}$$

It is assumed that $Q(R)$ is a strongly monotonic function having its optimum at $R = 0$. Because of the monotonicity, there are no further (local) optima.[3]

The extraordinary role played by the sphere model in ES theory is based on its R-dependency only. The evolution dynamics of the ES is completely described by the R state, having the progress rate φ (see Sect. 2.4). The progress rate, which is a function of R and σ,[4] i.e. $\varphi = \varphi(R, \sigma)$, is therefore scalable. After applying the normalization

$$\boxed{\text{NORMALIZATION:} \qquad \varphi^* := \varphi \frac{N}{R} \quad \text{and} \quad \sigma^* := \sigma \frac{N}{R},} \tag{2.22}$$

the progress rate $\varphi = \varphi(R, \sigma)$ becomes independent of R: $\varphi^* = \varphi^*(\sigma^*, N)$. Thus, the possibility to compare the performance of different strategies using analytical φ^*-formulae is opened up.

[3] Relaxation of this monotonicity requirement will open the chance for future investigations of the behavior of the ES in multimodal fitness landscapes. However, such models are not considered in this book.

[4] This refers to the case of isotropic Gaussian mutations according to Eq. (1.11).

2.2.2 The Hyperplane

The fitness landscape *hyperplane* is a linear function in $\mathbb{R}^N$, which is described by the equation

$$F(\mathbf{y}) := F_0 + \mathbf{c}^{\mathrm{T}}\mathbf{y}, \qquad \mathbf{c} = \mathrm{const} \tag{2.23}$$

It will not be particularly investigated in this book: provided that the ES algorithm is invariant as to coordinate rotations, the progress rate φ for finite mutations $\mathbf{z}$ ($\sigma < \infty$) on the hyperplane can be derived as a limit case of the sphere model for $R \to \infty$.

A proof of this asymptotic property is obtained in two steps. First, the transition of the sphere model to the plane equation using a Taylor expansion of the special sphere model $Q(\mathbf{R}+\mathbf{z}) = ||\mathbf{R}+\mathbf{z}||$ at $\mathbf{R}$ will be shown for large $R = ||\mathbf{R}||$

$$\begin{aligned}
Q(\mathbf{R}+\mathbf{z}) &= ||\mathbf{R}+\mathbf{z}|| = \sqrt{(\mathbf{R}+\mathbf{z})^2} = \sqrt{||\mathbf{R}||^2 + 2\mathbf{R}^{\mathrm{T}}\mathbf{z} + ||\mathbf{z}||^2} \\
&= ||\mathbf{R}||\sqrt{1 + \frac{2}{||\mathbf{R}||}\left(\mathbf{e}_R^{\mathrm{T}}\mathbf{z} + \frac{1}{2}\frac{||\mathbf{z}||^2}{||\mathbf{R}||}\right)} \\
&= R\left(1 + \frac{1}{R}\left(\mathbf{e}_R^{\mathrm{T}}\mathbf{z} + \frac{1}{2}\frac{||\mathbf{z}||^2}{R}\right) - \frac{1}{2}\frac{1}{R^2}\left(\mathbf{e}_R^{\mathrm{T}}\mathbf{z} + \frac{1}{2}\frac{||\mathbf{z}||^2}{R}\right)^2 + \ldots\right) \\
&= R + \mathbf{e}_R^{\mathrm{T}}\mathbf{z} + \frac{1}{2}\frac{||\mathbf{z}||^2 - (\mathbf{e}_R^{\mathrm{T}}\mathbf{z})^2}{R} + \mathcal{O}\left(\frac{1}{R^2}\right).
\end{aligned} \tag{2.24}$$

The mutation vector is denoted by $\mathbf{z}$, the unit vector in the $\mathbf{R}$-direction as $\mathbf{e}_R^{\mathrm{T}}$, and one uses the Taylor expansion

$$\sqrt{1+x} = 1 + \frac{1}{2}x - \frac{1}{8}x^2 + \frac{1}{16}x^3 - \ldots, \qquad |x| \le 1. \tag{2.25}$$

As one can see in Eq. (2.24), the special sphere model $Q(\mathbf{z}) = ||\mathbf{R}+\mathbf{z}||$ converges to the hyperplane similar to Eq. (2.23) for large values of R.

Now we expand φ^* in a Taylor series with respect to σ^*

$$\varphi^*(\sigma^*) = \varphi^*(0) + \left.\frac{\mathrm{d}\varphi^*}{\mathrm{d}\sigma^*}\right|_{\sigma^*=0}\sigma^* + \frac{1}{2}\left.\frac{\mathrm{d}^2\varphi^*}{\mathrm{d}\sigma^{*2}}\right|_{\sigma^*=0}\sigma^{*2} + \ldots. \tag{2.26}$$

If $\varphi^*(0) \equiv 0$ is taken into account, since $\sigma^* = 0 \;\to\; \sigma = 0$ is actually the "null mutation" $\mathbf{z} = \mathbf{0}$, one obtains using (2.22)

$$\varphi(\sigma) = \left.\frac{\mathrm{d}\varphi^*}{\mathrm{d}\sigma^*}\right|_{\sigma^*=0}\sigma + \frac{1}{2}\left.\frac{\mathrm{d}^2\varphi^*}{\mathrm{d}\sigma^{*2}}\right|_{\sigma^*=0}\sigma^2\,\frac{N}{R} + \ldots \tag{2.27}$$

and for $R \to \infty$ one gets the progress rate of the hyperplane

$$\varphi_{\text{hyperplane}}(\sigma) = \left.\frac{\mathrm{d}}{\mathrm{d}\sigma^*}\varphi^*_{\text{sphere}}\right|_{\sigma^*=0}\sigma, \tag{2.28}$$

where $\varphi^*_{\text{sphere}}$ denotes the normalized progress rate of the sphere model.

The geometrical interpretation of this derivation is the following: for very small mutations $\mathbf{z}$, i.e. for a very large radius R compared to the mutations, the hypersphere surface appears locally as a plane (tangential plane, to be exact). Very small mutations correspond to very small normalized mutation strengths σ^*. Therefore, one can approximate φ^* for $\sigma^* = 0$ using the linear term of the Taylor series only. Consequently, we obtain a φ-formula for the tangential plane of the hypersphere and therefore the formula for the hyperplane, if we omit thereafter the condition for small mutations.

2.2.3 Corridor and Discus – Hyperplanes with Restrictions

The *corridor model* (Rechenberg, 1973) defines a region $\mathcal{C}$ (see Fig. 2.3) similar to a corridor for the feasible parameter settings $\mathbf{y}$

$$\text{CORRIDOR} \qquad \mathcal{C}: \quad \begin{cases} -\infty \le y_1 \le \infty \\ -b \le y_i \le b, \quad 1 < i \le N. \end{cases} \tag{2.29}$$

The fitness function $F(\mathbf{y})$ is defined as

$$F(\mathbf{y}) := \begin{cases} f_{\mathcal{C}}(y_1), & \mathbf{y} \in \mathcal{C} \\ \text{“}\mathtt{lethal}\text{”}, & \mathbf{y} \notin \mathcal{C}, \end{cases} \tag{2.30}$$

where “`lethal`” stands for $+\infty$ (for the case of minimization) or $-\infty$ (maximization). Therefore, progress can be attained only in the y_1-direction, and $f_{\mathcal{C}}(y_1)$ is a monotonic function of y_1.

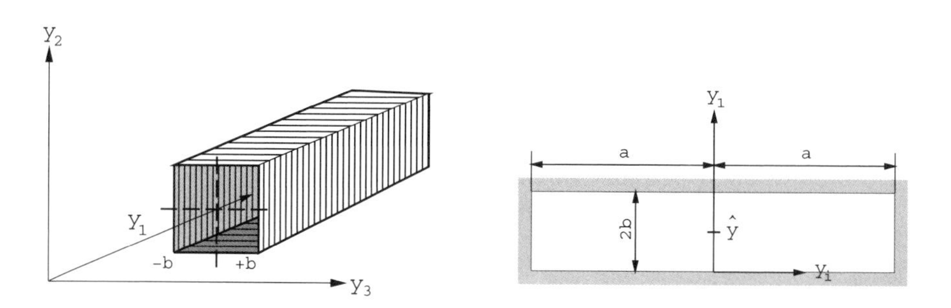

Fig. 2.3. The corridor model (left) and the discus (right)

In contrast to the corridor model, which is constrained in all directions but y_1, the *discus* model $\mathcal{D}$ has only one constraint which is in the y_1-direction. This is also the direction toward the optimum. All other y_i directions ($i = 2, \ldots, N$) are selectively neutral (Beyer, 1989a, 1990a)

$$\text{DISCUS} \qquad \mathcal{D}: \quad \begin{cases} 0 \le y_1 \le 2b \\ -a \le y_i \le a, \quad 1 < i \le N, \end{cases} \tag{2.31}$$

with the

$$\text{DISCUS CONDITION:} \qquad b \ll a. \tag{2.32}$$

The fitness function $F(\mathbf{y})$ is defined as follows

$$F(\mathbf{y}) := \begin{cases} f_{\mathcal{D}}(y_1), & \mathbf{y} \in \mathcal{D} \\ \text{“}\texttt{lethal}\text{”}, & \mathbf{y} \notin \mathcal{D}, \end{cases} \tag{2.33}$$

with the function $f_{\mathcal{D}}(y_1)$, which has its optimum at $y_1 = \hat{y}_1 = b$, and the boundary condition $f_{\mathcal{D}}(0) = f_{\mathcal{D}}(2b) = 0$.

Both models are introduced here for completeness. They will not be analyzed in this work.[5]

2.2.4 Quadratic Functions and Landscapes of Higher-Order

Particularly in the theory of the quality gain it is possible to analyze more complicated fitness landscapes than the sphere model. A natural approach is to expand the general fitness $F(\mathbf{y})$ in a Taylor series at the parental state $\mathbf{y}$. This approach is justified by the fact that relatively small mutations $\mathbf{x} = \mathbf{z}$ are applied to the parental state. Therefore, the first few terms – usually the first two – of the Taylor expansion suffice to approximate the original function locally

$$F(\mathbf{y}+\mathbf{x}) = F(\mathbf{y}) + \sum_{i=1}^{N} \frac{\partial F}{\partial y_i} x_i + \frac{1}{2} \sum_{i=1}^{N} \sum_{j=1}^{N} \frac{\partial^2 F}{\partial y_i \, \partial y_j} x_i x_j + \dots . \tag{2.34}$$

Thus, one obtains the *local fitness function* $Q_{\mathbf{y}}(\mathbf{x})$ (quality function, cf. Eq. (2.6))

$$F(\mathbf{y}+\mathbf{x}) - F(\mathbf{y}) = Q_{\mathbf{y}}(\mathbf{x}) := \mathbf{a}^{\mathrm{T}}\mathbf{x} - \mathbf{x}^{\mathrm{T}}\mathbf{Q}\,\mathbf{x}, \tag{2.35}$$

with vector $\mathbf{a}$ and matrix $\mathbf{Q}$ depending on the parental state $\mathbf{y}$

$$(\mathbf{a})_i = a_i(\mathbf{y}) := \frac{\partial F}{\partial y_i} \qquad \text{and} \qquad (\mathbf{Q})_{ij} = q_{ij}(\mathbf{y}) := -\frac{1}{2} \frac{\partial^2 F}{\partial y_i \, \partial y_j}. \tag{2.36}$$

Beyond the general quadratic model (2.35), it may be necessary to consider higher-order derivatives in (2.34). For example, if the $\mathbf{Q}$-matrix vanishes for a special $\mathbf{y}$, higher-order terms are left to consider. As an example, in Chap. 4, the local quality function

$$Q_{\mathbf{y}}(\mathbf{x}) := \sum_{i=1}^{N} a_i(\mathbf{y}) x_i - \sum_{i=1}^{N} c_i(\mathbf{y})\, x_i^4 \tag{2.37}$$

will be investigated.

[5] An analysis of the $(1, \lambda)$-ES on the corridor model can be found in Schwefel (1995).

2.2.5 The Bit Counting Function OneMax

The OneMax-function is a popular function in the newer theoretical works on the analysis of the GA (see, e.g., Mühlenbein & Schlierkamp-Voosen, 1993; 1994).[6,7] It takes the sum of all single bit positions in a bitstring $\mathbf{y} \in \mathbb{B}^\ell$ of length ℓ

$$F(\mathbf{y}) = \mathtt{OneMax}(\mathbf{y}) := \sum_{i=1}^{\ell} y_i \tag{2.38}$$

and has its maximum $\hat{F} = \ell$ at $y_i = 1$, $i = 1, \ldots, \ell$.

OneMax can be used also in the ES-analysis. In particular the quality gain of $(1 \overset{+}{,} \lambda)$ strategies can be calculated approximately very easily. The local quality function

$$Q_{\mathbf{y}}(\mathbf{x}) := \mathtt{OneMax}(\mathbf{x}) - \mathtt{OneMax}(\mathbf{y}) \tag{2.39}$$

is defined for this purpose, it measures the fitness change from the parent state $\mathbf{y}$ to an offspring state $\mathbf{x}$.

A peculiarity of OneMax is its "distance property" in $\mathbb{B}^\ell$, which is similar to the special sphere model $F(\mathbf{y}) = -\sqrt{\mathbf{y}^2}$ in $\mathbb{R}^N$, if the Hamming distance in $\mathbb{B}^\ell$ is used instead of the Euclidean distance (in $\mathbb{R}^N$). For both models one has $\varphi_{1 \overset{+}{,} \lambda} = \overline{Q}_{1 \overset{+}{,} \lambda}$, i.e. $\overline{Q}_{1 \overset{+}{,} \lambda}$ is the average change of the parental Hamming distance to the optimum.

2.2.6 Noisy Fitness Landscapes

Fitness landscapes at which the fitness measurements are superposed with statistical measurement errors represent an important problem class in the ES-analysis. Noisy fitness data can change the convergence behavior of the ES considerably, as will be shown in Chap. 3.

The essential behavior of the ES in the case of noisy fitness values can be investigated for simple landscapes, such as the sphere model. As a noise model, an *additive* noise on the local fitness $Q_{\mathbf{y}}(\mathbf{x})$ is assumed.[8] Thereby

[6] It should be mentioned that the investigation of the performance of simple EAs on the OneMax-bitcounting function has a long but mostly unknown history. It can be traced back to the late 1950s (!) and is associated with the work of Bremermann (1958) (see also Bremermann et al., 1966).

[7] Recently, investigations on the performance of the $(1+1)$-ES in binary search spaces experienced a revival. The reader is referred to the Ph.D. work of Rudolph (1997) and to Droste et al. (1998).

[8] Besides the case of noisy fitness as error source in optimization problems on material objects, there is also another source: the actuator realizing the state change $\mathbf{x}$ at the object may not work correctly. In such a case, the *realized* parameter vector $\mathbf{y} + \mathbf{x}$ is also noise perturbed. This noise type is not analyzed here. For the case of the $(1+1)$-ES the reader is referred to the work of Ott (1993) .

the noise ε_Q is realized using a normal distribution. Systematic errors are excluded, i.e. $\mathrm{E}\{\varepsilon_Q\} = 0$. The assumption of the normal distribution is not mandatory; however, the measurement errors in nature and technology can be modeled very often by using this distribution. Furthermore, the mean values of the stochastic processes mostly have this distribution (central limit theorem of statistics).

The measured fitness is represented by $\tilde{Q}_{\mathbf{y}}$. It is defined as

$$\tilde{Q}_{\mathbf{y}}(\mathbf{x}) := Q_{\mathbf{y}}(\mathbf{x}) + \varepsilon_Q(\mathbf{y}). \tag{2.40}$$

Here we denote by $\mathbf{y}$ in $\varepsilon_Q(\mathbf{y})$ that the magnitude of the error can depend on the parental state. The consideration of this parameter dependency is important, since, e.g., the analog measurement instruments have relative errors which depend on the magnitude of the measured quantity. The probability density of the error is given by the normal distribution

$$p(\varepsilon_Q) = \frac{1}{\sqrt{2\pi}\,\sigma_{\varepsilon_Q}} \exp\left(-\frac{1}{2}\left(\frac{\varepsilon_Q}{\sigma_{\varepsilon_Q}}\right)^2\right) \quad \text{with} \quad \sigma_{\varepsilon_Q} = \sigma_{\varepsilon_Q}(\mathbf{y}). \tag{2.41}$$

The dependency on the parental state is given consequently by the standard deviation of the error. Any dependency from $\mathbf{x}$ beyond these considerations is ignored in the model.

2.3 The Differential-Geometrical Model for Non-Spherical Fitness Landscapes

The effort to derive analytical formulae for the progress rate φ on the sphere model is cumbersome. One might argue that the practical use of such results is quite small, because they are only valid for functions of type $F(\mathbf{y}) = Q(\|\mathbf{y} - \hat{\mathbf{y}}\|)$. Therefore, there is a demand for generalizations of the results obtained on the sphere model.

Let us assume a sphere model which is deformed slightly. The model is defined by the local quality function $Q_{\mathbf{y}}(\mathbf{x})$, whereas the parental state $\mathbf{y}$ is considered as constant and the quality function value $Q_{\mathbf{y}}(\mathbf{0}) = 0$ is fixed independent of the degree of deformation. We are now interested in the change of the progress rate at the parental state $\mathbf{x} = \mathbf{0}$. In the case of small deformations we can expect that φ does not change much. The progress rate on the sphere model is determined essentially by the radius R of the hypersphere (which will be shown in Chap. 3, see also Sect. 2.2.1, p. 32). The radius is changed by the deformation, more precisely, the *local* curvature relations are changed. The basic idea here is to approximate the fitness landscape locally by a new hypersphere having a curvature $\varkappa = 1/R$ which is equal to the *mean curvature* $\langle\varkappa\rangle$ of the landscape in the state $\mathbf{x} = \mathbf{0}$.

The calculation of the mean curvature radius $R_\varkappa = 1/\langle\varkappa\rangle$ is a problem in N-dimensional Riemannian geometry. The derivation requires some familiarity with tensor calculus. Kreyszig (1968) is recommended for a good introduction into classical differential geometry. However, the principles of the derivations can be understood without the technical details by just following the geometrical interpretations.

The following considerations are mainly for the formal derivation of the scalar invariant $\langle\varkappa\rangle$ that describes the mean curvature of the landscape in the parental state $\mathbf{y}$. Thereafter, we postulate that the reciprocal value of this quantity can be used as the substitute radius in the φ-formulae. This *ad hoc* step is for the time being hypothetical, because it requires a probability theoretical substantiation, which follows in Chap. 4 by comparing the results of the analysis of the quality gain and of the normal progress rate.

In this subsection *Einstein's summation convention* will be used. That is, sums are taken from 1 to $N-1$, as long as the same lower case Greek summation indices appear both as subscripts *and* as superscripts, e.g.,

$$a_\alpha b^\alpha := \sum_{\alpha=1}^{N-1} a_\alpha b^\alpha. \tag{2.42}$$

This subsection comprises two parts. The first one, 2.3.1, introduces the theory of the mean curvature. In the second part, 2.3.2, this theory is applied to the local quality function $Q_\mathbf{y}(\mathbf{x})$.

2.3.1 Fundamentals of the Differential-Geometrical Model

2.3.1.1 The Local Hyperplane $\partial Q_\mathbf{y}$. The fitness landscape in the neighborhood of the parental state $\mathbf{y}$ is described locally by the quality function (2.6) $Q_\mathbf{y}(\mathbf{x}) = F(\mathbf{y}+\mathbf{x}) - F(\mathbf{y})$. The equation $Q_\mathbf{y}(\mathbf{x}) = 0$ thereby defines the surface $\partial Q_\mathbf{y}$ of the *local success domain*, i.e. the part of $\mathbb{R}^N$ that contains the offspring $\mathbf{x}$ which are better than their parent $\mathbf{y}$. The boundary $\partial Q_\mathbf{y}$: $Q_\mathbf{y}(\mathbf{x}) = 0$ is, geometrically interpreted, a hypersurface in $\mathbb{R}^N$.

If one introduces a Cartesian system $\{\mathbf{e}_i\}$, $i = 1, \ldots, N$, the hypersurface can be parameterized by the vector function

$$\mathbf{x}(u^1, u^2, \ldots, u^{N-1}) = \sum_{i=1}^{N} x_i(u^1, u^2, \ldots, u^{N-1})\, \mathbf{e}_i. \tag{2.43}$$

The vector components x_i depend on $N-1$ contravariant coordinates u^α ($\alpha = 1, \ldots, N-1$). Since we are interested in the local geometry of $\partial Q_\mathbf{y}$ in the parental state, the coordinate system

$$x_1 = u^1, \quad \ldots, \quad x_{N-1} = u^{N-1}, \quad x_N = f(u^1, u^2, \ldots, u^{N-1}) \tag{2.44}$$

appears to be a natural choice. The component x_N is obtained after formally resolving $Q_\mathbf{y}(\mathbf{x}(u^1, \ldots, u^{N-1})) = 0$ for x_N. It does not matter whether this is possible for the rest of the calculations. It suffices to assume that the inverse function $f(u^1, u^2, \ldots, u^{N-1})$ exists if one chooses the local system $\{\mathbf{e}_i\}$ appropriately and that x_N generally depends on the $N-1$ coordinates

u^α. In addition, we note that the parental state in the new coordinate system has the form

$$u^\alpha = 0 \quad (\alpha = 1, \ldots, N-1) \quad \Longleftrightarrow \quad \mathbf{x} = \mathbf{0}. \tag{2.45}$$

Equation (2.43) represents an $(N-1)$-dimensional manifold in $\mathbb{R}^N$. Its geometrical interpretation is the boundary $\partial Q_{\mathbf{y}}$ of the local success domain. Now, the differential geometry calculus can be applied to this manifold. To this end, one considers curves on $\partial Q_{\mathbf{y}}$ and measures their curvatures. The metric properties of the curved space $\partial Q_{\mathbf{y}}$ are needed for this purpose. These properties are given by the *metric tensor*, or more precisely, by the *first fundamental form*, whereas the curvature properties are described by the *second fundamental form.*

2.3.1.2 The Metric Tensor $g_{\alpha\beta}$. The local metric on the $(N-1)$-dimensional Riemannian manifold is given by the *first fundamental form*

$$(\mathrm{d}s)^2 = g_{\alpha\beta}\mathrm{d}u^\alpha\, \mathrm{d}u^\beta, \tag{2.46}$$

which is a quadratic form of the covariant metric tensor $g_{\alpha\beta}$. The metric tensor is obtained from the scalar product of the tangent vectors $\mathbf{x}_{u^\alpha}$ and $\mathbf{x}_{u^\beta}$

$$g_{\alpha\beta} = \mathbf{x}_{u^\alpha}^{\mathrm{T}} \mathbf{x}_{u^\beta} \qquad \text{with} \qquad \mathbf{x}_{u^\alpha} := \frac{\partial}{\partial u^\alpha}\mathbf{x}. \tag{2.47}$$

One obtains for the tangent vectors $\mathbf{x}_{u^\alpha}$, using (2.43) and (2.44),

$$\mathbf{x}_{u^\alpha} = \mathbf{e}_\alpha + \mathbf{e}_N \frac{\partial f}{\partial u^\alpha}. \tag{2.48}$$

The metric tensor follows as

$$g_{\alpha\beta} = \delta_{\alpha\beta} + \frac{\partial f}{\partial u^\alpha}\frac{\partial f}{\partial u^\beta}, \tag{2.49}$$

with Kronecker's $\delta_{\alpha\beta}$ symbol

$$\delta_{\alpha\beta} = \begin{cases} 1, & \alpha = \beta \\ 0, & \alpha \neq \beta. \end{cases} \tag{2.50}$$

To continue with the calculations, the *covariant* metric tensor is needed. Since

$$g_{\alpha\beta}g^{\beta\gamma} = g_\alpha{}^\gamma = \begin{cases} 1, & \alpha = \gamma \\ 0, & \alpha \neq \gamma \end{cases} \tag{2.51}$$

holds, $g^{\beta\gamma}$ is the inverse tensor $(g_{\beta\gamma})^{-1}$ to $g_{\alpha\beta}$. It can be formally obtained from a matrix power series, which is adapted from the scalar Taylor series

$$(1+c)^{-1} = 1 - c + c^2 - c^3 + c^4 - \ldots. \tag{2.52}$$

Let us consider the matrix $\mathbf{E} + \mathbf{C}$, where $\mathbf{E}$ stands for the identity matrix. Using (2.52) as a pattern one obtains

$$(\mathbf{E} + \mathbf{C})^{-1} = \mathbf{E} - \mathbf{C} + \mathbf{C}\,\mathbf{C} - \mathbf{C}\,\mathbf{C}\,\mathbf{C} + \mathbf{C}^4 - \ldots. \tag{2.53}$$

If the matrix entries of $\mathbf{C}$ are defined as

$$(\mathbf{C})_{\beta\gamma} := \frac{\partial f}{\partial u^\beta}\frac{\partial f}{\partial u^\gamma}, \qquad \text{considering} \qquad (\mathbf{E})_{\beta\gamma} = \delta_{\beta\gamma}, \tag{2.54}$$

one obtains from (2.53)

$$\begin{aligned}
(g_{\beta\gamma})^{-1} &= \left(\delta_{\beta\gamma} + \frac{\partial f}{\partial u^\beta}\frac{\partial f}{\partial u^\gamma}\right)^{-1} \\
&= \delta_{\beta\gamma} - \frac{\partial f}{\partial u^\beta}\frac{\partial f}{\partial u^\gamma} + \frac{\partial f}{\partial u^\beta}\left(\sum_\alpha \frac{\partial f}{\partial u^\alpha}\frac{\partial f}{\partial u^\alpha}\right)\frac{\partial f}{\partial u^\gamma} \\
&\quad - \frac{\partial f}{\partial u^\beta}\left(\sum_\alpha \frac{\partial f}{\partial u^\alpha}\frac{\partial f}{\partial u^\alpha}\right)\left(\sum_\alpha \frac{\partial f}{\partial u^\alpha}\frac{\partial f}{\partial u^\alpha}\right)\frac{\partial f}{\partial u^\gamma} + \ldots \\
(g_{\beta\gamma})^{-1} &= \delta_{\beta\gamma} - \frac{\partial f}{\partial u^\beta}\frac{\partial f}{\partial u^\gamma}\left[1 - \left(\sum_\alpha \frac{\partial f}{\partial u^\alpha}\frac{\partial f}{\partial u^\alpha}\right)\right. \\
&\quad \left. + \left(\sum_\alpha \frac{\partial f}{\partial u^\alpha}\frac{\partial f}{\partial u^\alpha}\right)^2 - \ldots\right].
\end{aligned} \tag{2.55}$$

Comparing (2.52) with the contents of the brackets in (2.55), the contravariant metric tensor follows as

$$g^{\beta\gamma} = (g_{\beta\gamma})^{-1} = \delta_{\beta\gamma} - \frac{\frac{\partial f}{\partial u^\beta}\frac{\partial f}{\partial u^\gamma}}{1 + \sum_{\alpha=1}^{N-1}\frac{\partial f}{\partial u^\alpha}\frac{\partial f}{\partial u^\alpha}}. \tag{2.56}$$

The correctness of this result can be easily verified by contracting (2.56) with (2.49) yielding (2.51) as the result.

2.3.1.3 The Second Fundamental Form. Let us consider curves $\mathfrak{c}$ on $\partial Q_\mathbf{y}$ which run through the parental point $\mathbf{x} = \mathbf{0}$ and are parameterized by their arc length s

$$\mathfrak{c}: \quad \mathbf{x}_\mathfrak{c} := \mathbf{x}\left(u^1(s),\ u^2(s),\ \ldots,\ u^{N-1}(s)\right). \tag{2.57}$$

The principal normal vector $\mathbf{h}$ of the curve $\mathfrak{c}$ is given by the second derivative of $\mathbf{x}$ with respect to its arc length s

$$\mathbf{h} := \frac{\mathbf{x}''}{\varkappa}, \qquad \varkappa := \|\mathbf{x}''\|, \qquad (\cdots)' = \frac{\mathrm{d}}{\mathrm{d}s}(\cdots), \tag{2.58}$$

where $\varkappa$ represents the curvature $\mathfrak{c}$. The second derivative of (2.57) is obtained by prepeated differentiation

$$\begin{aligned}
\mathbf{x}' &= \frac{\mathrm{d}\mathbf{x}}{\mathrm{d}s} = \mathbf{x}_{u^\alpha}\frac{\mathrm{d}u^\alpha}{\mathrm{d}s} = \mathbf{x}_{u^\alpha}u^{\alpha\prime} \\
\mathbf{x}'' &= \frac{\mathrm{d}}{\mathrm{d}s}\left(\mathbf{x}_{u^\alpha}u^{\alpha\prime}\right) = \mathbf{x}_{u^\alpha u^\beta}u^{\alpha\prime}u^{\beta\prime} + \mathbf{x}_{u^\alpha}u^{\alpha\prime\prime}.
\end{aligned} \tag{2.59}$$

If the curve $\mathfrak{c}$ is of such a form that the principal normal $\mathbf{h}$ is parallel to the surface normal $\mathbf{n}$ of $\partial Q_{\mathbf{y}}$ in the parental point, then one speaks of a *normal section curve* $\mathfrak{c}_n$ with *normal curvature* $\varkappa_n$. Based on the definition, the scalar product of the principal normal vector $\mathbf{h}$ and the surface normal vector $\mathbf{n}$ must be one. Thus, using (2.58) and (2.59) one obtains

$$\mathfrak{c}_n: \quad \mathbf{h}^{\mathrm{T}}\mathbf{n} = 1 \quad \Rightarrow \quad \varkappa_n = \mathbf{n}^{\mathrm{T}}\mathbf{x}'' = \mathbf{n}^{\mathrm{T}}\mathbf{x}_{u^\alpha u^\beta} u^{\alpha\prime} u^{\beta\prime} + \mathbf{n}^{\mathrm{T}}\mathbf{x}_{u^\alpha} u^{\alpha\prime\prime}. \tag{2.60}$$

The surface normal vector $\mathbf{n}$ on $\partial Q_{\mathbf{y}}$ is given by

$$\mathbf{n} = \frac{\nabla_x Q}{\|\nabla_x Q\|}, \qquad \nabla_x Q := \sum_{i=1}^{N} \mathbf{e}_i \frac{\partial}{\partial x_i} Q. \tag{2.61}$$

The differentiation of $Q\left(\mathbf{x}(u^1, u^2, \ldots, u^{N-1})\right) = 0$ yields

$$\frac{\partial Q}{\partial u^\alpha} = 0 \quad \Longrightarrow \quad 0 = \sum_{i=1}^{N} \frac{\partial Q}{\partial x_i}\frac{\partial x_i}{\partial u^\alpha} = \nabla_x Q\, \mathbf{x}_{u^\alpha}, \tag{2.62}$$

and, as a result,

$$\mathbf{n}^{\mathrm{T}}\mathbf{x}_{u^\alpha} = 0. \tag{2.63}$$

Therefore, one obtains the normal curvature $\varkappa_n$ from (2.60)

$$\varkappa_n = b_{\alpha\beta} u^{\alpha\prime} u^{\beta\prime} \tag{2.64}$$

as the so-called *second fundamental form*, composed using the $b_{\alpha\beta}$-tensor

$$b_{\alpha\beta} := \mathbf{n}^{\mathrm{T}}\mathbf{x}_{u^\alpha u^\beta}. \tag{2.65}$$

Note that the normal curvature $\varkappa_n$ depends on the direction of the tangent of the normal section curve $\mathfrak{c}_n$ (see Eq. (2.64)). There are $N-1$ linearly independent tangent directions on $\partial Q_{\mathbf{y}}$. They form the basis for the calculation of the mean curvature $\langle\varkappa\rangle$.

2.3.1.4 The Mean Curvature $\langle\varkappa\rangle$. The normal curvature $\varkappa_n$ depends on the tangent direction of the curve $\mathfrak{c}_n$ in the parental point. However, we are interested in a *curve-independent* characterization of the hypersurface $\partial Q_{\mathbf{y}}$ at the point $\mathbf{x} = \mathbf{0}$. This can be accomplished by determining the $N-1$ principal curvatures $\varkappa_p$. They are obtained as the extreme values of $\varkappa_n$. Therefore, at first (2.64) is rewritten as

$$\varkappa_n = b_{\alpha\beta} u^{\alpha\prime} u^{\beta\prime} = b_{\alpha\beta}\frac{\mathrm{d}u^\alpha}{\mathrm{d}s}\frac{\mathrm{d}u^\beta}{\mathrm{d}s} = \frac{b_{\alpha\beta}\mathrm{d}u^\alpha \mathrm{d}u^\beta}{(\mathrm{d}s)^2}. \tag{2.66}$$

Applying the first fundamental form (2.46) one obtains

$$\varkappa_n = \frac{b_{\alpha\beta}\mathrm{d}u^\alpha \mathrm{d}u^\beta}{g_{\alpha\beta}\mathrm{d}u^\alpha \mathrm{d}u^\beta}. \tag{2.67}$$

As already mentioned, the normal curvature depends merely on the direction of the tangents $\mathrm{d}u^\alpha$, since both $g_{\alpha\beta}$ and $b_{\alpha\beta}$ are constant for a fixed $\mathbf{x}$.

Now one can ask for the tangent directions for which (2.67) attains its extreme values. In order to simplify the notation $l^\alpha := du^\alpha$ is substituted

$$\varkappa_n(l^1, \ldots, l^{N-1}) = \frac{b_{\alpha\beta} l^\alpha l^\beta}{g_{\alpha\beta} l^\alpha l^\beta} \tag{2.68}$$

and we apply the extreme value condition

$$\varkappa_p = \varkappa_n \quad \Longleftrightarrow \quad \frac{\partial \varkappa_n}{\partial l^\alpha} = 0 \tag{2.69}$$

on (2.68). The differentiation gives

$$\frac{2 b_{\alpha\beta} l^\beta}{g_{\gamma\beta} l^\gamma l^\beta} - \frac{b_{\gamma\eta} l^\gamma l^\eta}{(g_{\gamma\beta} l^\gamma l^\beta)^2} 2 g_{\alpha\beta} l^\beta \stackrel{!}{=} 0$$

$$b_{\alpha\beta} l^\beta = \varkappa_p g_{\alpha\beta} l^\beta. \tag{2.70}$$

Since $l^\beta = g^{\beta\gamma} l_\gamma$ and (2.51) we obtain

$$b_\alpha{}^\gamma \, l_\gamma = \varkappa_p l_\alpha, \tag{2.71}$$

which is obviously an eigenvalue problem. As a result of (2.65), $b_\alpha{}^\gamma$ is a symmetrical matrix. Therefore, the eigenvalues $\varkappa_1, \varkappa_2, \ldots, \varkappa_{N-1}$ of (2.71) are real, and there is an orthogonal system of eigenvectors that defines the $N-1$ principal curvature directions. For each of these principal directions there exists a specific normal curvature $\varkappa_n = \varkappa_p$, called the principal curvature.

The sum of the principal curvatures defines the so-called mean curvature M^* (see Kreyszig, 1968, p. 372). This definition is somewhat confusing, since one usually expects the sum to be divided by the number of addends. For our case, it is necessary, because we expect in the case of the hypersphere the curvature $1/R$ and not $(N-1)/R$. Therefore, we use the more geometrically clear and strictly necessary definition of the *mean curvature of the hypersurface* as the mean value over the $N-1$ principal curvatures

$$\langle \varkappa \rangle := \frac{1}{N-1} \sum_{p=1}^{N-1} \varkappa_p. \tag{2.72}$$

The sum over the eigenvalues can be easily computed. As is known from eigenvalue theory, the sum of the eigenvalues is the trace of the matrix

$$\sum_{p=1}^{N-1} \varkappa_p = \mathrm{Tr}\,[b_\alpha{}^\gamma] = b_\alpha{}^\alpha = b_{\alpha\beta} g^{\alpha\beta}. \tag{2.73}$$

This is an extraordinarily satisfactory result, since the trace of a second-order tensor is obtained through contraction by means of the metric tensor $g^{\alpha\beta}$. The trace is therefore a scalar in the sense of the tensor calculus, i.e. the mean curvature $\langle \varkappa \rangle$ is an invariant, that reflects certain properties of the inner geometry of the hypersurface. Consequently, it is *independent* of the specially chosen coordinate system (2.43, 2.44). Furthermore, $\langle \varkappa \rangle$ is independent of the curve $\mathfrak{c}$ on $\partial Q_{\mathbf{y}}$, and in the case of the hypersphere with radius R one simply obtains $\langle\varkappa\rangle = 1/R$. Hence, as a natural definition of the *mean radius* $R_{\langle\varkappa\rangle}$ the absolute value of the reciprocal of the mean curvature (2.72)[9]

$$R_{\langle\varkappa\rangle} := \frac{1}{|\langle\varkappa\rangle|} = \frac{N-1}{|b_{\alpha\beta}g^{\alpha\beta}|}. \tag{2.74}$$

is to be used. The calculation of this quantity for the local quality function $Q_{\mathbf{y}}(\mathbf{x})$ is the subject of the next section.

2.3.2 The Calculation of the Mean Radius $R_{\langle\varkappa\rangle}$

In order to calculate $R_{\langle\varkappa\rangle}$, one first needs $g^{\alpha\beta}$ and $b_{\alpha\beta}$ on the manifold (2.44) at the parental position $\mathbf{x} = \mathbf{0}$, i.e. for $u^1 = u^2 = \ldots = u^{N-1} = 0$. After these steps the calculation of (2.74) can be completed.

The bilinear form (2.35) will be used as a fitness model. This might be considered as a restriction; however, the coefficients in (2.35) are associated with arbitrary, differentiable fitness landscapes because of (2.36). Therefore, the result to be derived here is the general solution to the mean radius problem in arbitrary fitness landscapes.

2.3.2.1 The Metric Tensor. The metric tensor $g^{\alpha\beta}$ is defined by Eq. (2.56). One needs $\frac{\partial f}{\partial u^\alpha}$ at $u^\alpha = 0$ in order to calculate $g^{\alpha\beta}$. Since the direct solution of the equation $Q\left(\mathbf{x}(u^1, \ldots, u^{N-1})\right) = 0$ for $x_N = f(u^1, \ldots, u^{N-1})$ (see Eq. (2.44)) is difficult, the direct computation of $\frac{\partial f}{\partial u^\alpha}$ is excluded. Implicit differentiation, however, brings us to this aim in a more elegant way.

If $Q(u^1, \ldots, u^{N-1}) = 0$ is differentiated with respect to u^α taking (2.44) into account, one gets

$$\frac{\partial Q}{\partial u^\alpha} = \frac{\partial}{\partial u^\alpha} Q\left(x_1(u^1), x_2(u^2), \ldots, x_N(u^1, u^2, \ldots, u^{N-1})\right) = 0$$

$$\frac{\partial Q}{\partial u^\alpha} = \frac{\partial Q}{\partial x_\alpha} + \frac{\partial Q}{\partial x_N}\frac{\partial x_N}{\partial u^\alpha} = \frac{\partial Q}{\partial x_\alpha} + \frac{\partial Q}{\partial x_N}\frac{\partial f}{\partial u^\alpha} = 0 \tag{2.75}$$

and further

$$\left.\frac{\partial f}{\partial u^\alpha}\right|_{u^{(\cdot)}=0} = -\left(\left.\frac{\partial Q}{\partial x_\alpha}\right|_{\mathbf{x}=\mathbf{0}}\right) \Big/ \left(\left.\frac{\partial Q}{\partial x_N}\right|_{\mathbf{x}=\mathbf{0}}\right). \tag{2.76}$$

For the concrete case of the fitness bilinear form (2.35, 2.36) one obtains

$$\left.\frac{\partial f}{\partial u^\alpha}\right|_{u^{(\cdot)}=0} = -\frac{a_\alpha}{a_N}, \tag{2.77}$$

and for the components of the contravariant metric tensor (2.56)

$$g^{\alpha\beta} = \delta_{\alpha\beta} - \frac{a_\alpha a_\beta / a_N^2}{1 + \frac{1}{a_N^2}\sum_{\gamma=1}^{N-1} a_\gamma a_\gamma} = \delta_{\alpha\beta} - \frac{a_\alpha a_\beta}{\sum_{i=1}^{N}(a_i)^2}. \tag{2.78}$$

[9] The absolute value operator is applied in order to ensure the positiveness of $R_{\langle\varkappa\rangle}$.

The result obtained here is valid for all fitness landscapes $F(\mathbf{y})$, because according to (2.36) the a_i are obtained by differentiation of $F(\mathbf{y})$ by y_i at the parental state.

2.3.2.2 The $b_{\alpha\beta}$-Tensor. The $b_{\alpha\beta}$-tensor is defined by (2.65). First, the hypersurface normal vector $\mathbf{n}$ at the parental state $\mathbf{x} = \mathbf{0}$ is calculated. Using (2.61) one obtains for the fitness bilinear form (2.35)

$$\nabla_x Q\big|_{\mathbf{x}=\mathbf{0}} = \sum_{i=1}^{N} a_i \mathbf{e}_i \quad \Longrightarrow \quad \mathbf{n} = \frac{\sum_{i=1}^{N} a_i \mathbf{e}_i}{\sqrt{\sum_{i=1}^{N} (a_i)^2}}. \tag{2.79}$$

The determination of $\mathbf{x}_{u^\alpha u^\beta}$ requires further considerations and the application of implicit differentiation. The application of the $\partial/\partial u^\beta$ operator on (2.48) gives at first

$$\mathbf{x}_{u^\alpha u^\beta} = \frac{\partial}{\partial u^\beta} \mathbf{x}_{u^\alpha} = \mathbf{e}_N \frac{\partial^2 f}{\partial u^\alpha \partial u^\beta}. \tag{2.80}$$

The second-order derivatives of f are obtained implicitly. Differentiation of (2.75) considering (2.44) gives

$$\frac{\partial^2 Q}{\partial u^\beta \partial u^\alpha} = \frac{\partial^2 Q}{\partial x_\beta \partial x_\alpha} + \frac{\partial^2 Q}{\partial x_N \partial x_\alpha} \frac{\partial f}{\partial u^\beta} + \frac{\partial^2 Q}{\partial x_\beta \partial x_N} \frac{\partial f}{\partial u^\alpha}$$

$$+ \frac{\partial^2 Q}{\partial x_N \partial x_N} \frac{\partial f}{\partial u^\beta} \frac{\partial f}{\partial u^\alpha} + \frac{\partial Q}{\partial x_N} \frac{\partial^2 f}{\partial u^\beta \partial u^\alpha} = 0. \tag{2.81}$$

At $\mathbf{x} = \mathbf{0}$, i.e. $u^{(\cdot)} = 0$, the computation using the fitness bilinear form (2.35) considering (2.77) yields

$$\begin{aligned} \frac{\partial^2 Q}{\partial u^\beta \partial u^\alpha}\bigg|_{u^{(\cdot)}=0} &= -2q_{\alpha\beta} - 2q_{\alpha N}\left(-\frac{a_\beta}{a_N}\right) - 2q_{\beta N}\left(-\frac{a_\alpha}{a_N}\right) \\ &\quad -2q_{NN}\left(\frac{a_\alpha a_\beta}{a_N^2}\right) + a_N \frac{\partial^2 f}{\partial u^\beta \partial u^\alpha}\bigg|_{u^{(\cdot)}=0} = 0, \end{aligned} \tag{2.82}$$

and for (2.80) it follows

$$\mathbf{x}_{u^\alpha u^\beta} = \mathbf{e}_N \frac{2}{a_N}\left(q_{\alpha\beta} - \frac{q_{\alpha N} a_\beta + q_{\beta N} a_\alpha}{a_N} + \frac{q_{NN}}{a_N^2} a_\alpha a_\beta\right). \tag{2.83}$$

Now one can combine these partial results to obtain the $b_{\alpha\beta}$-tensor (2.65). Using (2.79) and (2.83) one gets

$$\begin{aligned} b_{\alpha\beta} &= \mathbf{n}^T \mathbf{x}_{u^\alpha u^\beta} \\ &= \frac{2}{\sqrt{\sum_{i=1}^{N} a_i^2}}\left(q_{\alpha\beta} - \frac{q_{\alpha N} a_\beta + q_{\beta N} a_\alpha}{a_N} + \frac{q_{NN}}{a_N^2} a_\alpha a_\beta\right). \end{aligned} \tag{2.84}$$

2.3.2.3 The Computation of the Mean Radius $R_{\langle\varkappa\rangle}$. According to (2.74) the contraction of the $b_{\alpha\beta}$-tensor with the metric tensor $g^{\alpha\beta}$ is the main step toward the mean radius. Using the intermediate results (2.78) and (2.84), one obtains

$$
\begin{aligned}
b_{\alpha\beta}g^{\alpha\beta} = \frac{2}{\sqrt{\sum_{i=1}^{N} a_i^2}} \Bigg\{ & \sum_{\alpha=1}^{N-1} \left(q_{\alpha\alpha} - 2\frac{q_{\alpha N} a_\alpha}{a_N} + \frac{q_{NN}}{a_N^2} a_\alpha^2 \right) \\
& - \frac{1}{\sum_{i=1}^{N} a_i^2} \left[\sum_{\alpha=1}^{N-1} \sum_{\beta=1}^{N-1} a_\alpha q_{\alpha\beta} a_\beta - \frac{2}{a_N} \left(\sum_{\alpha=1}^{N-1} q_{\alpha N} a_\alpha \right) \left(\sum_{\beta=1}^{N-1} a_\beta^2 \right) \right. \\
& \left. + \frac{q_{NN}}{a_N^2} \left(\sum_{\alpha=1}^{N-1} a_\alpha^2 \right) \left(\sum_{\beta=1}^{N-1} a_\beta^2 \right) \right] \Bigg\} .
\end{aligned}
\tag{2.85}
$$

This result seems rather complicated; however, it can be simplified. The upper limit for the summation in (2.85) is extended from $N-1$ to N. After some reordering one gets

$$
\begin{aligned}
b_{\alpha\beta}g^{\alpha\beta} = \frac{2}{\sqrt{\sum_{i=1}^{N} a_i^2}} \Bigg\{ & \sum_{i=1}^{N} q_{ii} - q_{NN} - \frac{2}{a_N} \sum_{i=1}^{N} q_{iN} a_i + 2q_{NN} \\
& + \frac{q_{NN}}{a_N^2} \sum_{i=1}^{N} a_i^2 - q_{NN} \\
& + \frac{1}{\sum_{i=1}^{N} a_i^2} \left[- \sum_{i=1}^{N} \sum_{k=1}^{N} a_i q_{ik} a_k + 2a_N \sum_{i=1}^{N} q_{iN} a_i - q_{NN} a_N^2 \right. \\
& + \frac{2}{a_N} \left(\sum_{i=1}^{N} q_{iN} a_i \right) \left(\sum_{k=1}^{N} a_k^2 \right) - 2a_N \left(\sum_{i=1}^{N} q_{iN} a_i \right) \\
& - 2q_{NN} \left(\sum_{i=1}^{N} a_i^2 \right) + 2q_{NN} a_N^2 \\
& \left. - \frac{q_{NN}}{a_N^2} \left(\sum_{i=1}^{N} a_i^2 \right) \left(\sum_{k=1}^{N} a_k^2 \right) + 2q_{NN} \left(\sum_{i=1}^{N} a_i^2 \right) - q_{NN} a_N^2 \right] \Bigg\}
\end{aligned}
$$

$$
b_{\alpha\beta}g^{\alpha\beta} = \frac{2}{\sqrt{\sum_{i=1}^{N} a_i^2}} \left[\sum_{i=1}^{N} q_{ii} - \frac{\sum_{i=1}^{N} \sum_{k=1}^{N} a_i q_{ik} a_k}{\sum_{i=1}^{N} a_i^2} \right] . \tag{2.86}
$$

One can write alternatively

$$
b_{\alpha\beta}g^{\alpha\beta} = \frac{2}{\left(\sum_{i=1}^{N} a_i^2 \right)^{3/2}} \sum_{i=1}^{N} \sum_{k=1}^{N} q_{ik} \left[\left(\sum_{j=1}^{N} a_j^2 \right) \delta_{ik} - a_i a_k \right] . \tag{2.87}
$$

This is an astonishingly simple result; especially when formulated in matrix notation (see (2.36))

$$(\mathbf{Q})_{ik} = q_{ik} = -\frac{1}{2}\frac{\partial^2 F}{\partial y_i\, \partial y_j}, \qquad (\mathbf{a})_i = a_i = \frac{\partial F}{\partial y_i}, \qquad \|\mathbf{a}\| = \sqrt{\sum_{i=1}^{N} a_i^2}. \tag{2.88}$$

Thus, Eq. (2.86) can be rewritten as[10]

$$b_{\alpha\beta}g^{\alpha\beta} = \frac{2}{\|\mathbf{a}\|}\left(\mathrm{Tr}\,[\mathbf{Q}] - \frac{\mathbf{a}^{\mathrm{T}}\mathbf{Q}\,\mathbf{a}}{\|\mathbf{a}\|^2}\right). \tag{2.89}$$

Consequently, one obtains for the mean radius $R_{\langle\varkappa\rangle}$ (2.74) of the local success domain of the fitness landscape $F(\mathbf{y})$ in the parental state $\mathbf{y}$

MEAN RADIUS:
$$R_{\langle\varkappa\rangle} = \frac{N-1}{2}\frac{\|\mathbf{a}\|}{\left|\mathrm{Tr}\,[\mathbf{Q}] - \frac{\mathbf{a}^{\mathrm{T}}\mathbf{Q}\,\mathbf{a}}{\|\mathbf{a}\|^2}\right|}. \tag{2.90}$$

The mean radius can be used for the approximative analysis of the ES in nonspherical fitness landscapes. The idea here is to use $R_{\langle\varkappa\rangle}$ as the substitute radius in the progress rate formulae for the sphere model. Such formulae exist for the case of the sphere model in normalized form. They must be renormalized for the calculation of the progress rate φ. The insertion of the substitute radius $R_{\langle\varkappa\rangle}$ gives (using the normalization (2.22))

$$\varphi(\sigma) = \varphi^*_{\text{sphere}}\left(\sigma\frac{N}{R_{\langle\varkappa\rangle}}\right)\frac{R_{\langle\varkappa\rangle}}{N}. \tag{2.91}$$

For sufficiently large N, $N \to \infty$, one has $N/(N-1) \to 1$. Therefore, from (2.91) it follows using (2.90)

$$\varphi(\sigma) = \varphi^*_{\text{sphere}}\left(\sigma^*_{\text{sphere}}\right)\frac{1}{2}\frac{\|\mathbf{a}\|}{\left|\mathrm{Tr}\,[\mathbf{Q}] - \frac{\mathbf{a}^{\mathrm{T}}\mathbf{Q}\,\mathbf{a}}{\mathbf{a}^{\mathrm{T}}\mathbf{a}}\right|} \tag{2.92}$$

with

$$\sigma^*_{\text{sphere}} = \sigma\,\frac{2}{\|\mathbf{a}\|}\left|\mathrm{Tr}\,[\mathbf{Q}] - \frac{\mathbf{a}^{\mathrm{T}}\mathbf{Q}\,\mathbf{a}}{\mathbf{a}^{\mathrm{T}}\mathbf{a}}\right|. \tag{2.93}$$

Provided that the hypothesis is correct, the progress rate formulae from the sphere model can be applied to more complicated fitness landscapes. The probability-theoretical analysis for the $(1 \overset{+}{,} \lambda)$-ES will show that this assumption is approximately correct and gives results which are comparable to the normal progress rate φ_R (see Sect. 2.1.3, p. 30).

The applicability of the $R_{\langle\varkappa\rangle}$-approximation depends on the local fitness function $Q_{\mathbf{y}}(\mathbf{x})$. A further discussion will follow in Sect. 4.3. The quality of the approximation is mainly influenced by the eigenvalue spectrum of the

[10] Note, $\mathrm{Tr}\,[\mathbf{Q}]$ is the *trace* of the matrix $\mathbf{Q}$: $\mathrm{Tr}\,[\mathbf{Q}] = \sum_i q_{ii}$.

matrix $\mathbf{Q}$. If the eigenvalues are concentrated around $\mathrm{Tr}[\mathbf{Q}]/(N-1)$, excellent results are to be obtained. Assuming that $\mathbf{Q}$ is definite, one can even omit the second term in (2.89) since this term (also known as the Rayleigh-quotient) is bounded by the maximum/minimum of the spectrum of $\mathbf{Q}$. Under these assumptions, the formula

$$\boxed{\varphi(\sigma) = \varphi^*_{\text{sphere}}\left(\sigma^*_{\text{sphere}}\right) \frac{\|\mathbf{a}\|}{2\,|\mathrm{Tr}\,[\mathbf{Q}]|} \quad \text{with} \quad \sigma^*_{\text{sphere}} = \sigma\, \frac{2\,|\mathrm{Tr}\,[\mathbf{Q}]|}{\|\mathbf{a}\|}} \tag{2.94}$$

suffices for the calculations.[11] These formulae can be expressed by vector calculus notation. Using the N-dimensional Laplace operator Δ_y and the Nabla operator ∇_y, one obtains for the fitness function $F(\mathbf{y})$ considering (2.88) and $N \to \infty$: $R_{\langle\varkappa\rangle} \sim \|\nabla_y F\|/|\Delta_y F|$. This leads finally to

$$\varphi(\sigma) = \varphi^*_{\text{sphere}}\left(\sigma^*_{\text{sphere}}\right) \frac{\|\nabla_y F\|}{|\Delta_y F|} \qquad \text{with} \qquad \sigma^*_{\text{sphere}} = \sigma\, \frac{|\Delta_y F|}{\|\nabla_y F\|}\,. \tag{2.95}$$

2.4 The ES Dynamics

In general, the ES dynamics describe the evolution of the population over the generations g in the state space $\mathcal{A}$. In particular, it considers certain aspects of the evolution regarding the optimization behavior of the ES. At this point, we are interested especially in the expected values of certain functions on the state space $\mathcal{A}$. For example, it is interesting to compute the expected value of the parental parameter vector $\overline{\mathbf{y}}^{(g)} = \mathrm{E}\{\mathbf{y}\}$ (in the case of "$\mu=1$"-strategies) or of the centroid $\langle \mathbf{y}_p \rangle^{(g)}$. Formally, these quantities can be extracted from the stochastic dynamics (2.1, 2.2), e.g. for "$\mu=1$"-strategies one gets

$$\overline{\mathbf{y}}^{(g+1)} = \int \cdots \int \mathbf{y}' p^{(g+1)}\left(\mathfrak{P}'_1, \mathfrak{P}^{(g)}_1\right) \mathrm{d}\mathfrak{P}'_1. \tag{2.96}$$

Thus, the dynamics of the mean value of the parental vector is governed by an iterative mapping of the form $\overline{\mathbf{y}}^{(g+1)} = \mathbf{f}^{(g)}(\mathbf{y}^{(g)})$. Such an approach has not been taken in ES theory up to now. It could acquire importance for certain questions, e.g., the selection behavior in multimodal fitness landscapes. However, it is clear that such an analysis requires the determination of the transition densities in (2.2, 2.96).

An important and – at least for the sphere model – tractable analysis concerns the change of the distance $R(g)$ to the optimum over time (see Fig. 2.4). In the remaining part of this book, the term ES dynamics usually refers to this special dynamics.

[11] The applicability of this formula will also be discussed in Sect. 4.3. Another application concerning the estimation of the final optimum location error in noisy environments has been reported in Beyer and Arnold (1999).

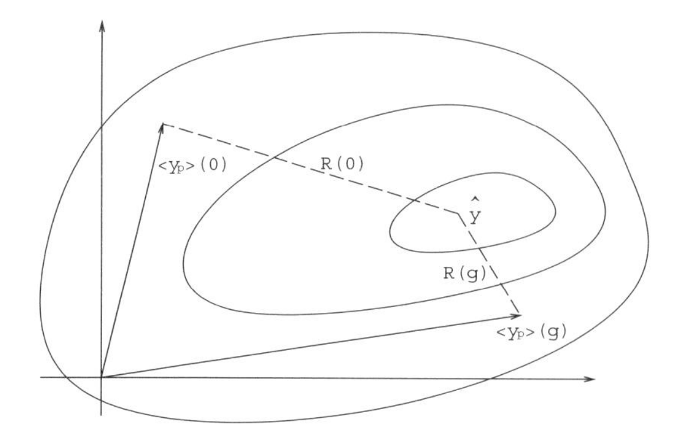

Fig. 2.4. On the dynamics of the residual distance $R(g)$ to the optimum. The evolution of the center of mass $\langle \mathbf{y}_p \rangle$ of the parental individuals is considered as an example. $\langle \mathbf{y}_p \rangle^{(0)}$ denotes the centroid of the parents at time 0 and $\langle \mathbf{y}_p \rangle^{(g)}$ at time g.

2.4.1 The *R*-Dynamics

Because the residual distance R is a function of $\mathbf{y}$, the calculation of the R-dynamics is principally not easier than, e.g., the determination of $\overline{\mathbf{y}}^{(g+1)}$ according to (2.96). In other words, complete knowledge of $\mathbf{y}$ at time g is necessary to determine $R(g)$. Exceptions are highly symmetrical fitness landscapes, such as the parabolic ridge and the sphere model. Only the sphere model will be considered in this work since the calculation of the progress rate φ (not φ_R) for the parabolic ridge is still under investigation.

Due to the sphere symmetry, the state of an individual $\mathfrak{a}$ in the parameter space $\mathcal{Y}$ is given completely by the residual distance $R(g)$ to the optimum.[12] Hence, the progress rate φ can be used directly for the derivation of the dynamics.

The expected distance $R(g+1)$ is given by (2.10, 2.15) or (2.13) according to

$$R(g+1) = R(g) - \varphi(g) \tag{2.97}$$

depending on the selection type used. Thus, $R(g+1)$ depends on $R(g)$ and φ. In the case of the sphere model, $\varphi(g)$ depends only on $R(g)$ and on the mutation strength $\sigma(g)$, that is $\varphi(g) = \varphi(R(g), \sigma(g))$. If one considers the scaling property of the sphere model, expressed by the normalization (2.22), one obtains

$$\begin{aligned} & R(g+1) = R(g) - \frac{R(g)}{N}\, \varphi^*(\sigma^*(g)) \\ \Longrightarrow \quad & \frac{R(g+1)}{R(g)} = 1 - \frac{1}{N}\, \varphi^*(\sigma^*(g)). \end{aligned} \tag{2.98}$$

This is a first-order difference equation. Progress, as the reduction of the distance to the optimum, is given for positive φ^*. As will be shown for fixed

[12] Up to and including Chap. 6 it is assumed that the strategy parameters, especially the mutation strength σ and σ^*, respectively, are given constants (exogenous parameters).

λ, φ^* has an upper bound; thus, φ^*/N goes to zero for $N \to \infty$. Hence, one can assume that φ^*/N is small with respect to one, i.e.,

$$\text{for progress:} \qquad 0 < \frac{1}{N}\,\varphi^*(\sigma^*(g)) \ll 1. \tag{2.99}$$

Therefore, one can expand $R(g+1)$ into a Taylor series at g (for sufficiently large g) and break off after the linear term

$$R(g+1) = R(g) + \frac{\mathrm{d}R(g)}{\mathrm{d}g}\,1 + \ldots. \tag{2.100}$$

Using (2.98) one obtains the differential equation

$$\frac{\mathrm{d}R(g)}{\mathrm{d}g} = -R(g)\,\frac{1}{N}\,\varphi^*(\sigma^*(g)), \tag{2.101}$$

which can be used as an approximation to the difference equation (2.98). One easily proves by differentiation that

$$R(g) = R(g_0)\exp\left[-\frac{1}{N}\int_{g'=g_0}^{g'=g} \varphi^*(\sigma^*(g'))\,\mathrm{d}g'\right] \tag{2.102}$$

is the solution to (2.101), where $R(g_0)$ stands for the residual distance at generation g_0.

2.4.2 The Special Case $\sigma^*(g) = \text{const}$

If the $\sigma^*(g)$-change over time is known, one can perform the integration in Eq. (2.102). An analytical integration of (2.102) is of course only possible in some special cases. One of these special cases is practically relevant: the maximal progress rate. Here, the mutation strength $\sigma(g)$ is set by a control process[13] in such a manner that $\sigma^*(g)$ is almost constant ($\sigma^*(g) \approx \hat{\sigma}^* = \text{const}$) and $\varphi^*(\hat{\sigma}^*) > 0$ is as large as possible.

For $\sigma^* = \text{const} > 0$, (2.102) yields a simple result

$$R(g) = R(g_0)\,\exp\left[-\frac{\varphi^*(\sigma^*)}{N}\,(g-g_0)\right]. \tag{2.103}$$

The distance $R(g)$ reduces exponentially with g. If one takes the logarithm of (2.103), one gets

$$\ln(R(g)) = \ln(R(g_0)) - \frac{\varphi^*(\sigma^*)}{N}\,(g-g_0), \tag{2.104}$$

i.e. the logarithm of R is a linear function of g. Such a time behavior is called *linear convergence*.

[13] Such control processes can be accomplished by, e.g., Rechenberg's 1/5th-rule, σ-self-adaptation (by mutative step-size control or cumulative path-length control), or Meta-ES techniques. The first two cases, i.e. the 1/5th-rule and the mutative step-size control, are analyzed in depth in Chap. 3 and Chap. 7, respectively.

Equation (2.104) can be used for estimating the number of generations G necessary to attain a relative reduction $R(g)/R(g_0)$ of the distance to the optimum. With $G := g - g_0$ one obtains

$$G = \frac{N}{\varphi^*(\sigma^*)} \ln\left(\frac{R(g_0)}{R(g)}\right). \tag{2.105}$$

As one can see, the EA time complexity[14] of the ES is of the order

$$\boxed{\text{EA TIME COMPLEXITY:} \qquad G = \mathcal{O}(N).} \tag{2.106}$$

It is interesting to observe that the ES often shows a linear convergence in empirical studies even for multimodal fitness landscapes in real-valued parameter spaces without constraints.[15] The prerequisite for that is adaptation of the mutation strength σ to the *local* topology by a suitable control mechanism. A possible explanation to the empirical result of linear convergence may be that many fitness landscapes behave locally similar to the sphere model with a radius chosen appropriately. The investigation of the ES dynamics in multimodal fitness landscapes is required for a more rigorous explanation of this observation. Such investigations should relate the empirically obtained performance measures to the local curvature of the landscape (differential-geometrical model!).

There are also other cases beyond the constant normalized mutation strength σ^* investigated in this section. The most important among those which allow to solve (2.102) analytically is the constant σ case in comma strategies. This case will be investigated in Chap. 3. Further questions that are related to ES dynamics are considered in Chap. 7.

---

[14] The term "EA time complexity" should be understood as the number of generations necessary to reach a certain relative convergence.

[15] It should be emphasized that the proof of linear convergence is *not* equivalent to the proof of *global* convergence. The linear convergence can also end up in a local optimum; linear convergence describes the *local convergence behavior*.

3. The Progress Rate of the $(1 \overset{+}{,} \lambda)$-ES on the Sphere Model

This chapter analyzes the progress rate of the $(1 \overset{+}{,} \lambda)$-ES. The first section is devoted to the exact theory of the $(1+1)$-ES on the sphere model, including the results considering the N-dependency and noisy fitness data. In the second and third section, analytical progress rate formulae are derived for the asymptotic case $N \to \infty$; moreover, the dynamics of the ES-process is investigated. The second section is dedicated to the ES without noise, whereas the third is concerned with the ES under the presence of noise on the fitness measurement process (fitness noise). The consequences are discussed, as well as ideas for the improvement of the convergence property. In the fourth section, the $(1, \lambda)$-ES is investigated again with the aim of finding a simpler approach to the derivation of N-dependent analytical progress rate formulae.

3.1 The Exact $(1+1)$-ES Theory

The analysis of the $(1+1)$-ES on the sphere model is a good starting point for the theoretical investigation of the progress rate φ of the ES. *Exact* integral representations of φ will be derived in this section for finite parameter space dimension N and isotropic normal mutations $\mathbf{z}$ with density (1.11). The case of deterministic fitness measurement is considered first. The case of noise-perturbed fitness is investigated in the second part, Sect. 3.1.2.

The benefit of such an analysis of the $(1+1)$-ES is twofold. Firstly, it allows the user a relatively simple, geometrically oriented approach to the $(1+1)$ theory, in contrast to the following, more technical chapters. Secondly, the numerical evaluation of the integrals for different N allows for a comparison with the analytical (approximative) formulae which will be derived later on. Particularly, the graphs of $\varphi^*(\sigma^*, N)$ clarify that the asymptotic φ^* formula ($N \to \infty$) of the $(1+1)$-ES yields practically usable approximations for $N \gtrsim 30$.

The approach to the $(1+1)$ theory will be given in a geometrically oriented manner. It is assumed that the ES system is at the parental state[1] $\mathbf{P} = \mathbf{y}_p$, and the state of the descendant

$$\tilde{\mathbf{y}} := \mathbf{y}_p + \mathbf{z} \qquad \text{with} \qquad \mathbf{z} := \sigma\,(\mathcal{N}(0,1), \mathcal{N}(0,1), \ldots, \mathcal{N}(0,1))^{\mathrm{T}} \tag{3.1}$$

[1] The index "$_p$" stands for parent.

Procedure (1 + 1)-ES-minimization;	line
Begin	1
$g := 0$;	2
`initialize`$(\mathbf{y}_p, \sigma)$;	3
Repeat	4
$F_p := F(\mathbf{y}_p)$;	5
$\tilde{\mathbf{y}} := \mathbf{y}_p + \sigma\,(\mathcal{N}(0,1), \mathcal{N}(0,1), \ldots, \mathcal{N}(0,1))^{\mathrm{T}}$;	6
$\tilde{F} := F(\tilde{\mathbf{y}})$;	7
If $\tilde{F} \le F_p$ **Then** $\mathbf{y}_p := \tilde{\mathbf{y}}$;	8
$g := g + 1$;	9
Until stop_criterion	10
End	11

Algorithm 3. The (1 + 1)-ES for the minimization of a fitness function $F(\mathbf{y})$. The implementation of the usual and practically necessary control of the mutation strength σ is omitted here.

is generated by a mutation $\mathbf{z}$ according to (1.9, 1.10). This is performed in Algorithm 3 at line 6.

The optimum is assumed to be at $\hat{\mathbf{y}}$. A "snapshot" of the $(1+1)$ process is shown in Fig. 3.1. The mutation $\mathbf{z}$ and the vector

$$\mathbf{R} := \mathbf{y}_p - \hat{\mathbf{y}} \tag{3.2}$$

(line segment $\overline{\mathbf{OP}}$) span a two-dimensional plane. The state of the descendant

$$\tilde{\mathbf{y}} = \mathbf{y}_p + \mathbf{z} = \hat{\mathbf{y}} + \tilde{\mathbf{R}} = \tilde{\mathbf{y}}_{1;1} \tag{3.3}$$

has the distance $\tilde{R} := \|\tilde{\mathbf{R}}\|$

$$\tilde{R} = \|\tilde{\mathbf{y}}_{1;1} - \hat{\mathbf{y}}\| = \|\hat{\mathbf{y}} - \tilde{\mathbf{y}}_{1;1}\| \tag{3.4}$$

to the optimum, and the parent $\mathbf{y}_p$ has the distance $R = \|\mathbf{y}_p - \hat{\mathbf{y}}\|$, respectively. According to definition (2.15), the progress rate is obtained as

$$\varphi = \mathrm{E}\left\{R - \tilde{R}\right\}. \tag{3.5}$$

In order to calculate the distance $\tilde{R}$, the mutation vector $\mathbf{z}$ is decomposed with respect to an orthogonal coordinate system $\{x_i\}$ having its origin at the parental state $\mathbf{y}_p$. The $x := x_1$ coordinate is chosen in the direction $\mathbf{e}_{\mathrm{opt}}$ toward the optimum $\hat{\mathbf{y}}$

$$\mathbf{e}_{\mathrm{opt}} := -\frac{\mathbf{R}}{\|\mathbf{R}\|} =: -\mathbf{e}_R. \tag{3.6}$$

Regarding this new coordinate system one has (see Fig. 3.1)

$$\mathbf{z} = x\mathbf{e}_{\mathrm{opt}} + \mathbf{h} \quad \text{with} \quad \mathbf{e}_{\mathrm{opt}}^{\mathrm{T}}\mathbf{h} = 0, \quad \mathbf{h} := (x_2, x_3, \ldots, x_N)^{\mathrm{T}}. \tag{3.7}$$

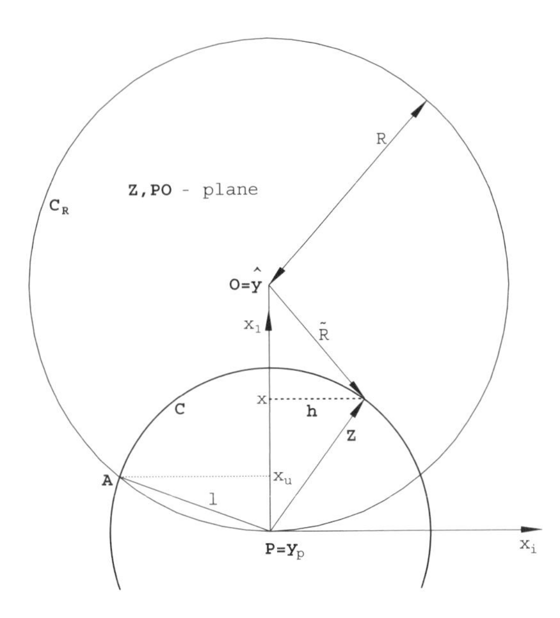

Fig. 3.1. The geometry of the sphere model. The vector $\mathbf{R} = \mathbf{y}_p - \hat{\mathbf{y}}$ (line segment $\overline{\mathbf{OP}}$) is not shown explicitly.

The distance $\tilde{R}$ is calculated using (3.3, 3.2, 3.7, 3.6)

$$\tilde{\mathbf{R}} = \mathbf{z} + \mathbf{y}_p - \hat{\mathbf{y}} \;=\; \mathbf{z} + \mathbf{R} \;=\; x\mathbf{e}_{\text{opt}} + \mathbf{h} + \mathbf{R}$$

$$\tilde{\mathbf{R}} = (R - x)\mathbf{e}_R + \mathbf{h} \tag{3.8}$$

and for the absolute value of $\tilde{\mathbf{R}}$

$$\tilde{R} = \sqrt{\tilde{\mathbf{R}}^{\mathrm{T}}\tilde{\mathbf{R}}} = \sqrt{(R - x)^2 + \mathbf{h}^2}. \tag{3.9}$$

Hence, the progress rate (3.5) follows as

$$\varphi_{1+1} = \mathrm{E}\left\{R - \sqrt{(R - x)^2 + \mathbf{h}^2}\right\}. \tag{3.10}$$

This formula is valid independent of the mutation type and the fitness model analyzed. The expected value is calculated in the space spanned by the random variables x and $\mathbf{h}$.

Isotropic Gaussian mutations are considered here in order to put the progress rate formula in concrete terms. As indicated by the adjective "isotropic," such mutations are invariant with respect to the orthogonal rotation group $\mathbf{O}(N)$. This can be shown immediately by considering the exponent part of the probability density (1.11), which is a function of the length of the vector $\mathbf{z}$. The length of $\mathbf{z}$ is not changed by any rotation transformation (see Beyer, 1989a, 1993). Therefore, the mutations are sphere-symmetrical (isotropic, they do not distinguish any direction in the space). Consequently, for the probability density of the x_i components of the $\mathbf{z}$ vector in the case of the coordinate rotation in Fig. 3.1 one has

$$p(x_1, x_2, \ldots, x_N) = \frac{1}{\left(\sqrt{2\pi}\,\sigma\right)^N} \exp\left(-\frac{1}{2}\frac{1}{\sigma^2}\sum_{i=1}^{N} x_i^2\right), \tag{3.11}$$

or for x and $\mathbf{h}$ in (3.7) and (3.10)

$$p_x(x) = \frac{1}{\sqrt{2\pi}\,\sigma} \exp\left(-\frac{1}{2}\left(\frac{x}{\sigma}\right)^2\right), \tag{3.12}$$
$$p(\mathbf{h}) = \frac{1}{\left(\sqrt{2\pi}\,\sigma\right)^{N-1}} \exp\left(-\frac{1}{2}\frac{\mathbf{h}^{\mathrm{T}}\mathbf{h}}{\sigma^2}\right).$$

If (3.10) is thoroughly examined and the symmetry of the sphere model is considered (see Fig. 3.1), one can notice that the $(N-1)$-dimensional density $p(\mathbf{h})$ is not required for the calculation of (3.10), but merely the probability density of the random variate

$$u := \mathbf{h}^2 = \sum_{i=2}^{N} x_i^2. \tag{3.13}$$

Obviously, the variate u obeys a χ^2 distribution $p_{h^2}(u)$ with $N-1$ degrees of freedom (see Fisz, 1971, p. 398)

$$p_{h^2}(u) = \frac{1}{\sigma^2} \frac{\left(\frac{u}{\sigma^2}\right)^{\frac{N-1}{2}-1} \mathrm{e}^{-\frac{u}{2\sigma^2}}}{2^{\frac{N-1}{2}}\,\Gamma\left(\frac{N-1}{2}\right)}, \tag{3.14}$$

where $\Gamma(x)$ stands for the complete gamma function (as to its definition, see (3.49), p. 62).

The case of deterministic fitness values (exact fitness measurements) will be considered next.

3.1.1 The Progress Rate without Noise in Fitness Measurements

Because of the symmetry properties of the sphere model and the isotropy of the mutations, the original N-dimensional expected value problem (3.5) in the space of the $\mathbf{z}$-mutations reduces to a two-dimensional problem in the space spanned by the random variables x and u, as expressed by (3.10). Therefore, the expected value integral of the progress rate attains the form

$$\varphi_{1+1} = \int\!\!\int \left(R - \sqrt{(R-x)^2 + u}\right) p_{h^2}(u) p_x(x)\, \mathrm{d}u\, \mathrm{d}x. \tag{3.15}$$

Hereby the integration limits are still left open. They are determined by the selection criterion of the (1 + 1)-ES: a mutation can only replace its parent if it is better than its parent (or as good as its parent), i.e. $\tilde{F} \le F_p$ (see Algorithm 3, line 8) and therefore $\tilde{R} \le R$.[2] This condition yields, together with (3.9) and (3.13), $\sqrt{(R-x)^2 + u} \le R$. After squaring the inequality and reordering terms one gets $u \le 2Rx - x^2$. Taking into account that the success

[2] The equal sign stands for the case of the acceptance of the offspring with equal fitness $F(\mathbf{y})$. It does not have further relevance for the sphere model case; however, for the ES algorithm at real applications, the convergence behavior may be improved by allowing a *random walk* on the "equifitness hypersurface."

region of the sphere model is bounded by $0 \le x \le 2R$ in the x direction, one obtains

$$\left.\begin{array}{l} 0 \le x \le 2R \\ 0 \le u \le 2Rx - x^2 \end{array}\right\} \tag{3.16}$$

for the acceptance region and therefore

$$\varphi_{1+1} = \int\limits_{x=0}^{x=2R} \int\limits_{u=0}^{u=2Rx-x^2} \left(R - \sqrt{(R-x)^2 + u}\right) p_{h^2}(u) p_x(x) \, \mathrm{d}u \, \mathrm{d}x. \tag{3.17}$$

Since (3.16) defines the region of successful mutations, the *success probability* P_s can also be expressed in a similar manner[3]

$$P_{\mathrm{s}1+1} = \int_{x=0}^{x=2R} \int_{u=0}^{u=2Rx-x^2} p_{h^2}(u) p_x(x) \, \mathrm{d}u \, \mathrm{d}x. \tag{3.18}$$

In order to proceed, the integral (3.17) is transformed using the substitutions

$$t = \frac{u}{2\sigma^2} \qquad \text{and} \qquad y = \frac{x}{\sigma}. \tag{3.19}$$

Considering (3.12) and (3.14) one obtains

$$\varphi_{1+1} = \frac{1}{\sqrt{2\pi}} \int_{y=0}^{y=2\frac{R}{\sigma}} \int_{t=0}^{t=\frac{R}{\sigma}y - \frac{y^2}{2}} \left(R - \sqrt{(R-\sigma y)^2 + 2\sigma^2 t}\right) \times \frac{\mathrm{e}^{-t}\, t^{\frac{N-1}{2}-1}}{\Gamma\left(\frac{N-1}{2}\right)} \, \mathrm{d}t \, \mathrm{e}^{-\frac{1}{2}y^2} \, \mathrm{d}y. \tag{3.20}$$

After introducing the normalized quantities $\varphi^* = \varphi N/R$ and $\sigma^* = \sigma N/R$, according to (2.22), one gets

$$\varphi^*_{1+1}(\sigma^*, N) = \frac{1}{\sqrt{2\pi}} \int_{y=0}^{y=\frac{2N}{\sigma^*}} \mathrm{e}^{-\frac{1}{2}y^2} \frac{N}{\Gamma\left(\frac{N-1}{2}\right)} \int_{t=0}^{t=\frac{N}{\sigma^*}y - \frac{y^2}{2}} \times \left(1 - \sqrt{\left(1 - \frac{\sigma^*}{N}y\right)^2 + 2\left(\frac{\sigma^*}{N}\right)^2 t}\right) \mathrm{e}^{-t}\, t^{\frac{N-1}{2}-1} \, \mathrm{d}t \, \mathrm{d}y. \tag{3.21}$$

In contrast to (3.20), this integral depends only on two variables: the normalized mutation strength[4] σ^* and the parameter space dimension N. Although we have obtained consequently an integral to express the progress rate of the (1 + 1)-ES on the sphere model, the further analytical evaluation of it is unfortunately not possible. However, the inner integration of the P_s integral (3.18) is tractable. Analogous to (3.21) one obtains

---

[3] The success probability P_s can be interpreted as the parental *death probability*, however, the subscript "$_s$" stands for "success."

[4] Note the definition of the normalized mutation strength σ^*: it does *not* explicitly depend on either R or N.

$$P_{s1+1} = \frac{1}{\sqrt{2\pi}} \int_{y=0}^{y=\frac{2N}{\sigma^*}} e^{-\frac{1}{2}y^2} \frac{1}{\Gamma\left(\frac{N-1}{2}\right)} \int_{t=0}^{t=\frac{N}{\sigma^*}y-\frac{y^2}{2}} e^{-t}\, t^{\frac{N-1}{2}-1}\, dt\, dy, \quad (3.22)$$

and taking the definition of the incomplete gamma function (see Abramowitz & Stegun, 1984, p. 81)

$$P(a, x) := \frac{1}{\Gamma(a)} \int_{t=0}^{t=x} e^{-t}\, t^{a-1}\, dt \quad (3.23)$$

into account, then it follows that

$$P_{s1+1}(\sigma^*, N) = \frac{1}{\sqrt{2\pi}} \int_{y=0}^{y=\frac{2N}{\sigma^*}} e^{-\frac{1}{2}y^2} P\left(\frac{N-1}{2}, \frac{N}{\sigma^*}y - \frac{y^2}{2}\right) dy. \quad (3.24)$$

As for the φ^* integral (3.21), the analytical integration of (3.24) is impossible. The significance of P_s for the $(1+1)$-ES algorithm will be discussed in Sect. 3.3.3.3.

The primary use of the exact integral formula (3.21) for the normalized progress rate φ^* lies in the fact that it allows for the investigation of the N-dependency of φ^*. The graphs of the function $\varphi^* = \varphi^*(\sigma^*, N)$ are generated by numerical integration for this purpose using `Mathematica` (a computer algebra system). Figure 3.2 shows $\varphi^* = \varphi^*(\sigma^*)$ curves for some selected N values. The common characteristic of all these graphs is the steep increase at

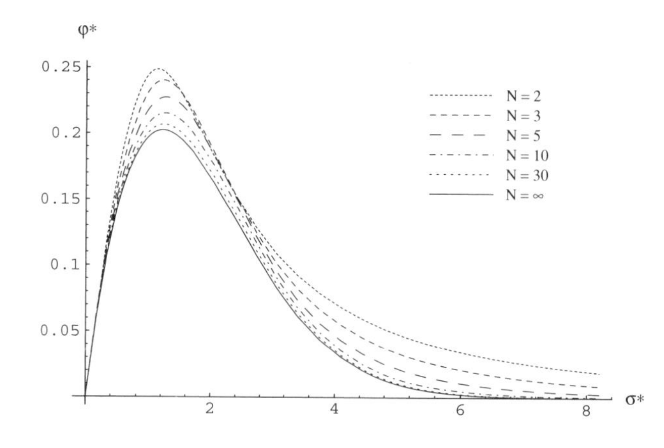

Fig. 3.2. The normalized progress rate φ^* versus the normalized mutation strength σ^* for $(1+1)$-ES on the sphere model for various values of the parameter space dimension N

first, then a maximum, and later a gradual decrease down to the asymptotic line $\varphi^* = 0$. There exists a $\hat{\sigma}^* = f(N)$ at which the maximal progress rate per generation is attained

$$\hat{\sigma}^*(N) := \arg\max_{\sigma^*} \left[\varphi^*(\sigma^*, N)\right]. \quad (3.25)$$

The relatively small dependency of the φ^* curves and the position of the optimum $\hat{\sigma}^*$ on the parameter space dimension N is remarkable. Furthermore, there is a limit curve for $N \to \infty$. It will be calculated in Sect. 3.2

analytically. It is important here to note that the limit curve $\varphi^* = \varphi^*(\sigma^*)$ already yields good and practically usable results as an approximation for the cases with $N \gtrapprox 30$. Therefore, the analytical derivation of the N-dependent approximation formulae for the (1 + 1)-ES on the sphere model is actually not mandatory. Hence, the following discussions on the (1 + 1)-ES as well as on the $(1 \overset{+}{,} \lambda)$ strategies will follow for the case $N \to \infty$ in Sect. 3.2.4 and 3.3, respectively. We will come back to the N-dependency in Sect. 3.4 for the $(1, \lambda)$-strategies.

3.1.2 The Progress Rate at Disturbed Fitness Measurements

According to the convention in Beyer (1993), the individuals having noise in the fitness data are signified with a tilde on the respective symbol in the strategy parentheses. In this special case, the (1 + 1)-ES with perturbed fitness calculation for the parents as well as for the offspring is denoted as $(\tilde{1} + \tilde{1})$-ES.[5]

The model used for the noise in the sphere model is as follows: the local fitness Q is rotation-symmetric, it only depends on the radius R, i.e. $Q = Q(R)$. The fitness value which is actually measured differs from the $Q(R)$ value by the superposition of an additive noise source ε_Q (see Sect. 2.2.6)

$$\tilde{Q}(R) := Q(R) + \varepsilon_Q(R). \tag{3.26}$$

The noise term $\varepsilon_Q(R)$ is assumed to be normally distributed as $\sigma_{\varepsilon_Q}(R)\mathcal{N}(0,1)$ (see (2.41)). Equation (3.26) describes the fitness measurement of the parent $\mathbf{y}_p$; for the descendant $\tilde{\mathbf{y}}$, having the distance $\tilde{R}$ to the optimum, one has, analogous to (3.26)

$$\tilde{Q}(\tilde{R}) := Q(\tilde{R}) + \varepsilon_Q(\tilde{R}). \tag{3.27}$$

Provided that the states of the parent and its descendant do not differ much from each other, one can assume as a good approximation that the standard deviations of the measurement errors of the descendant and parent are equal

$$\sigma_{\varepsilon_Q}(\tilde{R}) = \sigma_{\varepsilon_Q}(R), \tag{3.28}$$

and only depend on the parental state. Therefore

$$\left.\begin{array}{ll} \text{Parent:} & \tilde{Q}(R) = Q(R) + \sigma_{\varepsilon_Q}(R)\mathcal{N}(0,1) \\ \text{Descendant:} & \tilde{Q}(\tilde{R}) = Q(\tilde{R}) + \sigma_{\varepsilon_Q}(R)\mathcal{N}(0,1) \end{array}\right\}. \tag{3.29}$$

We consider without loss of generality (w.l.o.g.) the case of ES-minimization. The (1+1)-ES in Algorithm 3 is used here, whereas the fitness values F_p and $\tilde{F}$ are represented in this case by $\tilde{Q}(R)$ and $\tilde{Q}(\tilde{R})$, respectively. The selection

[5] Actually, the $(\tilde{1}+\tilde{1})$ notation does not introduce a new strategy, rather it indicates the disturbance in the fitness measurement process. Algorithm 3 is used without any alteration.

(see Algorithm 3, line 8) results from the noisy fitness values. A descendant is accepted if the condition

$$\tilde{Q}(\tilde{R}) \le \tilde{Q}(R) \tag{3.30}$$

is fulfilled. As a result of the perturbed fitness (3.29), this selection criterion is insecure as to the convergence behavior, i.e.

$$\tilde{Q}(\tilde{R}) \le \tilde{Q}(R) \quad \not\Rightarrow \quad \tilde{R} \le R. \tag{3.31}$$

Therefore, a fitness improvement does not necessarily imply a reduction of the remaining distance to the optimum. This has consequences on the progress rate φ. In the unperturbed case, the non-negativity of φ is guaranteed by definition (3.5); however, φ can be negative under the existence of noise. In the unperturbed case, no differentiation is needed between the selection criterion $\tilde{Q}(\tilde{R}) \le \tilde{Q}(R)$ and $\tilde{R} \le R$. For the perturbed case, we have to distinguish between them, and use the selection criterion (3.30). Therefore, the progress rate is obtained as a conditional expected value

$$\varphi_{\tilde{1}+\tilde{1}} = \mathrm{E}\left\{R - \tilde{R} \;\middle|\; \tilde{Q}(\tilde{R}) \le \tilde{Q}(R)\right\}. \tag{3.32}$$

The *acceptance probability* is introduced for the calculation of this expected value. A mutation $\mathbf{z}$ will be accepted if (3.30) is fulfilled. The probability P_a for that is

$$P_\mathrm{a}(\tilde{R}, R) := \Pr\left(\tilde{Q}(\tilde{R}) \le \tilde{Q}(R)\right). \tag{3.33}$$

In the special case of the normally distributed noise (3.29) it follows that

$$\begin{aligned} P_\mathrm{a}(\tilde{R}, R) &:= \Pr\left(\tilde{Q}(\tilde{R}) - \tilde{Q}(R) \le 0\right) \\ &= \Pr\left(Q(\tilde{R}) - Q(R) + \sigma_{\varepsilon_Q}(R)\left(\mathcal{N}_2(0,1) - \mathcal{N}_1(0,1)\right) \le 0\right). \end{aligned} \tag{3.34}$$

The subscripts of $\mathcal{N}$ indicate that there are two independent random variables (measurements). Since $\mathcal{N}_1$ and $\mathcal{N}_2$ are $(0,1)$ normally distributed, their difference is $\sqrt{2}\mathcal{N}(0,1)$ normally distributed. Therefore, the complete random function in (3.34) has a normal distribution with mean $Q(\tilde{R}) - Q(R)$ and standard deviation $\sqrt{2}\,\sigma_{\varepsilon_Q}(R)$. Hence, the acceptance probability is

$$P_\mathrm{a}(\tilde{R}, R) = \Phi\left(\frac{0 - \left(Q(\tilde{R}) - Q(R)\right)}{\sqrt{2}\,\sigma_{\varepsilon_Q}(R)}\right) = \Phi\left(\frac{Q(R) - Q(\tilde{R})}{\sqrt{2}\,\sigma_{\varepsilon_Q}(R)}\right), \tag{3.35}$$

where $\Phi(x)$ denotes the cumulative distribution function of the standard normal distribution (see Appendix C.1, Eq. (C.1)).

A mutation $\mathbf{z}$ with the actual distance $\tilde{R}$ is accepted with the acceptance probability $P_\mathrm{a}(\tilde{R}, R)$. $\tilde{R}$ is given according to (3.9); after the substitution $u = \mathbf{h}^2$, one gets $\tilde{R} = \sqrt{(R-x)^2 + u}$. Hence, P_a is a function of the random variates x and u. Similar to the expected value integral (3.17) for the progress rate φ_{1+1}, one obtains $\varphi_{\tilde{1}+\tilde{1}}$ as

$$\varphi_{\tilde{1}+\tilde{1}} = \int_{x=-\infty}^{x=\infty} \int_{u=0}^{u=\infty} \left(R - \sqrt{(R-x)^2 + u}\right) \times P_{\mathrm{a}}(\tilde{R}(x,u), R)\; p_{h^2}(u) p_x(x)\, \mathrm{d}u\, \mathrm{d}x. \tag{3.36}$$

The effect of a mutation $\mathbf{z}$ is completely determined by u and x. Unlike (3.17), the integral is taken here over all possible values for x and u

$$\left.\begin{array}{r} -\infty < x < \infty \\ 0 \le u < \infty \end{array}\right\}. \tag{3.37}$$

Note, the effect of the selection is implicitly included in the acceptance probability P_{a}. Considering (3.35), applying the substitutions (3.19) and the normalization (2.22) for φ^* and σ^*, one gets, analogously to (3.21),

$$\varphi^*_{\tilde{1}+\tilde{1}} = \frac{1}{\sqrt{2\pi}} \int_{y=-\infty}^{y=\infty} \mathrm{e}^{-\frac{1}{2}y^2} \frac{N}{\Gamma\left(\frac{N-1}{2}\right)} \int_{t=0}^{t=\infty} \left(1 - \sqrt{\left(1 - \frac{\sigma^*}{N} y\right)^2 + 2\left(\frac{\sigma^*}{N}\right)^2 t}\right)$$
$$\times\, \Phi\left(\frac{Q(R) - Q\left(R\sqrt{\left(1 - \frac{\sigma^*}{N}y\right)^2 + 2\left(\frac{\sigma^*}{N}\right)^2 t}\right)}{\sqrt{2}\,\sigma_{\varepsilon_Q}(R)}\right) \mathrm{e}^{-t}\, t^{\frac{N-1}{2}-1}\, \mathrm{d}t\, \mathrm{d}y. \tag{3.38}$$

If the fitness function $Q(R)$ is a simple power of R

$$Q(R) = cR^\alpha, \tag{3.39}$$

and after simplifying the argument in the Φ-function in (3.38), one gets

$$\varphi^*_{\tilde{1}+\tilde{1}}(\sigma^*, \varepsilon_r, N)$$
$$= \frac{1}{\sqrt{2\pi}} \int_{y=-\infty}^{y=\infty} \mathrm{e}^{-\frac{1}{2}y^2} \frac{N}{\Gamma\left(\frac{N-1}{2}\right)} \int_{t=0}^{t=\infty} \left(1 - \sqrt{\left(1 - \frac{\sigma^*}{N} y\right)^2 + 2\left(\frac{\sigma^*}{N}\right)^2 t}\right)$$
$$\times\, \Phi\left(\frac{1 - \left(\left(1 - \frac{\sigma^*}{N}y\right)^2 + 2\left(\frac{\sigma^*}{N}\right)^2 t\right)^{\alpha/2}}{\sqrt{2}\,\varepsilon_r}\right) \mathrm{e}^{-t}\, t^{\frac{N-1}{2}-1}\, \mathrm{d}t\, \mathrm{d}y. \tag{3.40}$$

The *relative measurement error* ε_r

$$\varepsilon_r(R) := \frac{\sigma_{\varepsilon_Q}(R)}{Q(R)} \tag{3.41}$$

was introduced here. Alternatively, instead of ε_r the *normalized noise strength* can be defined as (see also (2.22), p. 32)

$$\boxed{\sigma^*_\varepsilon := \sigma_{\varepsilon_Q}(R)\, \frac{N}{R}\, \frac{1}{Q'(R)} \qquad \text{with} \qquad Q'(R) := \frac{\mathrm{d}Q}{\mathrm{d}R}.} \tag{3.42}$$

The advantage of this normalization will become clear in Sect. 3.3, where the asymptotic case for large N is considered. For that case one obtains $\varphi^*_{\tilde{1}+\tilde{1}}(\sigma^*, \varepsilon_r, N) \to \varphi^*_{\tilde{1}+\tilde{1}}(\sigma^*, \sigma^*_\varepsilon)$.[6]

The case (3.39) is considered for further simplifications. It yields

$$\sigma^*_\varepsilon = \frac{N}{\alpha} \frac{\sigma_{\varepsilon_Q}(R)}{Q(R)} = \frac{N}{\alpha} \varepsilon_r, \tag{3.43}$$

and the normalized progress rate reads

$$\begin{aligned}
&\varphi^*_{\tilde{1}+\tilde{1}}(\sigma^*, \sigma^*_\varepsilon, N) \\
&\quad = \frac{1}{\sqrt{2\pi}} \int_{y=-\infty}^{y=\infty} \mathrm{e}^{-\frac{1}{2}y^2} \frac{N}{\Gamma\left(\frac{N-1}{2}\right)} \int_{t=0}^{t=\infty} \left(1 - \sqrt{\left(1 - \frac{\sigma^*}{N} y\right)^2 + 2\left(\frac{\sigma^*}{N}\right)^2 t}\right) \\
&\qquad \times \Phi\left(\frac{1 - \left(\left(1 - \frac{\sigma^*}{N} y\right)^2 + 2\left(\frac{\sigma^*}{N}\right)^2 t\right)^{\alpha/2}}{\sqrt{2}\,\alpha\sigma^*_\varepsilon / N}\right) \mathrm{e}^{-t}\, t^{\frac{N-1}{2}-1}\, \mathrm{d}t\, \mathrm{d}y.
\end{aligned} \tag{3.44}$$

Some plots of the progress rate $\varphi^*_{\tilde{1}+\tilde{1}}$ are shown in Fig. 3.3. They are generated using numerical integration. The influence of the normalized noise strength σ^*_ε on the (qualitative) change of φ^* can be observed very easily. In general, the following is valid:

$$\text{if} \quad \sigma^{*(1)}_\varepsilon > \sigma^{*(2)}_\varepsilon \quad \text{then} \quad \varphi^*_{\tilde{1}+\tilde{1}}(\sigma^*, \sigma^{*(1)}_\varepsilon, N) < \varphi^*_{\tilde{1}+\tilde{1}}(\sigma^*, \sigma^{*(2)}_\varepsilon, N). \tag{3.45}$$

The occurrence of *negative* φ^* values – in spite of the elitist selection – is related to this fact. Consequently, the $(1+1)$-ES as given by Algorithm 3 on p. 52 is *not* convergent under unfavorable conditions, e.g. if σ^*_ε is too large. The convergence security and first rudimentary approaches for its improvement are investigated in Sect. 3.3.2.2 (p. 92) and Sect. 3.3.4.

The performance of the $(\tilde{1}+\tilde{1})$-ES depends strongly on σ^*_ε. If its performance is compared with that of the nondisturbed $(1+1)$-ES (see Fig. 3.2), one notices a more significant influence of the parameter space dimension N on the qualitative behavior of the progress rate $\varphi^*_{\tilde{1}+\tilde{1}}$. The asymptotic limit $\varphi^*(\sigma^*, \sigma^*_\varepsilon, N \to \infty)$, which can be calculated analytically (see Sect. 3.3.3.3), is in contrast to the $\varphi^*(\sigma^*, \sigma^*_\varepsilon = 0, N \to \infty)$ case (Fig. 3.2) not always the worst case. In the low-dimensional case with larger σ^*_ε values it is meaningful to use smaller mutation strengths σ^*. The investigation of the corresponding optimal success probability P_s has not been performed yet.[7]

[6] Note that the normalized noise strength σ^*_ε is defined in such a manner that it does not explicitly depend on R or N.

[7] See also Sect. 3.3.3.3, p. 98, for the case $N \to \infty$, where the concept of σ-control based on the success probability P_s is discussed.

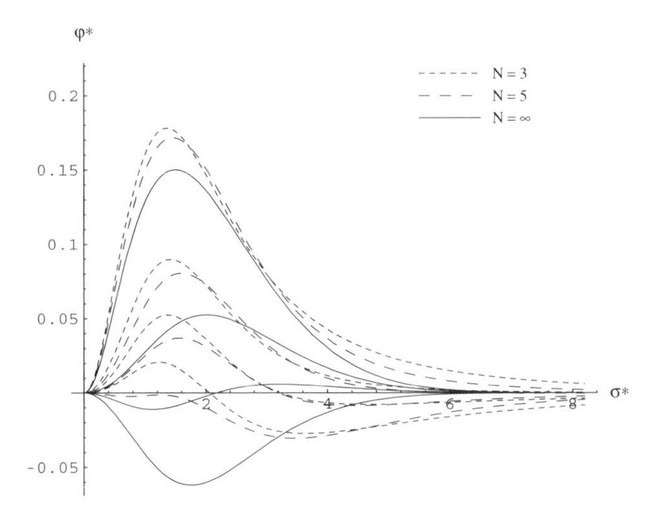

Fig. 3.3. The progress rate $\varphi^*_{\tilde{1}+\tilde{1}}$ of the (1 + 1)-ES according to Algorithm 3 on noisy fitness data. A quadratic fitness function is assumed for this diagram, i.e. $Q(R) = cR^\alpha$, $\alpha = 2$. In addition to the N-dependency (parameter space dimension), four different (normalized) noise strengths σ^*_ε are indicated for each case: $\sigma^*_\varepsilon = 0.5$, $\sigma^*_\varepsilon = 1.0$, $\sigma^*_\varepsilon = 1.25$, and $\sigma^*_\varepsilon = 1.5$. The respective curves lie on top of each other in the following order for σ^*_ε: 1.5, 1.25, 1.0, 0.5.

3.2 Asymptotic Formulae for the $(1 \stackrel{+}{,} \lambda)$-ES

3.2.1 A Geometrical Analysis of the (1 + 1)-ES

Several methods can be applied to the derivation of the analytical progress rate formulae for the case $N \to \infty$ on the (1 + 1)-ES at the sphere model. Some standard mathematical methods exist for the asymptotic derivation (see, e.g., Berg, 1968). An asymptotic expression for the double parameter integral (3.21) can be derived using these. This is a relatively simple approach; however, it is not very illuminating. Especially for a better understanding of the working mechanism of the ES, it is much more important to analyze the (1+1)-ES step by step and to develop a *geometrically* motivated model for the asymptotic case $N \to \infty$. Such a model was developed by Beyer (1989a) to analyze the evolution method for dissipative systems (Beyer, 1990a, 1990b). It was applied later (Beyer, 1992b, 1993) to the progress problem of the $(1 \stackrel{+}{,} \lambda)$-ES.

A central aspect of the geometrically oriented approach concerns the investigation of the properties of the mutation vector $\mathbf{z}$. One of these properties of $\mathbf{z}$, namely the spherical symmetry of the density $p(\mathbf{z})$, was already discussed in Sect. 3.1, p. 53, and was used for the computation of the densities of x and $\mathbf{h}$ (see Fig. 3.1). A second property is related to the asymptote of the length χ of the $\mathbf{z}$ vector and of the length h of the $\mathbf{h}$ vector, respectively.

They are investigated in the next section. The results obtained there will be used in the following section for a geometrically motivated derivation of the φ^*_{1+1} formula.

3.2.1.1 The Asymptote of the Mutation Vector z. The length χ

$$\chi := \sqrt{\sum_{i=1}^{N} z_i^2} \tag{3.46}$$

of an N-dimensional, isotropic Gaussian mutation vector $\mathbf{z}$ with the density (1.11) is a random variable which satisfies the so-called χ distribution

$$p(\chi) = \frac{2\chi^{N-1}\mathrm{e}^{-\frac{\chi^2}{2\sigma^2}}}{\sigma^N 2^{N/2}\Gamma\left(\frac{N}{2}\right)} \tag{3.47}$$

($\Gamma(x)$ is defined in (3.49)). This distribution can be derived easily from the χ^2 distribution (3.14) by substituting $N-1 \to N$ and using the transformation $\chi = \sqrt{u}$, i.e. $u(z) = \chi^2$, for the probability density $p(u)$ according to $p(\chi) = p_{h^2}(\chi)\frac{\mathrm{d}u}{\mathrm{d}\chi}$.

The expected value and the variance of χ are needed for the further investigation of (3.47). After a simple calculation one obtains the kth moment of the χ distribution

$$\overline{\chi^k} = \mathrm{E}\left\{\chi^k\right\} = \int_0^\infty \chi^k\, p(\chi)\,\mathrm{d}\chi = \left(\sqrt{2}\,\sigma\right)^k \frac{\Gamma\left(\frac{N}{2}+\frac{k}{2}\right)}{\Gamma\left(\frac{N}{2}\right)}, \tag{3.48}$$

considering the definition of the Γ function

$$\Gamma(x) := \int_0^\infty t^{x-1}\mathrm{e}^{-t}\,\mathrm{d}t. \tag{3.49}$$

The Γ-fraction in (3.48) can be expanded asymptotically (Abramowitz & Stegun, 1984, p. 78), yielding

$$\overline{\chi^k} = \left(\sqrt{N}\,\sigma\right)^k \left[1 + \frac{1}{N}\frac{k(k-2)}{4} + \mathcal{O}\left(\frac{1}{N^2}\right)\right]. \tag{3.50}$$

This formula is useful particularly for the odd k values; one obtains, e.g., for the expected value l of the length of χ: $l = \overline{\chi} = \mathrm{E}\{\chi\}$

$$l = \sqrt{N}\,\sigma\left[1 - \frac{1}{4N} + \mathcal{O}\left(\frac{1}{N^2}\right)\right]. \tag{3.51}$$

The second moment ($k=2$) of χ can be read directly from (3.48); because of the recurrence relation $\Gamma\left(\frac{N}{2}+1\right) = \frac{N}{2}\Gamma\left(\frac{N}{2}\right)$ one immediately finds

$$\overline{\chi^2} = N\sigma^2. \tag{3.52}$$

Consequently, the variance $\mathrm{D}^2\{\chi\} = \overline{(\chi-\overline{\chi})^2} = \overline{\chi^2} - \overline{\chi}^2 = \overline{\chi^2} - l^2$ is obtained as

$$\mathrm{D}^2\{\chi\} = \sigma^2 \left[\frac{1}{2} - \mathcal{O}\left(\frac{1}{N}\right)\right]. \tag{3.53}$$

Therefore, the variation coefficient $v\{\chi\}$ is of order

$$v\{\chi\} = \frac{\sqrt{\mathrm{D}^2\{\chi\}}}{\overline{\chi}} = \sqrt{\frac{1}{2N} - \mathcal{O}\left(\frac{1}{N\sqrt{N}}\right)} = \mathcal{O}\left(\frac{1}{\sqrt{N}}\right). \tag{3.54}$$

These results are very noteworthy. They indicate that the mean value of the length of the mutation vector $\mathbf{z}$ scales with the square root of the parameter space dimension N (cf. Eq. (3.51))

$$l \simeq \sigma\sqrt{N}, \qquad \mathrm{D}\{\chi\} \simeq \frac{\sigma}{\sqrt{2}}. \tag{3.55}$$

Furthermore, it can easily be estimated that the kth moment of the relative χ variation $(\chi - \overline{\chi})/\overline{\chi}$ is less than or equal to the order $\mathcal{O}(1/N)$

$$\forall k \in \mathbb{N}: \qquad \left|\mathrm{E}\left\{\left(\frac{\chi - \overline{\chi}}{\overline{\chi}}\right)^k\right\}\right| \le \mathcal{O}\left(\frac{1}{N}\right). \tag{3.56}$$

Therefore, the expected value expressions of the formulae containing the length of $\mathbf{z}$ or $\mathbf{h}$ can be estimated for the asymptotic limit $N \to \infty$ by substituting χ and u with their expected values l and $\overline{h}$, respectively.

Let us consider formula (3.9) as an important example, its asymptote reads

$$\tilde{R} \simeq \sqrt{(R-x)^2 + \overline{h}^2} \simeq \sqrt{(R-x)^2 + \sigma^2 N}. \tag{3.57}$$

Only an outline of the derivation will be given here. The basic idea is to expand the random variable $\tilde{R}$, Eq. (3.9), as a function of the random variables $h = \|\mathbf{h}\|$ in a Taylor series

$$\tilde{R} = \sqrt{(R-x)^2 + \mathbf{h}^2} = \overline{h}\sqrt{\left(\frac{R-x}{\overline{h}}\right)^2 + \left(1 + \frac{h - \overline{h}}{\overline{h}}\right)^2} \tag{3.58}$$

$$\tilde{R} = \overline{h}\sqrt{a + (1+y)^2} \quad \text{with} \quad a := \left(\frac{R-x}{\overline{h}}\right)^2 \quad \text{and} \quad y := \frac{h - \overline{h}}{\overline{h}}. \tag{3.59}$$

The expansion is done at $h = \overline{h}$. Therefore, (3.59) should be expanded at $y = 0$

$$\tilde{R} = \overline{h}\sqrt{a+1}\left\{1 + \frac{1}{a+1}y + \frac{1}{2}\left[\frac{1}{a+1} - \frac{1}{(a+1)^2}\right]y^2 + \ldots\right\}. \tag{3.60}$$

The moments of the left and right sides of (3.60) should be equal by definition. The moments of the right-hand side are obtained from the constant term $\overline{h}\sqrt{a+1}$ and from the expected values of y^k. The latter ones vanish due to (3.59) and (3.56) at least with $\mathcal{O}(1/N)$. Therefore, the error term in the braces of (3.60) vanishes at least with $\mathcal{O}(1/N)$.[8] After the backsubstitution and considering $\overline{h} \simeq \sigma\sqrt{N-1} \simeq \sigma\sqrt{N}$, the asymptotic formula (3.57) follows.

3.2.1.2 The Progress Rate of the (1 + 1)-ES on the Sphere Model. If the results from above are used, i.e. (3.57), the calculation of the progress rate (3.5) reduces to the one-dimensional integral

$$\begin{aligned}\varphi &= \int_{x_u}^{x_o} (R - \tilde{R}(x)) p_x(x)\,\mathrm{d}x \\ &= \int_{x_u}^{x_o} \left(R - \sqrt{(R-x)^2 + \overline{h}^2}\right) p_x(x)\,\mathrm{d}x \;+\; \mathcal{O}\!\left(\frac{R}{N^2}\right),\end{aligned} \tag{3.61}$$

where $p_x(x)$ stands for the density (3.12).[9] The limits x_u and x_o are given by the sphere boundary C_R (see Fig. 3.1). The condition $\tilde{R} \le R$ must be satisfied for a given $\overline{h}$; i.e. $\sqrt{(R-x)^2 + \overline{h}^2} \le R$. The limit condition is therefore $(R - x_{u/o})^2 + \overline{h}^2 = R^2$. Hence one obtains $x_{u/o}^2 - 2Rx_{u/o} + \overline{h}^2 = 0$, i.e. a quadratic equation in x. Its solutions are

$$x_{u/o} = R \mp R\sqrt{1 - \overline{h}^2/R^2}. \tag{3.62}$$

The closed analytical integration of (3.61) is impossible. However, since the limit case $N \to \infty$ is of interest anyway, the square root in (3.61) is expanded. The integral is separated into two parts for this purpose

$$\varphi = \frac{R}{\sqrt{2\pi}\,\sigma} \int_{x_u}^{l} \left(1 - \sqrt{\left(1 - \frac{x}{R}\right)^2 + \frac{\overline{h}^2}{R^2}}\right) \mathrm{e}^{-\frac{1}{2}\left(\frac{x}{\sigma}\right)^2} \mathrm{d}x \;+\; I_{\mathcal{R}}, \tag{3.63}$$

$$I_{\mathcal{R}} := \frac{R}{\sqrt{2\pi}\,\sigma} \int_{l}^{x_o} \left(1 - \frac{\tilde{R}(x)}{R}\right) \mathrm{e}^{-\frac{1}{2}\left(\frac{x}{\sigma}\right)^2} \mathrm{d}x. \tag{3.64}$$

The upper integration limit in the first part of (3.63) is chosen so that the square root in the integrand region differs only slightly from 1

$$\sqrt{\left(1 - \frac{x}{R}\right)^2 + \frac{\overline{h}^2}{R^2}} = \sqrt{1 - 2\underbrace{\left(\frac{x}{R} - \frac{\overline{h}^2}{2R^2} - \frac{x^2}{2R^2}\right)}_{\ll 1}}. \tag{3.65}$$

This is obviously the case for $0 \le x \le l$ because $l \simeq \sigma\sqrt{N} = \sigma^* R/\sqrt{N}$ (see (3.55) and (2.22)), if $l/R \simeq \sigma^*/\sqrt{N} \ll 1$; i.e. if

---

[8] If one postulates for a (see Eq. (3.59)) $x \ll R$, then one obtains $a^{-1} = \overline{h}^2/R^2 \sim \sigma^2 N/R^2 = \sigma^{*2}/N$, i.e. $a = \mathcal{O}(N)$. The error term in (3.60) vanishes under these conditions with $\mathcal{O}(1/N^2)$.

[9] The error term $\mathcal{O}(R/N^2)$ in (3.61) comes from (3.60) provided that $a = \mathcal{O}(N)$ (see also footnote 8). Since it does not change the final result (3.83), it is omitted in the following derivations.

$$\frac{\sigma^{*2}}{N} \ll 1 \tag{3.66}$$

is satisfied. Interestingly, these conditions should also hold for $\overline{h}^2/2R^2$, since asymptotically $\overline{h}^2 \simeq \sigma^2(N-1) \simeq \sigma^{*2}R^2/N$ is valid,

$$\frac{\overline{h}^2}{R^2} \simeq \frac{\sigma^{*2}}{N}. \tag{3.67}$$

The error term $I_\mathcal{R}$ (3.64) is easily estimated. Since $0 \le 1 - \tilde{R}(x)/R \le 1$ (see the selection criterion of the (1 + 1)-ES) one gets

$$\begin{aligned} 0 \le I_\mathcal{R} &\le \frac{R}{\sqrt{2\pi}\,\sigma} \int_l^{x_o} 1\mathrm{e}^{-\frac{1}{2}\left(\frac{x}{\sigma}\right)^2} \mathrm{d}x \\ &\le \frac{R}{\sqrt{2\pi}\,\sigma} \int_l^{\infty} \mathrm{e}^{-\frac{1}{2}\left(\frac{x}{\sigma}\right)^2} \mathrm{d}x = R\left[1 - \Phi\left(\frac{l}{\sigma}\right)\right] \end{aligned} \tag{3.68}$$

using the distribution function $\Phi(x)$ of the standard normal distribution

$$\Phi(x) := \frac{1}{\sqrt{2\pi}} \int_{-\infty}^{x} \mathrm{e}^{-\frac{1}{2}t^2} \mathrm{d}t. \tag{3.69}$$

Because of $l/\sigma \simeq \sqrt{N}$ and applying the asymptotic expansion of $\Phi(x)$ for large x

$$\Phi(x) \simeq 1 - \frac{1}{\sqrt{2\pi}} \frac{\mathrm{e}^{-\frac{1}{2}x^2}}{x}, \tag{3.70}$$

(see, e.g., Abramowitz & Stegun, 1984, p. 408) one obtains

$$0 \le I_\mathcal{R} \le \frac{R}{\sqrt{2\pi}} \frac{\mathrm{e}^{-\frac{N}{2}}}{\sqrt{N}} \quad \text{and} \quad 0 \le I_\mathcal{R}^* \le \frac{\sqrt{N}}{\sqrt{2\pi}} \mathrm{e}^{-\frac{N}{2}} = \mathcal{O}\left(\sqrt{\frac{N}{\mathrm{e}^N}}\right). \tag{3.71}$$

The error terms $I_\mathcal{R}$ and $I_\mathcal{R}^* = I_\mathcal{R}N/R$, normalized according to (2.22), vanish at least for the order $\mathcal{O}(\sqrt{N/\mathrm{e}^N})$ and can be neglected in all derivations. In other words, the upper integration limit x_o (3.62) in (3.61) can be replaced by $x_o = l$.

The calculation of (3.63) can be continued after this intermediate step. The square root (3.65) in (3.63) is expanded first. Because of

$$1 - \sqrt{1-2f} = f + \frac{1}{2}f^2 + \frac{1}{2}f^3 + \frac{5}{8}f^4 + \mathcal{O}(f^5)$$

one obtains

$$\begin{aligned} R - \tilde{R}(x) &= R\left(1 - \sqrt{1 - 2\left(\frac{x}{R} - \frac{\overline{h}^2}{2R^2} - \frac{x^2}{2R^2}\right)}\right) \\ &= x - \frac{\overline{h}^2}{2R} + \mathcal{R} \end{aligned} \tag{3.72}$$

with

$$\mathcal{R} = R\left[-\frac{x^2}{2R^2} + \frac{1}{2}\left(\frac{x}{R} - \frac{\overline{h}^2}{2R^2} - \frac{x^2}{2R^2}\right)^2 + \frac{1}{2}\left(\frac{x}{R} - \frac{\overline{h}^2}{2R^2} - \frac{x^2}{2R^2}\right)^3 + \ldots\right]. \tag{3.73}$$

After inserting this in (3.63) one gets

$$\varphi = \frac{1}{\sqrt{2\pi}\,\sigma}\int_{x_u}^{l}\left(x - \frac{\overline{h}^2}{2R}\right)\mathrm{e}^{-\frac{1}{2}\left(\frac{x}{\sigma}\right)^2}\mathrm{d}x \;+\; \frac{R}{N}I_{\mathcal{R}}^* \tag{3.74}$$

with the residuum

$$I_{\mathcal{R}}^* = \frac{N}{\sqrt{2\pi}\,\sigma}\int_{x_u}^{l}\left[-\frac{x^2}{2R^2} + \frac{1}{2}\left(\frac{x}{R} - \frac{\overline{h}^2}{2R^2} - \frac{x^2}{2R^2}\right)^2 + \frac{1}{2}\Big(\ldots\Big)^3 + \ldots\right]\mathrm{e}^{-\frac{1}{2}\left(\frac{x}{\sigma}\right)^2}\mathrm{d}x. \tag{3.75}$$

The substitution together with (3.67) and (2.22) yield for the normalized residual term

$$\begin{aligned}
I_{\mathcal{R}}^* &= \frac{N}{\sqrt{2\pi}}\int_{x_u/\sigma}^{l/\sigma}\left[-\left(\frac{\sigma^*}{N}\right)^2\frac{t^2}{2} + \frac{1}{2}\left(\frac{\sigma^*}{N}\right)^2\left(t - \frac{\sigma^*}{2} - \frac{\sigma^*}{N}\frac{t^2}{2}\right)^2 + \frac{1}{2}\left(\frac{\sigma^*}{N}\right)^3\ldots\right]\mathrm{e}^{-\frac{1}{2}t^2}\,\mathrm{d}t \\
&= \frac{\sigma^{*2}}{N}\frac{1}{\sqrt{2\pi}}\int_{x_u/\sigma}^{l/\sigma}\left[-\frac{t^2}{2} + \frac{1}{2}\left(t - \frac{\sigma^*}{2} - \frac{\sigma^*}{N}\frac{t^2}{2}\right)^2 + \frac{1}{2}\left(\frac{\sigma^*}{N}\right)\Big(\ldots\Big)^3 + \ldots\right]\mathrm{e}^{-\frac{1}{2}t^2}\,\mathrm{d}t \\
&= \frac{\sigma^{*2}}{N}\frac{1}{\sqrt{2\pi}}\int_{x_u/\sigma}^{l/\sigma}\left[\frac{1}{2}\sigma^*\left(\frac{\sigma^*}{4} - t\right) + \mathcal{O}\left(\frac{1}{N}\right)\right]\mathrm{e}^{-\frac{1}{2}t^2}\,\mathrm{d}t \\
&= \mathcal{O}\left(\frac{1}{N}\right).
\end{aligned} \tag{3.76}$$

Since the integral in (3.76) is finite, as can be shown for any l/σ and x_u/σ, the error term is of order $\mathcal{O}(1/N)$. Therefore, $I_{\mathcal{R}}^*$ can be neglected for $N \to \infty$. The practically relevant case $N < \infty$ is discussed further below in this chapter. After neglecting $I_{\mathcal{R}}^*$ in (3.74), the progress rate integral

$$\varphi \simeq \int_{x_u}^{l}\left(x - \frac{\sigma^*\sigma}{2}\right)p_x(x)\,\mathrm{d}x \tag{3.77}$$

remains. Considering (3.67), the lower limit is obtained from (3.62)

$$x_u = R\left(1 - \sqrt{1 - \frac{\sigma^{*2}}{N}}\right). \tag{3.78}$$

Because of Condition (3.66), one can perform the Taylor expansion of the square root. One obtains

$$\begin{aligned} x_u &= R\left[\frac{1}{2}\frac{\sigma^{*2}}{N} + \frac{1}{8}\left(\frac{\sigma^{*2}}{N}\right)^2 + \ldots\right] \\ &= \frac{\sigma^*\sigma}{2}\left[1 + \frac{\sigma^{*2}}{4N} + \ldots\right] = \frac{\sigma^*\sigma}{2}\left[1 + \mathcal{O}\left(\frac{1}{N}\right)\right]. \end{aligned} \tag{3.79}$$

Hence, for the asymptotic limit case one has

$$x_u = \frac{\sigma^*\sigma}{2}. \tag{3.80}$$

This result is consistent with the parentheses in the integrand of (3.77): the (+) selection strategy accepts only positive values for $R - \tilde{R}(x)$. The parentheses contain an asymptotically exact approximation, its positiveness is guaranteed by the lower limit given in (3.80).

In the last step of the derivation for φ^*, (3.12) and (2.22) are taken into account and $t = x/\sigma$ is substituted in (3.74). As a result one obtains

$$\varphi^* = \frac{1}{\sqrt{2\pi}} \int_{t=\sigma^*/2}^{t=\sqrt{N}} \left(\sigma^* t - \frac{\sigma^{*2}}{2}\right) \mathrm{e}^{-\frac{t^2}{2}} \, \mathrm{d}t \; + \; I_{\mathcal{R}}^*. \tag{3.81}$$

The simple integration thereafter yields

$$\varphi^* = \frac{\sigma^*}{\sqrt{2\pi}}\left[\mathrm{e}^{-\frac{\sigma^{*2}}{8}} - \mathrm{e}^{-\frac{N}{2}}\right] - \frac{\sigma^{*2}}{2}\left[\Phi\left(\sqrt{N}\right) - \Phi\left(\frac{\sigma^*}{2}\right)\right] \; + \; I_{\mathcal{R}}^*. \tag{3.82}$$

The terms containing N yield asymptotically $\mathrm{e}^{-N/2} = 0 + \mathcal{O}(1/\sqrt{\mathrm{e}^N})$ and $\Phi(\sqrt{N}) = 1 - \mathcal{O}(1/\sqrt{N\mathrm{e}^N})$, respectively. They can be neglected when compared to (3.76). The remaining result is the *asymptotic progress rate formula* of the (1 + 1)-ES on the sphere model which was derived by Rechenberg (1973) in another way

$$\boxed{\varphi_{1+1}^*(\sigma^*) = \frac{\sigma^*}{\sqrt{2\pi}}\,\mathrm{e}^{-\frac{1}{8}\sigma^{*2}} - \frac{\sigma^{*2}}{2}\left[1 - \Phi\left(\frac{\sigma^*}{2}\right)\right] \; + \; \mathcal{O}\left(\frac{1}{N}\right).} \tag{3.83}$$

Naturally, the $\mathcal{O}$ notation in (3.83) does not give any information on the approximation quality of the formula for practically interesting case $N < \infty$. The validity region (in the sense of usable theoretical predictions) is dictated by the condition (3.66), as far as the derivation is concerned. A more precise estimation of the error term is omitted here. Instead, the results of the exact theory of the (1 + 1)-ES are shown in Fig. 3.2, p. 56, by some examples. The formula (3.83) for $N \to \infty$ already gives good approximation results for

$N \gtrapprox 30$. N-dependent φ^* formulae of the comma strategies will be derived in Sect. 3.4 as well as in Chaps. 5 and 6.

3.2.1.3 The Success Probability P_{s1+1} and the EPP. In addition to the progress rate φ^*_{1+1}, the success probability P_s (3.18) is of certain interest. It measures the probability of a successful mutation, i.e. that the condition $\tilde{R} \le R$ is fulfilled according to Fig. 3.1. In the asymptotic $\overline{h} = \text{const}$ approximation, $\tilde{R} \le R$ is fulfilled for the range $x \in [x_u, x_o]$. Thus the calculation of P_s reduces to the calculation of the probability that the random variable x with density (3.12) lies in the interval $[x_u, x_o]$

$$P_{s1+1} = \Pr(x_u \le x \le x_o) = \int_{x_u}^{x_0} p_x(x)\,dx. \tag{3.84}$$

This integral is very similar to the φ-integral in (3.61). Therefore, (3.84) is decomposed in a manner similar to (3.63, 3.64). The error term is estimated analogously to (3.68) $\mathcal{O}(1/\sqrt{N e^N})$, where only the factor R is missing here

$$P_{s1+1} = \int_{x_u}^{l} p_x(x)\,dx \;+\; \mathcal{O}\left(\frac{1}{\sqrt{N e^N}}\right). \tag{3.85}$$

After inserting (3.12), considering $l \simeq \sigma\sqrt{N}$ and (3.79), as well as the substitution $t = x/\sigma$ thereafter, P_s becomes

$$P_{s1+1} = \frac{1}{\sqrt{2\pi}} \int_{t=\frac{\sigma^*}{2}\left(1+\mathcal{O}\left(\frac{1}{N}\right)\right)}^{t=\sqrt{N}} e^{-\frac{1}{2}t^2}\,dt \;+\; \mathcal{O}\left(\frac{1}{\sqrt{N e^N}}\right)$$

$$P_{s1+1} = \Phi\left(\sqrt{N}\right) - \Phi\left(\frac{\sigma^*}{2} + \frac{\sigma^*}{2}\mathcal{O}\left(\frac{1}{N}\right)\right). \tag{3.86}$$

The asymptotic expansion yields (applying (3.70) to the first term and Taylor expansion to the second term)

$$\boxed{P_{s1+1}(\sigma^*) = 1 - \Phi\left(\frac{\sigma^*}{2}\right) \;-\; \mathcal{O}\left(\frac{1}{N}\right).} \tag{3.87}$$

This result can be read directly from (3.81, 3.83), but without the order term. If we consider (3.81, 3.83), and (3.87) thoroughly, we see that the asymptotic progress rate can also be expressed as

$$\varphi^*_{1+1}(\sigma^*) = \frac{\sigma^*}{\sqrt{2\pi}} e^{-\frac{1}{8}\sigma^{*2}} \;-\; \frac{\sigma^{*2}}{2} P_{s1+1}(\sigma^*) \;+\; \mathcal{O}\left(\frac{1}{N}\right). \tag{3.88}$$

A first contact to the EPP (evolutionary progress principle, see Sect. 1.4.2, p. 20) can be established using this representation.

The decomposition of (3.88) in gain and loss terms is trivial in the sense of the EPP, and agrees with the order of the terms in (3.88). The backtracing of the derivation to Eqs. (3.77) and (3.74), respectively, is even more interesting: the gain term is affiliated with the (positive) x values, whereas the loss term is caused by $\overline{h}^2$. It can easily be seen that the x part is the component of the mutation vector $\mathbf{z}$ in the direction toward the optimum (see Fig. 3.1, p. 53); $\overline{h}$ measures the length of the component $\mathbf{h}$ perpendicular to it. If the mutation strength σ (or σ^*, since R and N are constant) is increased, the lengths of both x and $\mathbf{h}$ components increase. The gain term in (3.88) increases at first linearly in σ^*, and then is dampened by the exponential term at larger σ^* values. The loss term increases for small σ^* quadratically; the success probability decreases with increasing σ^* and moderates the loss term. In summary, both terms converge to zero for larger σ^*. This fact can be explained geometrically: mutations very seldom fall in the success region; therefore, the parent survives many generations, the state is not changed, and the progress rate φ converges to zero. The effect observed is the same as for $\sigma^* = 0$ (which corresponds to the mutation vectors $\mathbf{z}$ of length $l = 0$). There exists a σ^* interval with positive progress rate, which is designated by Rechenberg (1973) as an *evolution window* (in German: Evolutionsfenster). In particular, there exists an optimal $\hat{\sigma}^* = \arg\max_{\sigma^*}[\varphi^*(\sigma^*)] \approx 1.224$ (which can be obtained by numerical maximization of (3.83)) at which one obtains the maximal progress rate $\hat{\varphi}^* \approx 0.202$. Hence, the mutation strength σ should be controlled properly in a practical implementation. This can be achieved by the measurement of the success probability P_s. This point is investigated further in Sect. 3.3.3.3.

3.2.2 The Asymptotic $\varphi^*_{1\overset{+}{,}\lambda}$ Integral

The characteristic property of the $(1 \overset{+}{,} \lambda)$-ES with $\lambda > 1$ is the generation of λ descendants $\tilde{\mathbf{y}}_l$ around the parent individual $\mathbf{y}_p$ by the addition of the mutation $\mathbf{z}$, generated anew according to (3.1) and the selection of the best individual $\tilde{\mathbf{y}}_{1;\gamma}$ from the set of these offspring (in the case of comma selection, $\gamma = \lambda$) or from the set of offspring *and* the parent (for plus selection, $\gamma = 1 + \lambda$). Since we assume the rotationally symmetric fitness function $Q(R)$ to be monotonic, the selection of the best individual $\tilde{\mathbf{y}}_{1;\gamma}$ (see Sect. 1.2.2, p. 8) is equivalent to choosing the smallest $\tilde{R}$-value $\tilde{R}_{1:\lambda}$ in the comma strategy case, or $\min[\tilde{R}_{1:\lambda}, R]$ in the plus strategy case. Therefore, the progress rate is defined according to (2.15)

$$\varphi_{1\overset{+}{,}\lambda} = \mathrm{E}_{\overset{+}{,}}\{R - \tilde{R}_{1:\lambda}\}, \tag{3.89}$$

whereas the difference between (+)- and (,)-selection is "hidden" in the manner in which the expected value is calculated (see below).

The derivation of the asymptotic φ formulae for the $(1 \overset{+}{,} \lambda)$ strategies is done principally in a similar manner as in the $(1+1)$-ES case. The basic idea is again to decompose each mutation $\mathbf{z}_l$ in a $x_l\mathbf{e}_{\mathrm{opt}}$ component and a component $\mathbf{h}_l$ perpendicular to it according to (3.7) (cf. Fig. 3.1). The lengths h_l of the $\mathbf{h}_l$ vectors are assumed to be asymptotically (for $N \to \infty$) constant, hence they can be substituted by the expected value $\overline{h}$. Therefore, analogous to (3.57) one gets for the lth descendant $\tilde{R}_l \simeq \sqrt{(R - x_l)^2 + \overline{h}^2}$. Here, $\tilde{R}$ is a function of x. The best offspring has the residual distance $\tilde{R}_{1:\lambda}$ with the corresponding x value $x_{1;\lambda}$. The progress rate (3.89) follows (asymptotically)

$$\varphi_{1 \stackrel{+}{,} \lambda} \simeq \mathrm{E}_{\stackrel{+}{,}}\left\{R - \sqrt{(R - x_{1;\lambda})^2 + \overline{h}^2}\right\}. \tag{3.90}$$

The structure of this expected value formula agrees with the formula (3.61). The only difference lies in the $x = x_{1;\lambda}$ and the corresponding probability density. Therefore, the method of derivation for the φ_{1+1} formula (3.77) can be adapted for the $(1 \stackrel{+}{,} \lambda)$ variants. Again, the primary idea is the expansion of the square root in (3.90) in a Taylor series for $x_{1;\lambda} \ll R$, assuming (3.66). Analogous to (3.72) one obtains

$$R - \tilde{R}(x_{1;\lambda}) = x_{1;\lambda} - \frac{\overline{h}^2}{2R} + \mathcal{R}(x_{1;\lambda}). \tag{3.91}$$

The error term $\mathcal{R}$ is neglected in the scope of the approximation made. According to (3.91), the smallest $\tilde{R}$ value $R_{1:\lambda}$ belongs to the largest x value, i.e. $x_{1;\lambda} = x_{\lambda:\lambda}$ (see (1.8) for the notation). This follows immediately if the geometry is considered: x measures the component of the $\mathbf{z}$ mutation in the direction toward the optimum. Hence, the formula analogous to (3.77) reads

$$\begin{aligned} \varphi_{1 \stackrel{+}{,} \lambda} &\simeq \int_{x_u}^{l} \left(x_{\lambda:\lambda} - \frac{\sigma^* \sigma}{2}\right) p(x_{\lambda:\lambda})\, \mathrm{d}x_{\lambda:\lambda} \\ &= \int_{x_u}^{l} \left(x - \frac{\sigma^* \sigma}{2}\right) p_{\lambda:\lambda}(x)\, \mathrm{d}x, \end{aligned} \tag{3.92}$$

with the probability density $p_{\lambda:\lambda}(x)$ for the distribution of the *largest* x value from a sample of λ independent, identically distributed random variables with density $p_x(x)$.

The calculation of $p_{\lambda:\lambda}(x)$ is a standard task in order statistics (see, e.g., Arnold et al., 1992). Nevertheless, a short derivation of the $p_{\lambda:\lambda}(x)$ density will be given here, since the same derivation methodology will be used in an adapted or enhanced form in several parts of this book.

We will first consider only one of the λ mutations. The x component of this mutation – selected arbitrarily from the pool of λ offspring – has the density $p_x(x)$. All other $(\lambda - 1)$ mutations must have smaller x values if this x value is the largest. The probability of having a mutation X (meaning the x component in this notation) with $X < x$ is denoted by $P_x(x) = \Pr(X < x)$. The probability that *all* other $(\lambda - 1)$ mutations fulfill this condition is $[P_x(x)]^{\lambda-1}$, since they are statistically independent. Consequently, the density for the special mutation considered is $p_x(x)[P_x(x)]^{\lambda-1}$. Since there are λ individuals and, thus, λ exclusive cases of being the best individual, the factor λ is to be taken into account in the final $p_{\lambda:\lambda}(x)$ formula. Thus, one obtains

$$p_{\lambda:\lambda}(x) = \lambda p_x(x) \left[P_x(x)\right]^{\lambda-1} \tag{3.93}$$

with

$$P_x(x) = \Pr(X < x) = \int_{x'=-\infty}^{x'=x} p_x(x')\,\mathrm{d}x'. \tag{3.94}$$

Therefore, considering (3.69) the density (3.12) reads

$$p_{\lambda:\lambda}(x) = \frac{\lambda}{\sqrt{2\pi}\,\sigma}\,\mathrm{e}^{-\frac{1}{2}\left(\frac{x}{\sigma}\right)^2}\left[\Phi\left(\frac{x}{\sigma}\right)\right]^{\lambda-1} \quad \text{and} \quad P_x(x) = \Phi\left(\frac{x}{\sigma}\right). \tag{3.95}$$

After inserting (3.95) into (3.92), substituting $t = x/\sigma$, considering the asymptotic relation (3.55), $l \simeq \sigma\sqrt{N}$, and the transformation to the normalized φ^*, Eq. (2.22), a progress rate integral of the following form remains

$$\varphi^*_{1\overset{+}{,}\lambda} \simeq \frac{\lambda}{\sqrt{2\pi}}\int_{t=t_u=x_u/\sigma}^{t=t_o=\sqrt{N}}\left(\sigma^* t - \frac{\sigma^{*2}}{2}\right)\mathrm{e}^{-\frac{1}{2}t^2}\,[\Phi(t)]^{\lambda-1}\,\mathrm{d}t. \tag{3.96}$$

The lower integration limit $t_u = x_u/\sigma$ depends on the strategy type. In the case of $(1+\lambda)$-ES, only improvements are accepted, i.e., the x_u value is determined as in the $(1+1)$-ES case by (3.78) and herewith – in the limits of the approximation – by (3.80); therefore $t_u = \sigma^*/2$. For the $(1,\lambda)$-strategy, the parental state is ignored. Selection merely occurs among the λ offspring, i.e. since the x value of the best descendant is not constrained, in the worst case this value can be $-\infty$. In short, the lower limits at the sphere model are

$$\begin{aligned} (1+\lambda)\text{-ES} \Rightarrow \quad & t_u = \tfrac{\sigma^*}{2} \\ (1,\lambda)\text{-ES} \Rightarrow \quad & t_u = -\infty. \end{aligned} \tag{3.97}$$

The integral (3.96) can also be written as

$$\varphi^*_{1\overset{+}{,}\lambda}(\sigma^*) = \sigma^*\,\frac{\lambda}{\sqrt{2\pi}}\int_{t=t_u}^{t=\infty}\left(t - \frac{\sigma^*}{2}\right)\mathrm{e}^{-\frac{1}{2}t^2}\,[\Phi(t)]^{\lambda-1}\,\mathrm{d}t \;+\; \mathcal{O}\!\left(\frac{1}{N}\right). \tag{3.98}$$

The estimation of the order of the error term follows analogously to the $(1+1)$-ES. The applicability of (3.98) as an approximation for $N < \infty$ is limited by the condition (3.66). The cases $(1,\lambda)$-ES and $(1+\lambda)$-ES will be discussed separately in the following two points.

3.2.3 The Analysis of the $(1,\lambda)$-ES

3.2.3.1 The Progress Rate $\varphi^*_{1,\lambda}$.

The progress rate formula. In order to calculate $\varphi^*_{1,\lambda}$, (3.98) is evaluated considering $t_u = -\infty$ (3.97). The result is

$$\begin{aligned} \varphi^*_{1,\lambda}(\sigma^*) = \sigma^*\,&\frac{\lambda}{\sqrt{2\pi}}\int_{-\infty}^{\infty} t\,\mathrm{e}^{-\frac{1}{2}t^2}\,[\Phi(t)]^{\lambda-1}\,\mathrm{d}t \\ &- \frac{\sigma^{*2}}{2}\int_{-\infty}^{\infty}\frac{\mathrm{d}}{\mathrm{d}t}\,[\Phi(t)]^{\lambda}\,\mathrm{d}t \;+\; \mathcal{O}\!\left(\frac{1}{N}\right). \end{aligned} \tag{3.99}$$

The second integral in (3.99) gives, because of $\Phi(\infty) = 1$ and $\Phi(-\infty) = 0$, the value one. The first integral defines the *progress coefficient* $c_{1,\lambda}$ (Rechenberg, 1984)

$$c_{1,\lambda} := \frac{\lambda}{\sqrt{2\pi}} \int_{-\infty}^{\infty} t\,\mathrm{e}^{-\frac{1}{2}t^2} \left[\Phi(t)\right]^{\lambda-1} \mathrm{d}t, \tag{3.100}$$

which is a special case of the *progress coefficients*[10] $d_{1,\lambda}^{(k)}$ or of *general progress coefficients*[11] $e_{\mu,\lambda}^{\alpha,\beta}$, respectively. A survey of the $c_{1,\lambda}$ coefficients follows in the next point (p. 74).

Using (3.100), Eq. (3.99) simplifies to:

$$\varphi_{1,\lambda}^{*}(\sigma^*) = c_{1,\lambda}\sigma^* \;-\; \frac{\sigma^{*2}}{2} \;+\; \mathcal{O}\left(\frac{1}{N}\right). \tag{3.101}$$

This progress law is asymptotically $(N \to \infty)$ exact and astonishingly simple. The normalized progress rate is a quadratic function of the normalized mutation strength. Figure 3.4 shows plots of $\varphi_{1,\lambda}^*/\lambda$ versus σ^*. Valuable in-

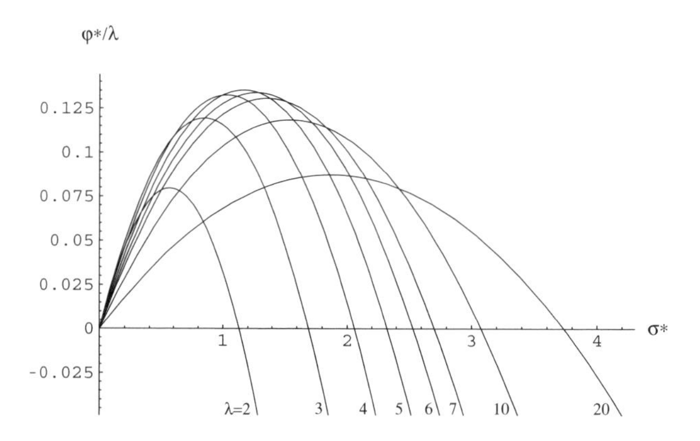

Fig. 3.4. The progress rate per *descendant* at the $(1,\lambda)$-ES for different numbers of descendants λ.

formation can be extracted from this picture. First of all, it is striking that φ^* can take negative values, in contrast to the (1+1)-ES (cf. Fig. 3.2, p. 56). This happens for $\sigma^* > 2c_{1,\lambda}$, as it can be easily read from (3.101) (ignore $\mathcal{O}(1/N)$, set φ^* to zero). The practical consequence is that the residual distance increases – and the *ES diverges* – for $\sigma^* > 2c_{1,\lambda} \;\Leftrightarrow\; \varphi^* < 0$ (see Eq. (2.101)). The necessary condition for the convergence of the ES is therefore the fulfillment of the

$$\text{Evolution Criterion } (1,\lambda)\text{-ES:} \qquad \sigma^* < 2c_{1,\lambda}. \tag{3.102}$$

[10] For $k = 1$, see Sect. 4.1.4, Eq. (4.42).

[11] For $\alpha = 1$, $\beta = 0$, $\mu = 1$, see Sect. 5.2.2, Eq. (5.112).

The mutation strength σ requires an *adaptation* aiming at the maximization of φ^*. Chapter 7 is devoted to this theme.

EPP. The *evolutionary progress principle* can be identified very easily in (3.101). The gain is the $c_{1,\lambda}\sigma^*$ term, whereas the loss is represented by $\sigma^{*2}/2$. Since the gain is linear in σ^* and the loss quadratic, φ^* is dominated for small mutation strengths by the positive, linear part. For increasing σ^*, the negative quadratic term becomes more influential. As a result, φ^* has necessarily a maximum $\hat{\varphi}^*$

$$\hat{\varphi}^* := \max_{\sigma^*}\left[\varphi^*(\sigma^*)\right] = \frac{c_{1,\lambda}^2}{2}, \qquad \hat{\sigma}^* := \arg\max_{\sigma^*}\left[\varphi^*(\sigma^*)\right] = c_{1,\lambda}, \tag{3.103}$$

immediately obtained from (3.101) (neglecting the $\mathcal{O}(1/N)$).

The $(1,\lambda)$ *fitness efficiency.* The *progress rate per fitness evaluation* (or per descendant), denoted as φ^*/λ, is shown in Fig. 3.4. If the maximal values $\hat{\varphi}^*_{1,\lambda}/\lambda$ are considered for the corresponding number of offspring λ, one observes that there is a "most efficient" $(1,\lambda)$-ES for this problem: the largest possible progress rate per descendant can be attained at $\lambda = 5$. In order to investigate this fact more thoroughly, the *fitness efficiency* $\eta_{1,\lambda}$ is defined

$$\text{FITNESS EFFICIENCY:} \qquad \eta_{1,\lambda} := \frac{\max_{\sigma^*}\left[\varphi^*(\sigma^*)\right]}{\lambda} \tag{3.104}$$

which measures the maximal progress per fitness evaluation depending on λ. Using (3.101) it can be calculated as

$$\eta_{1,\lambda} = \frac{c_{1,\lambda}^2}{2\lambda} \tag{3.105}$$

and is plotted in Fig. 3.5. Using this figure, it is possible to decide whether

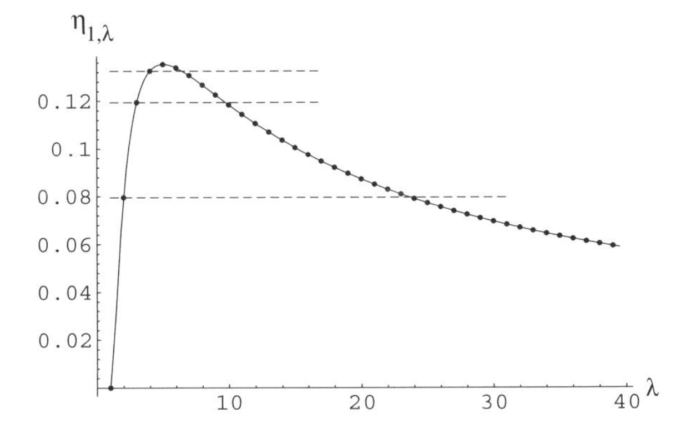

Fig. 3.5. The fitness efficiency $\eta_{1,\lambda} = c_{1,\lambda}^2/2\lambda$ is obtained by integrating (3.100) numerically. Its maximum lies around $\lambda \approx 5.02$; therefore, the $(1,5)$-ES is the most efficient $(1,\lambda)$-ES.

a $(1,G\lambda)$-ES or a $(1,\lambda)^G$-ES[12] should be used for a higher performance:

[12] The notation $(1,\lambda)^G$ stands for using $(1,\lambda)$-ES over G generations.

Assuming an optimal progress rate $\hat{\varphi}^*$, a $(1, G\lambda)$-ES yields an R-gain (2.97) of

$$R(g) - R(g+1) = \hat{\varphi}^*_{1,G\lambda} \frac{R(g)}{N} \tag{3.106}$$

per generation. If the time required for selection and the overhead of the ES algorithm are neglected as compared to the time cost for fitness calculations, one can consider a $(1, \lambda)$-ES which runs over G generations ($G\lambda$ fitness evaluations are required here again). The change of the residual distance after $G\lambda$ fitness evaluations is obtained iteratively as

$$\begin{aligned} R(g) - R(g+G) &= R(g) - R(g)\left(1 - \frac{\hat{\varphi}^*_{1,\lambda}}{N}\right)^G \\ &= R(g)\left[1 - \left(1 - G\frac{\hat{\varphi}^*_{1,\lambda}}{N} + \mathcal{O}\left(\frac{1}{N^2}\right)\right)\right] \\ &= G\,\hat{\varphi}^*_{1,\lambda}\frac{R(g)}{N}\left(1 - \mathcal{O}\left(\frac{1}{N}\right)\right). \end{aligned} \tag{3.107}$$

A $(1, G\lambda)$-ES is more efficient if (3.106) is greater than (3.107). Consequently, after neglecting the $\mathcal{O}(1/N)$ term in (3.107), $\hat{\varphi}^*_{1,G\lambda} > G\hat{\varphi}^*_{1,\lambda}$ should be valid, and after dividing by $G\lambda$ it follows that (considering (3.104))

$$\eta_{1,G\lambda} > \eta_{1,\lambda}. \tag{3.108}$$

This condition is fulfilled for $\lambda = 2$ and $\lambda = 3$. The points above the dashed horizontal lines in Fig. 3.5 fulfill the condition (3.108). For example, the $(1, 8)$-ES is more efficient than the $(1, 2)^4$-ES, and the $(1, 9)$-ES is more efficient than the $(1, 3)^3$-ES. However, for $\lambda \geq 4$, there is no $G > 1$ for which (3.108) is fulfilled. In other words, a $(1, \lambda)^G$-ES with $\lambda \geq 4$, $G \geq 1$ is always more efficient than a $(1, G\lambda)$-ES.

Therefore, increasing the number of offspring over $\lambda = 5$ (the most efficient offspring number for the sphere model) makes sense only if the computation time and not the number of fitness evaluations is to be reduced by parallelizing the algorithm. If there are ϖ processors available, then for $\varpi \geq 4$ the $(1, \lambda{=}\varpi)$-ES should be chosen.[13]

3.2.3.2 The Progress Coefficient $c_{1,\lambda}$.

Analytical and numerical calculation of $c_{1,\lambda}$. The progress coefficient (3.100) is the first moment of the $x_{\lambda:\lambda}$ order statistics from a sample of λ random numbers which are standard normally distributed, $\mathcal{N}(0, 1)$. The analytical calculation is only possible for $\lambda = 1, 2, 3, 4,$ and 5 (see, e.g., David, 1970, pp. 30ff)

[13] The Communication overhead is not considered in these considerations. The derivation of a more realistic model of the ES efficiency remains as a task for the future.

$$c_{1,1} = 0, \qquad c_{1,2} = \frac{1}{\sqrt{\pi}},$$
$$c_{1,3} = \frac{3}{2\sqrt{\pi}}, \qquad c_{1,4} = \frac{6}{\pi^{3/2}} \arctan\sqrt{2}, \tag{3.109}$$
$$c_{1,5} = \frac{5}{2\sqrt{\pi}}\left(\frac{6}{\pi}\arctan(\sqrt{2}) - 1\right).$$

For $\lambda > 5$ the results can only be obtained numerically. In Table 3.1, some $c_{1,\lambda}$ values are provided, a larger table of progress coefficients can be found in Appendix D.2.

Table 3.1. The first 18 $c_{1,\lambda}$ progress coefficients

λ	1	2	3	4	5	6	7	8	9
$c_{1,\lambda}$	0.000	0.564	0.846	1.029	1.163	1.267	1.352	1.424	1.485
λ	10	11	12	13	14	15	16	17	18
$c_{1,\lambda}$	1.539	1.586	1.629	1.668	1.703	1.736	1.766	1.794	1.820

The $c_{1,\lambda}$ progress coefficient is a slowly increasing function of λ, as can be seen in Fig. 3.6.

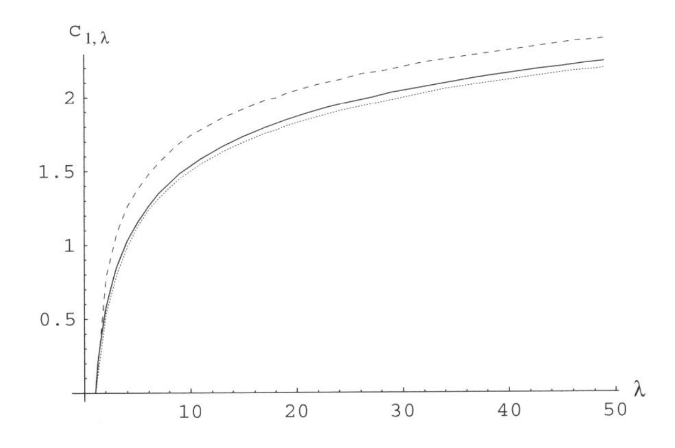

Fig. 3.6. The progress coefficient $c_{1,\lambda}$ as a function of λ, obtained by numerical integration and displayed as a solid line, compared with the approximation formula (3.110) ("...." dotted line) and the asymptotically exact $c_{1,\lambda}$ formula (3.113) ("- - -" dashed curve)

An approximate formula for $c_{1,\lambda}$. Approximation formulae are alternatives to the numerical integration approach. Rechenberg (1978) proposed the approximation

$$c_{1,\lambda} \approx \Phi^{-1}\left(\sqrt[\lambda]{\frac{1}{2}}\right). \tag{3.110}$$

Here, $\Phi^{-1}(.)$ stands for the inverse function to the cumulative distribution function $\Phi(.)$ of the Gaussian distribution (also called quantile function, see

Appendix C.1). The formula (3.110) yields usable results, as the comparison in Fig. 3.6 shows.

A geometrically motivated derivation for (3.110) can be attained easily. To this end, (3.100) is integrated by parts

$$c_{1,\lambda} = t\,[\Phi(t)]^\lambda \Big|_{-\infty}^{\infty} - \int_{-\infty}^{\infty} [\Phi(t)]^\lambda \,\mathrm{d}t = \lim_{t\to\infty} t - \int_{-\infty}^{\infty} [\Phi(t)]^\lambda \,\mathrm{d}t$$

$$c_{1,\lambda} = \lim_{w\to\infty} \int_{-w}^{w} \frac{1}{2}\,\mathrm{d}t - \int_{-\infty}^{\infty} [\Phi(t)]^\lambda \,\mathrm{d}t$$

$$= \lim_{w\to\infty} \int_{t=-w}^{t=w} \underbrace{\left(\frac{1}{2} - [\Phi(t)]^\lambda\right)}_{:=\, f(t)} \mathrm{d}t. \qquad (3.111)$$

The integrand function $f(t)$ is shown in Fig. 3.7. If geometrically interpreted,

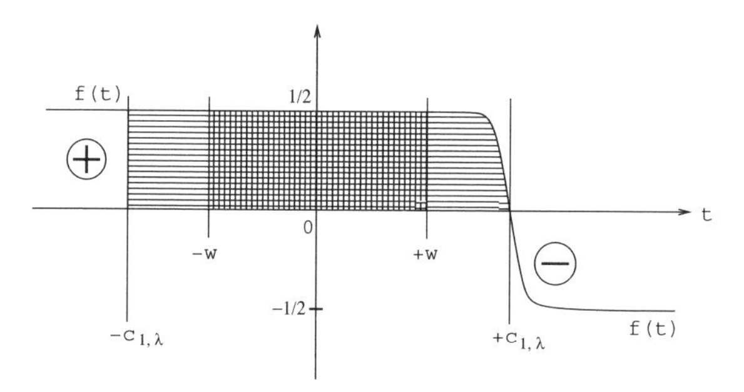

Fig. 3.7. On the graphically approximate calculation of $c_{1,\lambda}$

the integral in (3.111) expresses the area under the curve $f(t)$ from $t = -w$ to $t = w$. For large λ, the transition from positive to negative values of f is relatively sharp because of the exponentiation of $\Phi(t)$ (in the limit case, $\lambda \to \infty$, $f(t)$ represents a step function). If the integration interval $[-w, w]$ in Fig. 3.7 is increased gradually, the area increases until w reaches the zero point t_0 of the function $f(t)$, $w = t_0 \Leftrightarrow f(t_0) = 0$. If w is increased beyond t_0, the net area remains (approximately) constant, since the positive part "⊕" and the negative part "⊖" compensate each other. Thus, for $w \to \infty$, the integral value is represented by the area of the approximately rectangular-shaped region from $w = -t_0$ to $w = t_0$, i.e. $c_{1,\lambda} = 2t_0\,\frac{1}{2}$. Hence, the progress coefficient is determined approximately by

$$f(c_{1,\lambda}) \approx 0 \qquad \Rightarrow \qquad \frac{1}{2} \approx [\Phi(c_{1,\lambda})]^\lambda\,; \qquad (3.112)$$

its solution gives the formula (3.110).

On the asymtotes of the $c_{1,\lambda}$ values. Besides the derivation of $c_{1,\lambda}$ approximation formulae, the asymptotic behavior of the progress coefficients is also of interest. The asymptotes of the $c_{\mu/\mu,\lambda}$ progress coefficients are investigated in Sect. 6.3.1 and defined in the same chapter. Therefore $c_{1,\lambda}$ is a special case for $\mu = 1$. If one specifies $\mu = 1$ in (6.113) on p. 249, one obtains (see also Appendix D.1.3)

$$c_{1,\lambda} \simeq \frac{\lambda}{\sqrt{2\pi}} \exp\left[-\frac{1}{2}\left(\Phi^{-1}\left(1 - \frac{1}{\lambda}\right)\right)^2\right]. \tag{3.113}$$

The quality of (3.113) as a numerical approximation for small λ values is not very satisfactory, as can be seen in Fig. 3.6. However, Eq. (3.113) is the starting point for the derivation of simple expressions on the λ-order of $c_{1,\lambda}$. The investigation of this also follows in Chap. 6. Using (6.117), one obtains for $\mu = 1$

$$c_{1,\lambda} = \mathcal{O}\left(\sqrt{\ln \lambda}\right) \tag{3.114}$$

and as a result for the maximal possible progress rate[14] $\hat{\varphi}^*_{1,\lambda} = c^2_{1,\lambda}/2$ and for the fitness efficiency (3.105)

$$\hat{\varphi}^*_{1,\lambda} = \mathcal{O}(\ln \lambda) \qquad \text{and} \qquad \eta_{1,\lambda} = \mathcal{O}\left(\frac{\ln \lambda}{\lambda}\right). \tag{3.115}$$

As one can easily see, the progress rate scales with $\mathcal{O}(\ln \lambda)$. An increase in the number of offspring of the $(1, \lambda)$-ES yields a logarithmic increase of the progress rate. Consequently, the time complexity of the EA (2.105) (see p. 50) is of the order

$$\text{ES TIME COMPLEXITY } (1, \lambda)\text{-ES:} \qquad G_{1,\lambda} = \mathcal{O}\left(\frac{N}{\ln \lambda}\right). \tag{3.116}$$

3.2.4 The Analysis of the $(1 + \lambda)$-ES

3.2.4.1 The Progress Rate. For computing the progress rate $\varphi^*_{1+\lambda}$, the lower integration limit $t_u = \sigma^*/2$ (3.97) in (3.98) is to be used. The corresponding integral reads

$$\varphi^*_{1+\lambda}(\sigma^*) = \sigma^* \int_{t=\sigma^*/2}^{t=\infty} \left(t - \frac{\sigma^*}{2}\right) \frac{\mathrm{d}}{\mathrm{d}t} [\Phi(t)]^\lambda \,\mathrm{d}t \;+\; \mathcal{O}\left(\frac{1}{N}\right). \tag{3.117}$$

Its solution can be written thus

$$\varphi^*_{1+\lambda}(\sigma^*) = \sigma^* d^{(1)}_{1+\lambda}\left(\frac{\sigma^*}{2}\right) - \frac{\sigma^{*2}}{2}\left[1 - \left(\Phi\left(\frac{\sigma^*}{2}\right)\right)^\lambda\right] + \mathcal{O}\left(\frac{1}{N}\right). \tag{3.118}$$

[14] Note, while the normalized progress rate φ^* has an upper bound, i.e., due to Definition (2.22) $\varphi^* \leq N$, the correctness of the asymptotic relation $\hat{\varphi}^*_{1,\lambda} \sim \ln \lambda$ is still guaranteed as long as $\ln \lambda / N \to 0$.

Hereby, the $d^{(1)}_{1+\lambda}(x)$ symbol stands for the *first-order progress function* (Beyer, 1992b)

$$d^{(1)}_{1+\lambda}(x) = \frac{\lambda}{\sqrt{2\pi}} \int_x^\infty t\,\mathrm{e}^{-\frac{1}{2}t^2} \left[\Phi(t)\right]^{\lambda-1} \mathrm{d}t. \tag{3.119}$$

For the cases $\lambda = 1$ and $\lambda = 2$, analytical solutions are available

$$d^{(1)}_{1+1}(x) = \frac{\mathrm{e}^{-\frac{1}{2}x^2}}{\sqrt{2\pi}}, \tag{3.120a}$$

$$d^{(1)}_{1+2}(x) = \frac{1}{\sqrt{\pi}}\left(1 - \Phi\left(\sqrt{2}\,x\right)\right) + \frac{2}{\sqrt{2\pi}}\,\mathrm{e}^{-\frac{1}{2}x^2}\,\Phi(x). \tag{3.120b}$$

One can prove their correctness immediately by differentiation. The case $\lambda = 1$ was already discussed in Sect. 3.2.1, p. 67. Figure 3.8 shows the plots of (3.118) for $N \to \infty$ for various $\lambda \geq 1$. Similar to the $(1, \lambda)$-ES case, the progress is shown per descendant/fitness evaluation. A comparison with the

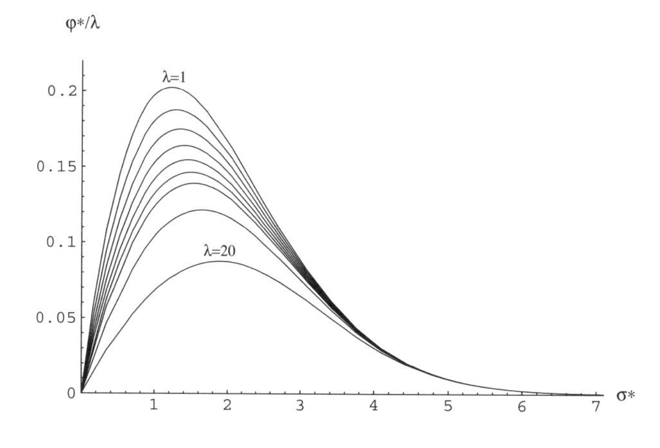

Fig. 3.8. The progress rate per descendant at the $(1 + \lambda)$-ES for the number of descendants λ = 1, 2, 3, 4, 5, 6, 7, 10, and 20. The curves for $\lambda > 2$ are obtained by numerical integration.

$(1, \lambda)$-ES, Fig. 3.4, p. 72, where the most efficient variant was $(1, 5)$ – as far as the number of generated offspring is concerned – shows that the $(1 + \lambda)$-ES has the largest progress per descendant for $\lambda = 1$, i.e. for the $(1 + 1)$ variant. It does not make sense to implement $\mu = 1$, $\lambda > 1$ strategies in order to obtain larger progress rates,[15] unless a multiprocessor system with ϖ processors is available. In such a case the $(1 + \lambda = \varpi)$-ES should be chosen, if the *wall-clock-time* per generation is to be minimized (see also footnote 13, p. 74).

Further properties of the $\varphi^*_{1+\lambda}$ function are

[15] This observation is valid for the sphere model, and it assumes that the ES works optimally using an appropriate control mechanism for the mutation strength σ, i.e. that $\varphi^* = \hat{\varphi}^*$ is realized approximately.

$$\varphi^*_{1+\lambda}(\sigma^*) \geq 0 \qquad \text{and} \qquad \varphi^*_{1+\lambda}(\sigma^*) \geq \varphi^*_{1,\lambda}(\sigma^*). \tag{3.121}$$

Their proof can be obtained easily by investigating (3.98): because of (3.97), $t_u = \sigma^*/2$ is valid for the plus strategy; therefore, the integrand in (3.98) cannot be negative. Hence, $\varphi^*_{1+\lambda}(\sigma^*) \geq 0$. Since $t_u = -\infty$ for comma strategies, the $\varphi^*_{1,\lambda}$-integral can be decomposed as

$$\varphi^*_{1,\lambda} = \underbrace{\cdots \int_{t=-\infty}^{t=\sigma^*/2} \left(t - \frac{\sigma^*}{2}\right) \cdots \mathrm{d}t}_{\leq 0} + \underbrace{\cdots \int_{t=\sigma^*/2}^{t=\infty} \left(t - \frac{\sigma^*}{2}\right) \cdots \mathrm{d}t}_{= \varphi^*_{1+\lambda}}. \tag{3.122}$$

The first integral is not positive, since $(t - \sigma^*/2) \leq 0$ does hold for $t \leq \sigma^*/2$, consequently $\varphi^*_{1,\lambda} - \varphi^*_{1+\lambda} \leq 0$ must hold.

3.2.4.2 The Success Probability $P_{s1+\lambda}$. The success probability $P_{s1+\lambda}$ measures the probability that the best offspring replaces its parent. For the approximation used here, this case occurs if $x_{\lambda:\lambda} \geq \overline{h}^2/2R = x_u$ is valid (cf. Eq. (3.92))

$$P_{s1+\lambda} = \Pr\left(x_{\lambda:\lambda} \geq x_u\right) = \int_{x=x_u}^{x=l} p_{\lambda:\lambda}(x)\,\mathrm{d}x. \tag{3.123}$$

The insertion of (3.95) and the calculation similar to (3.96) $\rightarrow$ (3.98) yields

$$P_{s1+\lambda} = \frac{\lambda}{\sqrt{2\pi}} \int_{t=\sigma^*/2}^{t=\infty} \mathrm{e}^{-\frac{t^2}{2}} \left[\Phi(t)\right]^{\lambda-1} \mathrm{d}t \;-\; \mathcal{O}\left(\frac{1}{N}\right), \tag{3.124}$$

with the result

$$\boxed{P_{s1+\lambda}(\sigma^*) = 1 - \left[\Phi\left(\frac{\sigma^*}{2}\right)\right]^{\lambda} - \mathcal{O}\left(\frac{1}{N}\right).} \tag{3.125}$$

The success probability can also be found – as for the $(1+1)$-ES, Eq. (3.88) – in the progress formula (3.118)

$$\varphi^*_{1+\lambda}(\sigma^*) = \sigma^* d^{(1)}_{1+\lambda}\left(\frac{\sigma^*}{2}\right) - \frac{\sigma^{*2}}{2} P_{s1+\lambda}(\sigma^*) + \mathcal{O}\left(\frac{1}{N}\right), \tag{3.126}$$

as one can see by comparing (3.117, 3.118) with (3.124, 3.125).

3.2.4.3 The Progress Function $d^{(1)}_{1+\lambda}(x)$. Analytical expressions for $d^{(1)}_{1+\lambda}(x)$ were given in (3.120a) for $\lambda = 1$ and in (3.120b) for $\lambda = 2$. The general tendency for larger λ values can be seen in Fig. 3.9. For increasing λ, $d^{(1)}_{1+\lambda}$ exhibits a saturation behavior for small values of x. In the asymptotic limit case $\lambda \rightarrow \infty$, one has

$$d^{(1)}_{1+\lambda}(0) \simeq c_{1,\lambda}. \tag{3.127}$$

Even for $\lambda \gtrapprox 20$ one has a satisfactory accordance $d^{(1)}_{1+\lambda}(x) \approx c_{1,\lambda}$, as long as $x \lessapprox c_{1,\lambda}/2$ is ensured. Hence, for $\sigma^* \lessapprox c_{1,\lambda}$, $d^{(1)}_{1+\lambda}$ can be substituted in

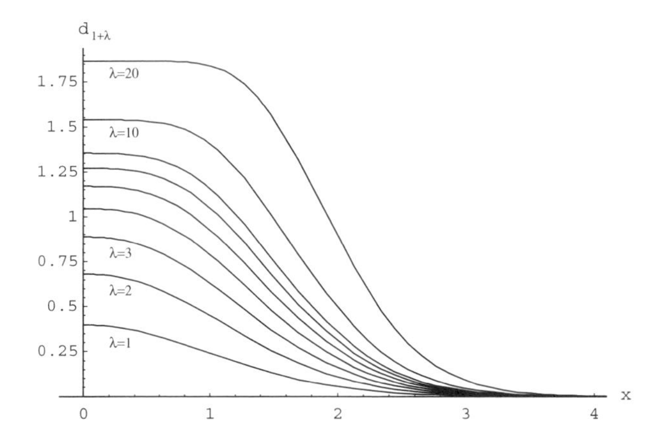

Fig. 3.9. The first-order progress function $d_{1+\lambda} := d^{(1)}_{1+\lambda}(x)$. The curves for $\lambda > 2$ are obtained by numerical integration. The cases $\lambda = 1, 2, 3, 4, 5, 6, 7, 10$, and 20 are shown here.

(3.118) by $c_{1,\lambda}$ (λ must be sufficiently large for that). Under these conditions (namely $\sigma^* \lessapprox c_{1,\lambda}$, $\lambda \gtrapprox 20$) one first obtains $P_{s1+\lambda} \approx 1$, and the progress rate of the $(1+\lambda)$-ES, Eq. (3.118), turns into the $(1,\lambda)$ formula. As an essential consequence, the maximal performance of the $(1+\lambda)$-ES can be estimated using the maximum of the progress rate of the $(1,\lambda)$-ES for large λ values

$$\hat{\varphi}^*_{1+\lambda} \approx \hat{\varphi}^*_{1,\lambda} = \frac{c^2_{1,\lambda}}{2} \qquad \text{and} \qquad \hat{\sigma}^*_{1+\lambda} \approx c_{1,\lambda}. \tag{3.128}$$

3.3 The Asymptotic Analysis of the $(\tilde{1} \overset{+}{,} \tilde{\lambda})$-ES

This section comprises four parts. The theory of the asymptotic $(\tilde{1} \overset{+}{,} \tilde{\lambda})$ progress rate will be developed in Sect. 3.3.1. In Sects. 3.3.2 and 3.3.3, the $(\tilde{1}, \tilde{\lambda})$-ES and the $(\tilde{1} + \tilde{\lambda})$-ES are analyzed, respectively. Section 3.3.4 deals with a method of convergence improvement by inheritance of down-scaled mutations. The case of minimization is considered w.l.o.g. in this section in order to simplify the notation.

3.3.1 The Theory of the $(\tilde{1} \overset{+}{,} \tilde{\lambda})$ Progress Rate

3.3.1.1 Progress Integrals and Acceptance Probabilities. Because of the noise in the measured fitness data, it is not possible to use the expected value of $R - \tilde{R}_{1:\lambda}$ (in (3.89)) for progress measurements. The descendants $\tilde{\mathbf{y}}_l$, $\tilde{\mathbf{y}}_l = \mathbf{y}_p + \mathbf{z}_l$ ($l = 1, \ldots, \lambda$) and the parents $\mathbf{y}_p$ are rather evaluated with respect to the noisy fitness $\tilde{Q}(\mathbf{y})$, i.e. $\tilde{Q}(R)$. The *seemingly* best descendant $\tilde{\mathbf{y}}_{\tilde{1};\tilde{\lambda}}$ with the residual distance $\tilde{R}_{\tilde{1};\tilde{\lambda}}$ has (in the case of minimization) the measured fitness value $\tilde{Q}_{\tilde{1}:\tilde{\lambda}}$ (the smallest descendant value); *however*, $\tilde{R}_{\tilde{1}:\tilde{\lambda}} \equiv \tilde{R}_{\tilde{1};\tilde{\lambda}}$ is not valid (see also (3.31)). Since we are interested in the effective distance change, instead of (3.89) we now have to calculate the expected value $\varphi_{\tilde{1}\overset{+}{,}\tilde{\lambda}} = \mathrm{E}_{\overset{+}{,}}\{R - \tilde{R}_{\tilde{1};\tilde{\lambda}}\}$, with $\tilde{R}_{\tilde{1};\tilde{\lambda}} = \|\tilde{\mathbf{R}}_{\tilde{1};\tilde{\lambda}}\| = \|\mathbf{R} + \mathbf{z}_{\tilde{1};\tilde{\lambda}}\|$ (see also Fig. 3.1 and Eq. (3.8)).

The approach to the progress rate is similar to the scheme in Sect. 3.1.2. For the asymptotic limit case $N \to \infty$, the expression for the expected value

$$\begin{aligned}\varphi_{\tilde{1}\overset{+}{,}\tilde{\lambda}} &= \mathrm{E}_{\overset{+}{,}}\left\{R - \left\|\mathbf{R} + \mathbf{z}_{\tilde{1};\tilde{\lambda}}\right\|\right\} \\ &= \mathrm{E}_{\overset{+}{,}}\left\{R - \sqrt{\left(R - x_{\tilde{1};\tilde{\lambda}}\right)^2 + \left\|\mathbf{h}_{\tilde{1};\tilde{\lambda}}\right\|^2}\right\}\end{aligned} \tag{3.129}$$

can be simplified. Taking $\|\mathbf{h}_{\tilde{1};\tilde{\lambda}}\|^2 \simeq \overline{h}^2 \simeq \sigma^2 N = R\sigma^*\sigma$ into account, one obtains

$$\begin{aligned}\varphi_{\tilde{1}\overset{+}{,}\tilde{\lambda}} &\simeq \int_{-\infty}^{\infty} \left(x_{\tilde{1};\tilde{\lambda}} - \frac{\sigma^*\sigma}{2}\right) p_{\overset{+}{,}}\left(x_{\tilde{1};\tilde{\lambda}}\right) \mathrm{d}x_{\tilde{1};\tilde{\lambda}} \\ &\simeq \int_{-\infty}^{\infty} \left(x - \frac{\sigma^*\sigma}{2}\right) p_{\tilde{1}\overset{+}{,}\tilde{\lambda}}(x)\,\mathrm{d}x\end{aligned} \tag{3.130}$$

analogous to (3.92), where the $p_{\tilde{1}\overset{+}{,}\tilde{\lambda}}(x)$ density remains to be calculated. The integration limits $-\infty$ and ∞ account for the fact that the x projection of $\mathbf{z}$ is not directly bounded by the $\tilde{Q}$ selection (in contrast to the $(1+\lambda)$-ES, where $x \ge \sigma^*\sigma/2$ is valid).

The error term produced by the simplification from (3.129) to (3.130) is of the order $\mathcal{O}(R/N^2)$ (cf. the considerations for Eq. (3.60) and footnote 8 on p. 64), so that one obtains for the normalized progress rate (2.22) from (3.130)

$$\varphi^*_{\tilde{1}\overset{+}{,}\tilde{\lambda}} = \sigma^* \int_{-\infty}^{\infty} \left(\frac{x}{\sigma} - \frac{\sigma^*}{2}\right) p_{\tilde{1}\overset{+}{,}\tilde{\lambda}}(x)\,\mathrm{d}x \;+\; \mathcal{O}\left(\frac{1}{N}\right). \tag{3.131}$$

The density $p_{\tilde{1}\overset{+}{,}\tilde{\lambda}}(x)$ must be derived next.

The determination of $p_{\tilde{1}\overset{+}{,}\tilde{\lambda}}(x)$ proceeds stepwise. Because of the $(1 \overset{+}{,} \lambda)$ selection, we have λ descendants, where any of them can be the best individual. If an individual is selected arbitrarily, it has an x value with density $p_x(x)$, Eq. (3.12). This x value appears with the acceptance probability $P_\mathrm{a}(x)$ to belong to the best individual. Consequently, the probability density for the individual considered to be the best is $p_x(x)P_\mathrm{a}(x)$. Since there are λ exclusive cases, $p_{\tilde{1}\overset{+}{,}\tilde{\lambda}}(x)$ results in

$$p_{\tilde{1}\overset{+}{,}\tilde{\lambda}}(x) = \lambda p_x(x) P_{\mathrm{a}\tilde{1}\overset{+}{,}\tilde{\lambda}}(x). \tag{3.132}$$

The $(\tilde{1},\tilde{\lambda})$ *case.* The case $(\tilde{1},\tilde{\lambda})$ is considered first. An individual with a certain x value is accepted if the noisy fitness $\tilde{Q}_{|x}$ generated by this x value is smaller than (or equal to) the fitness values of the other $\lambda - 1$ fitness values $\tilde{Q}$, i.e.

$$(\tilde{1},\tilde{\lambda})\text{-Selection Condition:} \qquad \tilde{Q}_{|x} \le \tilde{Q}_{1:\lambda-1}. \tag{3.133}$$

The random variable $\tilde{Q}_{|x}$ is assumed to have the conditional probability density function (pdf) $p(\tilde{Q}_{|x}|x) = p(\tilde{Q}_{|x}|Q(\tilde{R}(x)))$. At this point, $p(\tilde{Q}_{|x}|Q)$ stands for the distribution density of the fitness noise according to the noise model (2.40, 2.41, 3.27)

$$p\Big(\tilde{Q}_{|x}\Big|\,Q\Big) = \frac{1}{\sqrt{2\pi}\,\sigma_{\varepsilon_Q}} \exp\left(-\frac{1}{2}\left(\frac{\tilde{Q}_{|x} - Q}{\sigma_{\varepsilon_Q}}\right)^2\right). \tag{3.134}$$

The second random variable (which is independent of $\tilde{Q}_{|x}$) in the selection condition (3.133) is the first-order statistic of the $\lambda - 1$ remaining terms. From now on, it will be denoted as $\hat{Q} := \tilde{Q}_{1:\lambda-1}$. The density of $\hat{Q}$ is denoted as $p_{1:\lambda-1}(\hat{Q})$. It is obtained analogously to (3.93) from the distribution $P_1(\tilde{Q})$ of a single fitness trial $\tilde{Q}$ (the distribution $P_1(\tilde{Q})$ is therefore the cumulative fitness distribution of a single individual)

$$\begin{aligned} p_{1:\lambda-1}(\hat{Q}) &= (\lambda - 1)\, p_1(\hat{Q}) \left[\Pr(\tilde{Q} > \hat{Q})\right]^{\lambda-2} \\ &= (\lambda - 1)\, p_1(\hat{Q}) \left[1 - P_1(\hat{Q})\right]^{\lambda-2} \end{aligned} \tag{3.135}$$

$$p_{1:\lambda-1}(\hat{Q}) = -\frac{\mathrm{d}}{\mathrm{d}\hat{Q}}\left[1 - P_1(\hat{Q})\right]^{\lambda-1}, \quad P_1(\hat{Q}) = \int_{\tilde{Q}=-\infty}^{\tilde{Q}=\hat{Q}} p_1(\tilde{Q})\,\mathrm{d}\tilde{Q}. \tag{3.136}$$

The distribution $P_1(\tilde{Q})$ and the conditional density $p(\tilde{Q}_{|x}|x)$ will be calculated below. Continuing with the derivation of the $P_{\mathrm{a}\tilde{1},\tilde{\lambda}}(x)$, a fixed x value is considered. The probability of $\tilde{Q}_{|x}$ lying in the interval $[\tilde{Q}_{|x}, \tilde{Q}_{|x} + \mathrm{d}\tilde{Q}_{|x})$ is $p(\tilde{Q}_{|x}|x)\mathrm{d}\tilde{Q}_{|x}$, and of being accepted is

$$\mathrm{d}P_{\mathrm{a}\tilde{1},\tilde{\lambda}}\Big(x\,\Big|\tilde{Q}_{|x}\Big) = p\Big(\tilde{Q}_{|x}\Big|\,x\Big)\,\mathrm{d}\tilde{Q}_{|x}\,\Pr\Big(\tilde{Q}_{|x} \le \hat{Q}\Big). \tag{3.137}$$

The probability $\Pr(\tilde{Q}_{|x} \le \hat{Q}) = \Pr(\hat{Q} \ge \tilde{Q}_{|x})$ follows immediately from (3.136)

$$\begin{aligned} \Pr(\hat{Q} \ge \tilde{Q}_{|x}) &= \int_{\hat{Q}=\tilde{Q}_{|x}}^{\infty} p_{1:\lambda-1}(\hat{Q})\mathrm{d}\hat{Q} = -\left[1 - P_1(\hat{Q})\right]^{\lambda-1}\Bigg|_{\tilde{Q}_{|x}}^{\infty} \\ \Pr(\hat{Q} \ge \tilde{Q}_{|x}) &= \left[1 - P_1(\tilde{Q}_{|x})\right]^{\lambda-1}. \end{aligned} \tag{3.138}$$

If (3.137) is integrated over all $\tilde{Q}_{|x}$, one gets the acceptance probability

$$P_{\mathrm{a}\tilde{1},\tilde{\lambda}}(x) = \int_{-\infty}^{\infty} p\Big(\tilde{Q}_{|x}\Big|\,x\Big)\left[1 - P_1(\tilde{Q}_{|x})\right]^{\lambda-1} \mathrm{d}\tilde{Q}_{|x} \tag{3.139}$$

and for (3.132)

$$p_{\tilde{1},\tilde{\lambda}}(x) = \lambda p_x(x) \int_{-\infty}^{\infty} p\Big(\tilde{Q}_{|x}\Big|\,x\Big)\left[1 - P_1(\tilde{Q}_{|x})\right]^{\lambda-1} \mathrm{d}\tilde{Q}_{|x}. \tag{3.140}$$

The $(\tilde{1}+\tilde{\lambda})$ *case.* This case differs from the $(\tilde{1},\tilde{\lambda})$ variant in the selection condition. Instead of (3.133), the condition reads

$$(\tilde{1}+\tilde{\lambda})\text{-Selection:} \qquad \left(\tilde{Q}_{|x} \le \tilde{Q}_{1:\lambda-1}\right) \wedge \left(\tilde{Q}_{|x} \le \tilde{Q}_p\right), \tag{3.141}$$

where the second condition ensures that only the states with the measured fitness values less than (or equal to) the measured parental fitness will survive. Since these two conditions are statistically independent, Eq. (3.137) can be directly multiplied with the probability $\Pr(\tilde{Q}_{|x} \le \tilde{Q}_p)$ to get the result

$$\mathrm{d}P_{\mathrm{a}\tilde{1}+\tilde{\lambda}}\left(x \,\middle|\, \tilde{Q}_{|x}\right) = p\left(\tilde{Q}_{|x} \,\middle|\, x\right)\mathrm{d}\tilde{Q}_{|x}\ \Pr(\tilde{Q}_{|x} \le \hat{Q})\Pr(\tilde{Q}_{|x} \le \tilde{Q}_p). \tag{3.142}$$

The noise model (3.26) with the density

$$p(\tilde{Q}_p) = \frac{1}{\sqrt{2\pi}\,\sigma_{\varepsilon_Q}} \exp\left(-\frac{1}{2}\left(\frac{\tilde{Q}_p - Q_p}{\sigma_{\varepsilon_Q}}\right)^2\right), \qquad Q_p := Q(R) \tag{3.143}$$

is used to determine $\Pr(\tilde{Q}_{|x} \le \tilde{Q}_p) = \Pr(\tilde{Q}_p \ge \tilde{Q}_{|x})$

$$\Pr(\tilde{Q}_p \ge \tilde{Q}_{|x}) = \int_{\tilde{Q}_p=\tilde{Q}_{|x}}^{\infty} p(\tilde{Q}_p)\mathrm{d}\tilde{Q}_p = 1 - \Phi\left(\frac{\tilde{Q}_{|x} - Q_p}{\sigma_{\varepsilon_Q}}\right). \tag{3.144}$$

Consequently, one obtains after integrating (3.142)

$$P_{\mathrm{a}\tilde{1}+\tilde{\lambda}}(x) = \int_{-\infty}^{\infty} p\left(\tilde{Q}_{|x} \,\middle|\, x\right)\left[1 - P_1(\tilde{Q}_{|x})\right]^{\lambda-1} \times \left[1 - \Phi\left(\frac{\tilde{Q}_{|x} - Q_p}{\sigma_{\varepsilon_Q}}\right)\right]\mathrm{d}\tilde{Q}_{|x}. \tag{3.145}$$

For further treatment, $p(\tilde{Q}_{|x}|x)$ and $P_1(\tilde{Q})$ must be determined, which follows immediately.

3.3.1.2 The Asymptotic Fitness Model and the $p(\tilde{Q}_{|x}|x)$ Density. Assuming normally distributed measurement errors in the fitness evaluations, the density of the measurement values $\tilde{Q}_{|x}$ for a given x value is determined by (3.134). $Q = Q(\tilde{R}(x))$ is a function of the x projection of the $\mathbf{z}$ mutation vector. For $\tilde{R}$, the asymptotic relation (3.57) is used. Taking into account $\sigma^2 N/R = \sigma^{*2}/N$, it follows from (3.134) that

$$p\left(\tilde{Q}_{|x} \,\middle|\, x\right) = \frac{1}{\sqrt{2\pi}\,\sigma_{\varepsilon_Q}} \exp\left[-\frac{1}{2}\frac{1}{\sigma_{\varepsilon_Q}^2}\left(\tilde{Q}_{|x} - Q\left(R\sqrt{\left(1-\frac{x}{R}\right)^2 + \frac{\sigma^{*2}}{N}}\right)\right)^2\right]. \tag{3.146}$$

Assuming $x/R \ll 1$ and $\sigma^{*2}/N \ll 1$, a Taylor expansion for $Q(\tilde{R}(x))$ is applied with the intention of neglecting the nonlinear terms in the $(\cdots)^2$ parentheses of (3.146)

$$Q\left(\tilde{R}(x)\right) = Q\left(R\sqrt{1+\delta}\right), \quad \delta = 2\left[-\frac{x}{R}\left(1 - \frac{1}{2}\frac{x}{R}\right) + \frac{\sigma^{*2}}{2N}\right]. \tag{3.147}$$

The expansion $\sqrt{1+\delta} = 1 + \frac{1}{2}\delta - \frac{1}{2}\left(\frac{1}{2}\delta\right)^2 + \frac{1}{2}\left(\frac{1}{2}\delta\right)^3 + \mathcal{O}\left(\delta^4\right)$ yields

$$\begin{aligned}\tilde{R}(x) &= R + R\left[-\frac{x}{R}\left(1 - \frac{1}{2}\frac{x}{R}\right) + \frac{\sigma^{*2}}{2N}\right] - \frac{R}{2}\left[\cdots\right]^2 + \frac{R}{2}\left[\cdots\right]^3 + \ldots \\ &= R + R\frac{\sigma^{*2}}{2N}\left[1 - \frac{1}{4}\frac{\sigma^{*2}}{N} + \frac{1}{8}\left(\frac{\sigma^{*2}}{N}\right)^2 + \left(\frac{x}{R}\right)^2 + \ldots\right] \\ &\quad - R\frac{x}{R}\left[1 - \frac{1}{2}\frac{\sigma^{*2}}{N} + \frac{3}{4}\frac{\sigma^{*2}}{N}\frac{x}{R} + \ldots\right].\end{aligned} \tag{3.148}$$

If $\tilde{R}(x)$ is inserted into Q and $Q(\tilde{R})$ is expanded at $\tilde{R} = R$, and furthermore, taking

$$Q(R + R\,h) = Q(R) + Q'\,R\,h + \frac{1}{2}Q''R^2h^2 + \mathcal{O}\left((R\,h)^3\right),$$

into account, it follows that

$$\begin{aligned}Q(\tilde{R}) = Q(R) &+ Q'R\left[\frac{\sigma^{*2}}{2N} - \frac{x}{R} + \mathcal{O}\left(\left(\frac{\sigma^{*2}}{N}\right)^2\right) + \mathcal{O}\left(\frac{\sigma^{*2}}{N}\right)\mathcal{O}\left(\frac{x}{R}\right)\right] \\ &+ \frac{1}{2}Q''R^2\left[\mathcal{O}\left(\left(\frac{\sigma^{*2}}{N}\right)^2\right) + \mathcal{O}\left(\frac{\sigma^{*2}}{N}\right)\mathcal{O}\left(\frac{x}{R}\right) + \mathcal{O}\left(\left(\frac{x}{R}\right)^2\right)\right] \\ &+ \frac{1}{3!}Q'''R^3\left[\mathcal{O}\left(\left(\frac{\sigma^{*2}}{N}\right)^3\right) + \mathcal{O}\left(\left(\frac{\sigma^{*2}}{N}\right)^2\right)\mathcal{O}\left(\frac{x}{R}\right)\right. \\ &\qquad\left. + \mathcal{O}\left(\frac{\sigma^{*2}}{N}\right)\mathcal{O}\left(\left(\frac{x}{R}\right)^2\right) + \mathcal{O}\left(\left(\frac{x}{R}\right)^3\right)\right]\end{aligned} \tag{3.149}$$

with

$$Q' := \left.\frac{\mathrm{d}Q(\tilde{R})}{\mathrm{d}\tilde{R}}\right|_{\tilde{R}=R}, Q'' := \left.\frac{\mathrm{d}^2Q(\tilde{R})}{\mathrm{d}\tilde{R}^2}\right|_{\tilde{R}=R}, Q''' := \left.\frac{\mathrm{d}^3Q(\tilde{R})}{\mathrm{d}\tilde{R}^2}\right|_{\tilde{R}=R}. \tag{3.150}$$

The error terms in (3.149) can be summed up, provided that there is a $0 \le M < \infty$ for the fitness function $Q(R)$, such that

$$\frac{1}{|Q|}\left|\frac{\mathrm{d}^kQ}{\mathrm{d}R^k}R^k\right| \le M \tag{3.151}$$

is fulfilled, that is

$$\frac{\mathrm{d}^kQ}{\mathrm{d}R^k}R^k = \mathcal{O}(Q) \tag{3.152}$$

is valid. This condition is fulfilled by the power function (3.39), and therefore also for fitness functions of the form (for the case of minimization)

$$Q(R) = \sum_{j=1}^{K} a_j R^j, \qquad a_j \ge 0. \tag{3.153}$$

Assuming (3.152), (3.149) reduces to

$$Q(\tilde{R}) = Q(R) \,+\, Q' \frac{R}{N} \left(\frac{\sigma^{*2}}{2} - x \frac{N}{R} \right) \,+\, \mathcal{R}_Q, \tag{3.154}$$

with the error term $\mathcal{R}_Q$

$$\mathcal{R}_Q = \mathcal{O}(Q) \left[\mathcal{O}\!\left(\left(\frac{\sigma^{*2}}{N} \right)^2 \right) + \mathcal{O}\!\left(\frac{\sigma^{*2}}{N} \right) \mathcal{O}\!\left(\frac{x}{R} \right) + \mathcal{O}\!\left(\left(\frac{x}{R} \right)^2 \right) \right]. \tag{3.155}$$

Consequently, one obtains for (3.146) (note, $Q_p := Q(R)$)

$$p\Big(\tilde{Q}_{|x} \Big| \, x \Big) =$$

$$\frac{1}{\sqrt{2\pi}\, \sigma_{\varepsilon_Q}} \exp\left[-\frac{1}{2} \frac{1}{\sigma_{\varepsilon_Q}^2} \left(\tilde{Q}_{|x} - Q_p - Q' \frac{R}{N} \left(\frac{\sigma^{*2}}{2} - x \frac{N}{R} \right) - \mathcal{R}_Q \right)^2 \right]. \tag{3.156}$$

$P_1(\tilde{Q})$ can be calculated now after these preparatory steps.

3.3.1.3 The Calculation of $P_1(\tilde{Q})$. Assume that the measured (and therefore noisy) fitness $\tilde{Q}$ of a descendant has the distribution $P_1(\tilde{Q})$. $\tilde{Q}$ is obtained by a mutation of the state $\mathbf{y}_p \to \mathbf{y}_p + \mathbf{z}$ yielding the fitness $Q = Q(\mathbf{y}_p + \mathbf{z})$ and adding a noise term yielding $\tilde{Q}$ thereafter. In the $\tilde{R}$-picture, the R state is mutated first with the density $p_x(x)$. The resulting x value is a *parameter* in the conditional probability density $p(\tilde{Q}_{|x}|x)$. Consequently, the density of the random variable $\tilde{Q}_{|x}$ reads

$$p(\tilde{Q}_{|x}) = \int_{-\infty}^{\infty} p\Big(\tilde{Q}_{|x} \Big| \, x \Big)\, p_x(x)\, \mathrm{d}x \tag{3.157}$$

and the distribution function $P_1(\tilde{Q}) = \Pr(\tilde{Q}_{|x} \le \tilde{Q})$ becomes, using (3.12) and (3.146),

$$P_1(\tilde{Q}) = \int_{-\infty}^{\tilde{Q}} p\Big(\tilde{Q}_{|x} \Big)\, \mathrm{d}\tilde{Q}_{|x}$$

$$= \int_{-\infty}^{\infty} \frac{\mathrm{e}^{-\frac{1}{2}\left(\frac{x}{\sigma}\right)^2}}{\sqrt{2\pi}\,\sigma}\, \Phi \left[\frac{\tilde{Q} - Q\left(R\sqrt{\left(1 - \frac{x}{R}\right)^2 + \frac{\sigma^{*2}}{N}} \right)}{\sigma_{\varepsilon_Q}} \right] \mathrm{d}x. \tag{3.158}$$

Since this integral cannot be solved analytically, an asymptotic expansion will be used. Substituting $x/\sigma = t$ and considering (3.154), it follows that

$$P_1(\tilde{Q}) = \frac{1}{\sqrt{2\pi}} \int_{-\infty}^{\infty} \mathrm{e}^{-\frac{1}{2}t^2}\, \Phi \Bigg[\underbrace{\frac{\tilde{Q} - Q_p}{\sigma_{\varepsilon_Q}} - \frac{Q'R}{\sigma_{\varepsilon_Q} N} \sigma^* \left(\frac{\sigma^*}{2} - t \right)}_{:=\, A} \; \underbrace{- \frac{\mathcal{R}_Q}{\sigma_{\varepsilon_Q}}}_{:=\, B} \Bigg] \mathrm{d}t. \tag{3.159}$$

The Taylor expansion of $\Phi(A+B)$ at A gives (see Appendix B.1, Eq. (B.6))

$$\Phi(A+B) = \Phi(A) + \frac{B}{\sqrt{2\pi}}\,\mathrm{e}^{-\frac{1}{2}A^2}\sum_{k=0}^{\infty}\frac{(-1)^k B^k}{(k+1)!}\mathrm{He}_k(A), \tag{3.160}$$

with $\mathrm{He}_k(x)$ denoting the kth degree Hermite polynomial (see also Appendix B.2)

$$\mathrm{He}_k(x) := (-1)^k \mathrm{e}^{\frac{1}{2}x^2}\,\frac{\mathrm{d}^k}{\mathrm{d}x^k}\left(\mathrm{e}^{-\frac{1}{2}x^2}\right). \tag{3.161}$$

Using the substitutions (consider Definition (3.42) and (3.155))

$$A = at + b, \quad a = \frac{Q' R \sigma^*}{\sigma_{\varepsilon_Q} N} = \frac{\sigma^*}{\sigma^*_\varepsilon}, \quad b = \frac{\tilde{Q} - Q_p}{\sigma_{\varepsilon_Q}} - \frac{\sigma^{*2}}{2}\frac{R}{N}\frac{Q'}{\sigma_{\varepsilon_Q}} \tag{3.162}$$

and since

$$B = \mathcal{O}(N)\left[\mathcal{O}\!\left(\left(\frac{\sigma^{*2}}{N}\right)^2\right) + \mathcal{O}\!\left(\frac{\sigma^{*2}}{N}\right)\mathcal{O}\!\left(\frac{\sigma^*}{N}\right)t + \mathcal{O}\!\left(\left(\frac{\sigma^*}{N}\right)^2\right)t^2\right] \tag{3.163}$$

one obtains for (3.159)

$$P_1(\tilde{Q}) = \frac{1}{\sqrt{2\pi}}\int_{-\infty}^{\infty}\mathrm{e}^{-\frac{1}{2}t^2}\,\Phi(at+b)\,\mathrm{d}t$$
$$+\frac{1}{2\pi}\sum_{k=0}^{\infty}\frac{(-1)^k}{(k+1)!}\int_{-\infty}^{\infty}B(t)\,\mathrm{e}^{-\frac{1}{2}t^2}\mathrm{e}^{-\frac{1}{2}(at+b)^2}B(t)^k\,\mathrm{He}_k(at+b)\,\mathrm{d}t. \tag{3.164}$$

The solution of the first integral is given in Appendix A.1, Eq. (A.10). The factor $B(t)^k \mathrm{He}_k(at+b)$ is of the order $\mathcal{O}(1)$ (wrt. parameter space dimension N).[16] The remaining $B(t)$ in the integrand is of order $\mathcal{O}(1/N)$. The integration over t does not change the order. Consequently the whole second integral and therefore the sum is of order $\mathcal{O}(1/N)$. As a result we obtain

$$P_1(\tilde{Q}) = \Phi\left(\frac{\tilde{Q} - Q_p - Q'\,\frac{R}{N}\,\frac{\sigma^{*2}}{2}}{\sigma_{\varepsilon_Q}\sqrt{1+(\sigma^*/\sigma^*_\varepsilon)^2}}\right) + \mathcal{O}\left(\frac{1}{N}\right). \tag{3.165}$$

3.3.2 On the Analysis of the $(\tilde{1}, \tilde{\lambda})$-ES

3.3.2.1 The Asymptotic $\varphi^*_{\tilde{1},\tilde{\lambda}}$ Formula. To obtain $\varphi^*_{\tilde{1},\tilde{\lambda}}$, integral (3.131) is to be evaluated taking (3.140) into account

[16] Motivation: He_k is a kth degree polynomial. Since $(at+b)^k = \mathcal{O}(N^k)$ (because of (3.162) and (3.42)) as well as $B^k = \mathcal{O}(1/N^k)$ (because of (3.163)), their product $B^k \mathrm{He}_k = \mathcal{O}(1)$.

$$\varphi^*_{\tilde{1},\tilde{\lambda}} = \sigma^* \lambda \int_{-\infty}^{\infty} \left(\frac{x}{\sigma} - \frac{\sigma^*}{2} \right) p_x(x) \times \int_{-\infty}^{\infty} p(\tilde{Q}_{|x}|x) \left[1 - P_1(\tilde{Q}_{|x}) \right]^{\lambda-1} \mathrm{d}\tilde{Q}_{|x} \, \mathrm{d}x + \mathcal{O}\left(\frac{1}{N} \right). \quad (3.166)$$

Firstly, the integration order is exchanged and the substitution $\tilde{Q} := \tilde{Q}_{|x}$ is introduced to simplify the notation

$$\varphi^*_{\tilde{1},\tilde{\lambda}} = \sigma^* \lambda \int_{-\infty}^{\infty} \int_{-\infty}^{\infty} \frac{x}{\sigma} \, p(\tilde{Q}|x) \, p_x(x) \, \mathrm{d}x \left[1 - P_1(\tilde{Q}) \right]^{\lambda-1} \mathrm{d}\tilde{Q} - \frac{\sigma^{*2}}{2} \lambda \int_{-\infty}^{\infty} \underbrace{\int_{-\infty}^{\infty} p(\tilde{Q}|x) \, p_x(x) \, \mathrm{d}x}_{= \frac{\mathrm{d}}{\mathrm{d}\tilde{Q}} P_1(\tilde{Q})} \left[1 - P_1(\tilde{Q}) \right]^{\lambda-1} \mathrm{d}\tilde{Q} + \mathcal{O}\left(\frac{1}{N} \right). \quad (3.167)$$

The inner integral of the second double integral can be simplified because of (3.157, 3.158); thus, the integrand of the outer integral is $-\frac{1}{\lambda} \frac{\mathrm{d}}{\mathrm{d}\tilde{Q}} [1 - P_1(\tilde{Q})]^\lambda$. Therefore, the whole double integral yields $1/\lambda$. The result of the second line of (3.167) is $-\sigma^{*2}/2 + \mathcal{O}(1/N)$.

The integration of the first line in (3.167) needs further treatment. First (3.12, 3.156) and (3.165) are inserted into the first line of (3.167). Thereafter, the substitutions

$$t = \frac{x}{\sigma} \qquad \text{and} \qquad s = \frac{\tilde{Q} - Q_p - Q' \, \frac{R}{N} \frac{\sigma^{*2}}{2}}{\sigma_{\varepsilon_Q} \sqrt{1 + (\sigma^*/\sigma^*_\varepsilon)^2}} \quad (3.168)$$

yield for the first line of (3.167), abbreviated as Z_1,

$$Z_1 = \frac{\sigma^* \lambda}{2\pi} \sqrt{1 + \left(\frac{\sigma^*}{\sigma^*_\varepsilon} \right)^2} \int_{-\infty}^{\infty} \times \int_{-\infty}^{\infty} t \mathrm{e}^{-\frac{t^2}{2}} \exp\left[-\frac{1}{2} \Bigg(\underbrace{\sqrt{1 + \left(\frac{\sigma^*}{\sigma^*_\varepsilon} \right)^2} \, s + \frac{\sigma^*}{\sigma^*_\varepsilon} t}_{:= C} - \underbrace{\frac{\mathcal{R}_Q(t)}{\sigma^*_\varepsilon}}_{:= B} \Bigg)^2 \right] \mathrm{d}t \times \left[1 - \Phi(s) - \mathcal{O}\left(\frac{1}{N} \right) \right]^{\lambda-1} \mathrm{d}s. \quad (3.169)$$

For the next calculation, take

$$[1 - \Phi(s) - \mathcal{O}(1/N)]^{\lambda-1} = [1 - \Phi(s)]^{\lambda-1} - \mathcal{O}(1/N)$$

into account; furthermore, the Taylor expansion of $\exp[-(C+B)^2/2]$ yields

$$\mathrm{e}^{-\frac{1}{2}(C+B)^2} = \mathrm{e}^{-\frac{1}{2}C^2} (1 - CB + \ldots) = \mathrm{e}^{-\frac{1}{2}C^2} (1 - C\mathcal{O}(B)),$$

because of $C \neq f(N)$ and (3.163), $B = \mathcal{O}(1/N)$. Using the abbreviations

$$C := at + b, \quad a := \frac{\sigma^*}{\sigma^*_\varepsilon}, \quad \text{and} \quad b := \sqrt{1 + (\sigma^*/\sigma^*_\varepsilon)^2}\, s,$$

it follows for (3.169) that

$$Z_1 = \frac{\sigma^* \lambda}{2\pi} \sqrt{1 + \left(\frac{\sigma^*}{\sigma^*_\varepsilon}\right)^2} \int_{-\infty}^{\infty} \int_{-\infty}^{\infty} t\, \mathrm{e}^{-\frac{1}{2}t^2} \mathrm{e}^{-\frac{1}{2}(at+b(s))^2}$$
$$\times \left[1 - (at + b(s))\, \mathcal{O}(B(t))\right] \mathrm{d}t \left[(1 - \Phi(s))^{\lambda - 1} - \mathcal{O}\left(\frac{1}{N}\right)\right] \mathrm{d}s. \quad (3.170)$$

Since the inner integral in (3.170) is convergent, the error term is of order $\mathcal{O}(B) = \mathcal{O}(1/N)$. Likewise, the outer integral yields an error term of $\mathcal{O}(1/N)$, therefore Z_1 becomes

$$Z_1 = \frac{\sigma^* \lambda}{2\pi} \sqrt{1 + \left(\frac{\sigma^*}{\sigma^*_\varepsilon}\right)^2} \int_{-\infty}^{\infty} \int_{-\infty}^{\infty} t\, \mathrm{e}^{-\frac{t^2}{2}} \mathrm{e}^{-\frac{1}{2}(at+b(s))^2} \mathrm{d}t\, \left[1 - \Phi(s)\right]^{\lambda - 1} \mathrm{d}s$$
$$+ \mathcal{O}\left(\frac{1}{N}\right). \quad (3.171)$$

The inner integral can be evaluated using (A.8) from Appendix A.1. One obtains

$$Z_1 = -\frac{\sigma^* \lambda}{\sqrt{2\pi}} \frac{\sigma^*}{\sigma^*_\varepsilon} \frac{1}{\sqrt{1 + (\sigma^*/\sigma^*_\varepsilon)^2}} \int_{-\infty}^{\infty} s\, \mathrm{e}^{-\frac{1}{2}s^2} \left[1 - \Phi(s)\right]^{\lambda - 1} \mathrm{d}s + \mathcal{O}\left(\frac{1}{N}\right). \quad (3.172)$$

After substituting $s = -t$ one gets

$$Z_1 = \frac{\sigma^*}{\sqrt{1 + (\sigma^*_\varepsilon/\sigma^*)^2}} \frac{\lambda}{\sqrt{2\pi}} \int_{-\infty}^{\infty} t\, \mathrm{e}^{-\frac{1}{2}t^2} \left[1 - \Phi(-t)\right]^{\lambda - 1} \mathrm{d}t + \mathcal{O}\left(\frac{1}{N}\right).$$

Considering (3.100) and the symmetry property $1 - \Phi(-t) = \Phi(t)$ (see Appendix C.1), and collecting all the terms thereafter, the *asymptotic* formula for the progress rate of the $(\tilde{1}, \tilde{\lambda})$-ES follows

$$\boxed{\varphi^*_{\tilde{1},\tilde{\lambda}}(\sigma^*, \sigma^*_\varepsilon) = c_{1,\lambda} \sigma^* \frac{1}{\sqrt{1 + (\sigma^*_\varepsilon/\sigma^*)^2}} - \frac{\sigma^{*2}}{2} + \mathcal{O}\left(\frac{1}{N}\right).} \quad (3.173)$$

As one can see, the special case, $(1, \lambda)$-ES without noise, is obtained for $\sigma^*_\varepsilon = 0$ in (3.173) (see (3.101)).

If this result is interpreted in the sense of the *EPP* (see Sect. 1.4.2.1 and also p. 73), it becomes clear that the noise-disturbed fitness has *no* effect on the loss term $\sigma^{*2}/2$ of φ^*. However, the gain term gets worse by a factor of $1/\sqrt{1 + (\sigma^*_\varepsilon/\sigma^*)^2}$. Since the selection (and its result, respectively) is responsible for the progress attained, this means that noise deceives the selection. This can be demonstrated clearly: a fitness improvement does not necessarily result in a decrease of the residual distance. Conversely, another case is imaginable: a mutation which would yield a decrease in the residual distance is interpreted as a worsening because of the random noise on the measured value, i.e. the offspring gets a higher probability of being eliminated. Therefore, noisy fitness causes a deterioration of the *selection sharpness* S, which can be defined as

$$\text{SELECTION SHARPNESS:} \qquad S := \frac{1}{\sqrt{1 + (\sigma_\varepsilon^* / \sigma^*)^2}}. \tag{3.174}$$

For $\sigma_\varepsilon^* = 0$ we have $S = 1$, the selection accuracy is maximal. If noise exists, e.g. $\sigma_\varepsilon^* \neq 0$, the selection sharpness decreases. In order to take countermeasures one must increase σ^* – the mutation strength. Unfortunately, this leads to a quadratic increase in the loss term $\sigma^{*2}/2$ in (3.173). Methods for improving the selection sharpness and decreasing the loss term in $(\tilde{1} \overset{+}{,} \tilde{\lambda})$ strategies will be discussed in the next subsection and in Sect. 3.3.4.

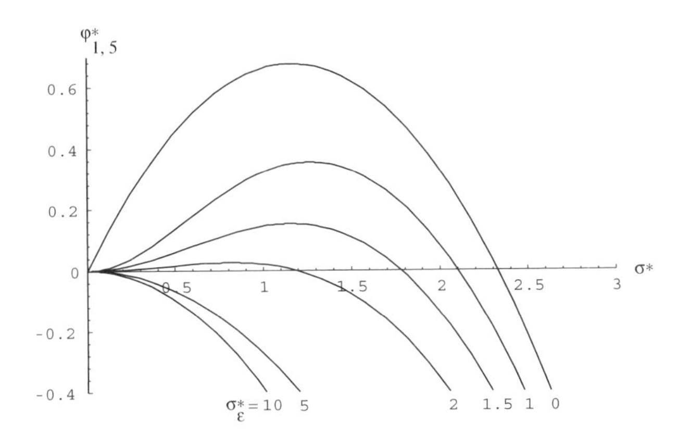

Fig. 3.10. The normalized progress rate of the $(\tilde{1}, \tilde{5})$-ES using noisy fitness data with various normalized noise strengths σ_ε^*

A qualitative feeling for the performance decrease of the ES under the presence of fitness noise can be obtained from Fig. 3.10. Plots of Eq. (3.173) are shown for the case of the $(\tilde{1}, \tilde{5})$-ES. As one can see, there is clearly a critical σ_ε^* beyond which a positive φ^* is not possible. Below this critical σ_ε^* there is a $\hat{\sigma}^* = \arg\max_{\sigma^*}[\varphi_{\tilde{1},\tilde{\lambda}}^*(\sigma^*, \sigma_\varepsilon^*)] = f(\sigma_\varepsilon^*)$, where φ^* becomes maximal. Because of (3.103), one has $\hat{\sigma}^* = c_{1,\lambda}$ for the noise-free case $\sigma_\varepsilon^* = 0$. For $\sigma_\varepsilon^* > 0$, $\hat{\sigma}^*$ can be obtained from (3.173) by setting the first derivative to zero. However, by this one obtains a third-order equation; its analytical solution is not very informative. The principal properties of $\hat{\sigma}^* = f(\sigma_\varepsilon^*)$ can also be read directly from Fig. 3.10: for increasing σ_ε^*, $\hat{\sigma}^*$ increases at first; however, it does so with a decreasing tendency. Beyond a certain σ_ε^* value, $\hat{\sigma}^* = f(\sigma_\varepsilon^*)$ becomes a decreasing function, which takes the value $\hat{\sigma}^* = 0$ at the stability boundary (see below).

3.3.2.2 The Dynamics of the $(\tilde{1}, \tilde{\lambda})$-ES. The differential equation of the ES dynamics (2.101) attains with (3.173) in the asymptotic limit $(N \to \infty)$

the form

$$\frac{\mathrm{d}R(g)}{\mathrm{d}g} = -\frac{R(g)}{N}(\sigma^*(g))^2\left[\frac{c_{1,\lambda}}{\sqrt{(\sigma^*(g))^2+(\sigma^*_\varepsilon(g))^2}}-\frac{1}{2}\right] \tag{3.175}$$
$$\frac{\mathrm{d}R(g)}{\mathrm{d}g} = f_R(R(g),\sigma^*(g),\sigma^*_\varepsilon(g))\,.$$

Since this equation contains not only $R(g)$ but additionally also $\sigma^*(g)$ and $\sigma^*_\varepsilon(g)$, its solution is not defined uniquely. The complete ES dynamics requires in general two additional differential equations for $\sigma^*(g)$ and $\sigma^*_\varepsilon(g)$

$$\begin{aligned}\frac{\mathrm{d}\sigma^*(g)}{\mathrm{d}g} &= f_\sigma(R(g),\sigma^*(g),\sigma^*_\varepsilon(g))\,,\\ \frac{\mathrm{d}\sigma^*_\varepsilon(g)}{\mathrm{d}g} &= f_{\sigma_\varepsilon}(R(g),\sigma^*(g),\sigma^*_\varepsilon(g))\,.\end{aligned} \tag{3.176}$$

For the special case of the $(1,\lambda)$-ES (without noise), a theory will be developed in Chap. 7, which also includes the σ control. At this point, only certain basic properties of (3.175) can be investigated, which hold independently of (3.176).

Evolution criteria. Evolution criteria are mathematical relations ensuring the convergence of the ES algorithm, i.e. that for any residual distance $R = R(g) > 0$ there exists a number of generations g_c beyond which

$$\forall g,\ 0 \le g_c \le g < \infty: \qquad \frac{\mathrm{d}R(g)}{\mathrm{d}g} < 0 \tag{3.177}$$

is fulfilled. Provided that $\sigma^*(g) < \infty$, this is fulfilled for (3.175) when the quantity in the brackets is positive. After a simple calculation one obtains the *sufficient evolution criterion* of the $(\tilde{1},\tilde{\lambda})$-ES

$$\boxed{(\sigma^*(g))^2 + (\sigma^*_\varepsilon(g))^2 < 4c^2_{1,\lambda}.} \tag{3.178}$$

This criterion is graphically illustrated in Fig. 3.11. The $(\tilde{1},\tilde{\lambda})$-ES is stable (in the sense of convergence (3.177)) if the two sides of the triangle, composed of the normalized mutation strength σ^* and the normalized noise level σ^*_ε, are *inside* the Thales circle with radius $c_{1,\lambda}$. Consequently, as a *necessary* evolution condition for the Comma-ES under noise one obtains the *necessary evolution criterion* of the $(\tilde{1},\tilde{\lambda})$-ES

$$\boxed{\sigma^*_\varepsilon(g) < 2c_{1,\lambda}.} \tag{3.179}$$

The *complementary* σ^* criterion to that is $\sigma^* < 2c_{1,\lambda}$ (see also (3.102)). The stability limit of the $(\tilde{1},\tilde{\lambda})$-ES is defined by the Thales circle

$$(\sigma^*(g))^2 + (\sigma^*_\varepsilon(g))^2 = 4c^2_{1,\lambda}. \tag{3.180}$$

The points on the circle descibe solutions for (3.175) with $\frac{\mathrm{d}R}{\mathrm{d}g} = 0$, i.e. *stationary* states. For such $(\sigma^*,\sigma^*_\varepsilon)$ pairs, the ES does not converge to $R(g) \xrightarrow{g\to\infty} 0$,

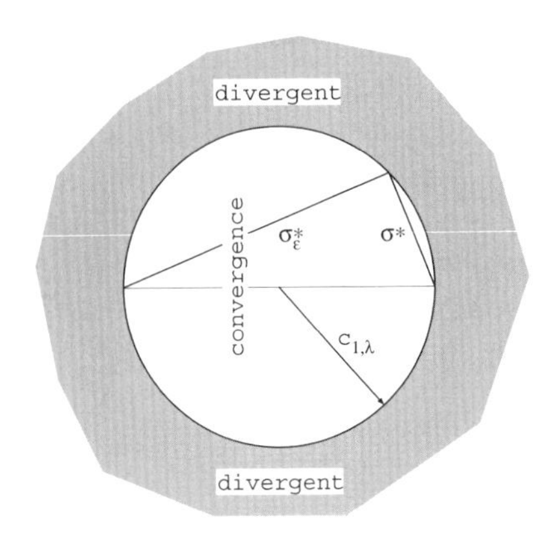

Fig. 3.11. The stability card of the $(\tilde{1}, \tilde{\lambda})$-ES – the Thales criterion

but to a constant, finite residual distance $R(g) \xrightarrow{g\to\infty} R_\infty \neq 0$ with $R_\infty < \infty$. The question now concerns the relevance of these stationary states.

Stationary states – evolution under σ = const *and* σ_{ϵ_Q} = const. If an ES-algorithm is considered with σ^* = const and σ^*_ε = const, whether or not this ES converges can be tested immediately by inserting σ^* and σ^*_ε into (3.178). However, other operating regimes are imaginable and of practical relevance.

The first case to be investigated is σ^* = const, σ_{ε_Q} = const. Since (3.42), one obtains from (3.180) with (3.39), $Q = cR^\alpha$,

$$4c_{1,\lambda}^2 - \sigma^{*2} = \sigma_{\varepsilon_Q}^2 \frac{N^2}{\alpha^2\,(Q(R_\infty))^2}$$

$$\Longrightarrow \quad R_\infty = \sqrt[\alpha]{\frac{\sigma_{\varepsilon_Q} N}{c\,\alpha} \frac{1}{\sqrt{4c_{1,\lambda}^2 - \sigma^{*2}}}}. \tag{3.181}$$

As one can easily see, the stationary state R_∞ is *by no means* a "singular event," since it is numerically observable. This state rather occurs inevitably for $\sigma_{\varepsilon_Q} > 0$. The stationary state R_∞ is influenced by different factors. R_∞ increases with the parameter space dimension N, a similar effect holds for σ_{ε_Q} and σ^*. In the framework of the standard $(1,\lambda)$-ES algorithm, a decrease is only possible by increasing $c_{1,\lambda}$, i.e. by increasing λ (c, α, and N are problem specific and therefore cannot be altered). Beyond the $(1,\lambda)$ standard algorithm, there are additional ways of decreasing R_∞, to be discussed below and in Sect. 3.3.4, respectively.

Besides the case σ^* = const, σ_{ε_Q} = const there is also the case with σ = const, σ_{ε_Q} = const For the $(1,\lambda)$-ES without noise ($\sigma_{\varepsilon_Q} = 0$) one obtains with (2.22) and (3.180)

$$(1,\lambda)\text{-ES}, \quad \sigma = \text{const} > 0: \qquad R_\infty = \frac{\sigma N}{2c_{1,\lambda}}, \tag{3.182}$$

independent of the fitness model $Q = Q(R)$ considered. The $(1, \lambda)$-ES is therefore *not* convergent at constant mutation strength σ and yields for $g \to \infty$ a stationary residual distance R_∞. Equation (3.182) expresses the importance of adapting the mutation strength to the local topology by an appropriate σ-control mechanism and to drive σ to zero as the optimum is approached.

The noisy case, i.e. $\sigma_{\varepsilon_Q} = \text{const} > 0$ and $\sigma > 0$, yields with (3.180) a nonlinear equation for R_∞ which cannot be solved for the general $Q = cR^\alpha$ case. Only for $\alpha = 2$, one obtains a quadratic equation with the solution

$$(\tilde{1}, \tilde{\lambda})\text{-ES}, \quad \alpha = 2: \quad R_\infty = \frac{1}{\sqrt{2}} \frac{\sigma N}{2c_{1,\lambda}} \sqrt{1 + \sqrt{1 + \left(\frac{2c_{1,\lambda}\sigma_{\varepsilon_Q}}{cN\sigma^2}\right)^2}}. \tag{3.183}$$

The case $\sigma_{\varepsilon_Q} = 0$, Eq. (3.182), is included in (3.183) as a special case.

On improving the convergence property at noisy fitness. The prerequisite for the convergence of the $(\tilde{1}, \tilde{\lambda})$-ES is the fulfillment of (3.179). With (3.42) it follows that

$$\sigma_{\varepsilon_Q} < \frac{2c_{1,\lambda}RQ'}{N} \qquad \text{and} \qquad \sigma_{\varepsilon_Q} < \frac{2c_{1,\lambda}\alpha Q}{N} \quad (\text{if } Q = cR^\alpha). \tag{3.184}$$

The evolution criterion can always be fulfilled by an appropriate choice of $c_{1,\lambda}$ (i.e. λ). Unfortunately, as a result of (3.114) the computation time increases super-exponentially in λ

$$\lambda \sim \exp\left(\frac{N\sigma_{\varepsilon_Q}}{2RQ'}\right)^2, \tag{3.185}$$

and this method is only practicable for relatively small σ_{ε_Q}.

A more effective method for convergence improvement (Beyer, 1993) aims at the decrease of σ_{ε_Q}. Instead of the original $(1, \lambda)$ algorithm, where the fitness $\tilde{Q}_l$ of the descendant states $\tilde{\mathbf{y}}_l$ are measured *once*, m (independent) fitness measurements per descendant are carried out, and their mean values $\langle \tilde{Q}_l \rangle$ enter the $(1, \lambda)$ selection

$$\langle \tilde{Q}_l \rangle := \frac{1}{m} \sum_{i=1}^{m} \tilde{Q}(\tilde{\mathbf{y}}_l), \qquad \tilde{\mathbf{y}}_l = \text{const}. \tag{3.186}$$

Because the fitness fluctuations are statistically independent, the standard deviation D of the mean values becomes

$$\mathrm{D}\left\{\langle \tilde{Q}_l \rangle\right\} = \frac{\sigma_{\varepsilon_Q}}{\sqrt{m}}. \tag{3.187}$$

The evolution condition (3.184) is improved accordingly

$$\sigma_{\varepsilon_Q} < \sqrt{m}\,\frac{2c_{1,\lambda}RQ'}{N}. \tag{3.188}$$

For the $Q = cR^\alpha$ case, one obtains after introducing the *relative measurement error* ε_r (3.41)

$$\varepsilon_r < \sqrt{m}\,\frac{2c_{1,\lambda}\alpha}{N}. \qquad (3.189)$$

Compared to (3.185), the computational cost is "only" of quadratic order

$$m\lambda \sim \lambda\left(\frac{N\sigma_{\varepsilon_Q}}{2c_{1,\lambda}RQ'}\right)^2. \qquad (3.190)$$

Therefore, the m times averaged $(\tilde{1}, \tilde{\lambda})$-ES should be preferred to the $(\tilde{1}, \widetilde{\lambda m})$-ES for large m.[17] For small λ and m values, however, this need not be the case. In the interval $m\lambda \le 22$ there are variants for which the $(\tilde{1}, \widetilde{\lambda m})$-ES fulfills the evolution criterion better than the $(\tilde{1}, \tilde{\lambda})$-ES with m times averaged fitness values. For this case,

$$\sqrt{m}\,\frac{2c_{1,\lambda}RQ'}{N} < \frac{2c_{1,\lambda m}RQ'}{N} \qquad \Rightarrow \qquad \sqrt{m}c_{1,\lambda} < c_{1,\lambda m} \qquad (3.191)$$

must be valid because of (3.184) and (3.188). The numerical investigation using Tables 3.1 and D.1 gives admissible combinations fulfilling (3.191). They are shown in Table 3.2.

Table 3.2. Strategy parameter combinations for which the $(\tilde{1}, \widetilde{\lambda m})$-ES fulfills the necessary evolution criterion better than the $(\tilde{1}, \tilde{\lambda})$-ES with m measurement repetitions (resampling) and subsequent fitness averaging

λ	2	3
m	< 12	< 4

3.3.3 On the Analysis of the $(\tilde{1} + \tilde{\lambda})$-ES

3.3.3.1 The Asymptotic $\varphi^*_{\tilde{1}+\tilde{\lambda}}$ Integral and $P_{\mathrm{s}\tilde{1}+\tilde{\lambda}}$. The progress rate integral $\varphi^*_{\tilde{1}+\tilde{\lambda}}$ is obtained from (3.131, 3.132) and (3.145). Because of (3.139, 3.145), it differs from the $\varphi^*_{\tilde{1},\tilde{\lambda}}$ integral (3.166) only by an additional factor (3.144) in the $\tilde{Q}_{|x}$ integrand. Therefore, one can write instead of (3.167)

$$\varphi^*_{\tilde{1}+\tilde{\lambda}} = \sigma^*\lambda \int_{-\infty}^{\infty}\int_{-\infty}^{\infty} \frac{x}{\sigma}\, p(\tilde{Q}|x)\, p_x(x)\,\mathrm{d}x \left[1 - P_1(\tilde{Q})\right]^{\lambda-1}\left[1 - \Phi\left(\frac{\tilde{Q}-Q_p}{\sigma_{\varepsilon_Q}}\right)\right]\mathrm{d}\tilde{Q}$$
$$- \frac{\sigma^{*2}}{2}\int_{-\infty}^{\infty}\lambda\frac{\mathrm{d}P_1(\tilde{Q})}{\mathrm{d}\tilde{Q}}\left[1 - P_1(\tilde{Q})\right]^{\lambda-1}\left[1 - \Phi\left(\frac{\tilde{Q}-Q_p}{\sigma_{\varepsilon_Q}}\right)\right]\mathrm{d}\tilde{Q} + \mathcal{R}. \qquad (3.192)$$

[17] For sufficiently large populations with noisy fitness evaluation, the "democracy principle" (computing the mean value) is more suitable than the "elitist principle" (selecting the seemingly best individual) for a fast system evolution/adaptation.

The estimation of the error term $\mathcal{R}$ is also completely analogous to 3.3.2.1, it is not repeated here. The error term is of the order $\mathcal{R} = \mathcal{O}(1/N)$. Since the first line of (3.192) differs only in the $[1-\Phi((\tilde{Q}-Q_p)/\sigma_{\varepsilon_Q})]$ part from the first line in (3.166), the result (3.172) can be transferred directly to (3.192). Only the s substitution (3.168) must be considered in the $[1-\Phi((\tilde{Q}-Q_p)/\sigma_{\varepsilon_Q})]$ part

$$\frac{\tilde{Q}-Q_p}{\sigma_{\varepsilon_Q}} = \sqrt{1+(\sigma^*/\sigma^*_\varepsilon)^2}\, s + \frac{\sigma^{*2}}{2\sigma^*_\varepsilon}. \tag{3.193}$$

One obtains

$$\varphi^*_{\tilde{1}+\tilde{\lambda}} = -\frac{\sigma^*\lambda}{\sqrt{2\pi}}\frac{\sigma^*}{\sigma^*_\varepsilon}\frac{1}{\sqrt{1+\frac{\sigma^{*2}}{\sigma^{*2}_\varepsilon}}}\int_{-\infty}^{\infty} s\,\mathrm{e}^{-\frac{1}{2}s^2}\left[1-\Phi(s)\right]^{\lambda-1} \left[1-\Phi\left(\sqrt{1+\frac{\sigma^{*2}}{\sigma^{*2}_\varepsilon}}\, s + \frac{\sigma^{*2}}{2\sigma^*_\varepsilon}\right)\right]\mathrm{d}s$$

$$-\frac{\sigma^{*2}}{2}\int_{-\infty}^{\infty} -\frac{\mathrm{d}}{\mathrm{d}s}\left[1-\Phi(s)\right]^{\lambda}\left[1-\Phi\left(\sqrt{1+\frac{\sigma^{*2}}{\sigma^{*2}_\varepsilon}}\, s + \frac{\sigma^{*2}}{2\sigma^*_\varepsilon}\right)\right]\mathrm{d}s + \mathcal{O}\left(\frac{1}{N}\right).$$

After the substitution $s=-t$ and using the relation $1-\Phi(-a)=\Phi(a)$ (see Eq. (C.8) on p. 352), the asymptotic φ^* integral follows

$$\boxed{\begin{aligned}\varphi^*_{\tilde{1}+\tilde{\lambda}} = {} & \frac{\sigma^*}{\sqrt{1+\frac{\sigma^{*2}_\varepsilon}{\sigma^{*2}}}}\frac{\lambda}{\sqrt{2\pi}}\int_{-\infty}^{\infty} t\,\mathrm{e}^{-\frac{1}{2}t^2}\left[\Phi(t)\right]^{\lambda-1}\Phi\left(\sqrt{1+\frac{\sigma^{*2}}{\sigma^{*2}_\varepsilon}}\, t - \frac{\sigma^{*2}}{2\sigma^*_\varepsilon}\right)\mathrm{d}t \\ & -\frac{\sigma^{*2}}{2}\frac{\lambda}{\sqrt{2\pi}}\int_{-\infty}^{\infty}\mathrm{e}^{-\frac{1}{2}t^2}\left[\Phi(t)\right]^{\lambda-1}\Phi\left(\sqrt{1+\frac{\sigma^{*2}}{\sigma^{*2}_\varepsilon}}\, t - \frac{\sigma^{*2}}{2\sigma^*_\varepsilon}\right)\mathrm{d}t + \mathcal{O}\left(\frac{1}{N}\right).\end{aligned}} \tag{3.194}$$

The integral in the second line of (3.192) contains the success probability $P_{\mathrm{s}\tilde{1}+\tilde{\lambda}}$. It can also be interpreted as the death probability of the parent. This follows immediately since

$$\lambda\frac{\mathrm{d}P_1(\tilde{Q})}{\mathrm{d}\tilde{Q}}\left[1-P_1(\tilde{Q})\right]^{\lambda-1} = p_{1:\lambda}(\tilde{Q})$$

is the probability density for the smallest $\tilde{Q}$ value $\tilde{Q}_{1:\lambda}$ in λ trials, and $[1-\Phi((\tilde{Q}-Q_p)/\sigma_{\varepsilon_Q})]$ is the probability that $\tilde{Q}=\tilde{Q}_{1:\lambda}$ is additionally smaller than the noisy fitness value Q_p of the parent (i.e. $\tilde{Q}_p \ge \tilde{Q}_{1:\lambda}$). In other words, the best descendant $\tilde{Q}_{1:\lambda}$ with the density $p(\tilde{Q}_{1:\lambda}) = p_{1:\lambda}(\tilde{Q})$ is successful (since the parent dies) if for the noisy parental fitness $\tilde{Q}_p \ge \tilde{Q}_{1:\lambda}$ $(=\tilde{Q})$ holds. This happens with probability

$$\Pr(\tilde{Q}_p \ge \tilde{Q}) = 1-\Pr(\tilde{Q}_p < \tilde{Q}) = 1-\Phi((\tilde{Q}-Q_p)/\sigma_{\varepsilon_Q}).$$

Therefore, the success probability $\mathrm{d}P_\mathrm{s}(\tilde{Q})$ that the best descendant is in the interval $[\tilde{Q}, \tilde{Q} + \mathrm{d}\tilde{Q})$ and survives is

$$\mathrm{d}P_\mathrm{s}(\tilde{Q}) = p_{1:\lambda}(\tilde{Q})\mathrm{d}\tilde{Q}\,[1 - \Pr(\tilde{Q}_p < \tilde{Q})].$$

The integration over all $\tilde{Q}$ yields

$$P_{\mathrm{s}\tilde{1}+\tilde{\lambda}} = \int_{-\infty}^{\infty} p_{1:\lambda}(\tilde{Q}) \left[1 - \Pr(\tilde{Q}_p < \tilde{Q})\right] \mathrm{d}\tilde{Q}$$

$$P_{\mathrm{s}\tilde{1}+\tilde{\lambda}} = \lambda \int_{-\infty}^{\infty} \frac{\mathrm{d}P_1(\tilde{Q})}{\mathrm{d}\tilde{Q}} \left[1 - P_1(\tilde{Q})\right]^{\lambda-1} \left[1 - \Phi\left(\frac{\tilde{Q} - Q_p}{\sigma_{\varepsilon_Q}}\right)\right] \mathrm{d}\tilde{Q}. \quad (3.195)$$

Consequently, the comparison of (3.195) with (3.192) and (3.194) gives for the success probability

$$P_{\mathrm{s}\tilde{1}+\tilde{\lambda}} = \frac{\lambda}{\sqrt{2\pi}} \int_{-\infty}^{\infty} \mathrm{e}^{-\frac{1}{2}t^2} \left[\Phi(t)\right]^{\lambda-1} \Phi\left(\sqrt{1 + \frac{\sigma^{*2}}{\sigma_\varepsilon^{*2}}}\, t - \frac{\sigma^{*2}}{2\sigma_\varepsilon^*}\right) \mathrm{d}t \;+\; \mathcal{O}\left(\frac{1}{N}\right). \quad (3.196)$$

The evaluation of the integrals (3.196) and (3.194) is cumbersome. Analytical results only exist for the $(\tilde{1} + \tilde{1})$ case, which will be discussed in Sect. 3.3.3.3. For $\lambda > 1$ numerical integration methods must be used. As an example, Fig. 3.12 shows the progress rate of the $(\tilde{1} + \tilde{5})$-ES.

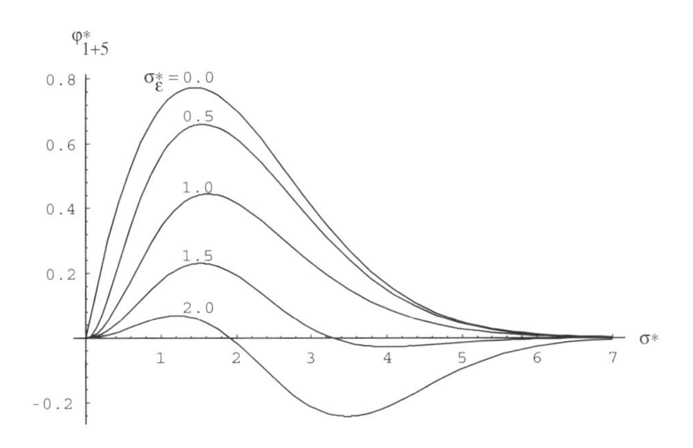

Fig. 3.12. The normalized progress rate of the $(\tilde{1}+\tilde{5})$-ES with noisy fitness data for different normalized noise strengths σ_ε^*

3.3.3.2 An (Almost) Necessary Evolution Criterion for $(\tilde{1} + \tilde{\lambda})$-ES. As can be seen from the $(\tilde{1} + \tilde{5})$-ES example in Fig. 3.12, there are strategy parameter ranges for which $\varphi^* < 0$ holds, even though elitist selection is used in these strategies. The extraction of the evolution criteria is therefore – as in the case of the $(\tilde{1}, \tilde{\lambda})$-ES – an important part of the analysis of the $(\tilde{1}+\tilde{\lambda})$-ES. However, since an analytical calculation of the progress rate integral (3.194), which can be written for $N \to \infty$ also in terms of

$$\varphi^*_{\tilde{1}+\tilde{\lambda}} = \frac{\sigma^{*2}\lambda}{\sqrt{2\pi}} \int_{-\infty}^{\infty} \left(\frac{t}{\sqrt{\sigma^{*2}+\sigma_\varepsilon^{*2}}} - \frac{1}{2} \right) \mathrm{e}^{-\frac{1}{2}t^2} \left[\Phi(t)\right]^{\lambda-1} \times \Phi\left(\frac{\sqrt{\sigma^{*2}+\sigma_\varepsilon^{*2}}}{\sigma_\varepsilon^*} t - \frac{\sigma^{*2}}{2\sigma_\varepsilon^*} \right) \mathrm{d}t, \tag{3.197}$$

is almost always excluded, such investigations are rather difficult. Nevertheless, it is possible to derive a geometrically motivated *necessary* evolution criterion for sufficiently small σ^*. For this purpose, Figs. 3.10 and 3.12 are considered for the optimal σ^* value $\hat{\sigma}^* = \arg\max_{\sigma^*}[\varphi^*(\sigma^*, \sigma_\varepsilon^*)] = f(\sigma_\varepsilon^*)$. For sufficiently large σ_ε^*, $\hat{\sigma}^*$ is a monotonically decreasing function of σ_ε^*. As long as the maximum $\hat{\varphi}^* > 0$, (and therefore $\hat{\sigma}^* > 0$), φ^* exhibits at $\sigma^* = 0$ a non-negative curvature $\left.\frac{\mathrm{d}^2\varphi^*}{\mathrm{d}\sigma^{*2}}\right|_{\sigma^*=0} \geq 0$. However, as soon as $\hat{\varphi}^* = 0$, the curvature changes its sign from positive to negative. The limit case is at the parameter value σ_ε^* for which the curvature vanishes at $\sigma^* = 0$ (and $\sigma_\varepsilon^* > 0$)

$$\left.\frac{\mathrm{d}^2\varphi^*}{\mathrm{d}\sigma^{*2}}\right|_{\sigma^*=0} = f(\sigma_\varepsilon^*) \overset{!}{=} 0. \tag{3.198}$$

Equation (3.197) is of the type $\sigma^{*2}h(\sigma^*)$; therefore, one obtains for the second derivative $[\sigma^{*2}h(\sigma^*)]'' = 2h(\sigma^*) + 4\sigma^* h'(\sigma^*) + \sigma^{*2}h''(\sigma^*)$. For $\sigma^* = 0$ it follows that $\left.[\sigma^{*2}h(\sigma^*)]''\right|_{\sigma^*=0} = 2h(0)$. Therefore one gets for Condition (3.198), applied to (3.197),

$$\begin{aligned} 0 &\overset{!}{=} \frac{1}{\sigma_\varepsilon^*} \frac{\lambda}{\sqrt{2\pi}} \int_{-\infty}^{\infty} \left(t - \frac{\sigma_\varepsilon^*}{2} \right) \mathrm{e}^{-\frac{1}{2}t^2} \left[\Phi(t)\right]^{\lambda} \mathrm{d}t \\ &= \frac{1}{\sigma_\varepsilon^*} \frac{\lambda}{\lambda+1} \left(c_{1,1+\lambda} - \frac{\sigma_\varepsilon^*}{2} \right). \end{aligned} \tag{3.199}$$

The definition of the progress coefficient (3.100) was used here. The limit case of the vanishing curvature is given as a result of (3.199) by $\sigma_\varepsilon^* = 2c_{1,1+\lambda}$. Larger σ_ε^* values yield negative φ^* values for $\sigma^* \to 0$, since the curvature is negative at $\sigma^* = 0$. Consequently, we have found a *necessary evolution criterion* for the $(\tilde{1}+\tilde{\lambda})$-ES

$$\boxed{\sigma_\varepsilon^*(g) < 2c_{1,1+\lambda} \quad (\text{for } \sigma^* \to 0).} \tag{3.200}$$

In the heading of this point I used "(almost)." The reason for this is related to the condition $\sigma^* \to 0$. This condition is trivial for the $(\tilde{1}, \tilde{\lambda})$-ES as can be seen by looking at (3.178).[18] For the $(\tilde{1}+\tilde{\lambda})$ strategy, however, the condition $\sigma_\varepsilon^*(g) < 2c_{1,1+\lambda}$ need not to be stringently fulfilled if one drops the prerequisite $\sigma^* \to 0$. As an example, consider the $(\tilde{1}+\tilde{1})$-ES in Fig. 3.3, p. 61, for the $N \to \infty$ case: the parameter value $\sigma_\varepsilon^* = 1.25$ yields a negative curvature for $\sigma^* \to 0$; even though there is a $\sigma^* > 0$ yielding $\hat{\varphi}^* > 0$. If the criterion (3.200)

[18] The necessary evolution criterion (3.179) could also be obtained by applying (3.198) to (3.173).

were used in this case, one would obtain with (3.109): $\sigma^*_\varepsilon(g) < 2/\sqrt{\pi} \approx 1.128$. However, the $(\tilde{1}+\tilde{1})$-ES can still converge, provided that one is able to find a σ^* with $\varphi^*(\sigma^*) > 0$. This is indeed possible as one can see in Fig. 3.3 on p. 61 (see also Fig. 3.14).

3.3.3.3 Some Aspects of the $(\tilde{1}+\tilde{1})$-ES.

The progress rate formula. Analytical expressions can be derived for the $(\tilde{1}+\tilde{1})$-ES. With the integral formulae (A.10) and (A.12) from Appendix A.1, the success probability reads

$$\boxed{P_{\mathrm{s}\tilde{1}+\tilde{1}}(\sigma^*,\sigma^*_\varepsilon) = 1 - \Phi\left(\frac{\frac{\sigma^*}{2}}{\sqrt{1+2\frac{\sigma^{*2}_\varepsilon}{\sigma^{*2}}}}\right) + \mathcal{O}\left(\frac{1}{N}\right)} \tag{3.201}$$

and the progress rate (3.194) becomes

$$\boxed{\begin{aligned}\varphi^*_{\tilde{1}+\tilde{1}} = \frac{\sigma^*}{\sqrt{2\pi}}\frac{1}{\sqrt{1+2\frac{\sigma^{*2}_\varepsilon}{\sigma^{*2}}}}\exp\left[-\frac{1}{2}\left(\frac{\frac{\sigma^*}{2}}{\sqrt{1+2\frac{\sigma^{*2}_\varepsilon}{\sigma^{*2}}}}\right)^2\right] \\ - \frac{\sigma^{*2}}{2}P_{\mathrm{s}\tilde{1}+\tilde{1}}(\sigma^*,\sigma^*_\varepsilon) + \mathcal{O}\left(\frac{1}{N}\right).\end{aligned}} \tag{3.202}$$

The success probability is shown in Fig. 3.13. The curve of $\varphi^*_{\tilde{1}+\tilde{1}}$ was already

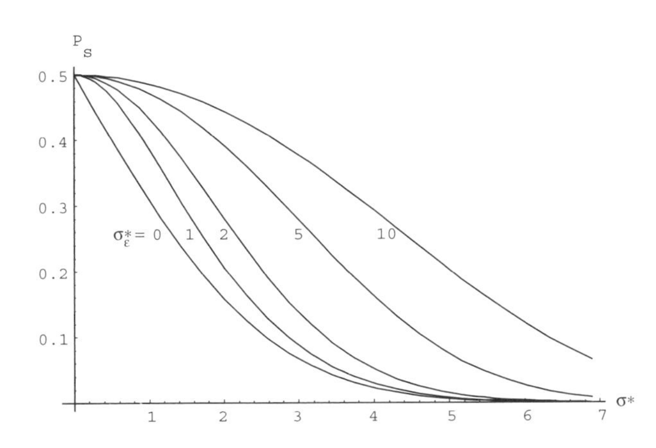

Fig. 3.13. Success probability $P_{\mathrm{s}\tilde{1}+\tilde{1}}$ of the $(\tilde{1}+\tilde{1})$-ES with noisy fitness data is a monotonically decreasing function of σ^*.

displayed in Fig. 3.3 on p. 61 for several noise strengths σ^*_ε. It is important to note that the curve shown on p. 61 for $N = \infty$ indicates a $\hat{\sigma}^*$ (value of σ^* at maximum $\hat{\varphi}^*$) that increases monotonically in σ^*_ε, which is in contrast to the properties of the $(\tilde{1} \overset{+}{,} \tilde{5})$-ES (Figs. 3.10 and 3.12). Accordingly, the mutation strength should be increased for increasing σ^*_ε at the $(\tilde{1}+\tilde{1})$-ES. However, it can be suspected that this recommendation has only limited validity because for sufficiently large σ^* and $N < \infty$ the condition $\sigma^{*2}/N \ll 1$ is violated. A look at Fig. 3.3 for the $N = 3$ case agrees with this conjecture. Therefore, we directly arrive at the next topic: control of the mutation strength σ.

The σ control using the 1/5th rule. All σ control mechanisms aim to control the ES in an operation region with $\varphi^* \approx \text{const} > 0$ and thus obtain a linear convergence order (see Sect. 2.4.2). For maximal possible progress, φ^* should be in the vicinity of its maximum $\hat{\varphi}^*$.

In the $(1+1)$-ES-Algorithm 3 on p. 52, a mutation strength control was not incorporated. However, in real applications σ should be controlled by the algorithm so that $\hat{\sigma}^*$ is approximately realized and φ^* is in the vicinity of its maximum. The problem is to decide whether an initially chosen $\sigma^{(0)} = \sigma_{\text{init}}$ should be increased or decreased after a certain number of generations. Since the local curvature radius $R^{(g)}$ of the fitness landscape is not known at time g in the state $\mathbf{y}^{(g)}$, one cannot simply calculate $\sigma^{*(g)} = \sigma^{(g)} N/R$ and compare it with the optimal value of the $(1+1)$-ES, $\hat{\sigma}^* \approx 1.224$. However, there is the possibility of obtaining the $\sigma^{*(g)}$ information indirectly – by measuring the success probability P_s. Assuming that $\sigma_\varepsilon^* = \text{const}$, σ^* can be obtained by resolving (3.201). This is possible because $P_{\text{s}\tilde{1}+\tilde{1}}$ is an invertible function of σ^*, which can be easily verified using Fig. 3.13. That is, knowing P_s, one can determine σ^* and therefore $\varphi^* = \varphi^*(P_\text{s}, \sigma_\varepsilon^*)$. This dependency is shown in Fig. 3.14.

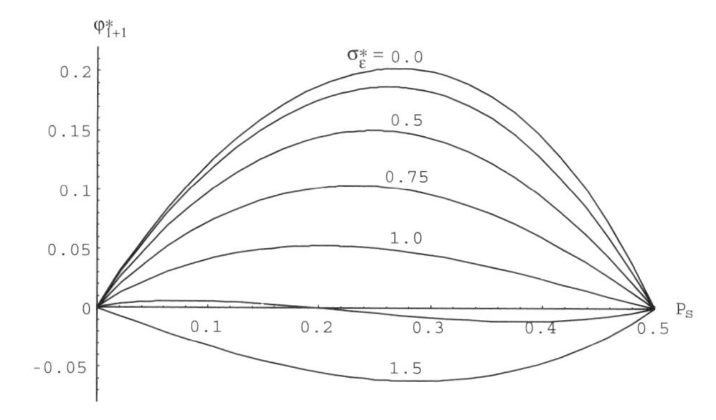

Fig. 3.14. One the dependency of the progress rate φ^* of the $(\tilde{1}+\tilde{1})$-ES on the success probability P_s for various normalized noise levels $\sigma_\varepsilon^* = 0$, 0.25, 0.5, 0.75, 1.0, 1.25, and 1.5

It can be seen that the progress rate at the sphere model uniquely depends on the success probability P_s (for $\sigma_\varepsilon^* = \text{const}$). The optimal $P_\text{s} = P_\text{opt}$ at which φ^* becomes maximal is a function of σ_ε^*. Figure 3.15 shows this interesting dependency.

For the noiseless case, i.e. $\sigma_\varepsilon^* = 0$, one has $P_\text{opt} \approx 0.27$, in other words the $(1+1)$-ES works optimally if in each generation on average each $1/0.27 \approx 3.7$th descendant is successful (better than or equal to the parent). With increasing noise level P_opt decreases; at first slowly, but for $\sigma_\varepsilon^* \gtrapprox 1$ drastically.

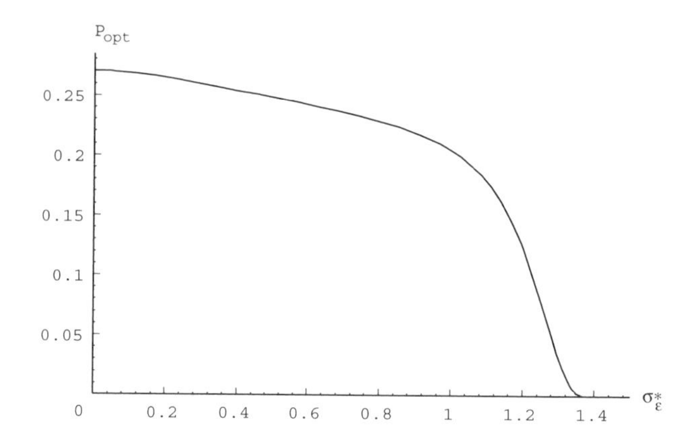

Fig. 3.15. The optimal success probability of the $(\tilde{1} + \tilde{1})$-ES for the sphere model

How should P_s be chosen if nothing is known about σ_ε^*? As Rechenberg (1973) suggested the idea of the σ control based on the success probability, he was faced with a similar problem. The starting point was a fitness landscape with $\sigma_\varepsilon^* = 0$. Rechenberg investigated the fitness models sphere and corridor (see Sects. 2.2.1 and 2.2.3) and found $P_{opt} \approx 0.27$ (see above) for the sphere and $P_{opt} \approx 0.184$ for the corridor. If one assumes that the real fitness landscapes are somehow "mixed forms" of the sphere and corridor, regarding these two as "extreme cases," then a suitable compromise should be found by taking the mean value of the reciprocals of both P_{opt} values. Therefore, Rechenberg suggested the $P_{opt} = 1/5$ rule. For the noisy fitness landscapes he suggested, using plausibility arguments, $P_{opt} = 1/10$ (Rechenberg, 1973, p. 123, a $\varphi_{\tilde{1}+\tilde{1}}^*$ formula was only available for the corridor model). Based on the results from the sphere model under noise, this suggestion will be evaluated quantitatively below.

The success probability $P_s = 1/5 \ldots 1/10$ is optimal for noise strengths in the interval $\sigma_\varepsilon^* = 1 \ldots 1.2$, as can be seen in Fig. 3.15. After decreasing the success probability from 1/5 to 1/10, a relatively small change in σ_ε^* is achieved. For $\sigma_\varepsilon^* \gtrapprox 1$, P_{opt} depends strongly on the noise strength σ_ε^*, moreover for $\sigma_\varepsilon^* \gtrapprox 1.3$ it approaches zero very quickly. The calibration of an operation point in this noise interval is difficult and numerically not very stable.[19] Additionally, for $P_{opt} = 1/10$ only small progress rates are to be expected (see Fig. 3.14). Therefore, it does not seem to be reasonable to use $P_{opt} = 1/10$; rather the 1/5th rule should be applied also for the case with noisy fitness data.

The convergence behavior of the $(\tilde{1}+\tilde{1})$-ES should be improved preferably by methods aimed at a decrease of σ_ε^*, e.g., by averaging over m fitness

[19] Because of the method used for measuring P_s (for the implementation of the σ control, see below), by counting the number of successful mutations over a finite number of generations, small P_s values have a larger relative measurement error than the large P_s values.

measurements (resampling, analogous to the method suggested on p. 92f; see also Sect. 3.3.4).

On the application of the 1/5th rule. As for the practical realization of the 1/5th rule, the actual success probability P_s must be estimated first. This is done by measuring the relative frequency of the successful mutations over a certain period of generations G.[20] If there are G_s successful mutations (i.e. mutations substituting their parents), then $P_s \approx G_s/G$. Using this estimate, the actual σ^* can be determined by Eq. (3.201) $\sigma^* = \sigma^*(P_s, \sigma_\varepsilon^*)$ (assuming that $\sigma_\varepsilon^* = \text{const}$). Since the mutation strength is known, the local curvature radius R can be determined using Eq. (2.22)

$$R = N\sigma/\sigma^*(P_s, \sigma_\varepsilon^*). \tag{3.203}$$

For maximal performance $\sigma^* = \hat{\sigma}^* = \sigma^*(P_{opt}, \sigma_\varepsilon^*)$ should be valid, that is $\sigma^*(P_{opt}, \sigma_\varepsilon^*) = \sigma_{opt}\frac{N}{R}$. If (3.203) is substituted, one obtains

$$\sigma_{opt} = \frac{\sigma^*(P_{opt}, \sigma_\varepsilon^*)}{\sigma^*(P_s, \sigma_\varepsilon^*)}\,\sigma. \tag{3.204}$$

If σ_ε^* is known, Eq. (3.204) can directly be used for the control of σ. The actual σ is simply multiplied by a correction factor, and one obtains σ_{opt}, which can serve as the constant mutation strength in the (1 + 1)-ES for the next G generations.

If σ_ε^* is unknown, (3.204) cannot be realized directly. Rechenberg's 1/5th heuristics should be used instead. Because of the monotonicity of $P_s(\sigma^*)$ (see Fig. 3.13) one has

$$\begin{aligned} \frac{\sigma^*(P_{opt}, \sigma_\varepsilon^*)}{\sigma^*(P_s, \sigma_\varepsilon^*)} < 1 \quad &\Leftrightarrow \quad P_s < P_{opt}, \\ \frac{\sigma^*(P_{opt}, \sigma_\varepsilon^*)}{\sigma^*(P_s, \sigma_\varepsilon^*)} > 1 \quad &\Leftrightarrow \quad P_s > P_{opt}. \end{aligned} \tag{3.205}$$

If $P_s < P_{opt}$, σ must be decreased accordingly; if not, it must be increased. As a compromise $P_{opt} = 1/5$ has been suggested, resulting in the

$$\boxed{\text{1/5TH RULE:} \qquad \sigma_{new} := \begin{cases} \sigma_{old}\, a, & \text{if} \quad P_s > 1/5 \\ \sigma_{old}/a, & \text{if} \quad P_s < 1/5\,. \\ \sigma_{old}, & \text{if} \quad P_s = 1/5 \end{cases}} \tag{3.206}$$

The factor a is usually between 1.1 and 1.5. It depends particularly on the measurement period G (where σ remains constant). For $\sigma_\varepsilon^* = 0$ and $G = N$, Schwefel (1995, p. 112) calculated $1/a \approx 0.817$ and recommended using $1/a = 0.85$. The choice of a is relatively uncritical. Values for a with $G \neq N$ can easily be obtained from (3.204) with (3.203), taking the R-change (2.103) into account.

[20] The mutation strength σ is kept constant during the measurement period G.

Limits of the 1/5th rule. The 1/5th rule has a remarkable validity domain. Although it has been derived considering the sphere and the corridor model only, this *heuristic* also works for nonspherical fitness landscapes as long as the landscape can be approximated locally using a substitute sphere with a mean curvature radius $R_{\langle\varkappa\rangle}$ (see Sect. 2.3, Eq. (2.90)). However, if this approximation assumption cannot be fulfilled, the 1/5th rule may fail. This is particularly the case when discrete or combinatorial optimization problems are considered, e.g., the traveling salesman problem (see, e.g., Fogel & Ghozeil, 1996). This rule can also fail for continuous parameter optimization when the neighborhood of the parental state $\mathbf{y}_p$ is not continuously differentiable, e.g., at the sharp ridge. Similar problems can also arise in the vicinity of active restrictions (Schwefel, 1995, p. 215).

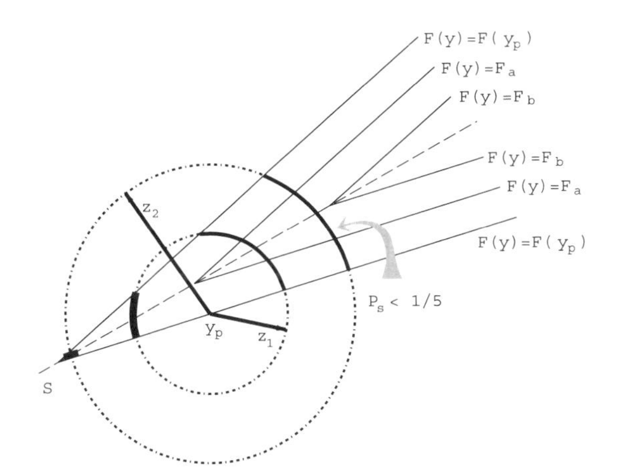

Fig. 3.16. On the failure of the 1/5th rule at the sharp ridge. The picture of isometric lines of the sharp ridge in $N = 2$ dimensions is shown here. $\mathbf{z}_1$ and $\mathbf{z}_2$ are two mutation vectors with different mutation strengths $\sigma_1 < \sigma_2$.

The sharp ridge case is shown in Fig. 3.16. The desired progress is in the tangential direction (in the ridge direction). The success probability P_s of the mutations is obtained (in the figure) from the ratio of the length of the boldly emphasized parts of the mutation radius to the corresponding circumference of the circle (of course, only for the $N = 2$ case considered here). The 2D-section of the equifitness surfaces (isometric lines) shows that the mutation strength σ should be decreased in order to fulfill the 1/5th rule. These smaller mutations produce in the statistical mean a progress in the direction toward S. Therefore, the individuals get closer and closer to the ridge axis. However, due to the scaling invariance of the ridge geometry, the success probability can only increase by a further decrease of σ. Hence, the object parameter components in the desired progress direction (tangential to the ridge axis S) obtain continuously smaller increases. It is geometrically clear that in order to obtain a significant progress in the direction of the ridge axis, σ *must not* be decreased in this fitness landscape.

3.3.4 Convergence Improvement by Inheriting Scaled Mutations

3.3.4.1 Theoretical Fundamentals. Besides the method of fitness averaging (resampling) mentioned on p. 92 which aims at the reduction of σ_ε^*, there is also the possibility to improve the convergence reliability of the $(\tilde{1} \overset{+}{,} \tilde{\lambda})$-ES by using the principle "*vary large and inherit small*" (Rechenberg, 1994).

The analysis of the $(\tilde{1}, \tilde{\lambda})$-ES showed that the noise in the fitness values decreases the gain part in the EPP; however, it has no influence on the loss part (see Eq. (3.173) and the related discussion of the EPP).[21] Averaging the fitness measurements of the same setting increases the gain part. Is there any way to decrease the loss part? This is indeed possible because of the nonlinearity in σ^{*2}. The $(\tilde{1} \overset{+}{,} \tilde{\lambda})$-ES is used in the classical way. The descendants $\tilde{\mathbf{y}}_l$ are generated by the parent $\mathbf{y}_p^{(g)}$ according to $\tilde{\mathbf{y}}_l := \mathbf{y}_p^{(g)} + \mathbf{z}_l$, and selected using the $(1 \overset{+}{,} \lambda)$ rule. However, the best descendant $\tilde{\mathbf{y}}_{1;\lambda}$ does not become the parent $\mathbf{y}_p^{(g+1)}$ of the next generation, but rather the mutation $\mathbf{z}_{1;\lambda}$ which produced $\tilde{\mathbf{y}}_{1;\lambda}$ is *inherited after down-scaling*

$$\boxed{\mathbf{y}_p^{(g+1)} := \mathbf{y}_p^{(g)} + \frac{1}{\kappa}\mathbf{z}_{1;\lambda}, \qquad \kappa \ge 1} \tag{3.207}$$

(of course, for the $(\tilde{1} + \tilde{\lambda})$-ES only if the mutation $\mathbf{z}_{1;\lambda}$ gave a better fitness value $\tilde{Q}$ than the parental $\tilde{Q}_p$).

The ES algorithm altered by (3.207) can be analyzed easily by the theory developed in Sect. 3.3.1. Almost no changes are necessary in the derivation. The only difference is in the distance formula $R - \|\mathbf{R} + \mathbf{z}_{\tilde{1};\tilde{\lambda}}\|$ which is to be altered to $R - \|\mathbf{R} + \frac{1}{\kappa}\mathbf{z}_{\tilde{1};\tilde{\lambda}}\|$ caused by (3.207) and having consequences in the expected value expression (3.129). Instead of (3.129), one has to evaluate

$$\varphi_{\tilde{1}\overset{+}{,}\tilde{\lambda}} = \mathrm{E}_{\overset{+}{,}}\left\{ R - \sqrt{\left(R - \frac{1}{\kappa}x_{\tilde{1};\tilde{\lambda}}\right)^2 + \left\|\frac{1}{\kappa}\mathbf{h}_{\tilde{1};\tilde{\lambda}}\right\|^2} \right\} \tag{3.208}$$

and obtains analogous to (3.130) the expected value integral

$$\varphi_{\tilde{1}\overset{+}{,}\tilde{\lambda}} \simeq \int_{-\infty}^{\infty} \left(\frac{x}{\kappa} - \frac{\sigma^*\sigma}{2\kappa^2}\right) p_{\tilde{1}\overset{+}{,}\tilde{\lambda}}(x)\,\mathrm{d}x. \tag{3.209}$$

The gain term is decreased by a factor of κ. In comparison, the loss term decreases much more, by a factor of κ^2. The relative contribution of the gain term to the progress rate increases; consequently, the convergence reliability is increased.

Since (3.209) differs from (3.130) only in the first parentheses, the derivation and results (3.173) and (3.197) can be used directly. The asymptotic limit case reads $(N \to \infty)$

---

[21] Unfortunately, in the case of the $(\tilde{1} + \tilde{\lambda})$-ES, the loss part increases as well, since for increasing σ_ε^* the success probability P_s increases, as one can see in the example of the $(\tilde{1} + \tilde{1})$-ES (Fig. 3.13 and Eq. (3.202)).

$$\varphi^*_{\tilde{1},\tilde{\lambda}}(\sigma^*, \sigma^*_\varepsilon) = \frac{1}{\kappa}\left[c_{1,\lambda}\sigma^* \frac{1}{\sqrt{1+(\sigma^*_\varepsilon/\sigma^*)^2}} - \frac{\sigma^{*2}}{2\kappa}\right], \tag{3.210}$$

$$\varphi^*_{\tilde{1}+\tilde{\lambda}} = \frac{1}{\kappa}\frac{\sigma^{*2}\lambda}{\sqrt{2\pi}} \int_{-\infty}^{+\infty} \left(\frac{t}{\sqrt{\sigma^{*2}+\sigma^{*2}_\varepsilon}} - \frac{1}{2\kappa}\right) \mathrm{e}^{-\frac{1}{2}t^2} \left[\Phi(t)\right]^{\lambda-1} \times \Phi\left(\frac{\sqrt{\sigma^{*2}+\sigma^{*2}_\varepsilon}}{\sigma^*_\varepsilon}\, t - \frac{\sigma^{*2}}{2\sigma^*_\varepsilon}\right) \mathrm{d}t. \tag{3.211}$$

The success probability (3.196) does not change.

One finds for the necessary evolution criteria of the $(\tilde{1} \overset{+}{,} \tilde{\lambda})$-ES analogous to (3.179) and (3.200) the *necessary criterion*

$$(\tilde{1}, \tilde{\lambda})\text{-ES:} \qquad \sigma^*_\varepsilon(g) < \kappa\, 2c_{1,\lambda} \tag{3.212}$$

and the *(almost) necessary criterion*

$$(\tilde{1} + \tilde{\lambda})\text{-ES:} \qquad \sigma^*_\varepsilon(g) < \kappa\, 2c_{1,\lambda+1}. \tag{3.213}$$

These criteria show that the convergence of the $(\tilde{1} \overset{+}{,} \tilde{\lambda})$-ES can always be forced by choosing κ sufficiently large (in the framework of the asymptotic theory). The $(\tilde{1}, \tilde{\lambda})$-ES will be investigated in detail next.

3.3.4.2 Discussion of the $(\tilde{1}, \tilde{\lambda})$-ES. Down-scaling the mutations improves the convergence properties of the $(\tilde{1}, \tilde{\lambda})$-ES as a result of (3.212) considerably. Because of (3.210), the absolute value of φ^* decreases (about) by the factor κ. In order to obtain practically useful progress rates, it is necessary to increase σ^* accordingly. Figure 3.17 shows the effect of σ^* and κ on the progress rate φ^* for the $(\tilde{1}, \tilde{5})$ case. $\sigma^*_\varepsilon = 5$ was chosen as the normalized mutation strength. In order to satisfy (3.212), $\kappa \gtrapprox 2.15$ must be fulfilled.

By choosing σ^* and κ appropriately, values for φ^* can be theoretically realized which are comparable with the maximal value $\hat{\varphi}^* = c^2_{1,5}/2 \approx 0.676$ of the $(1, 5)$-ES without noise. In practice, however, this is only conditionally the case, since the $\sigma^{*2}/N \ll 1$ condition must hold for the derivations made. A satisfactory treatment of the problem of choosing σ^* *and* κ requires the formulation of an N-dependent theory of the $(\tilde{1}, \tilde{\lambda})$-ES – a task for future research. Here, only a recommendation can be given: choose σ^* so that $\sigma^{*2}/N \ll 1$ holds. As one can see in Fig. 3.17, for any $\sigma^* = \text{const}$ there exists a κ for which φ^* becomes maximal. By considering the extreme value of (3.210), one obtains

$$\kappa_{\text{opt}} = \frac{1}{c_{1,\lambda}}\sqrt{\sigma^{*2}+\sigma^{*2}_\varepsilon} \qquad \text{and} \qquad \varphi^* = \frac{1}{1+(\sigma^*_\varepsilon/\sigma^*)^2}\frac{c^2_{1,\lambda}}{2}. \tag{3.214}$$

The formulae obtained in that way should be used only if $\sigma^{*2}/N \ll 1$. The limits of the applicability cannot be estimated by the asymptotic theory. An N-dependent φ^* formula should be derived for that purpose. For the case

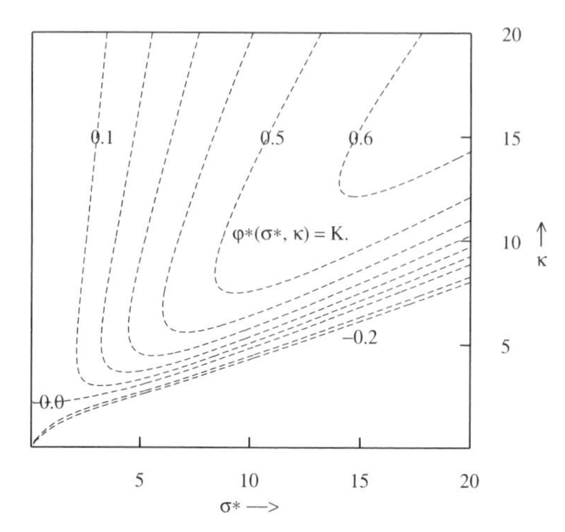

Fig. 3.17. Lines of constant progress rate for the $(\tilde{1}, \tilde{5})$-ES with $\sigma_\varepsilon^* = 5$

of the $(1, \lambda)$-ES without noise, such a formula will be derived in the next section.

3.4 The N-Dependent $(1, \lambda)$ Progress Rate Formula

3.4.1 Motivation

A deficiency of the asymptotic theory developed in Sects. 3.2 and 3.3 is that its validity is limited for the real case $N < \infty$ by the condition $\sigma^{*2}/N \ll 1$ (3.66). This limitation is usually not noticeable for the investigation of the $(1, \lambda)$-ES since the region of interest is $\varphi^* \gtrapprox 0$, i.e. $\sigma^* \lessapprox 2c_{1,\lambda}$. However, there are questions as well as ES algorithms which require the consideration of larger mutation strengths σ^*. For instance, in Sect. 3.3.4.2 the validity of (3.214) had to be limited by $\sigma^{*2}/N \ll 1$. The same is true for the applicability in the recombinant ES case (Chap. 6). Principally, all algorithms which reduce the loss term in the EPP (see Sect. 1.4.2.1) by appropriate techniques as compared to the standard $(\mu \stackrel{+}{,} \lambda)$-ES are affected.

The aim of the φ^* formula to be derived here is to extend the validity region of the approximations beyond the "magic" $\sigma^{*2}/N \ll 1$ condition. As we will see in the derivations, this approximation does not yield any statement for the case with very small N values. A practical limit should be $N \gtrapprox 30$, although this does not imply that useful results cannot be obtained for smaller N values (see also footnote 23). It is difficult to give corresponding error limits. However, the practical benefit of such estimations is rather small. Instead, the analytical results will be verified in any case by results obtained from numerical ES simulations. Therefore, all important results obtained from the analytical derivations in this section and in the following chapters will be supported by ES simulations.

The derivation of the progress rate formula for the $(1, \lambda)$-ES starts with Eq. (3.89): for the comma strategy we have $\varphi_{1,\lambda} = \mathrm{E}\left\{R - \tilde{R}_{1:\lambda}\right\}$. However, instead of taking $\tilde{R}_{1:\lambda}$ as a function of the random variable $x_{1;\lambda}$ as in (3.90), $\tilde{R}_{1:\lambda}$ itself is taken as an independent random variable. Provided that the density $p(\tilde{R})$ of a single descendant $\tilde{\mathbf{y}}_l$ is known, $\mathrm{E}\left\{R - \tilde{R}_{1:\lambda}\right\}$ can be calculated

$$\varphi_{1,\lambda} = \mathrm{E}\left\{R - \tilde{R}_{1:\lambda}\right\} = \int_0^\infty (R - r)\, p_{1,\lambda}(r)\, \mathrm{d}r. \tag{3.215}$$

The $p_{1,\lambda}(r)$ density stands for the first-order statistics $1 : \lambda$ of $\tilde{R}$ (see Eq. (3.135))

$$p_{1,\lambda}(r) = p_{1:\lambda}(r) = \lambda\, p(r) \left[\Pr(\tilde{R} > r)\right]^{\lambda - 1}, \tag{3.216}$$

and $p(r)$ is the density of the random variable $r := \tilde{R}_{1:1}$. It will be calculated in the next subsection.

3.4.2 The $p(r)$ Density

According to the decomposition in (3.7), the square of the residual distance $r = \tilde{R}$, Eq. (3.9), can be calculated

$$r^2 = (R - x)^2 + \mathbf{h}^2 = R^2 - 2Rx + x^2 + \mathbf{h}^2 \approx R^2 - 2Rx + \mathbf{h}^2. \tag{3.217}$$

The x^2 is neglected with respect to $\mathbf{h}^2$. This is tolerable since

$$\mathrm{E}\left\{\mathbf{h}^2\right\} = \overline{\mathbf{h}^2} = (N-1)\sigma^2 \qquad \text{and} \qquad \mathrm{E}\left\{x^2\right\} = \overline{x^2} = \sigma^2 \tag{3.218}$$

is valid because of (3.52) using the density (3.12). Therefore, the quotient vanishes with $\mathrm{E}\left\{x^2\right\} / \mathrm{E}\left\{\mathbf{h}^2\right\} = 1/(N-1)$.[22] After neglecting x^2 one obtains a sum consisting of the constant R^2 and two *independent* random variables

$$v_1 := -2Rx \qquad \text{and} \qquad v_2 := \mathbf{h}^2. \tag{3.219}$$

Due to (3.12), v_1 is a normally distributed random variable with zero mean and standard deviation $2R\sigma$, i.e. $v_1 = 2R\sigma\mathcal{N}(0,1)$. It is desirable to approximate v_2 by a normal distribution, as a result one would obtain a normal distribution for r^2. The approximation of v_2 using a normal distribution is based on the central limit theorem in statistics (Fisz, 1971). For this purpose, the expected value $\overline{v_2}$ and the standard deviation $\mathrm{D}\left\{v_2\right\}$ of the random variable v_2 must be calculated.[23]

[22] For the kth order moments one can show using (3.50) for the quotient $\mathrm{E}\left\{x^k\right\} / \mathrm{E}\left\{\|\mathbf{h}\|^k\right\} = \mathcal{O}\left(N^{-k/2}\right)$.

[23] The central limit theorem is only applicable for sufficiently large N, $N \gtrsim 30$. For the approximation in smaller parameter space dimensions, the density function

As to the normal approximation of $\mathbf{h}^2$, note that $\mathbf{h}^2$ is the sum of $N-1$ i.i.d. random variables x_i^2 with $\mathrm{D}\{x_i\} = \sigma$ (see (3.13)). Because of (3.11), the expected value of x_i^2 is $\overline{x_i^2} = \sigma^2$ (can also be read from (3.52)). Consequently, the expected value of the sum reads

$$\overline{v_2} = \mathrm{E}\{v_2\} = \mathrm{E}\{\mathbf{h}^2\} = \sum_{i=2}^{N} \overline{x_i^2} = (N-1)\sigma^2 \sim N\sigma^2. \tag{3.220}$$

In order to calculate the standard deviation $\mathrm{D}\{v_2\} = \sqrt{\overline{v_2^2} - \overline{v_2}^2}$, $\overline{v_2^2}$ is to be determined. Since x_i are statistically independent,

$$\overline{v_2^2} = \sum_{i=2}^{N} \overline{x_i^4} + \sum_{i=2}^{N} \sum_{j \neq i} \overline{x_i^2}\,\overline{x_j^2} = 3(N-1)\sigma^4 + (N-1)(N-2)\sigma^4 \tag{3.221}$$

follows, where the fourth moment of the normal distribution $\overline{x_i^4} = 3\sigma^4$ has been taken into account (see Table A.1 in Appendix A.1). Using (3.220) and (3.221), one obtains for the standard deviation

$$\mathrm{D}\{v_2\} = \sqrt{\overline{v_2^2} - \overline{v_2}^2} = \sqrt{2(N-1)}\,\sigma^2 \simeq \sqrt{2N}\,\sigma^2. \tag{3.222}$$

After combining the partial results one obtains for the densities of v_1 and v_2

$$p_{v_1}(v_1) = \frac{1}{\sqrt{2\pi}\,2R\sigma} \mathrm{e}^{-\frac{1}{2}\left(\frac{v_1}{2R\sigma}\right)^2} \tag{3.223}$$

and

$$p_{h^2}(v_2) = \frac{1}{\sqrt{2\pi}\,\sqrt{2N}\,\sigma^2} \mathrm{e}^{-\frac{1}{2}\left(\frac{v_2 - N\sigma^2}{\sqrt{2N}\,\sigma^2}\right)^2}. \tag{3.224}$$

Since r^2 is given by (3.217) one obtains for the expected value of r^2 using (3.218)

$$\overline{r^2} = R^2 + N\sigma^2. \tag{3.225}$$

For the calculation of the standard deviation $\mathrm{D}\{r^2\}$ it shall be recalled from (3.217) that r^2 is constitued – within the approximations used – of the sum of v_1, v_2, and R^2. Thus, $\mathrm{D}\{r^2\} = \sqrt{\mathrm{D}^2\{v_1\} + \mathrm{D}^2\{v_2\}}$ and one gets

$$\mathrm{D}\{r^2\} = \sqrt{4R^2\sigma^2 + 2N\sigma^4} = 2\sigma R\sqrt{1 + \frac{\sigma^2 N}{2R^2}}. \tag{3.226}$$

Hence, the distribution function of r^2 reads

$$\Pr(\tilde{R}^2 \le r^2) = \Phi\left(\frac{r^2 - R^2 - \sigma^2 N}{2\sigma R\sqrt{1 + \frac{\sigma^2 N}{2R^2}}}\right). \tag{3.227}$$

---

of v_2 should be estimated accordingly, and a "correction" to the normal distribution is necessary. The approximation technique for such cases is developed in Appendix B.3.2. Applications can be found in Chap. 4.

Finally one obtains, considering

$$\Pr(\tilde{R}^2 \le r^2) = \Pr(\tilde{R} \le r) =: P(r) \quad \text{and} \quad p(r) = \frac{\mathrm{d}}{\mathrm{d}r} P(r), \tag{3.228}$$

for the probability density of the random variable r

$$p(r) = \frac{r}{\sqrt{2\pi}\,\sigma R \sqrt{1 + \frac{\sigma^2 N}{2R^2}}} \exp\left[-\frac{1}{2} \left(\frac{r^2 - R^2 - \sigma^2 N}{2\sigma R \sqrt{1 + \frac{\sigma^2 N}{2R^2}}} \right)^2 \right]. \tag{3.229}$$

Note that r^2 is normally distributed, but not r. A normal approximation for r is derived in Sect. 3.4.5.

3.4.3 The Derivation of the φ^* Progress Rate Formula

The progress rate is obtained from the integral (3.215). With (3.216) and (3.228) one gets

$$\varphi_{1,\lambda} = R - \lambda \int_0^\infty r\, p(r) \left[1 - P(r)\right]^{\lambda - 1} \mathrm{d}r. \tag{3.230}$$

Inserting (3.227) and (3.229), substituting

$$\begin{aligned} t &= \left(\frac{r^2 - R^2 - \sigma^2 N}{2\sigma \sqrt{R^2 + \frac{\sigma^2 N}{2}}} \right) \\ \Longrightarrow \quad r &= \sqrt{R^2 + \sigma^2 N} \sqrt{1 + 2\sigma \frac{\sqrt{R^2 + \sigma^2 N/2}}{R^2 + \sigma^2 N} t} \end{aligned} \tag{3.231}$$

thereafter and using the normalization scheme (2.22) one gets the integral

$$\begin{aligned} \varphi^*_{1,\lambda} = N - N\sqrt{1 + \frac{\sigma^{*2}}{N}} \frac{\lambda}{\sqrt{2\pi}} \int_{t_u}^\infty & \sqrt{1 + 2\frac{\sigma^*}{N} \frac{\sqrt{1 + \frac{\sigma^{*2}}{2N}}}{1 + \frac{\sigma^{*2}}{N}} t} \\ & \times \mathrm{e}^{-\frac{1}{2}t^2} \left[1 - \Phi(t)\right]^{\lambda - 1} \mathrm{d}t. \end{aligned} \tag{3.232}$$

The lower limit t_u reads, using (3.231),

$$t_u = t(r)|_{r=0} = -\frac{N}{2\sigma^*} \frac{1 + \sigma^{*2}/N}{\sqrt{1 + \sigma^{*2}/2N}} \le -\sqrt{\frac{N}{2}}. \tag{3.233}$$

The last relation can be verified by direct calculation. The largest value of the lower limit in (3.232) is $t_u = -\sqrt{N/2}$.

Considering the factor at the integration variable in the square root of the integrand in (3.232) one verifies

$$2a := 2\frac{\sigma^*}{N} \frac{\sqrt{1 + \sigma^{*2}/2N}}{1 + \sigma^{*2}/N} \le \sqrt{\frac{2}{N}}. \tag{3.234}$$

For $t \in (-\sqrt{N/2}, \sqrt{N/2})$, the square root in the integrand of (3.233) can be expanded into a Taylor series $\sqrt{1+2at} = 1 + at - (at)^2/2 + (at)^3/2 - \ldots$

$$\sqrt{1 + 2\frac{\sigma^*}{N}\frac{\sqrt{1+\frac{\sigma^{*2}}{2N}}}{1+\frac{\sigma^{*2}}{N}}t}$$
$$= 1 + \frac{\sigma^*}{N}\frac{\sqrt{1+\frac{\sigma^{*2}}{2N}}}{1+\frac{\sigma^{*2}}{N}}\left[t - \frac{1}{2}\frac{\sigma^*}{N}\frac{\sqrt{1+\frac{\sigma^{*2}}{2N}}}{1+\frac{\sigma^{*2}}{N}}t^2 + \ldots\right]. \qquad (3.235)$$

The application of this expansion for (3.232) limits first the integration to the boundaries $t \in (-\sqrt{N/2}, \sqrt{N/2})$

$$\varphi^*_{1,\lambda} = N - N\sqrt{1+\frac{\sigma^{*2}}{N}}\frac{\lambda}{\sqrt{2\pi}}\int_{-\sqrt{N/2}}^{\sqrt{N/2}} \sqrt{1+2at}\ \mathrm{e}^{-\frac{1}{2}t^2}\left[1-\Phi(t)\right]^{\lambda-1}\mathrm{d}t$$
$$+\mathcal{R}. \qquad (3.236)$$

The error term $\mathcal{R}$ can be written as

$$\mathcal{R} = -N\sqrt{1+\frac{\sigma^{*2}}{N}}\frac{\lambda}{\sqrt{2\pi}}\int_{\sqrt{N/2}}^{\infty} \sqrt{1+2at}\ \mathrm{e}^{-\frac{1}{2}t^2}\left[1-\Phi(t)\right]^{\lambda-1}\mathrm{d}t. \quad (3.237)$$

Considering (3.234), $\sqrt{1+2at} \le 1 + t/\sqrt{2N}$ (for $t \ge \sqrt{N/2}$), the error term can be estimated as

$$|\mathcal{R}| \le N\sqrt{1+\frac{\sigma^{*2}}{N}}\frac{\lambda}{\sqrt{2\pi}}\int_{\sqrt{N/2}}^{\infty}\left(1+\sqrt{\frac{1}{2N}}\,t\right)\mathrm{e}^{-\frac{1}{2}t^2}\left[1-\Phi(t)\right]^{\lambda-1}\mathrm{d}t$$
$$\le N\sqrt{1+\frac{\sigma^{*2}}{N}}\frac{\lambda}{\sqrt{2\pi}}\int_{\sqrt{N/2}}^{\infty}\left(1+\sqrt{\frac{1}{2N}}\,t\right)\mathrm{e}^{-\frac{1}{2}t^2}\mathrm{d}t$$
$$= N\sqrt{1+\frac{\sigma^{*2}}{N}}\lambda\left[1-\Phi\left(\sqrt{\frac{N}{2}}\right)+\frac{1}{\sqrt{2\pi}}\frac{\mathrm{e}^{-N/4}}{\sqrt{2N}}\right]$$
$$\simeq \sqrt{1+\frac{\sigma^{*2}}{N}}\frac{\lambda}{\sqrt{2\pi}}\left(\sqrt{2}+\frac{1}{\sqrt{2}}\right)\sqrt{N}\mathrm{e}^{-N/4}. \qquad (3.238)$$

Equation (3.70) was used in order to get the last line. As one can see, the error term is of the order $\mathcal{R} = \mathcal{O}\left(\sqrt{N}\mathrm{e}^{-N/4}\right)$. It can be neglected for $N \to \infty$, anyway. However, we consider mainly the case of relatively small N, $N \gtrsim 30$. It is evident that the error term decreases very fast with increasing N because of the exponential term. For example, for $\sqrt{N}\mathrm{e}^{-N/4}$, $N = 30$, one gets $\mathcal{R} \approx 0.00303$. This behavior is caused by the $\mathrm{e}^{-t^2/2}$ term in the integrand of (3.232) and (3.237), respectively. The probability mass of the standard normal distribution is concentrated more than 99.99% in the interval $t \in [-4, 4]$ around zero. The contribution to the integral beyond these limits are therefore very small. As a result, one can restrict the integration interval to $t \in (-\sqrt{N/2}, \sqrt{N/2})$ without harm. With (3.235) one obtains from (3.236)

$$\varphi^*_{1,\lambda} = N - N\sqrt{1+\frac{\sigma^{*2}}{N}}\frac{\lambda}{\sqrt{2\pi}}\int_{-\sqrt{N/2}}^{\sqrt{N/2}}\left\{1+\frac{\sigma^*}{N}\frac{\sqrt{1+\frac{\sigma^{*2}}{2N}}}{1+\frac{\sigma^{*2}}{N}}\right.$$

$$\left.\times\left[t-\frac{\sigma^*}{2N}\frac{\sqrt{1+\frac{\sigma^{*2}}{2N}}}{1+\frac{\sigma^{*2}}{N}}t^2+\ldots\right]\right\}\mathrm{e}^{-\frac{1}{2}t^2}\left[1-\Phi(t)\right]^{\lambda-1}\mathrm{d}t + \mathcal{R}. \quad (3.239)$$

After successfully expanding the square root in (3.236), one can go the opposite way: the integration limits in (3.239) can be extended to $\pm\infty$. The argumentation and the estimation of the error term is analogous to (3.238), and therefore not repeated here. For further calculation, the substitution $t \to -t$ is used; taking the symmetry relation $1-\Phi(-t)=\Phi(t)$ into account, it follows that

$$\varphi^*_{1,\lambda} = N - N\sqrt{1+\frac{\sigma^{*2}}{N}}\frac{\lambda}{\sqrt{2\pi}}\int_{-\infty}^{\infty}\left\{1+\frac{\sigma^*}{N}\frac{\sqrt{1+\frac{\sigma^{*2}}{2N}}}{1+\frac{\sigma^{*2}}{N}}\right.$$

$$\left.\times\left[-t-\frac{\sigma^*}{2N}\frac{\sqrt{1+\frac{\sigma^{*2}}{2N}}}{1+\frac{\sigma^{*2}}{N}}t^2+\ldots\right]\right\}\mathrm{e}^{-\frac{1}{2}t^2}\left[\Phi(t)\right]^{\lambda-1}\mathrm{d}t + \tilde{\mathcal{R}}. \quad (3.240)$$

Considering (3.100) and referencing forward to (4.41), the integration yields

$$\varphi^*_{1,\lambda} = N\left(1-\sqrt{1+\frac{\sigma^{*2}}{N}}\right)$$

$$+\sigma^*\sqrt{\frac{1+\frac{\sigma^{*2}}{2N}}{1+\frac{\sigma^{*2}}{N}}}\left[c_{1,\lambda}+\frac{d^{(2)}_{1,\lambda}}{2}\frac{\sigma^*}{N}\frac{\sqrt{1+\frac{\sigma^{*2}}{2N}}}{1+\frac{\sigma^{*2}}{N}}+\ldots\right]. \quad (3.241)$$

For the most relevant cases it is sufficient to consider the $c_{1,\lambda}$ term in (3.241), since the $d^{(2)}_{1,\lambda}$ term is of order $\mathcal{O}(1/\sqrt{N})$ (see (3.234)). Consequently, we obtain the N *dependent progress rate formula*

$$\boxed{\varphi^*_{1,\lambda}(\sigma^*,N) = c_{1,\lambda}\sigma^*\sqrt{\frac{1+\frac{\sigma^{*2}}{2N}}{1+\frac{\sigma^{*2}}{N}}} - N\left(\sqrt{1+\frac{\sigma^{*2}}{N}}-1\right) + \mathcal{O}\left(\frac{1}{\sqrt{N}}\right).} \quad (3.242)$$

3.4.4 Comparison with Experiments

As already pointed out, it is difficult to determine the error term in the progress rate formulae exactly by theoretical analysis. Giving the order of the error is only conditionally helpful. For instance, the error term in (3.242) is of $\mathcal{O}(1/\sqrt{N})$, but in (3.101) only of $\mathcal{O}(1/N)$. However, the former is derived under the condition $\sigma^{*2}/N \ll 1$, whereas (3.242) does *not* have such a restriction. Particularly for large σ^* values, differences between the approximations are expected (see below).

The approximation quality of (3.242) and (3.101) can be evaluated best by plotting these formulae and comparing it with results obtained from ES experiments. The experiments are realized as follows: at a randomly chosen state $\mathbf{y}$ having the distance R to the optimum $\hat{\mathbf{y}}$, G "*one-generation experiments*" are performed. In other words, in any of these experiments, λ descendants $\tilde{\mathbf{y}}_l$ are generated from the state $\mathbf{y}$ by mutations with $\sigma = \text{const}$; the best descendant $\tilde{\mathbf{y}}_{1;\lambda}$ among them is determined. The resulting progress rate $(R - \tilde{R}_{1:\lambda})$ is averaged over G experiments, and the normalization (2.22) is applied thereafter. In that manner one obtains a (σ^*, φ^*) pair for a given σ.

The $(1,8)$-ES is investigated for $N = 100$ as an example, with $G = 20,000$ for each (σ^*, φ^*) pair. The results are shown in Fig. 3.18. The standard deviations of the experiments are smaller than the diameter of the data points. As can be seen in the figure, the simulation results conform with (3.242). The small deviation in the left figure can be decreased further by using the $d^{(2)}$ term in (3.241). However, since the formulae become less manageable, the more accurate formula will not be used in this book.

The fundamental difference between (3.242) and (3.101) lies in the σ^* asymptotic. For large σ^*, the progress rate φ^* is *not* a quadratic function of the (normalized) mutation strength σ^*, but a *linearly decreasing function*

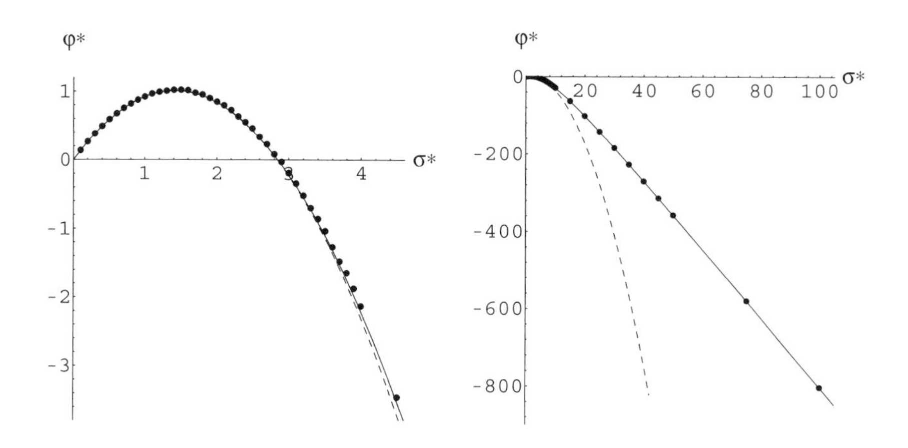

Fig. 3.18. The comparison of the theory (shown as curves) and simulations (as dots) for the $(1,8)$-ES on the sphere model ($N = 100$). The dashed curves belong to Eq. (3.101).

$$\varphi_{1,\lambda}^*(\sigma^*, N) \simeq N - \sigma^* \left(\sqrt{N} - \frac{c_{1,\lambda}}{\sqrt{2}} \right), \tag{3.243}$$

as one can easily see in Fig. 3.18 (right). However, both approximations yield comparable results for sufficiently small σ^* values. Equation (3.242) can be converted to (3.101) by expanding the square roots in (3.242) using a Taylor expansion with respect to σ^{*2}/N. Consequently, (3.101) can be used for $\sigma^{*2} \lessapprox N$.

3.4.5 A Normal Approximation for $p(r)$

The approximation (3.229) derived for the density $p(r)$ in Sect. 3.4.2 is *not* a Gaussian distribution. However, it is desirable to have $p(r)$ in the form of a normal distribution for the calculations in Chaps. 5 and 6, since the derivations will be simpler in this case. Such an approximation is possible by neglecting the $\mathrm{d}^{(2)}$ term in (3.241), i.e. by allowing a (maximal) error of $\mathcal{O}(1/\sqrt{N})$. In this case, the Taylor expansion (3.235) can be cut off after the linear term. When performing this series cut off in (3.231) one gets

$$r = \sqrt{R^2 + \sigma^2 N} \; + \; \sigma \sqrt{\frac{R^2 + \frac{\sigma^2 N}{2}}{R^2 + \sigma^2 N}} \; t \; + \; \mathcal{O}\left(\frac{\sigma \, t^2}{\sqrt{N}} \right). \tag{3.244}$$

After the back-substitution of t, one obtains the normal approximation of $p(r)$ as

$$p(r) \simeq \frac{1}{\sqrt{2\pi}\,\tilde{\sigma}} \exp\left[-\frac{1}{2} \left(\frac{r - \sqrt{R^2 + \sigma^2 N}}{\tilde{\sigma}} \right)^2 \right] \tag{3.245}$$

with

$$\tilde{\sigma} := \sigma \sqrt{\frac{R^2 + \frac{\sigma^2 N}{2}}{R^2 + \sigma^2 N}}. \tag{3.246}$$

The advantage of this approximation is that one can obtain the progress rate formula (3.242) immediately (without Taylor expansion) by inserting (3.245) in (3.230). The intermediate calculation steps are not shown here. Merely the treatment of the lower integration limit should be mentioned again. After the respective substitution, one gets for the lower limit

$$t_u = -\frac{1}{\tilde{\sigma}} \sqrt{R^2 + \sigma^2 N}. \tag{3.247}$$

The reasoning given for the extension of the lower limit t_u to $-\infty$ in Sect. 3.4.3 is also valid for the density (3.245). Therefore, as far as the comma strategy is concerned, $t_u = -\infty$ is used as the lower limit in all following chapters.

4. The $(1 \overset{+}{,} \lambda)$ Quality Gain

The analysis of the quality gain for the $(1 \overset{+}{,} \lambda)$ strategies can be carried out for a large number of fitness landscapes and mutation operators. This chapter will provide the theoretical background as well as first application examples. Firstly, an integral expression will be derived for $\overline{Q}_{1 \overset{+}{,} \lambda}$ that can be properly approximated. The central problem in the derivation of analytical $\overline{Q}_{1 \overset{+}{,} \lambda}$ formulae lies in the approximation of the distribution function of the mutation-induced fitness distribution $Q(\mathbf{z})$. The Hermite polynomials are used for this purpose. Fourier coefficients occur in these derivations. They are expressed by statistical parameters, such as mean value, standard deviation, and the cumulants of kth order. These parameters are calculated for some special fitness landscapes, and the respective analytical $\overline{Q}_{1 \overset{+}{,} \lambda}$ formulae are obtained, serving as application examples. These formulae are further compared with the results of ES simulations. The connection to the differential-geometric model is established. As a result, it is shown why two kinds of performance measures are necessary for the ES analysis.

4.1 The Theory of the $(1 \overset{+}{,} \lambda)$ Quality Gain

4.1.1 The $\overline{Q}_{1 \overset{+}{,} \lambda}$ Integral

The case of maximization is considered in this chapter w.l.o.g. The local quality function $Q(\mathbf{x}) = Q_{\mathbf{y}}(\mathbf{x})$ (see Eq. (2.6)) is assumed to have its maximum $\hat{Q}$ at $\hat{\mathbf{x}}$, i.e.

$$\hat{\mathbf{x}} = \arg\max_{\mathbf{x}} \left[Q(\mathbf{x})\right], \qquad \hat{Q} = \max_{\mathbf{x}} \left[Q(\mathbf{x})\right]. \tag{4.1}$$

Let the parent be at $\mathbf{x} = \mathbf{0}$.[1] Therefore, the offspring are defined by the states $\mathbf{x}_l = \mathbf{0} + \mathbf{z}_l$, and have the fitness values $Q_l = Q(\mathbf{z}_l)$. Note that $Q(\mathbf{0}) = 0$ (see Eq. (2.6)). For the case of fitness maximization, one obtains by specializing (2.7) the quality gain

$$\overline{Q}_{1 \overset{+}{,} \lambda} = \int_{Q=Q_u}^{Q=\hat{Q}} Q \, p_{\lambda:\lambda}(Q) \, \mathrm{d}Q, \tag{4.2}$$

[1] The notation and definitions related to the quality gain can be found in Sect. 2.1.1.

with

$$Q_u = \begin{cases} -\infty, & \text{for } (1,\lambda)\text{-ES} \\ 0, & \text{for } (1+\lambda)\text{-ES}. \end{cases} \tag{4.3}$$

The probability density for the $Q := Q_{1;\lambda} = Q_{\lambda:\lambda}$ value, i.e. the fitness value of the *best* descendant (and therefore the one with the largest fitness) is denoted hereby as $p_{\lambda:\lambda}(Q)$. If the $p(Q)$ density for a single offspring is known, $p_{\lambda:\lambda}$ can be determined as the λth order statistic (see Eq. (3.93))

$$p_{\lambda:\lambda}(Q) = \lambda p(Q) \left[P(Q)\right]^{\lambda-1}, \qquad P(Q) = \int_{-\infty}^{Q} p(Q') \, \mathrm{d}Q'. \tag{4.4}$$

Therefore, one obtains

$$\overline{Q}_{1\stackrel{+}{,}\lambda} = \lambda \int_{Q=Q_u}^{Q=\hat{Q}} Q \, p(Q) \left[P(Q)\right]^{\lambda-1} \mathrm{d}Q. \tag{4.5}$$

$p(Q)$ is needed to calculate (4.5). The density $p(Q)$ is obtained from the *mutation-induced* distribution of the function $Q = Q(\mathbf{z})$. Since the distribution $p(\mathbf{z})$ of the mutation vectors $\mathbf{z}$ is known, $p(Q(\mathbf{z}))$ can principally be calculated. However, such an analytical calculation has its limitations. Therefore, analytic approximations for $p(Q)$ are developed further below. As preparation, $p(Q)$ in (4.5) is transformed to a standard density function $p_z(z)$ with the variable

$$z := \frac{Q - M_Q}{S_Q} \tag{4.6}$$

with

$$M_Q := \mathrm{E}\{Q\} = \overline{Q} \qquad \text{and} \qquad S_Q := \sqrt{\mathrm{E}\left\{\left(Q - \overline{Q}\right)^2\right\}}. \tag{4.7}$$

Using $Q = M_Q + S_Q z$ in (4.5), one gets

$$\overline{Q}_{1\stackrel{+}{,}\lambda} = \lambda \int_{z=\frac{Q_u - M_Q}{S_Q}}^{z=\frac{\hat{Q}-M_Q}{S_Q}} (M_Q + S_Q z) \, p_z(z) \left[P_z(z)\right]^{\lambda-1} \mathrm{d}z, \tag{4.8}$$

considering $p(Q)\,\mathrm{d}Q = p_z(z)\,\mathrm{d}z$. After ordering terms one obtains

$$\overline{Q}_{1\stackrel{+}{,}\lambda} = \underbrace{M_Q \int_{z=\frac{Q_u - M_Q}{S_Q}}^{z=\frac{\hat{Q}-M_Q}{S_Q}} \frac{\mathrm{d}}{\mathrm{d}z} \left[P_z(z)\right]^{\lambda} \mathrm{d}z}_{= I^{(1)}_{1\stackrel{+}{,}\lambda}} + \underbrace{S_Q \lambda \int_{z=\frac{Q_u - M_Q}{S_Q}}^{z=\frac{\hat{Q}-M_Q}{S_Q}} z \, p_z(z) \left[P_z(z)\right]^{\lambda-1} \mathrm{d}z}_{= I^{(2)}_{1\stackrel{+}{,}\lambda}}. \tag{4.9}$$

The first integral in (4.9) gives

$$I^{(1)}_{1 \overset{+}{,} \lambda} = M_Q \left[1 - \left(P_z\left(\frac{Q_u - M_Q}{S_Q} \right) \right)^{\lambda} \right], \tag{4.10}$$

using the equality $P_z\Big((\hat{Q} - M_Q)/S_Q\Big) = P(\hat{Q}) = 1$. Since $\hat{Q}$ is the maximal possible Q value, $P(\hat{Q}) = \Pr(Q \le \hat{Q}) = 1$ is valid for the random variable Q. The lower limit Q_u depends on the selection strategy. For the comma strategy, one obtains by (4.3) $P_z((-\infty - M_Q)/S_Q) = P(-\infty) = 0$, thus

$$I^{(1)}_{1,\lambda} = M_Q, \tag{4.11}$$

and for the plus strategy

$$I^{(1)}_{1+\lambda} = M_Q \left[1 - \left(P_z\left(\frac{-M_Q}{S_Q} \right) \right)^{\lambda} \right]. \tag{4.12}$$

The brackets in (4.12) contain the success probability $P_{\mathrm{s}1+\lambda}$, introduced in Sect. 3.2

$$\boxed{P_{\mathrm{s}1+\lambda} = 1 - \left(P_z\left(\frac{-M_Q}{S_Q} \right) \right)^{\lambda},} \tag{4.13}$$

since the integration in $I^{(1)}_{1+\lambda}$, Eq. (4.9), is taken over Q values greater than (or equal to) the parental $Q = 0$.

The substitution $f = P_z(z)$, $\frac{\mathrm{d}f}{\mathrm{d}z} = p_z(z)$, and $z = P_z^{-1}(f)$ (inverse function to P, also known as the quantile function) is used to handle the second integral in (4.9)

$$I^{(2)}_{1 \overset{+}{,} \lambda} = S_Q \lambda \int_{f=f_u}^{f=1} P_z^{-1}(f)\, f^{\lambda-1}\, \mathrm{d}f. \tag{4.14}$$

The upper limit $f = 1$ follows from $P(\hat{Q}) = P_z\Big((\hat{Q} - M_Q)/S_Q\Big) = P_z(\hat{z}) = 1$. The lower limit is obtained by inserting the lower limit of (4.9) in $P_z(z)$

$$f_u = \begin{cases} 0, & \text{for } (1,\lambda)\text{-ES} \\ P_z\left(-\frac{M_Q}{S_Q}\right), & \text{for } (1+\lambda)\text{-ES.} \end{cases} \tag{4.15}$$

4.1.2 On the Approximation of $P_z(z)$

In order to calculate (4.10) and (4.14), one needs M_Q, S_Q, $P_z(z)$, and $P_z^{-1}(f)$. Due to Eq. (4.6) these quantities depend on the function $Q = Q(\mathbf{z})$ and the mutation density $p(\mathbf{z})$ (e.g., Eq. (1.12)). It is difficult to find expressions for $P_z(z)$ and $P_z^{-1}(f)$. Therefore, it is advisable to choose a general, approximative approach to these two expressions. Principally, this is *always* possible, provided that one can expand the density $p_z(z)$ or the distribution function

$P_z(z)$ of the standardized Q variables, $z = (Q - M_Q)/S_Q$, in a *uniformly convergent* function series. However, the vital question is on which (*orthogonal and complete*) function system one should decide. Of course, a fast convergence is desired, i.e. the series should yield even with a few terms a good approximation.

A potential hint comes from the investigation of the fitness function

$$Q(\mathbf{z}) = \sum_{i=1}^{N} \left(a_i z_i - q_i z_i^2\right) \tag{4.16}$$

with *isotropic* Gaussian mutations (1.11). Since the variates z_i are not correlated, we have $Q(\mathbf{z}) = \sum_{i=1}^{N} Q_i(z_i)$, with $Q_i(z_i) = a_i z_i - q_i z_i^2$, that is a sum of N *independent* and identically distributed random variables Q_i. For large N and moderate values of a_i and q_i the *central limit theorem*[2] is applicable

$$\Pr\left(\tfrac{Q - \mathrm{E}\{Q\}}{\mathrm{D}\{Q\}} \le z\right) \xrightarrow{N \to \infty} \Phi(z). \tag{4.17}$$

Therefore, the sum Q of the random variables Q_i is asymptotically normal-distributed. In the concrete case of (4.16) one immediately gets (cf. Appendix A.1, Table A.1)

$$\mathrm{E}\{Q\} = \sum_{i=1}^{N} \mathrm{E}\{Q_i\} = \sum_{i=1}^{N} -\sigma^2 q_i = -\sigma^2 \sum_{i=1}^{N} q_i \tag{4.18}$$

and

$$\begin{aligned}
\mathrm{D}^2\{Q\} &= \sum_{i=1}^{N} \mathrm{D}^2\{Q_i\} = \sum_{i=1}^{N} \left(\mathrm{E}\{Q_i^2\} - (\mathrm{E}\{Q_i\})^2\right) \\
&= \sum_{i=1}^{N} \left(\sigma^2 a_i^2 + 3\sigma^4 q_i^2 - \sigma^4 q_i^2\right) \\
\mathrm{D}\{Q\} &= \sigma \sqrt{\sum_i a_i^2 + 2\sigma^2 \sum_i q_i^2}.
\end{aligned} \tag{4.19}$$

As a result, one obtains for $P_z(z)$

$$\Pr\left(\tfrac{Q - M_Q}{S_Q} \le z\right) \xrightarrow{N \to \infty} \Phi(z), \tag{4.20}$$

using the notation

$$M_Q = \mathrm{E}\{Q\} = -\sigma^2 \sum_{i=1}^{N} q_i \tag{4.21}$$

and

[2] For this, a_i and q_i must be of the same order so that the variables Q_i fulfill the Lindeberg-Feller condition (Fisz, 1971, p. 245).

$$S_Q = \mathrm{D}\{Q\} = \sigma \sqrt{\sum_i a_i^2 + 2\sigma^2 \sum_i q_i^2}. \tag{4.22}$$

For the $N < \infty$ case, the relation $P_z(z) \simeq \Phi(z)$ is only approximate. Therefore, "correction terms" should be added to $\Phi(z)$, such that

$$P_z(z) = \Phi(z) + \sum_{m=1}^{\infty} c_m f_m(z) \tag{4.23}$$

and

$$p_z(z) = \frac{\mathrm{e}^{-\frac{1}{2}z^2}}{\sqrt{2\pi}} \left[1 + \sum_{m=1}^{\infty} a_m f_m(z)\right]. \tag{4.24}$$

The series approach to the $p_z(z)$ density in (4.24) proves suitable for this idea. The remaining question concerns the choice of $f_m(z)$. Because $p_z(z)$ should be expanded as an orthogonal function series, the desired function system $\{f_m\}$ should fulfill the orthogonality relation with respect to the scalar product weighted by $\mathrm{e}^{-z^2/2}$:

$$\int_{-\infty}^{\infty} \mathrm{e}^{-\frac{1}{2}z^2} f_m(z)\, f_n(z)\, \mathrm{d}z = a_m \delta_{mn}. \tag{4.25}$$

This is the case for the *Hermite polynomials* $\{\mathrm{He}_k(z)\}$. Thus, one obtains the function series

$$p_z(z) = \frac{\mathrm{e}^{-\frac{1}{2}z^2}}{\sqrt{2\pi}} \sum_{m=0}^{\infty} \frac{a_m}{m!} \mathrm{He}_m(z), \tag{4.26}$$

where the Fourier coefficients a_m still remain to be determined. The first four Hermite polynomials (see Appendix B.2) are

$$\begin{aligned} \mathrm{He}_0(z) &= 1, \qquad & \mathrm{He}_1(z) &= z, \\ \mathrm{He}_2(z) &= z^2 - 1, \qquad & \mathrm{He}_3(z) &= z^3 - 3z, \end{aligned} \tag{4.27}$$

yielding the approximation $p_z^{(3)}(z)$ to the $p_z(z)$ density

$$p_z^{(3)}(z) = \frac{\mathrm{e}^{-\frac{1}{2}z^2}}{\sqrt{2\pi}} \left[a_0 + a_1 z + \frac{a_2}{2!}\left(z^2 - 1\right) + \frac{a_3}{3!}\left(z^3 - 3z\right)\right]. \tag{4.28}$$

The coefficients a_m can be expressed by the moments of the exact $p_z(z)$ density

$$\overline{z^k} = \int_{-\infty}^{\infty} z^k p_z(z)\, \mathrm{d}z \stackrel{!}{=} \int_{-\infty}^{\infty} z^k p_z^{(3)}(z)\, \mathrm{d}z, \qquad k = 0, 1, 2, 3. \tag{4.29}$$

Because x is a standardized random variate (see Eq. (4.6)), one has $\overline{z^0} = 1$, $\overline{z^1} = 0$, and $\overline{z^2} = 1$. After evaluating the $p_z^{(3)}(z)$ integrals using (A.2), one obtains

$$a_0 = 1, \quad a_1 = 0, \quad a_2 = 0 \quad \text{and} \quad a_3 = \overline{z^3}. \tag{4.30}$$

Therefore, the $p_z(z)$ in Eq. (4.26) can be approximated by using the first four Hermite terms ($m \le 3$)

$$p_z(z) = \frac{e^{-\frac{1}{2}z^2}}{\sqrt{2\pi}} \left[1 + \frac{\overline{z^3}}{3!} \mathrm{He}_3(z) + \ldots \right]. \tag{4.31}$$

$P_z(z)$ is obtained by integrating (4.31)

$$P_z(z) = \Phi(z) - \frac{e^{-\frac{1}{2}z^2}}{\sqrt{2\pi}} \left[\frac{\overline{z^3}}{3!} \mathrm{He}_2(z) + \ldots \right], \tag{4.32}$$

which can be verified by differentiation.

Higher-order terms can be obtained using the idea of (4.29), see Appendix B.3.2 for details, where the relationship between a_m and the *cumulants* $\kappa_k\{z\}$ is derived. As shown in (B.59), one has

$$P_z(z) = \Phi(z) - \frac{e^{-\frac{1}{2}z^2}}{\sqrt{2\pi}} \left[\frac{\kappa_3}{3!} \mathrm{He}_2(z) + \left(\frac{\kappa_4}{4!} \mathrm{He}_3(z) + \frac{\kappa_3^2}{2 \cdot 3! \cdot 3!} \mathrm{He}_5(z) \right) \right.$$
$$\left. + \left(\frac{\kappa_5}{5!} \mathrm{He}_4(z) + \frac{\kappa_3 \, \kappa_4}{3! \cdot 4!} \mathrm{He}_6(z) + \frac{\kappa_3^3}{(3!)^4} \mathrm{He}_8(z) \right) + \ldots \right]. \tag{4.33}$$

The statistical parameters $\kappa_k = \kappa_k\{z\}$ in (4.33) are called cumulants or *semi-invariants*. They can be calculated from the $\overline{z^k}$ moments. The comparison of (4.32) and (4.33) gives $\kappa_3 = \overline{z^3}$. For the first κ_k, $k \ge 3$ one gets (see Appendix B.3.1, Eq. (B.34))

$$\kappa_3\{z\} = \overline{z^3}, \qquad \kappa_4\{z\} = \overline{z^4} - 3 \quad \text{and} \quad \kappa_5\{z\} = \overline{z^5} - 10\overline{z^3}. \tag{4.34}$$

The two groups in the parentheses in the brackets of (4.33) cover the terms of certain $\mathcal{O}((1/\sqrt{N})^{k-2})$ error orders (see Appendix B.3.2, p. 346), if Q is a sum of N independent identically distributed variables Q_i. In such a case, the κ_3 term is of error order $\mathcal{O}(1/\sqrt{N})$, the first parentheses group is of order $\mathcal{O}(1/N)$ and the second is of order $\mathcal{O}(1/N\sqrt{N})$.

4.1.3 On Approximating the Quantile Function $P_z^{-1}(f)$

A proper approximation of the quantile function $P_z^{-1}(f)$ is of the utmost importance for approximating the $I^{(2)}_{1 \stackrel{+}{,} \lambda}$ integral (4.14). The basic idea is to approximate $P_z^{-1}(f)$ also in a series of Hermite polynomials

$$P_z^{-1}(f) = \sum_{k=0}^{\infty} c_k \mathrm{He}_k\left(\Phi^{-1}(f)\right). \tag{4.35}$$

These polynomials have as argument $\Phi^{-1}(f)$ (the inverse function to the standard normal distribution Φ, see also Appendix C.1). This "strange" choice of argument is easily explained: to this end, let $\kappa_k \equiv 0$ ($k \ge 3$) in

(4.33) with the result $P_z(z) = \Phi(z)$. For this special case, (4.35) must read $P_z^{-1}(f) = c_1 \mathrm{He}_1(\Phi^{-1}(f))$, i.e. $\kappa_k \equiv 0$ $(k \geq 3)$ implies $c_1 = 1$ and $c_k = 0$ $(k \neq 1)$. Conversely, if $\kappa_k \not\equiv 0$, "corrections" in (4.35) are necessary, leading to values $c_k \neq 0$. The calculation of c_k according to the $\kappa_k\{z\}$ parameters is lengthy, as is sketched in Appendix B.4. Using (B.61, B.62, B.63) and (B.66), one obtains an expression in terms of (4.35). However, in order to ease the connection to the concepts of the ES analysis already developed, an expansion in exponents of $\Phi^{-1}(f)$ is preferred as an alternative to (4.35)

$$P_z^{-1}(f) = \sum_{k=0}^{\infty} b_k \left[\Phi^{-1}(f)\right]^k. \tag{4.36}$$

This expansion is obtained from Eq. (B.68) in Appendix B.4. One gets

$$\begin{aligned} P_z^{-1}(f) = & -\frac{\kappa_3}{6} + \left[1 + \left(\frac{5}{36}\kappa_3^2 - \frac{\kappa_4}{8}\right)\right] \Phi^{-1}(f) \\ & + \frac{\kappa_3}{6}\left[\Phi^{-1}(f)\right]^2 + \left(\frac{\kappa_4}{24} - \frac{\kappa_3^2}{18}\right)\left[\Phi^{-1}(f)\right]^3 + \ldots . \end{aligned} \tag{4.37}$$

The corresponding b_k coefficients are obtained

$$\begin{aligned} b_0 &= -\frac{\kappa_3}{6} + \ldots, & b_1 &= 1 + \left(\frac{5}{36}\kappa_3^2 - \frac{\kappa_4}{8}\right) + \ldots, \\ b_2 &= \frac{\kappa_3}{6} + \ldots, & b_3 &= \left(\frac{\kappa_4}{24} - \frac{\kappa_3^2}{18}\right) + \ldots . \end{aligned} \tag{4.38}$$

Further terms can be calculated using (B.67).

4.1.4 The $\overline{Q}_{1,\lambda}$ Formula

Considering (4.11, 4.14) and (4.36), the quality gain formula (4.9) becomes

$$\overline{Q}_{1,\lambda} = M_Q + S_Q \sum_{k=0}^{\infty} b_k \lambda \int_{f=0}^{f=1} \left[\Phi^{-1}(f)\right]^k f^{\lambda-1} \mathrm{d}f. \tag{4.39}$$

The substitutions $\Phi^{-1}(f) = t$, $f = \Phi(t)$, and $\mathrm{d}f = \frac{1}{\sqrt{2\pi}} \mathrm{e}^{-t^2/2} \mathrm{d}t$ yield

$$\overline{Q}_{1,\lambda} = M_Q + S_Q b_0 + S_Q \sum_{k=1}^{\infty} b_k \frac{\lambda}{\sqrt{2\pi}} \int_{-\infty}^{\infty} t^k \mathrm{e}^{-\frac{1}{2}t^2} \left[\Phi(t)\right]^{\lambda-1} \mathrm{d}t. \tag{4.40}$$

Introducing the *higher-order progress coefficients*

$$\boxed{d_{1,\lambda}^{(k)} := \frac{\lambda}{\sqrt{2\pi}} \int_{-\infty}^{\infty} t^k \, \mathrm{e}^{-\frac{1}{2}t^2} \left[\Phi(t)\right]^{\lambda-1} \mathrm{d}t} \tag{4.41}$$

and considering

$$\boxed{d_{1,\lambda}^{(1)} = c_{1,\lambda}} \tag{4.42}$$

(see also Sect. 3.2.3.1), the quality gain of the $(1, \lambda)$-ES follows

$$\overline{Q}_{1,\lambda} = M_Q + S_Q b_0 + c_{1,\lambda} b_1 S_Q + S_Q \sum_{k=2}^{\infty} b_k d_{1,\lambda}^{(k)}. \tag{4.43}$$

Inserting the b_k coefficients from (4.38) one obtains

$$\boxed{\begin{aligned} \overline{Q}_{1,\lambda} = M_Q \; - \; S_Q \frac{\kappa_3}{6} \; + \; c_{1,\lambda} S_Q \left[1 + \left(\frac{5}{36} \kappa_3^2 - \frac{\kappa_4}{8} \right) \right] \\ + \; d_{1,\lambda}^{(2)} S_Q \frac{\kappa_3}{6} \; + \; d_{1,\lambda}^{(3)} S_Q \left(\frac{\kappa_4}{24} - \frac{\kappa_3^2}{18} \right) \; + \; \ldots . \end{aligned}} \tag{4.44}$$

The approximate $\overline{Q}_{1,\lambda}$ formula is sufficient for most practical purposes. The progress coefficients appearing in (4.44) and (4.43) are tabled in Appendix D.2. The remaining quantities (M_Q, S_Q, and κ_k) depend on the fitness function and the mutation distribution. Sample cases are handled in Sect. 4.2.

4.1.5 The $\overline{Q}_{1+\lambda}$ Formula

In case of the $(1 + \lambda)$-ES, (4.9) yields with (4.12 – 4.15) and (4.36)

$$\overline{Q}_{1+\lambda} = M_Q P_{s1+\lambda} + S_Q \sum_{k=0}^{\infty} b_k \lambda \int_{f=P_z(-M_Q/S_Q)}^{f=1} \left[\Phi^{-1}(f) \right]^k f^{\lambda-1} \mathrm{d}f. \tag{4.45}$$

Also inserting $\Phi^{-1}(f) = t$ in this case, one gets

$$\begin{aligned} \overline{Q}_{1+\lambda} &= (M_Q + S_Q b_0) \, P_{s1+\lambda} \\ &+ S_Q \sum_{k=1}^{\infty} b_k \frac{\lambda}{\sqrt{2\pi}} \int_{t=\Phi^{-1}[P_z(-M_Q/S_Q)]}^{t=\infty} t^k \mathrm{e}^{-\frac{1}{2}t^2} \left[\Phi(t) \right]^{\lambda-1} \mathrm{d}t. \end{aligned} \tag{4.46}$$

After introducing the kth order *progress function* $d_{1+\lambda}^{(k)}(x)$

$$\boxed{d_{1+\lambda}^{(k)}(x) := \frac{\lambda}{\sqrt{2\pi}} \int_{t=x}^{t=\infty} t^k \, \mathrm{e}^{-\frac{1}{2}t^2} \left[\Phi(t) \right]^{\lambda-1} \mathrm{d}t,} \tag{4.47}$$

one obtains for the quality gain

$$\begin{aligned} \overline{Q}_{1+\lambda} &= (M_Q + S_Q b_0) \left[1 - \left(P_z \left(-\frac{M_Q}{S_Q} \right) \right)^{\lambda} \right] \\ &+ S_Q \sum_{k=1}^{\infty} b_k d_{1+\lambda}^{(k)} \left(\Phi^{-1} \left[P_z \left(-\frac{M_Q}{S_Q} \right) \right] \right). \end{aligned} \tag{4.48}$$

P_z is given by Eq. (4.33)

$$
P_z\left(-\frac{M_Q}{S_Q}\right) = 1 - \Phi\left(\frac{M_Q}{S_Q}\right) - \frac{\exp\left[-\frac{1}{2}\left(\frac{M_Q}{S_Q}\right)^2\right]}{\sqrt{2\pi}} \left\{ \frac{\kappa_3}{3!} \mathrm{He}_2\left(\frac{M_Q}{S_Q}\right) \right.
$$

$$
- \left[\frac{\kappa_4}{4!} \mathrm{He}_3\left(\frac{M_Q}{S_Q}\right) + \frac{\kappa_3^2}{2 \cdot 3! \cdot 3!} \mathrm{He}_5\left(\frac{M_Q}{S_Q}\right)\right]
$$

$$
\left. + \left[\frac{\kappa_5}{5!} \mathrm{He}_4\left(\frac{M_Q}{S_Q}\right) + \frac{\kappa_3 \kappa_4}{3! \cdot 4!} \mathrm{He}_6\left(\frac{M_Q}{S_Q}\right) + \frac{\kappa_3^3}{(3!)^4} \mathrm{He}_8\left(\frac{M_Q}{S_Q}\right)\right] + \ldots \right\}. \tag{4.49}
$$

If only the b_k coefficients given in (4.38) are considered, one obtains the approximate $\overline{Q}_{1+\lambda}$ formula as

$$
\boxed{
\begin{aligned}
\overline{Q}_{1+\lambda} = {} & \left(M_Q - S_Q \frac{\kappa_3}{6}\right) \left[1 - \left(P_z\left(-\frac{M_Q}{S_Q}\right)\right)^\lambda\right] \\
& + S_Q \left[1 + \left(\frac{5}{36}\kappa_3^2 - \frac{\kappa_4}{8}\right)\right] d_{1+\lambda}^{(1)}\left(\Phi^{-1}\left[P_z\left(-\frac{M_Q}{S_Q}\right)\right]\right) \\
& + S_Q \frac{\kappa_3}{6}\, d_{1+\lambda}^{(2)}\left(\Phi^{-1}\left[P_z\left(-\frac{M_Q}{S_Q}\right)\right]\right) \\
& + S_Q \left(\frac{\kappa_4}{24} - \frac{\kappa_3^2}{18}\right) d_{1+\lambda}^{(3)}\left(\Phi^{-1}\left[P_z\left(-\frac{M_Q}{S_Q}\right)\right]\right) + \ldots.
\end{aligned}
} \tag{4.50}
$$

The corresponding $P_z(-M_Q/S_Q)$ approximation reads

$$
\boxed{
\begin{aligned}
P_z\left(-\frac{M_Q}{S_Q}\right) = {} & 1 - \Phi\left(\frac{M_Q}{S_Q}\right) - \frac{\exp\left[-\frac{1}{2}\left(\frac{M_Q}{S_Q}\right)^2\right]}{\sqrt{2\pi}} \left\{ \frac{\kappa_3}{3!} \mathrm{He}_2\left(\frac{M_Q}{S_Q}\right) \right. \\
& \left. - \left[\frac{\kappa_4}{4!} \mathrm{He}_3\left(\frac{M_Q}{S_Q}\right) + \frac{\kappa_3^2}{2 \cdot 3! \cdot 3!} \mathrm{He}_5\left(\frac{M_Q}{S_Q}\right)\right] + \ldots \right\}.
\end{aligned}
} \tag{4.51}
$$

The $\overline{Q}_{1+\lambda}$ formula in (4.48) and the approximation (4.50) contain progress functions defined by the integral (4.47). The limit of the analytically closed representation of (4.50) is reached at this point: analytical expressions for $d_{1+\lambda}^{(k)}(x)$ (Eq. 4.47) could only be derived for $\lambda = 1$ and $\lambda = 2$ up to now. For $\lambda = 1$, one obtains considering (A.13) and (A.17)

$$
d_{1+1}^{(2)}(x) = 1 - \Phi(x) + \frac{x}{\sqrt{2\pi}} \mathrm{e}^{-\frac{1}{2}x^2}, \qquad d_{1+1}^{(3)}(x) = \frac{2 + x^2}{\sqrt{2\pi}} \mathrm{e}^{-\frac{1}{2}x^2}. \tag{4.52}
$$

The first-order progress function $d_{1+1}^{(1)}(x)$ was already used in Sect. 3.2.4.1; also the $d_{1+2}^{(1)}(x)$ function can be found there in Eq. (3.120b). The functions $d_{1+2}^{(2)}(x)$ and $d_{1+2}^{(3)}(x)$ are obtained by using (A.21) and (A.25), and also (A.26)

$$d_{1+2}^{(2)}(x) = 1 - [\Phi(x)]^2 + \frac{2x}{\sqrt{2\pi}}\,\Phi(x)\,\mathrm{e}^{-\frac{1}{2}x^2} + \frac{1}{2\pi}\,\mathrm{e}^{-x^2}, \qquad (4.53)$$

$$d_{1+2}^{(3)}(x) = \frac{1}{\sqrt{\pi}}\frac{5}{2}\left(1 - \Phi(\sqrt{2}\,x)\right) + \frac{2}{\sqrt{2\pi}}\,(2+x^2)\,\Phi(x)\,\mathrm{e}^{-\frac{1}{2}x^2} + \frac{x}{2\pi}\,\mathrm{e}^{-x^2}.$$

For $\lambda > 2$, numerical integration methods can be applied to compute the integral in (4.47). See Sect. 3.2.4.3 for a discussion on the first-order progress function $d_{1,\lambda}^{(1)}(x)$.

As in the case of the $(1,\lambda)$-ES, the statistical parameters M_Q, S_Q, and κ_k remain to be determined, which is done in the following section.

4.2 Fitness Models and Mutation Operators

4.2.1 The General Quadratic Model and Correlated Mutations

For practical considerations, the statistical parameters M_Q, S_Q, and $\kappa_k\{z\}$ should be calculated for concrete model classes. In this section, the general quadratic fitness model (2.35) $Q(\mathbf{x}) = \mathbf{a}^{\mathrm{T}}\mathbf{x} - \mathbf{x}^{\mathrm{T}}\mathbf{Q}\mathbf{x}$ will be considered. It is assumed that $\mathbf{Q} = \mathbf{Q}^{\mathrm{T}}$ holds.[3] Further conditions on $\mathbf{Q}$ are not required. (Particularly, the positive-definiteness of $\mathbf{Q}$ is *not* a prerequisite.)

The most general form of Gaussian mutations is described by the density (1.12), which allows correlations between the components of the vector $\mathbf{z}$. In the local coordinate system we use $\mathbf{x}$ instead of $\mathbf{z}$ in order to avoid confusion with the standardized Q variable z. The density reads

$$p(\mathbf{x}) = \frac{1}{\left(\sqrt{2\pi}\right)^N}\,\frac{1}{\sqrt{\det[\mathbf{C}]}}\,\exp\left(-\frac{1}{2}\,\mathbf{x}^{\mathrm{T}}\mathbf{C}^{-1}\mathbf{x}\right). \qquad (4.54)$$

Calculating M_Q. The statistical parameter M_Q is the expected value of Q for a *single* mutation $\mathbf{x}$, thus $M_Q = \mathrm{E}\{Q(\mathbf{x})\} = \overline{Q(\mathbf{x})}$ (see (4.7)). Considering the results in Appendix C.3 for the product moments of correlated mutations (Eqs. (C.28) and (C.31)), one obtains

$$M_Q = \mathrm{E}\{\mathbf{a}^{\mathrm{T}}\mathbf{x} - \mathbf{x}^{\mathrm{T}}\mathbf{Q}\mathbf{x}\} = \mathrm{E}\left\{\sum_{i=1}^{N} a_i x_i - \sum_{i=1}^{N}\sum_{j=1}^{N} Q_{ij} x_i x_j\right\}$$

$$M_Q = \sum_{i=1}^{N} a_i\,\overline{x_i} - \sum_{i=1}^{N}\sum_{j=1}^{N} Q_{ij}\,\overline{x_i x_j} = -\sum_{i=1}^{N}\sum_{j=1}^{N} Q_{ij} C_{ij} = -\sum_{i=1}^{N}\sum_{j=1}^{N} Q_{ij} C_{ji}$$

$$M_Q = -\sum_{i=1}^{N} (\mathbf{QC})_{ii} = -\mathrm{Tr}[\mathbf{QC}]. \qquad (4.55)$$

Note, $\mathrm{Tr}[\mathbf{M}]$ denotes the trace $\mathrm{Tr}[\mathbf{M}] = \sum_i (\mathbf{M})_{ii}$ of matrix $\mathbf{M}$.

[3] This condition is not restrictive since the asymmetric part $\mathbf{M}^- = -(\mathbf{M})^{\mathrm{T}}$ of a matrix $\mathbf{M} = \mathbf{M}^+ + \mathbf{M}^-$ (with $\mathbf{M}^+ = (\mathbf{M}^+)^{\mathrm{T}}$) does not contribute to the bilinear form $\mathbf{x}^{\mathrm{T}}\mathbf{M}\mathbf{x}$. One has $\mathbf{x}^{\mathrm{T}}\mathbf{M}\mathbf{x} = \mathbf{x}^{\mathrm{T}}\mathbf{M}^+\mathbf{x}$.

Calculating S_Q. According to (4.7), we have $S_Q^2 = \overline{(Q(\mathbf{x}))^2} - \overline{Q(\mathbf{x})}^2$, i.e. the expected value of Q^2 shall be calculated next. Since

$$Q^2 = \left(\sum_i a_i x_i - \sum_{ij} Q_{ij} x_i x_j\right)\left(\sum_k a_k x_k - \sum_{kl} Q_{kl} x_k x_l\right)$$

$$Q^2 = \sum_{ik} a_i a_k x_i x_k - \sum_{ikl} a_i Q_{kl} x_i x_k x_l - \sum_{ijk} Q_{ij} a_k x_i x_j x_k + \sum_{ijkl} Q_{ij} Q_{kl} x_i x_j x_k x_l, \tag{4.56}$$

it follows using (C.31, C.33) and (C.35)

$$\overline{Q^2} = \sum_{ik} a_i a_k C_{ik} + \sum_{ijkl} Q_{ij} Q_{kl} \left(C_{il} C_{jk} + C_{jl} C_{ik} + C_{kl} C_{ij}\right). \tag{4.57}$$

Considering the symmetry of $\mathbf{C}$ and $\mathbf{Q}$, this becomes

$$\overline{Q^2} = \sum_{ik} a_i C_{ik} a_k + 2\sum_{ik}\left(\sum_j Q_{ij} C_{jk}\right)\left(\sum_l Q_{kl} C_{li}\right) + \left(\sum_{ij} Q_{ij} C_{ji}\right)^2.$$

This result can be written compactly in matrix notation as

$$\overline{Q^2} = \mathbf{a}^{\mathrm{T}} \mathbf{C} \mathbf{a} + 2\mathrm{Tr}\left[(\mathbf{QC})^2\right] + \left(\mathrm{Tr}[\mathbf{QC}]\right)^2. \tag{4.58}$$

Therefore, using (4.55) and (4.58), the standard deviation S_Q of the random variable Q becomes

$$S_Q = \sqrt{\mathbf{a}^{\mathrm{T}} \mathbf{C} \mathbf{a} + 2\mathrm{Tr}[(\mathbf{QC})^2]}. \tag{4.59}$$

Calculating $\kappa_3\{z\}$. The quantity $\kappa_3\{z\}$ is defined in (4.34) as the third moment of the standard variable z (given in (4.6)),

$$\kappa_3\{z\} = \overline{\left(\frac{Q - \overline{Q}}{S_Q}\right)^3} = \frac{1}{S_Q^3}\left(\overline{Q^3} - 3\overline{Q^2}\,\overline{Q} + 2\overline{Q}^3\right). \tag{4.60}$$

The term Q^3 will be calculated first. Considering the symmetry of $\mathbf{Q}$, this reads

$$Q^3 = \sum_{ijk} x_i x_j x_k \left(a_i - \sum_l Q_{il} x_l\right)\left(a_j - \sum_m Q_{jm} x_m\right)\left(a_k - \sum_n Q_{kn} x_n\right)$$

$$Q^3 = \sum_{ijk} a_i a_j a_k x_i x_j x_k - 3\sum_{ijkl} a_i a_j Q_{kl} x_i x_j x_k x_l + 3\sum_{ijklm} a_i Q_{jk} Q_{lm} x_i x_j x_k x_l x_m - \sum_{ijklmn} Q_{ij} Q_{kl} Q_{mn} x_i x_j x_k x_l x_m x_n.$$

According to (C.38), the product moments with odd numbers of factors vanish. Therefore, $\overline{Q^3}$ reduces to

$$\overline{Q^3} = -3 \sum_{ijkl} a_i a_j Q_{kl} \left(C_{il} C_{jk} + C_{jl} C_{ik} + C_{kl} C_{ij} \right) - \sum_{ijklmn} Q_{ij} Q_{kl} Q_{mn} \overline{x_i x_j x_k x_l x_m x_n}. \tag{4.61}$$

Equation (C.35) was inserted here, but not (C.36), since the formula would be relatively long. After collecting terms one gets

$$\overline{Q^3} = -3\mathrm{Tr}[\mathbf{QC}]\,\mathbf{a}^\mathrm{T}\mathbf{Ca} - 6\mathbf{a}^\mathrm{T}\mathbf{CQCa} - 8\mathrm{Tr}\big[(\mathbf{QC})^3\big] - 6\mathrm{Tr}[\mathbf{QC}]\,\mathrm{Tr}\big[(\mathbf{QC})^2\big] - (\mathrm{Tr}[\mathbf{QC}])^3. \tag{4.62}$$

Together with (4.58, 4.55) and (4.59), (4.60) becomes

$$\kappa_3\{z\} = -\frac{6\mathbf{a}^\mathrm{T}\mathbf{CQCa} + 8\mathrm{Tr}\big[(\mathbf{QC})^3\big]}{\left(\sqrt{\mathbf{a}^\mathrm{T}\mathbf{Ca} + 2\mathrm{Tr}[(\mathbf{QC})^2]}\right)^3}. \tag{4.63}$$

Calculating $\kappa_4\{z\}$. The related calculations are simple, but very long. Therefore, the whole derivation is not presented here. Instead, the most important steps given below can be compared with a recalculation left to the reader, if desired. Starting with (4.34), κ_4 reads

$$\kappa_4\{z\} = \overline{\left(\frac{Q - \overline{Q}}{S_Q}\right)^4} - 3 = \frac{1}{S_Q^4}\left[\underbrace{\left(\overline{Q^4} - 4\overline{Q^3}\,\overline{Q} + 6\overline{Q^2}\,\overline{Q}^2 - 3\overline{Q}^4\right)}_{=\overline{(Q-\overline{Q})^4}} - 3S_Q^4\right]. \tag{4.64}$$

Because of (C.38), $\overline{Q^4}$ only contains the product moments with an even number of factors

$$\overline{Q^4} = \overline{\left[\sum_i x_i \left(a_i - \sum_j Q_{ij} x_j\right)\right]^4}$$

$$\overline{Q^4} = \sum_{ijkl} a_i a_j a_k a_l \,\overline{x_i x_j x_k x_l} + 6 \sum_{ijklmn} a_i a_j Q_{kl} Q_{mn}\, \overline{x_i x_j x_k x_l x_m x_n} + \sum_{ijklmnop} Q_{ij} Q_{kl} Q_{mn} Q_{op}\, \overline{x_i x_j x_k x_l x_m x_n x_o x_p}. \tag{4.65}$$

After inserting (C.38) and very long calculations, one gets

$$\begin{aligned}\overline{Q^4} = {} & 3(\mathbf{a}^\mathrm{T}\mathbf{Ca})^2 + 48\mathbf{a}^\mathrm{T}\mathbf{CQCQCa} + 24\mathrm{Tr}[\mathbf{QC}]\,\mathbf{a}^\mathrm{T}\mathbf{CQCa} \\ & + 12\mathrm{Tr}\big[(\mathbf{QC})^2\big]\,\mathbf{a}^\mathrm{T}\mathbf{Ca} + 6\,(\mathrm{Tr}[\mathbf{QC}])^2\,\mathbf{a}^\mathrm{T}\mathbf{Ca} + 48\mathrm{Tr}\big[(\mathbf{QC})^4\big] \\ & + 32\mathrm{Tr}\big[(\mathbf{QC})^3\big]\,\mathrm{Tr}[\mathbf{QC}] + 12\left(\mathrm{Tr}\big[(\mathbf{QC})^2\big]\right)^2 + (\mathrm{Tr}[\mathbf{QC}])^4 \\ & + 12\mathrm{Tr}\big[(\mathbf{QC})^2\big]\,(\mathrm{Tr}[\mathbf{QC}])^2.\end{aligned} \tag{4.66}$$

Therefore, $\overline{(Q-\overline{Q})^4}$ reads

$$\overline{(Q-\overline{Q})^4} = 3(\mathbf{a}^{\mathrm{T}}\mathbf{C}\mathbf{a})^2 + 48\mathbf{a}^{\mathrm{T}}\mathbf{CQCQCa} + 12\mathrm{Tr}\big[(\mathbf{QC})^2\big]\,\mathbf{a}^{\mathrm{T}}\mathbf{C}\mathbf{a} \\ + 48\mathrm{Tr}\big[(\mathbf{QC})^4\big] + 12\left(\mathrm{Tr}\big[(\mathbf{QC})^2\big]\right)^2. \quad (4.67)$$

Finally, the insertion of (4.67) and (4.59) into (4.64) yields

$$\kappa_4\{z\} = 48\,\frac{\mathbf{a}^{\mathrm{T}}\mathbf{CQCQCa} + \mathrm{Tr}\big[(\mathbf{QC})^4\big]}{\left(\mathbf{a}^{\mathrm{T}}\mathbf{C}\mathbf{a} + 2\mathrm{Tr}[(\mathbf{QC})^2]\right)^2}. \quad (4.68)$$

This is a surprisingly simple result. One may assume that a simpler, yet to be determined, shorter derivation also exists.

4.2.2 The Special Case of Isotropic Gaussian Mutations

The case of a fully occupied $\mathbf{Q}$ *matrix.* The formulae derived for the statistical parameter in the previous subsection can also be used in the case of isotropic Gaussian mutations. In this case, the density function is given by (1.11); therefore, we get the covariance matrix $\mathbf{C}$

$$\text{ISOTROPIC MUTATIONS:} \qquad \mathbf{C} = \sigma^2\mathbf{E}. \quad (4.69)$$

Hence, M_Q and S_Q result from (4.55) and (4.59)

$$M_Q = -\sigma^2\mathrm{Tr}[\mathbf{Q}] \qquad \text{and} \qquad S_Q = \sigma\sqrt{||\mathbf{a}||^2 + 2\sigma^2\mathrm{Tr}[\mathbf{Q}^2]}. \quad (4.70)$$

The cumulants of third- and fourth-order, Eqs. (4.63) and (4.68), yield

$$\kappa_3 = -\sigma\,\frac{6\mathbf{a}^{\mathrm{T}}\mathbf{Q}\mathbf{a} + 8\sigma^2\mathrm{Tr}[\mathbf{Q}^3]}{\left(\sqrt{||\mathbf{a}||^2 + 2\sigma^2\mathrm{Tr}[\mathbf{Q}^2]}\right)^3} \quad (4.71)$$

and

$$\kappa_4 = 48\sigma^2\,\frac{||\mathbf{Q}\mathbf{a}||^2 + \sigma^2\mathrm{Tr}[\mathbf{Q}^4]}{\left(||\mathbf{a}||^2 + 2\sigma^2\mathrm{Tr}[\mathbf{Q}^2]\right)^2}. \quad (4.72)$$

The case with a diagonal $\mathbf{Q}$ *matrix.* If only the diagonal of the $\mathbf{Q}$ matrix is occupied, that is

$$(\mathbf{Q})_{ij} = q_i\delta_{ij}, \quad (4.73)$$

the formulae (4.70), (4.71), and (4.71) simplify further. One obtains

$$M_Q = -\sigma^2\sum_{i=1}^{N} q_i, \qquad S_Q = \sigma\sqrt{\sum_{i=1}^{N} a_i^2 + 2\sigma^2\sum_{i=1}^{N} q_i^2}, \quad (4.74)$$

$$\kappa_3 = -\sigma\,\frac{6\sum_i a_i^2 q_i + 8\sigma^2\sum_i q_i^3}{\left(\sqrt{\sum_{i=1}^N a_i^2 + 2\sigma^2\sum_{i=1}^N q_i^2}\right)^3}, \quad (4.75)$$

and

$$\kappa_4 = 48\sigma^2 \frac{\sum_i a_i^2 q_i^2 + \sigma^2 \sum_i q_i^4}{\left(\sum_{i=1}^N a_i^2 + 2\sigma^2 \sum_{i=1}^N q_i^2\right)^2}. \tag{4.76}$$

The $\overline{Q}_{1 \overset{+}{,} \lambda}$ formulae (4.44) and (4.50), which use (4.74) and (4.75) for their parameters, are compared with experimental results in Sect. 4.3.2.

4.2.3 Examples of Non-Quadratic Fitness Functions

The advantage of the quality gain formula is the fact that only the knowledge of the statistical parameters M_Q, S_Q, and $\kappa_k\{(Q(\mathbf{x}) - M_Q)/S_Q\}$ is necessary to make quantitative predictions which (usually) can be expressed in an analytically closed manner. Therefore, quality gain formulae are not limited to the sphere model or Gaussian mutations. In principle, the method works for any mutation distribution and for any fitness function, provided that the statistical parameter of the mutation-induced Q distribution can be determined. The spectrum of the analyzable fitness models and mutation types should be very large and should contain cases such as mixed-integer and combinatorial optimization problems. Until now this method has been applied in examples with biquadratic fitness and isotropic Gaussian mutations (Beyer, 1994a), and for the extreme case of the optimization in bitstrings – the `OneMax` function (Beyer, 1995a).

4.2.3.1 Biquadratic Fitness with Isotropic Gaussian Mutations. The fitness function

$$Q_{\mathbf{y}}(\mathbf{x}) = \sum_{i=1}^{N} Q_i(x_i), \qquad Q_i(x_i) = a_i x_i - c_i x_i^4 \tag{4.77}$$

introduced in Sect. 2.2.4, Eq. (2.37), with isotropic Gaussian mutations

$$p(x_i) = \frac{1}{\sqrt{2\pi}\,\sigma} \exp\left(-\frac{1}{2}\left(\frac{x_i}{\sigma}\right)^2\right) \tag{4.78}$$

will be investigated. Since Q_i are statistically independent, the calculation simplifies considerably. For the first three moments of Q_i, one obtains using Table A.1

$$\overline{Q_i} = -3\,\sigma^4 c_i, \qquad \overline{Q_i^2} = \sigma^2 a_i^2 + 105\,\sigma^8 c_i^2,$$
$$\overline{Q_i^3} = -45\,\sigma^6 a_i^2 c_i - 10\,395\,\sigma^{12} c_i^3, \tag{4.79}$$

so that, according to (4.7)

$$M_Q = \sum_{i=1}^{N} \overline{Q_i} = -3\,\sigma^4 \sum_{i=1}^{N} c_i \tag{4.80}$$

and

$$S_Q = \sqrt{\sum_{i=1}^{N} \left(\overline{Q_i^2} - \overline{Q_i}^2 \right)} = \sigma \sqrt{\sum_{i=1}^{N} a_i^2 + 96\,\sigma^6 \sum_{i=1}^{N} c_i^2} \tag{4.81}$$

is obtained. The cumulants $\kappa_k\{z\}$ of z, Eq. (4.6), are obtained considering the addition property (B.27) as well as (B.25)

$$\kappa_k\{z\} = \kappa_k \left\{ \frac{\sum_i (Q_i - \overline{Q_i})}{S_Q} \right\} = \frac{1}{S_Q^k} \sum_{i=1}^{N} \kappa_k\{Q_i - \overline{Q_i}\}\,. \tag{4.82}$$

Using (4.34), the case $k = 3$ yields

$$\kappa_3\{z\} = \sum_i \overline{(Q_i - \overline{Q_i})^3} / S_Q^3. \tag{4.83}$$

Inserting (4.79) in $\overline{(Q_i - \overline{Q_i})^3} = \overline{Q_i^3} - 3\overline{Q_i^2}\,\overline{Q_i} + 2\overline{Q_i}^3$, one gets

$$\overline{(Q_i - \overline{Q_i})^3} = -36\sigma^6 (a_i^2 c_i + 264 c_i^3), \tag{4.84}$$

and for κ_3, considering (4.81) and (4.82)

$$\kappa_3\{z\} = -36\sigma^3 \frac{\sum_i a_i^2 c_i + 264 \sum_i c_i^3}{\left(\sqrt{\sum_i a_i^2 + 96\,\sigma^6 \sum_i c_i^2} \right)^3}. \tag{4.85}$$

Since the derivation is clear and simple, the calculation of $\kappa_4\{z\}$ is omitted here. The biquadratic fitness function (4.77) will be investigated experimentally in Sect. 4.3.2, and the results of the quality gain analysis based on the statistical parameters (4.80, 4.81) will be discussed.

4.2.3.2 The Counting Ones Function `OneMax`. In the derivations of (4.44) and (4.50) it is assumed that the mutation-induced Q distribution $P_z((Q - M_Q)/S_Q)$ as well as the quantile function $P_z^{-1}(f) = z$ are smooth functions. Therefore, the quality gain analysis of the functions which map bitstrings $\mathbf{b}$ of length ℓ to the range of integers is an extreme application of the method developed.

As an example, the counting ones function `OneMax` with the fitness

$$F(\mathbf{b}) = \sum_{i=1}^{\ell} b_i, \qquad b_i \in \{0, 1\} \tag{4.86}$$

will be considered. Its local quality function (2.39)

$$Q_{\mathbf{y}}(\mathbf{x}) = F(\mathbf{x}) - F_0 = \sum_{i=1}^{\ell} x_i - \sum_{i=1}^{\ell} y_i, \qquad x_i, y_i \in \{0, 1\} \tag{4.87}$$

takes only discrete values, which leads to discontinuities in $P_z(Q)$. Nevertheless, the analysis produces usable results if ℓ is sufficiently large.

Calculation of the statistical parameters. Assume that the ES system is in a state $\mathbf{y}$, with the fitness value $F_0 := F(\mathbf{y})$. The bit vector $\mathbf{y}$ is mutated with probability density (1.13). For a single bit y_i (parent), the transition to a new state x_i (descendant) has the probability

$$p(x_i) = p_m\,\delta(x_i - 1 + y_i) \;+\; (1 - p_m)\,\delta(x_i - y_i), \tag{4.88}$$

i.e. y_i is negated with probability p_m (mutation rate) and remains unchanged with $(1-p_m)$. Due to the statistical independence of the mutations, the parts of the sum (4.87) can be treated separately

$$Q_{\mathbf{y}}(\mathbf{x}) = \sum_{i=1}^{\ell} Q_i(x_i), \qquad Q_i(x_i) = x_i - y_i. \tag{4.89}$$

Firstly, the moments $\overline{Q_i^k}$ will be calculated[4]

$$\begin{aligned}\overline{Q_i^k} &= p_m \int_{-\infty}^{\infty} (x_i - y_i)^k\,\delta(x_i - 1 + y_i)\,\mathrm{d}x_i \\ &\quad + (1 - p_m) \int_{-\infty}^{\infty} (x_i - y_i)^k\,\delta(x_i - y_i)\,\mathrm{d}x_i \\ &= p_m\,(1 - 2y_i)^k. \end{aligned} \tag{4.90}$$

Using (4.7), it follows that for M_Q

$$M_Q = \sum_{i=1}^{\ell} \overline{Q_i} = p_m \sum_{i=1}^{\ell} (1 - 2y_i) = p_m\,(\ell - 2F_0) = -p_m\,(2F_0 - \ell) \tag{4.91}$$

and for S_Q^2

$$S_Q^2 = \sum_{i=1}^{\ell}\left(\overline{Q_i^2} - \overline{Q_i}^2\right) = \sum_{i=1}^{\ell} p_m\,(1 - 2y_i)^2 - p_m^2\,(1 - 2y_i)^2$$

$$S_Q^2 = p_m\,(1 - p_m) \sum_{i=1}^{\ell} (1 - 2y_i)^2 = p_m\,(1 - p_m)\,\ell; \tag{4.92}$$

therefore, one obtains

$$S_Q = \sqrt{\ell}\,\sqrt{p_m\,(1 - p_m)}. \tag{4.93}$$

The cumulants for $k = 3$ and $k = 4$ can be determined using (4.82) and (4.34)

$$\kappa_3\{z\} = \frac{1}{S_Q^3} \sum_{i=1}^{\ell} \overline{(Q_i - \overline{Q_i})^3}, \qquad \kappa_4\{z\} = \frac{1}{S_Q^4} \sum_{i=1}^{\ell} \overline{(Q_i - \overline{Q_i})^4} \;-\; 3. \tag{4.94}$$

One obtains for $\overline{(Q_i - \overline{Q_i})^3} = \overline{Q_i^3} - 3\overline{Q_i^2}\,\overline{Q_i} + 2\overline{Q_i}^3$

[4] Note, for Dirac's delta function $\delta(\cdot)$, one has $\int_{-\infty}^{\infty} f(x)\,\delta(x - y)\,\mathrm{d}x = f(y)$.

$$\overline{(Q_i - \overline{Q_i})^3} = \left(p_m - 3p_m^2 + 2p_m^3\right)(1-2y_i)^3$$
$$= p_m\,(1-p_m)(1-2p_m)(1-2y_i)(1-2y_i)^2$$

and hence

$$\kappa_3\{z\} = -\frac{(1-2p_m)\left(2\,\frac{F_0}{\ell}-1\right)}{\sqrt{\ell}\,\sqrt{p_m\,(1-p_m)}}, \tag{4.95}$$

using $(1-2y_i)^2 \equiv 1$, and decomposing the p_m polynomial into its factors. Analogously, one finds

$$\overline{(Q_i - \overline{Q_i})^4} = \ell\, p_m\,(1-p_m)\left(1 - 3p_m + 3p_m^2\right) \tag{4.96}$$

and consequently

$$\kappa_4\{z\} = \frac{1}{\ell}\,\frac{1-6p_m+6p_m^2}{p_m\,(1-p_m)}. \tag{4.97}$$

Discussion of the $\overline{Q}_{1,\lambda}$ formula. Using (4.91, 4.93) and (4.95), the formula (4.44) up to and including the κ_3 term yields

$$\begin{aligned}\overline{Q}_{1,\lambda} &= c_{1,\lambda}\sqrt{p_m\ell}\,\sqrt{1-p_m} - p_m\ell\left(2\,\frac{F_0}{\ell}-1\right)\\ &\quad - \frac{d^{(2)}_{1,\lambda}-1}{6}\,(1-2p_m)\left(2\,\frac{F_0}{\ell}-1\right) + \ldots\end{aligned} \tag{4.98}$$

$$\begin{aligned}\overline{Q}_{1,\lambda} &= c_{1,\lambda}\sqrt{p_m\ell}\,\sqrt{1-p_m}\\ &\quad - p_m\ell\left(2\,\frac{F_0}{\ell}-1\right)\left(1+\frac{d^{(2)}_{1,\lambda}-1}{6}\,\frac{1-2p_m}{p_m\ell}\right) + \;\ldots .\end{aligned} \tag{4.99}$$

The terms with κ_3 in (4.99) can be neglected for mutation rates $p_m > 0$ and long bitstrings $\ell \to \infty$. One obtains the asymptotically exact formula

$$\boxed{\ell\to\infty:\quad \overline{Q}_{1,\lambda} = c_{1,\lambda}\sqrt{p_m\ell}\,\sqrt{1-p_m}\; - \; p_m\ell\left(2\,\frac{F_0}{\ell}-1\right).} \tag{4.100}$$

This formula is very similar to the $\varphi_{1,\lambda}$ formula (3.101) on the sphere model. Neglecting the $\sqrt{1-p_m}$ term (for sufficiently small p_m), one tends to identify the term $\sqrt{p_m\ell}$ with σ. Indeed, considering (4.93) $\sigma = S_Q$; σ is therefore a measure for the deviation strength of Q and at the same time a measure for the deviation strength of the change of the Hamming distance to the optimum $\hat{Q} = \ell$. In principle, it is possible to establish a fully formal correspondence to the sphere model, if one applies an appropriate normalization. Under the condition $F_0 > \ell/2$, the following is suitable:

$$\left.\begin{aligned} \sigma^* &:= \sigma\, 2\left(2\frac{F_0}{\ell} - 1\right), \quad \text{with} \quad \sigma := \sqrt{p_m \ell}, \\ \overline{Q}^*_{1,\lambda} &:= \overline{Q}_{1,\lambda}\, 2\left(2\frac{F_0}{\ell} - 1\right) \end{aligned}\right\} \quad \text{for } F_0 > \frac{\ell}{2}, \tag{4.101}$$

$$\overline{Q}^*_{1,\lambda} = c_{1,\lambda}\sigma^* - \frac{\sigma^{*2}}{2} + \ldots \qquad (\text{for } 0 < p_m \ll 1). \tag{4.102}$$

Equation (4.100) can be interpreted using the meaning of the EPP. To this end, we will first consider the $F_0 \geq \ell/2$ case. The gain term is given by the $c_{1,\lambda}S_Q$ term, and the loss term by the M_Q term.

In the $F_0 = \ell/2$ case, M_Q vanishes (and in (4.99) the κ_3 term). Hence, only evolutive progress exists (of course, in the statistical mean). This is immediately clear, since 50% of the bits are "1" for $F_0 = \ell/2$. After the generation of the offspring, statistically half of the descendants will have more "1" bits. The $(1, \lambda)$ selection guarantees even for $\lambda = 2$ (in a statistical mean), that an offspring with $F > \ell/2$ will be selected. That is, for $F_0 = \ell/2$ one can state always a non-negative $\overline{Q}_{1,\lambda}$. For $p_m = 0$ or $p_m = 1$ one obtains $\overline{Q}_{1,\lambda} = 0$. Both of these mutation probabilities do not change F: For $p_m = 0$, this is trivial; and for $p_m = 1$ all bits are inverted, yielding $F = \ell/2$ again.[5] The optimal mutation rate for $F_0 = \ell/2$ is $p_m = 1/2$, which can be shown simply by the maximization of $\sqrt{p_m(1 - p_m)}$.

An interesting phenomenon emerges for $F_0 < \ell/2$: the loss term M_Q becomes positive, independent of p_m. Considering the $F_0 = 0$ case, one can directly state the optimal mutation rate p_m: $p_m = 1$, all bits must be flipped, resulting in $\overline{Q}_{1,\lambda} = \ell$.

The optimal mutation rate $\hat{p}_m$. As one can see, $\overline{Q}_{1,\lambda}$ depends on F_0 and p_m. Equation (4.100) can principally be used to determine the optimal mutation rate $\hat{p}_m$; the $\sigma^* \leftrightarrow p_m$ correspondence to the sphere model (3.101) is also valid here. However, (4.102) will not be used for this purpose, but the more accurate formula (4.100) instead. By setting the first derivative with respect to p_m to zero, one obtains

$$\hat{p}_m = \frac{1}{2}\left(1 \pm \sqrt{1 - \frac{c^2_{1,\lambda}}{c^2_{1,\lambda} + \ell\,(2F_0/\ell - 1)^2}}\right), \qquad \begin{aligned} &+ \text{ for } F_0 \leq \ell/2 \\ &- \text{ for } F_0 > \ell/2 \end{aligned}. \tag{4.103}$$

One observes in Equation (4.103) that $\hat{p}_m$ depends on $c_{1,\lambda}$, ℓ, and the ratio $f_0 := F_0/\ell = F_0/\max[F]$. For large ℓ, the square root can be expanded using a Taylor series, therefore

$$\hat{p}_m \approx \frac{1}{4}\frac{c^2_{1,\lambda}}{c^2_{1,\lambda} + \ell\,(2F_0/\ell - 1)^2} \sim \frac{c^2_{1,\lambda}}{4\,(2f_0 - 1)^2}\frac{1}{\ell} \qquad \text{for } f_0 > \frac{1}{2}, \quad \ell \to \infty.$$

[5] Strictly speaking, this is only valid for even ℓ and/or $\ell \to \infty$.

That is, if the state of the `OneMax` function is given by f_0 ($f_0 = F_0/\ell$ is the ratio of the "1" bits), the optimal mutation rate is inversely proportional to ℓ. This qualitative statement is in accordance with the results of Mühlenbein (1992).[6] Different from Mühlenbein (1992) or Bäck (1992), however, the optimal mutation rate can be given *analytically* using (4.103), within the bounds of the approximation (4.100).

Outlook and open questions. The analysis of the quality gain at functions on bitstrings is still at an early stage. For $\ell \to \infty$, exact $\overline{Q}_{1\dot{+}\lambda}$ formulae can be derived, since in this case – at least for the `OneMax` function – the Moivre-Laplace limit theorem (Fisz, 1971) is valid; in other words, the P_z distribution (4.32, 4.33) is a normal distribution. For $\ell < \infty$, P_z has a certain "granularity," which can affect the approximation quality unfavorably. Particularly, it is an open question whether or not the derivation (4.33) is the most appropriate way of reordering (4.26) and its integral, respectively. (See the corresponding remarks to Eq. (B.50) in Appendix B.3, p. 347.) Alternatively, one might imagine a reordering based on the exponents of z, i.e., the series cutoff would be based on the $d_{1,\lambda}^{(k)}$ coefficients/functions.

4.3 Experiments and Interpretations of the Results

4.3.1 Normalization

4.3.1.1 Quadratic Fitness, Isotropic Gaussian Mutations and the Differential-Geometric Model.

Normalization and the correspondence principle. In the theory of the progress rate φ on the sphere model, the normalization (2.22) has played an important role. The normalization (2.22) led to a reduction of the free parameters in the φ formulae. The φ^* do not contain the local curvature radius. Therefore, the φ^* formulae are "state independent" to some extent, since they only depend on the normalized mutation strength σ^* (and on the exogenous strategy parameters μ, λ, and N, of course). As a result, the performance of the ES can be evaluated independently of R. A normalization of the quality gain has exactly the same intention.

Formally, normalizations always contain a certain degree of "arbitrariness," and reflect the preferences of the introducer.[7] The evident, plausible guideline is the *correspondence principle* . In other words, the normalization should yield for the sphere model case, i.e. for the quadratic fitness functions $Q(\mathbf{x}) = \mathbf{a}^{\mathrm{T}}\mathbf{x} - \mathbf{x}^{\mathrm{T}}\mathbf{Q}\mathbf{x}$ with $\mathbf{Q} = q\,\mathbf{E}$ ($q > 0$, $\mathbf{E}$-identity matrix), the normalization (2.22). This requirement determines the normalization to some extent:

---

[6] It should be mentioned here that the optimum mutation rate $p_m = 1/\ell$ for `OneMax`-like functions was already obtained by Bremermann (1958). See also Bremermann et al. (1966).

[7] For instance, a different normalization scheme has been used in Beyer (1994a).

1. $\overline{Q}_{1\stackrel{+}{,}\lambda}$ should be converted to a distance change in the parameter space $\mathcal{Y}$ equivalent to the corresponding fitness change
2. If possible, the local curvature relations should be characterized by a mean radius to be defined appropriately.

The first requirement leads us directly to the *normal progress* φ_R, Eq. (2.19), introduced in Sect. 2.1.3. The latter one suggests the usage of the mean radius $R_{\langle\varkappa\rangle}$, calculated using differential-geometric methods in Sect. 2.3.

Realization. Under the assumptions $|\mathrm{Tr}[\mathbf{Q}]| \gg |\mathbf{a}^{\mathrm{T}}\mathbf{Q}\mathbf{a}|/||\mathbf{a}||^2$ and $N \gg 1$, one obtains with (2.92), (2.93) and (2.19) for the fitness model (2.35) with isotropic Gaussian mutations (4.69), (4.54) the equations for the normalization

$$\sigma^* := \sigma \frac{2\,|\mathrm{Tr}[\mathbf{Q}]|}{||\mathbf{a}||} \qquad \text{and} \qquad \overline{Q}^*_{1\stackrel{+}{,}\lambda} := \overline{Q}_{1\stackrel{+}{,}\lambda} \frac{2\,|\mathrm{Tr}[\mathbf{Q}]|}{||\mathbf{a}||^2}. \tag{4.104}$$

Considering the κ_3 term, the $\overline{Q}_{1,\lambda}$ formula (4.44) reads

$$\overline{Q}_{1,\lambda} = c_{1,\lambda} S_Q + M_Q + S_Q \frac{\kappa_3}{6}\left(d^{(2)}_{1,\lambda} - 1\right) + \ldots. \tag{4.105}$$

Substituting (4.70, 4.71) and applying the normalization (4.104), it follows that[8]

$$\begin{aligned}\overline{Q}^*_{1,\lambda} = {} & c_{1,\lambda}\sigma^* \sqrt{1 + \frac{\sigma^{*2}}{2}\frac{\mathrm{Tr}[\mathbf{Q}^2]}{(\mathrm{Tr}[\mathbf{Q}])^2}} - \frac{\sigma^{*2}}{2}\,\mathrm{sign}(\mathrm{Tr}[\mathbf{Q}]) \\ & - \frac{\sigma^{*2}}{2}\left(d^{(2)}_{1,\lambda} - 1\right)\frac{1}{\mathrm{Tr}[\mathbf{Q}]}\,\frac{\frac{\mathbf{a}^{\mathrm{T}}\mathbf{Q}\mathbf{a}}{||\mathbf{a}||^2} + \frac{\sigma^{*2}}{3}\frac{\mathrm{Tr}[\mathbf{Q}^3]}{(\mathrm{Tr}[\mathbf{Q}])^2}}{1 + \frac{\sigma^{*2}}{2}\frac{\mathrm{Tr}[\mathbf{Q}^2]}{(\mathrm{Tr}[\mathbf{Q}])^2}} + \ldots. \end{aligned} \tag{4.106}$$

$\mathrm{sign}(\mathrm{Tr}[\mathbf{Q}])$ stands for the sign of $\mathrm{Tr}[\mathbf{Q}]$. If the $\mathbf{Q}$ matrix is positive definite, we have $\mathrm{sign}(\mathrm{Tr}[\mathbf{Q}]) = 1$ (maximization!). Therefore, $\mathrm{Tr}[\mathbf{Q}^2] \le (\mathrm{Tr}[\mathbf{Q}])^2$ and $\mathrm{Tr}[\mathbf{Q}^3] \le (\mathrm{Tr}[\mathbf{Q}])^3$, since the trace operator of a matrix is equal to the summation over the eigenvalues of the matrix. If the eigenvalues of the matrix $\mathbf{Q}$ are $q_i \ge 0$, then $\mathrm{Tr}[\mathbf{Q}] = \sum_i q_i$, $\mathrm{Tr}[\mathbf{Q}^2] = \sum_i q_i^2$ and $\mathrm{Tr}[\mathbf{Q}^3] = \sum_i q_i^3$. The inequalities above immediately follow from these facts. Now, we can easily formulate the condition under which (4.106) becomes formally equivalent to the asymptotic $\varphi^*_{1,\lambda}$ formula (3.101) on the sphere model:

$$\textsc{Sphere Model Condition:} \qquad \boxed{\sigma^{*2}\mathrm{Tr}\left[\mathbf{Q}^2\right] \ll (\mathrm{Tr}[\mathbf{Q}])^2.} \tag{4.107}$$

If (4.107) is valid, one obtains

$$\max[q_i^2] \ll (\mathrm{Tr}[\mathbf{Q}])^2$$

and

---

[8] Note that the structure of (4.106) reflects the EPP in a very beautiful manner: the gain term is associated with the $(1,\lambda)$ selection, whereas the loss term is dominated by the M_Q term.

$$\frac{\mathbf{a}^{\mathrm{T}}\mathbf{Q}\mathbf{a}}{||\mathbf{a}||^2} \le \max[q_i] \le \mathrm{Tr}[\mathbf{Q}],$$

as well as

$$\mathrm{Tr}[\mathbf{Q}^3] = \sum_i q_i^3 \le \sum_i q_i \sum_j q_j^2 = \mathrm{Tr}[\mathbf{Q}]\,\mathrm{Tr}[\mathbf{Q}^2] \ll (\mathrm{Tr}[\mathbf{Q}])^3.$$

Thus, one gets for sufficiently small σ^*

$$\boxed{\overline{Q}^*_{1,\lambda} = c_{1,\lambda}\sigma^* - \frac{\sigma^{*2}}{2}} \quad + \quad \text{higher-order terms.} \tag{4.108}$$

The result[9] (4.108) is remarkable: if Condition (4.107) is satisfied, the results of the asymptotic sphere model theory can directly be transferred to the case of the general quadratic fitness $Q(\mathbf{x}) = \mathbf{a}^{\mathrm{T}}\mathbf{x} - \mathbf{x}^{\mathrm{T}}\mathbf{Q}\mathbf{x}$, if the special normalization (4.104) is used. Since this normalization is based on the differential-geometric model of Sect. 2.3, (4.107) also expresses the condition for the applicability of the differential-geometric model.

The plus strategies. The considerations above are also valid for the $(1 + \lambda)$-ES: its quality gain is approximated by the formula (4.50). In this formula the distribution $P_z(-M_Q/S_Q)$ appears additionally. It is approximated by (4.51). Since (4.70), one gets for M_Q/S_Q

$$\frac{M_Q}{S_Q} = -\frac{\sigma^*}{2} \frac{\mathrm{sign}(\mathrm{Tr}[\mathbf{Q}])}{\sqrt{1 + \frac{\sigma^{*2}}{2}\frac{\mathrm{Tr}[\mathbf{Q}^2]}{(\mathrm{Tr}[\mathbf{Q}])^2}}} \tag{4.109}$$

and if (4.107) is satisfied and the $\mathbf{Q}$ matrix is positive definite, one obtains the asymptotic

$$\text{SPHERE MODEL APPROXIMATION:} \quad \boxed{\frac{M_Q}{S_Q} = -\frac{\sigma^*}{2}.} \tag{4.110}$$

Remarks for the case of correlated mutations. The normalization (4.104) cannot be applied to the case of correlated mutations. However, the scaling transformation (2.19) $\overline{Q}_{1\dagger\lambda} \mapsto \varphi_R$ can. In contrast to the single mutation strength σ in the isotropic case, one has now a (symmetric) correlation matrix $\mathbf{C}$. That is, the quality gain $\overline{Q}_{1\dagger\lambda}$ generally depends on $N(N+1)/2$ strategy parameters. A normalization does not appear to make sense. Particularly, a

---

[9] Rechenberg (1994) has suggested the formula $\Phi = \Delta - \Delta^2$ as "the central progress law." Within the framework developed here, this is a $\overline{Q}^*_{1,\lambda}$ formula equivalent to (4.108), with the normalization $\Phi := \overline{Q}_{1,\lambda}\Omega/(||\mathbf{a}||c^2_{1,\lambda})$ and $\Delta = \sigma\Omega/c_{1,\lambda}$ (Φ should not be confused with the normal distribution function). The quantity $\Omega = \sum_i q_i/||\mathbf{a}||$ is called by Rechenberg as the "quadratic complexity." Based on the investigations in Sect. 2.3, we know that $\Omega \approx N/2R_{\langle\varkappa\rangle}$, i.e. it is proportional to the mean curvature. Therefore, the Condition (4.107) determines the limit of the "central progress law."

reduction to the asymptotic sphere model is not possible. However, it is possible to decompose the mutation vector $\mathbf{x}$ multiplicatively in a component $\mathbf{x}_I$ with $\det[\mathbf{C}_I] = 1$ and a step size factor σ

$$\mathbf{x} := \sigma \mathbf{x}_I, \qquad \mathbf{x}_I = \mathcal{N}(\mathbf{0}, \mathbf{C}_I) \quad \text{with} \quad \det[\mathbf{C}_I] = 1. \tag{4.111}$$

This would be in the sense of Schwefel's ES variant (Schwefel, 1995) with self-adaptive correlation matrix. The σ could be normalized thereafter appropriately. Any related investigations remain to be done in the future.

4.3.1.2 The Normalization for the Biquadratic Case. The case introduced in Sect. 4.2.3.1 will be chosen for the analysis. It possesses some properties which are counterintuitive to the expectations known for the quadratic fitness landscapes. The first of these peculiarities is shown here, the second one in the section on the ES experiments.

The first property: there is *no* differential-geometric model for the special fitness landscape (4.77). The reason is that the local curvature in the parental state $\mathbf{x} = \mathbf{0}$ vanishes. Since the second derivatives also identically vanish, the concept of the mean curvature radius $R_{\langle \varkappa \rangle}$ (see Sect. 2.3) is not applicable, i.e. a normalization in the sense of (2.22) is in principle *not* possible. The correspondence principle postulated in Sect. 4.3.1.1, p. 131, is not applicable. However, the reduction to the normal progress (2.19) $\varphi_R = \overline{Q}_{1\overset{+}{,}\lambda}/\|\mathbf{a}\|$ remains applicable.

The σ normalization is chosen in order to produce the most formal similarity to the progress formula (4.108). Particularly, the gain term according to the EPP (see footnote 8, p. 132) should be asymptotically of the form $c_{1,\lambda}\sigma^*$, and the loss term $-\sigma^{*4}[1 + \ldots]$. These conditions are fulfilled by the normalization

$$\sigma^* := \sigma \sqrt[3]{\frac{3\sum_i c_i}{\|\mathbf{a}\|}} \qquad \text{and} \qquad \overline{Q}^*_{1\overset{+}{,}\lambda} := \frac{\overline{Q}_{1\overset{+}{,}\lambda}}{\|\mathbf{a}\|} \sqrt[3]{\frac{3\sum_i c_i}{\|\mathbf{a}\|}}. \tag{4.112}$$

Using (4.105, 4.80, 4.81) and (4.85), one obtains the normalized quality gain on the $(1, \lambda)$-ES

$$\overline{Q}^*_{1,\lambda} = c_{1,\lambda}\sigma^* \sqrt{1 + \frac{32}{3}\sigma^{*6}\frac{\sum_i c_i^2}{\left(\sum_i c_i\right)^2}} - \sigma^{*4}$$
$$- \sigma^{*4}\left(d^{(2)}_{1,\lambda} - 1\right)\frac{2}{\sum_i c_i}\frac{\dfrac{\sum_i a_i^2 c_i}{\|\mathbf{a}\|^2} + \dfrac{88}{3}\sigma^{*6}\dfrac{\sum_i c_i^3}{\left(\sum_i c_i\right)^2}}{1 + \dfrac{32}{3}\sigma^{*6}\dfrac{\sum_i c_i^2}{\left(\sum_i c_i\right)^2}} + \ldots . \tag{4.113}$$

This formula has a certain similarity to (4.106). Analogous to (4.107), postulating the

$$\text{UNIFORMITY CONDITION:} \qquad \boxed{\sigma^{*6}\sum_{i=1}^N c_i^2 \ll \left(\sum_{i=1}^N c_i\right)^2,} \tag{4.114}$$

which ensures the uniformity of the c_i, (4.113) is reduced for sufficiently small σ^* to

$$\boxed{\overline{Q}^*_{1,\lambda} = c_{1,\lambda}\sigma^* - \sigma^{*4}} \quad + \text{ higher-order terms.} \tag{4.115}$$

As to the M_Q/S_Q ratio in the $(1+\lambda)$ formulae (4.50) and (4.51) one obtains

$$\frac{M_Q}{S_Q} = -\sigma^{*3} \frac{1}{\sqrt{1 + \frac{32}{3}\sigma^{*6} \frac{\sum_i c_i^2}{\left(\sum_i c_i\right)^2}}} \quad \text{and} \quad \frac{M_Q}{S_Q} \xrightarrow{(4.114)} -\sigma^{*3}. \tag{4.116}$$

4.3.2 Experiments and Approximation Quality

As already pointed out, in general one cannot derive any practically usable, quantitative statements on the approximation error in the $\overline{Q}_{1 \overset{+}{,} \lambda}$ formulae. This does not exclude that the order of the error term for a given Q function cannot be estimated. However, comparisons with the experiments should be done to visualize this error. The quadratic fitness function (2.35) with

$$N = 100, \qquad (\mathbf{Q})_{ij} = i\delta_{ij}, \qquad a_i = 1 \tag{4.117}$$

and the biquadratic function (4.77) with

$$N = 100, \qquad c_i = 1, \qquad a_i = 1 \tag{4.118}$$

are investigated here as examples. Both fitness functions fulfill the sphere model and uniformity conditions, (4.107) and (4.114), respectively. This can be shown by simple calculations[10]

$$\frac{\mathrm{Tr}\left[\mathbf{Q}^2\right]}{\left(\mathrm{Tr}[\mathbf{Q}]\right)^2} = \frac{\sum_{i=1}^N i^2}{\left(\sum_{i=1}^N i\right)^2} = \frac{2}{3}\frac{2N+1}{N\,(N+1)} = \mathcal{O}\left(\frac{1}{N}\right) \tag{4.119}$$

$$\frac{\sum_i c_i^2}{\left(\sum_i c_i\right)^2} = \frac{\sum_{i=1}^N 1}{\left(\sum_{i=1}^N 1\right)^2} = \frac{1}{N}. \tag{4.120}$$

Therefore, the expressions (4.108, 4.110) and (4.115, 4.116) suffice for the approximation of the quality gain as long as σ^* is sufficiently small. The experimental verification is done by $(1 \overset{+}{,} \lambda)$ *"one-generation experiments"* after the scheme in Sect. 3.4.4. In Figs. 4.1 and 4.2, $G = 40{,}000$ "one-generation experiments" with the $(1,5)$-ES and the $(1+1)$-ES are performed for each data point, shown as "$\diamond$" symbols. The mean fitness change and the mutation strength σ obtained are normalized using (4.104) and (4.112), respectively. A very good agreement between theory and experiments can be observed.

---

[10] The summation formulae $\sum_{i=1}^N i = N\,(N{+}1)/2$, $\sum_{i=1}^N i^2 = N\,(N{+}1)(2N{+}1)/6$, and $\sum_{i=1}^N i^4 = N\,(N+1)(2N+1)(3N^2+3N-1)/30$ have been used in the intermediate steps (Bronstein & Semendjajew, 1981, p. 166).

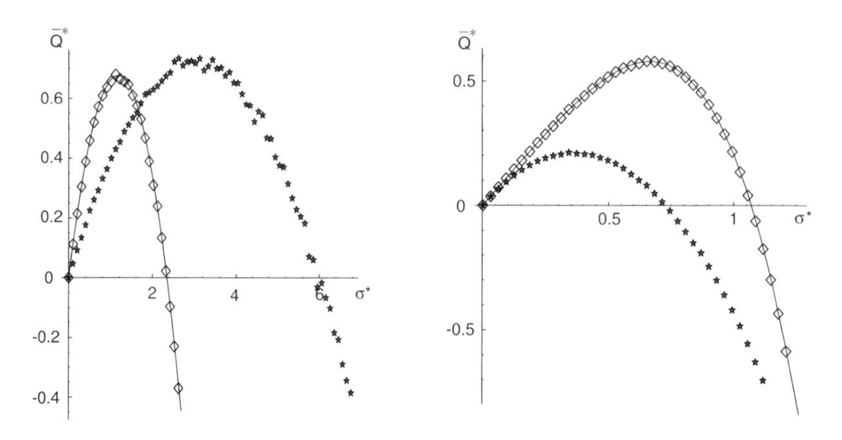

Fig. 4.1. The (1, 5)-ES experiments. Left: quadratic fitness, according to (4.117). Right: biquadratic fitness, according to (4.118). The results of the quality gain experiments are indicated by "◇." The curves are the predictions of (4.108) and (4.115), respectively. The "⋆" symbols show the progress rate φ^* (see Sect. 4.3.3).

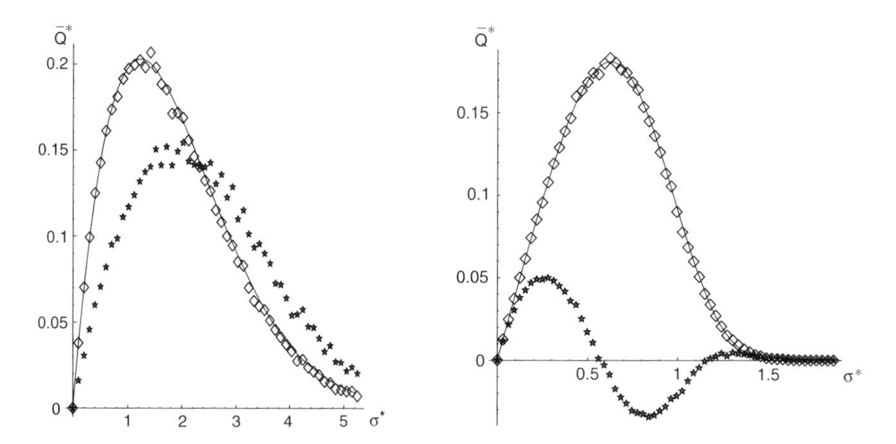

Fig. 4.2. The (1 + 1)-ES experiments. Left: quadratic fitness, according to (4.117). Right: biquadratic fitness, according to (4.118). See Fig. 4.1 for further explanations.

If the quadratic fitness function (4.117) is considered, one realizes that the hypersurfaces $Q(\mathbf{x}) = \text{const}$ define ellipsoids parallel to the axis. The maximal ratio of the half-axes is $\sqrt{1}/\sqrt{100} = 1/10$. For such a ratio, the assumption of the local approximate-ability of the fitness landscape seems still imaginable. Interestingly, the ratio of the half-axes, however, does not have a great influence on the approximation quality. A fitness model with $(\mathbf{Q})_{ij} = i^2\delta_{ij}$ yields comparable results, the Condition (4.107) is fulfilled in a similar manner as in (4.117, 4.119)

$$(\mathbf{Q})_{ij} = i^2\delta_{ij}: \quad \frac{\operatorname{Tr}\left[\mathbf{Q}^2\right]}{\left(\operatorname{Tr}[\mathbf{Q}]\right)^2} = \frac{6}{5}\,\frac{3N^2+3N-1}{N\,(N+1)(2N+1)} = \mathcal{O}\left(\frac{1}{N}\right). \tag{4.121}$$

Naturally, one can violate the conditions (4.107) and (4.114). In order to do so, one has to consider the "dominance" of some isolated eigenvalues. In the case of quadratic fitness, the violation of (4.107) can be achieved for example by the following choice

$$N = 100, \qquad a_i = 1, \qquad (\mathbf{Q})_{ij} = \begin{cases} 200, & \text{if } i = j = 1 \\ 1, & \text{if } i = j \neq 1 \\ 0, & \text{if } i \neq j. \end{cases} \tag{4.122}$$

The theoretical predictions and the experimental results (indicated with "$\diamond$")

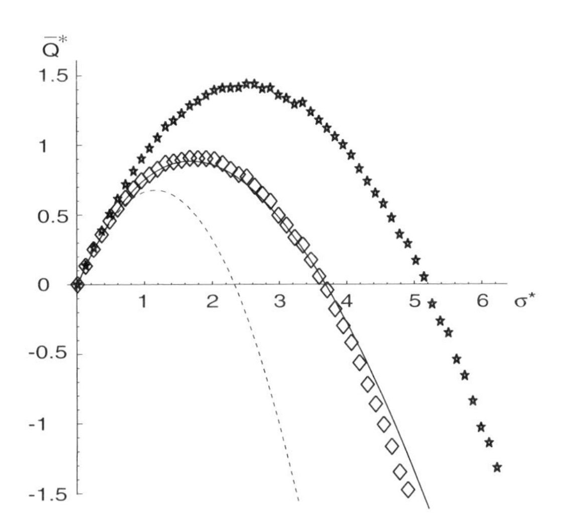

Fig. 4.3. The violation of the sphere model condition (4.107) on the $(1,5)$-ES. For the explanation of the data indicated with "$\star$" see Sect. 4.3.3.

are shown in Fig. 4.3. The approximation for $\overline{Q}_{1,\lambda}$ (4.108) is indicated by the dashed line. It differs considerably from the experimental data $\diamond$ (see also footnote 9, p. 133). Whereas, if (4.106) is used, shown by a solid curve in Fig. 4.3, one obtains a good accordance to the experiment in the interval $\overline{Q}_{1,\lambda} \gtrapprox 0$. For larger σ^* values, this approximation also deviates from the experimental results. This deviation can be treated by considering further cumulants in the quality gain formula (4.44).

4.3.3 Quality Gain or Progress Rate?

This chapter considered the calculation of the progress measure called quality gain. One might come to the idea that $\overline{Q}_{1\,\overset{+}{,}\,\lambda}$ can be converted to $\varphi_{1\,\overset{+}{,}\,\lambda}$. Especially the normal progress φ_R suggested by Rechenberg (1994) (see Sect. 2.1.3) may lead to this belief. But this is *not* the case. Both progress measures describe totally different aspects of the ES performance, and generally one has

$$\varphi_R \neq \varphi \qquad \text{and even} \qquad \varphi_R \not\approx \varphi. \tag{4.123}$$

The $\overline{Q}_{1\overset{+}{,}\lambda}$ measure describes the fitness change, where the φ measure quantifies the advancement toward the optimum. It is essential to note that for the nonspherical models $\varphi_R \not\approx \varphi$ is often observed, i.e., not even a similar behavior of the two quantities should be expected. All fitness examples in the previous subsection consider $\varphi_R \not\approx \varphi$. The actual progress rate φ^*, obtained from the simulation results and normalized thereafter, is marked by "$\star$" data points displayed in Figs. 4.1–4.3. The progress rate φ^* is obtained by ES experiments, analogous to the $\overline{Q}_{1\overset{+}{,}\lambda}$ data points. For this purpose, the change in the residual distance to the optimum is measured (statically) according to the definition of $\varphi_{1\overset{+}{,}\lambda}$ in (2.15). This is possible, since the optimum $\hat{\mathbf{y}} = \hat{\mathbf{x}}$ is known for the cases (4.117, 4.118) and (4.122)

$$\begin{aligned}
&(4.117){:} \quad \hat{x}_i = \frac{1}{2i}, \\
&(4.118){:} \quad \hat{x}_i = \sqrt[3]{\frac{1}{4}}, \\
&(4.122){:} \quad \hat{x}_1 = \frac{1}{400}, \quad \hat{x}_i = \frac{1}{2} \ (\text{for } i > 1).
\end{aligned}$$

The normalizations to get σ^* given in (4.104) and (4.112) are used accordingly, and the normalization scheme given to obtain $\overline{Q}^*_{1\overset{+}{,}\lambda}$ is used also to obtain the normalized progress rate φ^*. In general, it makes no sense to compare the quantities of $\overline{Q}^*_{1\overset{+}{,}\lambda}$ and $\varphi^*_{1\overset{+}{,}\lambda}$ as to their absolute values, since they are two different quantities by their nature.[11] Of course, one can discuss and compare the dependence of the respective performance on the mutation strength σ^*.

Discussion of the elliptic landscape. In the case of the (1, 5)-ES, Fig. 4.1, the left-hand figure shows the comparison of the $\overline{Q}^*_{1,5}$ measure "$\diamond$" with the $\varphi^*_{1,5}$ measure "$\star$." One can see that the measures are not similar. In the case of the φ^* data points, one can assume the validity of the EPP and its dependence on the mutation strength; however, the maximal progress rate $\hat{\varphi}^*$ is attained at a mutation strength σ^*, at which $\overline{Q}^*$ is *negative*! Or, conversely, an ES working optimally in $\overline{Q}^*$ does *not* yield the maximal possible progress rate $\hat{\varphi}^*$. A geometrical explanation for this observation is given in Fig. 4.4. The mutations $\mathbf{z}_1$ of strength σ_1 and $\mathbf{z}_2$ of strength σ_2 are generated around the parent $\mathbf{y}$. In the case of small mutations, $\mathbf{z}_1$, there is a high probability that the best offspring is in the local success region $Q(\mathbf{x}) \geq 0$ (fitness maximization). If, on average, at least one descendant fulfills the condition $Q(\mathbf{x}) \geq 0$, then we have $\overline{Q}_{1,\lambda} > 0$. The new parent obtained in this way has the distance

[11] The φ measure quantifies the progress in the direction toward the optimum, whereas the *normalized* quality gain provides an imaginary progress in the direction of the gradient.

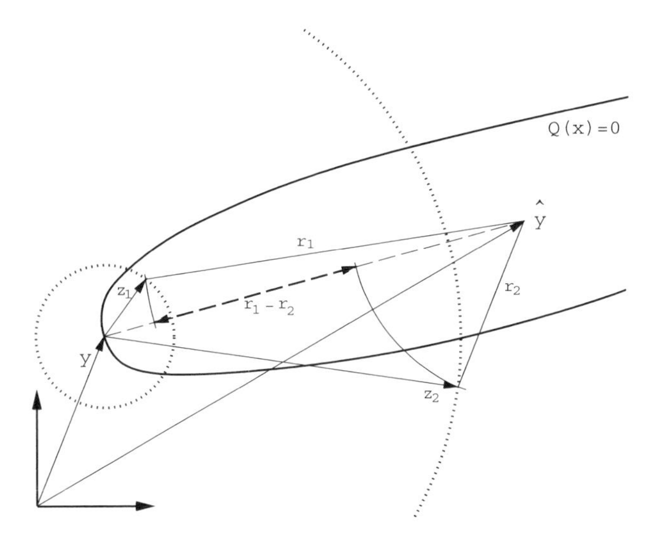

Fig. 4.4. The ES on a prolonged ellipsoid. On the geometrical explanation of the observations in Fig. 4.1 left. The length l of the mutations $\mathbf{z}$ is almost constant because of (3.55), $l \approx \sigma\sqrt{N}$. Therefore, they can be visualized as if they were on the shell of a hypersphere.

r_1 to the optimum $\hat{\mathbf{y}}$. Whereas, if $\sigma = \sigma_2$ is chosen, which is much larger than σ_1, the $(1, \lambda)$ selection gives on average only Q values which are smaller than zero. However, the selected vectors are not uniformly distributed on the corresponding hypersphere surface $||\mathbf{z}|| \approx \sigma_2\sqrt{N}$. Mostly the mutations having the tendency toward $Q(\mathbf{x}) = 0$ are favored by the $(1, \lambda)$-selection. This is indicated by the mutation $\mathbf{z}_2$ in Fig. 4.4. For σ_2, one therefore obtains a negative $\overline{Q}_{1,\lambda}$, but an r_2 much smaller than r_1. Consequently, the distance change toward the optimum and therefore φ is larger at σ_2 than at σ_1.

If σ is increased further so that $\sigma\sqrt{N} \gtrapprox ||\hat{\mathbf{y}} - \mathbf{y}||$, then also the φ performance measure yields negative values, i.e. the distance to the optimum increases. Hence, the graphs in Fig. 4.1 (left) are completely interpreted and explained.

The same explanation scheme is principally valid also for the $(1 + \lambda)$-ES on the same fitness model, shown in Fig. 4.2 left ($\lambda = 1$). However, the plus selection prohibits the survival of mutations with large σ but negative Q. As a result, the differences between $\overline{Q}^*_{1+\lambda}$ and $\varphi^*_{1+\lambda}$ are not as large as in the $(1, \lambda)$-ES.

Discussion of the biquadratic fitness landscape. The biquadratic model (4.77, 4.118) shows a different behavior from the quadratic one discussed above. A glance at Fig. 4.1 right shows the "inverse" relations. The maximum of the quality gain $\overline{Q}^*_{1,5}$ ("$\diamond$" data points) is reached for very small, almost negative progress rates $\varphi^*_{1,5}$ ("$\star$" data points). $\overline{Q}^*_{1,5}$ shows a long linear increase, followed by a relatively sharp decrease. Yet more astonishing is the behavior of the $(1 + 1)$-ES, shown in Fig. 4.2 right. Although the $(1 + 1)$-selection forces a positive quality gain $\overline{Q}^*_{1+1}$, in spite of the elitist selection, φ^*_{1+1} can take negative values. In our special case, φ^* is even negative for $\sigma^* = \hat{\sigma}^*$, where $\overline{Q}^*$ is maximal.

These observations can also be explained geometrically. To this end, consider Fig. 4.5, where the two-dimensional isometric picture of the fitness function (4.77) with the parameter values (4.118), for $x_i = 0$ $(i > 2)$ is shown. The fitness function $Q(\mathbf{x}) = \sum_i (x_i - x_i^4)$ has a symmetry axis (point symmetry) which goes through the origin and the optimum $\hat{\mathbf{x}}$. Therefore, all x_i-x_k-planes $(i \neq k)$ are the same. The parent $\mathbf{x} = \mathbf{0}$ has distance R to $\hat{\mathbf{x}}$. The points with the same distance $r = R$ to $\hat{\mathbf{x}}$ are shown by the dashed circle segment in Fig. 4.5. The mutations with $r < R$, i.e. which fall into this circle, positively contribute to the progress rate φ. As can be seen, there are mutations $\mathbf{z}$ in the local success region $Q(\mathbf{x}) \geq 0$ but which do *not* fulfill the condition $r < R$. Such points yield a negative contribution to φ. Thus, the principal shape of $\overline{Q}^*_{1,\lambda}$ and $\varphi^*_{1,\lambda}$ in Fig. 4.1 right is explained.

The relatively long linear increase of $\overline{Q}^*_{1,\lambda}$ can be explained by the disappearance of the local curvature at the parental state $\mathbf{x} = \mathbf{0}$ (also, see Sect. 4.3.1.2). Even for relatively large σ^*, the fitness landscape is only slightly curved in the vicinity of $\mathbf{x} = \mathbf{0}$. That is, one has an inclined hyperplane. However, if σ^* is increased further, the nonlinearity term x^4 dominates abruptly and causes the sharp drop in the $\overline{Q}^*_{1,\lambda}(\sigma^*)$ curve.

The astonishing behavior of φ^* in the case of the (1 + 1)-ES (Fig. 4.2 right, "$\star$" data points) now also becomes understandable. Because of the (1 + 1) selection, the successful mutations are captured by the local success region $Q(\mathbf{x}) \geq 0$. Let the mutation strength σ increase stepwise starting from zero. At the beginning, the tangential planes of $Q(\mathbf{x})$ and of the hypersphere (with radius R) coincide, where the latter defines the region with a smaller distance to the optimum. Therefore, mutations increasing the fitness $(Q > 0)$

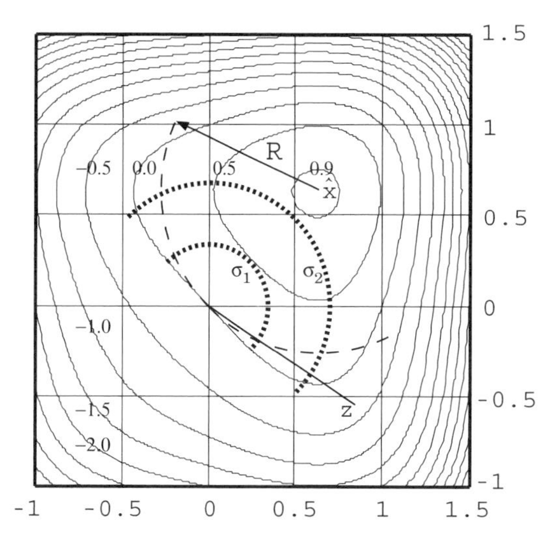

Fig. 4.5. The isometric plot $Q(\mathbf{x}) = \text{const}$ of the biquadratic fitness function $Q(\mathbf{x}) = \sum_i x_i - \sum_i x_i^4$ in the x_1-x_2-plane $(x_3, \ldots, x_N = 0)$.

will also satisfy the $r < R$ condition, thus[12] yielding a positive φ. However, if σ is increased further, such as to $\sigma = \sigma_1$ (see Fig. 4.5), the curvature of the hypersphere attains significance. That is, the linear φ^* increase becomes slower. If σ is increased further, such as to $\sigma = \sigma_2$, then almost all mutations are concentrated in the region $Q(\mathbf{x}) > 0 \wedge r > R$. The reason is that all mutations $\mathbf{z}$ lie almost at the sphere edge of length $l \approx \sigma_2\sqrt{N}$ (shown as a dotted line in Fig. 4.5); however, the component of $\mathbf{z}$ in the direction toward the optimum is only of order σ. Hence, the selected individuals concentrate in the region $r > R$, and φ becomes negative.

A further increase of σ results in an expected mutation step length $l \approx \sigma\sqrt{N}$, which is larger than the distance of the parental state $\mathbf{x} = \mathbf{0}$ to the intersection of the isometric line $Q = 0$ with the circle $r = R$. The probability for successful mutations become smaller and smaller. However, such mutations are in the vicinity of the $\varphi = 0$ line (or hypersurface) yielding $r < R$ which inevitably causes $\varphi > 0$. The "$\star$" plot in Fig. 4.2 right reflects exactly this behavior.

Conclusions, open problems, and outlook. It was the aim of this subsection to demonstrate that the two performance measures quality gain $\overline{Q}^*_{1 \overset{+}{,} \lambda}$ and progress rate $\varphi^*_{1 \overset{+}{,} \lambda}$ describe different aspects of the *microscopic* behavior of an EA.

The quality gain $\overline{Q}$ is an *observable* quantity; it can be measured in the progression of the evolution process. The control and change of the mutation strength σ or of the correlation matrix $\mathbf{C}$ can only be based on the direct Q information, i.e. on the local improvements in the fitness space (statistical analysis methods[13] applied to the population in the search space are disregarded in this discussion).

In contrast, the φ measure is a "hidden" measure. In real applications, φ cannot be measured; since it presupposes the *a priori* knowledge of the optimal state $\hat{\mathbf{y}}$ which is to be found by the EA. Hence, φ is a purely theoretical quantity.

Of course, it is desirable in practice to control the algorithm so that it operates with the largest possible φ. Ultimately, a large φ guarantees a fast convergence to the optimum. Herdy (1992) and Rechenberg (1994) investigated in this direction and introduced the hierarchical ES with isolation period γ, called Meta-ES, population ES, or nested ES. For example, such algorithms can exhibit better performance on the parabolic ridge

$$Q(\mathbf{x}) = ax_1 - \sum_{i=2}^{N} x_i^2 \tag{4.124}$$

[12] As a reminder, the quantity r measures the distance of a mutation $\mathbf{z}$ to optimum $\hat{\mathbf{x}}$: $r = \|\hat{\mathbf{x}} - \mathbf{z}\|$.

[13] Particularly, the approaches of Ostermeier et al. (1995) and Hansen et al. (1995) for the adaptation of the strategy parameters by using statistical inference methods belong to this category.

than the nonhierarchical ES. However, there is no general principle guaranteeing the superiority of Meta-ES in any case. As to the ridge function, the Meta-ES can take advantage of the shape of the local success domain. In this success domain, the performance measures behave in a similar manner to that in Fig. 4.1 left. Therefore, by using populations with different σ values, it seems possible to select the mutation strength σ which yields the largest progress. However, the diagnosis of a significant progress advantage requires a sufficiently long isolation of the populations. The isolation time γ is a new strategy parameter that must be properly chosen (as a first guess, γ should be of order N). A theory for γ (e.g., for elliptic fitness landscapes or ridge functions) presupposes the knowledge of φ in terms of analytic approximations. Up until now, such formulae are not available for models beyond the sphere, corridor, and discus model. This is yet another challenge for future research.[14]

It should be emphasized that the hierarchical ES is by no means the final solution to the problem of finding the optimal progress rate. These strategies use $\overline{Q}$ information for the control of σ. They just use the *nonlocal* $\overline{Q}$ information, i.e. the Q values cumulated over γ generations are used. This only makes sense if the dependence of the cumulated $\overline{Q}^*$ and φ^* functions are more similar to each other than in the case of a single generation. It is another open problem to determine for which classes of fitness landscapes this is the case.

The quality gain performance measure has been investigated only in a few fitness models. The application examples in this book should firstly demonstrate the capability of this approach. The spectrum of application possibilities has been speculated on at several points in this book. Beyond extreme cases, such as integer optimization or the optimization of bitstrings, also investigations in multimodal landscapes using the quality gain are imaginable. Especially, the selection behavior of strategies in the neighborhood of the saddle points and plateaus should be considered in future research.

[14] Recently, some progress was made in the ES performance analysis of ridge functions. The reader is referred to Oyman et al. (1998, 2000), Oyman and Beyer (2000), and Oyman (1999).

5. The Analysis of the (μ, λ)-ES

This chapter is dedicated to the analysis of the (μ, λ)-ES on the sphere model. It has three parts. The theoretical background is presented in the first section. In the second section, the program established in the previous section is carried out in the linear approximation. It will hereby be possible to perform the calculations for the final progress rate formulae analytically. The results obtained are compared with the simulation results in the third section, along with a discussion. For small σ^*, the progress rate formula for the hyperplane is obtained as a byproduct. This can be used for the empirical verification of the $c_{\mu,\lambda}$ progress coefficients. The exploration behavior of the (μ, λ)-ES is investigated on the sphere model. As a result, it is shown that the search behavior observed cannot *be characterized as a "diffusion along the gradient."*

5.1 Fundamentals and Theoretical Framework

5.1.1 Preliminaries

The analysis of the progress rate $\varphi_{\mu,\lambda}$ on the sphere model is the most complicated and technically most ambitious chapter of this book. In other chapters, the analysis is carried out *de facto* for strategies with one parent.[1] However, the parental population with μ parents must be taken into consideration in the analysis of the (μ, λ)-ES. The distribution of these parents $\mathbf{y}_m$ in the parameter space $\mathcal{Y}$ is an unknown function. Moreover, it changes in successive generations. If this function were known, the calculation of the progress rate would be a simple task, comparable to the difficulty of the quality gain analysis of the $(1, \lambda)$-ES.

The fundamental relations are shown and/or derived in this section. First of all, the relevant part of the (μ, λ)-ES algorithm is taken to a form which is appropriate for the φ definition on the sphere model. In Sect. 5.1.3, the *formal* expression of $\varphi_{\mu,\lambda}$ follows as an integral over the distribution of a *single*

[1] This statement should be clear for the $(1 \overset{+}{,} \lambda)$-ES. In Chap. 6, the intermediate $(\mu/\mu_I, \lambda)$-ES will be investigated: in this case, the descendants are generated from the (same) parental mean; therefore, the algorithm is similar to the $(1, \lambda)$-ES. The dominant $(\mu/\mu_D, \lambda)$-ES is an exceptional case: the distribution of the parents is described in the analysis using an *ad hoc* approach.

descendant denoted by $p(r)$. This distribution $p(r)$ is generation-dependent, and of course unknown. The actual challenge in this chapter is determining it approximately. The distribution of the parental population can be calculated using this $p(r)$; however, it is not explicitly required for further calculations.

In Sect. 5.1.4, $p(r)$ will be approximated formally using a function series of Hermite polynomials. Using this method, the problem of determining $p(r)$ is mapped to determining certain coefficients ($\overline{r}$, s, γ, see below) of the function series. The coefficients (s, γ) are functions of the moments of the offspring population. They are "statistically estimated" by the random selection process of the (μ, λ)-ES, i.e. *by the ES*. The mathematical modeling of this statistical estimation process is the main technical problem of the chapter. The expected values of the moments with respect to the parental distribution are to be determined. Sections 5.1.5–5.1.9 deal with several aspects of this estimation. Finally, Sect. 5.1.10 describes the idea of the *self-consistent method*, which will be used for the determination of the stationary $p(r)$ distribution.

5.1.2 The (μ, λ) Algorithm and the $\varphi_{\mu,\lambda}$ Definition

In the (μ, λ)-ES algorithm, see Sect. 1.2.1, each of the λ offspring $\tilde{\mathbf{y}}_l$ ($l = 1, \ldots, \lambda$) is generated by adding a random mutation vector $\mathbf{z}$ to the coordinates of a parent randomly chosen from the parental pool $\mathbf{y}_m$ ($m = 1, \ldots, \mu$). The offspring are marked with a tilde and they carry the subscript l. The subscript m is used for the parents. The descendant l is generated according to[2]

$$\tilde{\mathbf{y}}_l := \mathbf{y}_m + \mathbf{z}_l, \quad m := \mathtt{Random}\{1, \ldots, \mu\}, \quad \mathbf{z}_l := \sigma \left(\mathcal{N}(0,1), \ldots, \mathcal{N}(0,1)\right)^{\mathrm{T}}. \tag{5.1}$$

The characteristics of the algorithm will be formalized next. Equation (5.1) should be considered as a "moment exposure" at (generation) time $g+1$. The parents of generation $g+1$ are chosen by deterministic (μ, λ) selection as the best μ individuals from the offspring pool at generation g, $\mathbf{y}_m^{(g+1)} := \tilde{\mathbf{y}}_{m;\lambda}^{(g)}$ ($m = 1, \ldots, \mu$). This selection method was already introduced in Sect. 1.2.2.1. Consequently, the relevant part of the ES algorithm can be represented in compact form

$$\underline{(\mu, \lambda)\text{-ES:}} \qquad \forall\, l = 1, \ldots, \lambda: \quad \begin{cases} m_l := \mathtt{Random}\{1, \ldots, \mu\}, \\ \tilde{\mathbf{y}}_l^{(g+1)} := \tilde{\mathbf{y}}_{m_l;\lambda}^{(g)} + \mathbf{z}_l. \end{cases} \tag{5.2}$$

For the investigation of the (μ, λ)-ES in the following, it is reasonable to use a suitably adapted coordinate system. We will use the one introduced in Chap. 3, Fig. 3.1 (p. 53). The optimum $\hat{\mathbf{y}}$ (the center of the hypersphere) is

---
[2] $\mathtt{Random}\{1, \ldots, \mu\}$ is a random number generator. It produces discrete values from the set $\{1, \ldots, \mu\}$ in a uniformly distributed manner, i.e. with the probability $p = 1/\mu$ each.

placed at the origin. Consequently, the parents and offspring are specified by the radius vectors $\mathbf{R}_m$ and $\tilde{\mathbf{R}}_l$, respectively. Because of the sphere-symmetry, the fitness of the individuals only depends on the magnitude of their radius vectors, $R = \|\mathbf{R}\|$. This has essential consequences for the "search behavior" of the (μ, λ)-ES. That point is investigated in detail at the end of this chapter.

We use (2.11) for defining the progress rate $\varphi_{\mu,\lambda}$. By choosing a coordinate system so that the optimum is placed at the origin, it follows that

$$\varphi_{\mu,\lambda} = \frac{1}{\mu} \sum_{m=1}^{\mu} \mathrm{E}\left\{ \left\| \mathbf{R}_m^{(g)} \right\| - \left\| \tilde{\mathbf{R}}_{m;\lambda}^{(g)} \right\| \right\}. \tag{5.3}$$

This will be abbreviated as

$$\varphi_{\mu,\lambda} := \overline{\langle R \rangle} - \overline{\langle \tilde{R} \rangle} \tag{5.4}$$

with

$$\overline{\langle R \rangle} := \frac{1}{\mu} \sum_{m=1}^{\mu} \mathrm{E}\left\{ \left\| \mathbf{R}_m^{(g)} \right\| \right\} = \frac{1}{\mu} \sum_{m=1}^{\mu} \mathrm{E}\left\{ R_m \right\} = \frac{1}{\mu} \sum_{m=1}^{\mu} \int_{r=0}^{\infty} r\, p_m(r)\, \mathrm{d}r \tag{5.5}$$

and

$$\overline{\langle \tilde{R} \rangle} := \frac{1}{\mu} \sum_{m=1}^{\mu} \mathrm{E}\left\{ \left\| \tilde{\mathbf{R}}_{m;\lambda}^{(g)} \right\| \right\} = \frac{1}{\mu} \sum_{m=1}^{\mu} \mathrm{E}\left\{ \tilde{R}_{m:\lambda} \right\}$$

$$= \frac{1}{\mu} \sum_{m=1}^{\mu} \int_{r=0}^{\infty} r\, p_{m:\lambda}(r)\, \mathrm{d}r. \tag{5.6}$$

$R_m = \left\| \mathbf{R}_m^{(g)} \right\|$ denotes the distance of the parent m to the optimum (at generation g), and $p_m(r)$ denotes the corresponding probability density. The random variable $\tilde{R}_{m:\lambda} = \left\| \tilde{\mathbf{R}}_{m;\lambda}^{(g)} \right\|$ is the distance of the mth best offspring $\tilde{\mathbf{R}}_{m;\lambda}^{(g)}$ to the optimum; one has

$$\tilde{R}_{1:\lambda} \le \tilde{R}_{2:\lambda} \le \ldots \le \tilde{R}_{m:\lambda} \le \ldots \le \tilde{R}_{\mu:\lambda}, \tag{5.7}$$

i.e. $\tilde{R}_{m:\lambda}$ is the mth order statistics $m : \lambda$, with density $p_{m:\lambda}(r)$.

The progress rate (5.4) can be calculated if one explicitly knows the $p_{m:\lambda}(r)$ distribution at generation g or the distribution $p(r)$ of a single individual causing $p_{m:\lambda}(r)$. The explicit knowledge of the parental distributions $p_m(r)$ (see (5.5)) is not necessary. The only requirement is that they have certain expected values at generation g, for example $\overline{\langle R \rangle}$.

As in the case of the $(1 \stackrel{+}{,} \lambda)$-ES, it is reasonable to introduce normalized quantities. For the sphere model, the definition in (2.22) is appropriate. However, $\overline{\langle R \rangle}$ is used instead of R, denoting the expected value of the mean distance of the parents to the optimum:

$$\textsc{Normalization:} \qquad \varphi^*_{\mu,\lambda} := \varphi_{\mu,\lambda} \frac{N}{\overline{\langle R \rangle}} \quad \text{and} \quad \sigma^* := \sigma \frac{N}{\overline{\langle R \rangle}}. \tag{5.8}$$

5.1.3 The $\varphi_{\mu,\lambda}$ Integral

Considering (5.4) and (5.6), one obtains for $\varphi_{\mu,\lambda}$

$$\varphi_{\mu,\lambda} = \overline{\langle R \rangle} - \frac{1}{\mu} \sum_{m=1}^{\mu} \int_{r=0}^{\infty} r\, p_{m:\lambda}(r)\, \mathrm{d}r. \tag{5.9}$$

The density function of the mth order statistic is represented by $p_{m:\lambda}(r)$. This density is generated by λ descendants having density $p(r)$. The function $p(r)$ is the density of a *single* descendant, generated according to (5.1). In this section, it is assumed that $p(r)$ is known for generation g. The actual calculation of $p(r)$ follows after that. It takes almost the whole of the chapter.

The density function $p_{m:\lambda}(r)$, given the distribution $P(r) = \int_0^r p(r')\, \mathrm{d}r'$ of a single offspring, is needed to proceed. This is a standard task in order statistics (David, 1970)

$$p_{m:\lambda}(r) = \frac{\lambda!}{(m-1)!\,(\lambda-m)!}\, p(r)[P(r)]^{m-1}[1-P(r)]^{\lambda-m}. \tag{5.10}$$

Equation (5.10) is easy to understand. A single descendant is generated with the density $p(r)$. To be the mth best with its special r value, there must be $(m-1)$ samples (individuals) better than that, *and* $(\lambda-m)$ samples worse. Being "better" is formalized by having a distance $\tilde{R}$ which is smaller than (or equal to) r. This occurs with the distribution $\Pr(\tilde{R} \le r) =: P(r)$ for a single individual. Similarly, being "worse" means larger $\tilde{R}$ values with $\Pr(\tilde{R} > r) = 1 - P(r)$. Consequently, such a constellation has the density $p(r)[P(r)]^{m-1}[1-P(r)]^{\lambda-m}$. There are exactly "$\lambda$ *times* $[(m-1)$ *out of* $(\lambda-1)]$" different, exclusive cases of such constellations. Therefore, one gets the factor $\lambda\binom{\lambda-1}{m-1} = \lambda!/((m-1)!(\lambda-m)!)$ (see also (5.74)).

Inserting (5.10) into the $\varphi_{\mu,\lambda}$ formula (5.9) one obtains

$$\varphi_{\mu,\lambda} = \overline{\langle R \rangle} - \frac{\lambda}{\mu} \int_{r=0}^{\infty} r\, p(r) \sum_{m=1}^{\mu} \frac{(\lambda-1)!}{(m-1)!\,(\lambda-m)!} [P(r)]^{m-1}[1-P(r)]^{\lambda-m}\, \mathrm{d}r. \tag{5.11}$$

For most of the derivations, it is reasonable to express the sum in (5.11) by an integral

$$\begin{aligned} &\sum_{m=1}^{\mu} \frac{(\lambda-1)!}{(m-1)!\,(\lambda-m)!} P^{m-1}[1-P]^{\lambda-m} \\ &\quad = \sum_{k=0}^{\mu-1} \frac{(\lambda-1)!}{k!\,(\lambda-k-1)!} P^k [1-P]^{\lambda-k-1} \\ &\quad = \sum_{k=0}^{\mu-1} \binom{\lambda-1}{k} P^k[1-P]^{\lambda-k-1} = 1 - \sum_{k=\mu}^{\lambda-1} \binom{\lambda-1}{k} P^k[1-P]^{\lambda-k-1} \\ &\quad = 1 - \mathrm{I}_P(\mu, \lambda-\mu) = \mathrm{I}_{1-P}(\lambda-\mu, \mu). \end{aligned} \tag{5.12}$$

The binomial theorem was applied here as well as the relation to the *regularized incomplete beta function* $\mathrm{I}_y(a, b)$ (Abramowitz & Stegun, 1984, p. 83). $\mathrm{I}_y(a, b)$ has the integral representation

$$\mathrm{I}_{1-P(r)}(\lambda - \mu, \mu) = \frac{(\lambda - 1)!}{(\lambda - \mu - 1)!(\mu - 1)!} \int_0^{1-P(r)} x^{\lambda-\mu-1}(1 - x)^{\mu-1} \mathrm{d}x, \quad (5.13)$$

therefore one obtains

$$\sum_{m=1}^{\mu} \frac{P^{m-1}[1 - P]^{\lambda-m}}{(m - 1)!\,(\lambda - m)!} = \frac{1}{(\lambda - \mu - 1)!(\mu - 1)!} \int_0^{1-P(r)} x^{\lambda-\mu-1}(1 - x)^{\mu-1} \mathrm{d}x. \quad (5.14)$$

This relation can be proven by successive integration by parts (see Abramowitz & Stegun, 1984). Inserting (5.14) into (5.11) one gets

$$\varphi_{\mu,\lambda} = \overline{\langle R \rangle} - \frac{\lambda!}{(\lambda - \mu - 1)!\,\mu!} \int_{r=0}^{r=\infty} \int_{x=0}^{x=1-P(r)} rp(r) x^{\lambda-\mu-1}(1 - x)^{\mu-1} \mathrm{d}x\, \mathrm{d}r, \quad (5.15)$$

and exchanging the order of integration yields

$$\varphi_{\mu,\lambda} = \overline{\langle R \rangle} - (\lambda - \mu)\binom{\lambda}{\mu} \int_{x=0}^{x=1} \int_{r=0}^{r=P^{-1}(1-x)} rp(r) x^{\lambda-\mu-1}(1 - x)^{\mu-1} \mathrm{d}r\, \mathrm{d}x. \quad (5.16)$$

Here P^{-1} stands for the inverse function of $P(x)$ (quantile function). Using the substitution $x = 1 - P(t)$, $\mathrm{d}x = -p(t)\,\mathrm{d}t$ one finally gets

$$\varphi_{\mu,\lambda} = \overline{\langle R \rangle} - (\lambda - \mu)\binom{\lambda}{\mu} \int_{t=0}^{t=\infty} p(t)[1 - P(t)]^{\lambda-\mu-1}[P(t)]^{\mu-1} \int_{r=0}^{r=t} rp(r)\, \mathrm{d}r\, \mathrm{d}t. \quad (5.17)$$

5.1.4 Formal Approximation of the Offspring Distribution $p(r)$

The main problem in the analysis of the (μ, λ)-ES is the determination of the offspring distribution density $p(r)$ required in the integral (5.17). It changes from generation to generation, since the position of the parents and descendants change because of the distance changes to the optimum. If we start with parents having all the same object vector, so to say with the same "ancient forbear," the offspring distribution $p^{(g)}(r)$ will successively evolve driven by (5.1) to a (time-dependent) population distribution. However, it is expected that some properties of this generation-dependent distribution will become stationary for large g. The aim is to describe this distribution as simply as possible, and most importantly in an analytical (approximate) manner. Since the mutations are generated using the normal distribution, one expects that the descendants show a similar distribution behavior. Generally, considering the geometry, one expects that the descendants should exhibit a similar behavior to their parents, even if they would be generated by a distribution other than Gaussian. That is, they will be in the state space $\mathcal{Y}$ in the neighborhood of their parents with a greater probability than far away from them.

From the arguments given above, a bell-shaped distribution of the offspring population is also expected for non-Gaussian mutations. Thus, one can assume the Gaussian distribution as a first approximation. Accordingly, an approximate $p(r)$ should be possible by adding correction terms. A function series yielding a "natural" and complete approximation is given by the system of *Hermite polynomials* (see Appendix B.2). It was already used in Chap. 4 for the approximation of the $p(Q)$ distribution (see Sect. 4.1.2).

The standardized quantity z is introduced for the formal expansion of the $p(r)$ density

$$z := \frac{r - \overline{r}}{s}, \tag{5.18}$$

with the expected value $\overline{r}$ of r

$$\overline{r} := E\{r\} = \int_0^\infty r\, p(r)\, \mathrm{d}r \tag{5.19}$$

and the standard deviation $s = \sqrt{s^2}$

$$s^2 := \mathrm{E}\{(r - \overline{r})^2\} = \int_0^\infty r^2 p(r)\, \mathrm{d}r \ - \ \overline{r}^2. \tag{5.20}$$

The formal expansion of the *standardized* $p_z(z)$ distribution is given by the formula (B.47) in Appendix B.3

$$p_z(z) = \frac{\mathrm{e}^{-\frac{1}{2}z^2}}{\sqrt{2\pi}} \left[1 \ + \ \gamma \mathrm{He}_3(z) \ + \ \gamma_2 \mathrm{He}_4(z) \ + \ \gamma_3 \mathrm{He}_5(z) \ + \ \ldots\right], \tag{5.21}$$

with the coefficients[3]

$$\gamma := \frac{\kappa_3\{z\}}{3!}, \qquad \gamma_2 := \frac{\kappa_4\{z\}}{4!}, \qquad \gamma_3 := \frac{\kappa_5\{z\}}{5!}. \tag{5.22}$$

The *cumulants* in this expansion can be expressed by the central moments M_k of r

$$M_k := \int_0^\infty (r - \overline{r})^k p(r)\, \mathrm{d}r. \tag{5.23}$$

[3] Note, one *cannot* infer a generation rule for the terms γ_k with $k > 3$ from the expressions in (5.22); see Appendix B.3, p. 346.

Considering (4.34) one gets[4]

$$\gamma = \frac{1}{6}\frac{M_3}{s^3}, \qquad \gamma_2 = \frac{1}{24}\frac{M_4}{s^4} - \frac{1}{8}, \qquad \gamma_3 = \frac{1}{120}\frac{M_5}{s^5} - \frac{1}{12}\frac{M_3}{s^3}. \tag{5.24}$$

The expansion given in (5.21) is valid for the standard variable z. Because of the transformation rule for densities, $p(r) = p_z(z(r))\left|\frac{\mathrm{d}z}{\mathrm{d}r}\right|$, one can transform back to r

$$p(r) = \frac{\exp\left[-\frac{1}{2}\left(\frac{r-\overline{r}}{s}\right)^2\right]}{\sqrt{2\pi}\,s}\left[1 + \gamma\mathrm{He}_3\left(\frac{r-\overline{r}}{s}\right) + \gamma_2\mathrm{He}_4\left(\frac{r-\overline{r}}{s}\right) + \ldots\right] \tag{5.25}$$

and the distribution function becomes

$$P(r) = \Phi\left(\frac{r-\overline{r}}{s}\right) - \frac{\exp\left[-\frac{1}{2}\left(\frac{r-\overline{r}}{s}\right)^2\right]}{\sqrt{2\pi}}\left[\gamma\mathrm{He}_2\left(\frac{r-\overline{r}}{s}\right) + \gamma_2\mathrm{He}_3\left(\frac{r-\overline{r}}{s}\right) + \ldots\right]. \tag{5.26}$$

The correctness of (5.26) can easily be tested by differentiation. Alternatively, one can also use the expression in (B.59) with (5.22).

With the approach of series expansions (5.25, 5.26), we are now able to approximate the (unknown) offspring distribution in a suitable manner. In what follows, the determination of $p(r)$ will reduce to the self-consistent calculation of the coefficients s, γ, γ_2, etc., which are still undetermined. As one can see in (5.20) and (5.24), these coefficients *only* depend on the *central* moments M_k of the distribution density $p(r)$. In other words, they are *independent* of the expected value of the distribution. This translation invariance is an important and desired property of the chosen approach. Moreover, it makes the calculation of the stationary offspring population possible for the first time. We will come back to this point in Sect. 5.1.10.

In the following, the coefficients of the function series will be estimated statistically. The calculations necessary for that are very long. Therefore, the investigations are carried out only for $\overline{r}$, s, and γ. Such a series cut-off is admissible; see Sect. 5.3.1.2 for the experimental validation. A theoretical estimation of the error produced by this cut-off has not been done yet, it is expected to be very difficult to accomplish. Intuitively, the series cut-off expresses the approximation of the offspring population by a Gaussian distribution that is distorted by a certain *skewness* M_3/s^3, considering that $6\gamma = M_3/s^3$. Furthermore, using the γ_2 term, one would consider the excess of the distribution, where $24\gamma_2 = M_4/s^4 - 3$ is the distribution *excess*.

[4] A proof of the correctness of these coefficients can be obtained by calculating explicitly the moments $\overline{r}$, $M_2 = s^2$, and M_k, as given by the formulae (5.19, 5.20) and (5.23), respectively, using (5.25) as the density. This method is independent of the derivation given in Appendix B.3.2.

5.1.5 Estimation of $\overline{r}$, s, and γ, and of $\overline{r}_{|R_m}$, $M_{2|R_m}$, and $M_{3|R_m}$

The parameters $\overline{r}$, s, and γ for the series approximation can be interpreted as a statistical estimation of the distribution parameters of the actual offspring population produced by the algorithm (5.1, 5.2). We consider a "snapshot" of the ES process (5.1): a parent with the number m is chosen with probability $1/\mu$ from the set $\{\mathbf{R}_1, \dots, \mathbf{R}_\mu\}$. This parent has distance R_m to the optimum. The application of the mutation $\mathbf{z}$ gives a descendant $\tilde{\mathbf{R}}$ with a new distance $\tilde{R}$ to the optimum. The distribution $\Pr(\tilde{R} < r \mid R_m) =: P(r|R_m)$ of such a descendant was already calculated in Sect. 3.4.2 for the N-dependent theory of the $(1, \lambda)$-ES. The density of the distribution can be approximated very well by the normal distribution (3.245); this will be the starting point of the following derivations in this chapter

$$p(r|R_m) = \frac{1}{\sqrt{2\pi}\,\tilde{\sigma}_m} \exp\left[-\frac{1}{2}\left(\frac{r - \sqrt{R_m^2 + \sigma^2 N}}{\tilde{\sigma}_m}\right)^2\right], \tag{5.27}$$

with

$$\tilde{\sigma}_m := \sigma \sqrt{\frac{R_m^2 + \frac{\sigma^2 N}{2}}{R_m^2 + \sigma^2 N}}. \tag{5.28}$$

The index m specifies the parental origin. As one can see, the standard derivation $\tilde{\sigma}$ also depends on the parental state, i.e. on R_m. However, this dependency is relatively weak, as will be shown in the next section.

In (5.27), one has the density $p(r|R_m)$ under the condition that the parent R_m is chosen randomly. Since any of the μ parents can be selected with the same probability $1/\mu$, the offspring density $p(r)$ is obtained as a *mixture* of these μ densities (5.27)

$$p(r \mid R_1, R_2, \dots R_\mu) = \frac{1}{\mu} \sum_{m=1}^{\mu} p(r|R_m). \tag{5.29}$$

It is important to note the conditional nature of this density function. It depends on μ random variables R_m.

Now we can start with the calculation of the statistical parameters. Because of (5.19, 5.20) and (5.24), the moments $\overline{r}$ and M_k should be determined first. In a first step, the conditional moments $\overline{r}_{|R_m}$ and $M_{k|R_m}$ will be calculated. Using (5.27) and (5.29), one obtains

$$\begin{aligned}\overline{r}_{|R_m} &= \int_{r=0}^{\infty} r\, p(r \mid R_1, R_2, \dots, R_\mu)\, \mathrm{d}r \\ &= \int_0^\infty r\, \frac{1}{\mu} \sum_{m=1}^{\mu} \frac{1}{\sqrt{2\pi}\tilde{\sigma}_m} \exp\left[-\frac{1}{2}\left(\frac{r - \sqrt{R_m^2 + \sigma^2 N}}{\tilde{\sigma}_m}\right)^2\right] \mathrm{d}r. \end{aligned} \tag{5.30}$$

Substituting $r = \tilde{\sigma}_m t + \sqrt{R_m^2 + \sigma^2 N}$ one gets

$$\overline{r}_{|R_m} = \frac{1}{\mu}\sum_{m=1}^{\mu}\frac{1}{\sqrt{2\pi}}\int_{-\infty}^{\infty}\left(\tilde{\sigma}_m t + \sqrt{R_m^2+\sigma^2 N}\right)\mathrm{e}^{-\frac{1}{2}t^2}\,\mathrm{d}t. \tag{5.31}$$

The lower limit $t = -\sqrt{R_m^2+\sigma^2 N}/\tilde{\sigma}_m$ was shifted to $t = -\infty$. The discussion on the admissibility of this simplification was done in Sect. 3.4.5, p. 111. The principal idea is that the factor $\mathrm{e}^{-t^2/2}$ in the integrand of (5.31) goes very quickly to zero. As a result, the tails of the normal distribution do not play a role in the evaluation of the integral. Therefore, the lower integration limit is changed to $-\infty$ as long as N is sufficiently large in all following calculations.

The result of the integration of (5.31) reads

$$\overline{r}_{|R_m} = \frac{1}{\mu}\sum_{m=1}^{\mu}\sqrt{R_m^2+\sigma^2 N}. \tag{5.32}$$

For the kth order central moments $M_{k|R_m}$

$$M_{k|R_m} = \int_0^{\infty}(r-\overline{r}_{|R_m})^k p(r \mid R_1, R_2, \ldots, R_\mu)\,\mathrm{d}r, \tag{5.33}$$

one finds after similar calculations

$$M_{k|R_m} = \frac{1}{\mu}\sum_{m=1}^{\mu}\frac{1}{\sqrt{2\pi}}\int_{-\infty}^{\infty}\left[\tilde{\sigma}_m t + \left(\sqrt{R_m^2+\sigma^2 N} - \overline{r}_{|R_m}\right)\right]^k \mathrm{e}^{-\frac{1}{2}t^2}\,\mathrm{d}t.$$

Hence, one gets for the square of the standard deviation $s^2_{|R_m} = M_{2|R_m}$

$$M_{2|R_m} = \frac{1}{\mu}\sum_{m=1}^{\mu}\tilde{\sigma}_m^2 + \frac{1}{\mu}\sum_{m=1}^{\mu}\left(\sqrt{R_m^2+\sigma^2 N} - \overline{r}_{|R_m}\right)^2 \tag{5.34}$$

and for $M_{3|R_m}$ (5.23)

$$\begin{aligned} M_{3|R_m} = {} & \frac{3}{\mu}\sum_{m=1}^{\mu}\tilde{\sigma}_m^2\left(\sqrt{R_m^2+\sigma^2 N} - \overline{r}_{|R_m}\right) \\ & + \frac{1}{\mu}\sum_{m=1}^{\mu}\left(\sqrt{R_m^2+\sigma^2 N} - \overline{r}_{|R_m}\right)^3 . \end{aligned} \tag{5.35}$$

Here we have obtained the expressions of $\overline{r}_{|R_m}$, $s^2_{|R_m}$, and $M_{3|R_m}$ as functions of type $f_i = f_i(R_1, \ldots, R_\mu)$ depending on the parental R_m. Therefore, we are now closer to the estimation of $\overline{r}$, s, and γ. The relation to the offspring of the previous generation will be established next. The parental R_m is obtained as the mth order statistic of the offspring population of the previous generation $R_m^{(g+1)} = \tilde{R}_{m:\lambda}^{(g)}$. Thus, statistical estimation means taking expected values of the functions f_i (i.e. $\overline{r}_{|R_m}$, $M_{2|R_m}$, and $M_{3|R_m}$) with respect to the first μ order statistics. Hence, one has to deal with μ-dimensional expected value integrals over the corresponding domain $R_1 \le R_2 \le \ldots \le R_\mu$. The integrals are of the form

$$\mathrm{E}\{f_i\} = \int\limits_{R_1 \le} \int\limits_{R_2 \le} \cdots \int\limits_{\le R_\mu} f_i(R_1, \ldots, R_\mu)\, p_o(R_1, \ldots, R_\mu)\, \mathrm{d}R_1 \cdots \mathrm{d}R_\mu. \quad (5.36)$$

The joint density p_o reads[5]

$$p_o(R_1, \ldots, R_\mu) = \frac{\lambda!}{(\lambda - \mu)!} \left[1 - P(R_\mu)\right]^{\lambda-\mu} \prod_{m=1}^{\mu} p(R_m), \quad (5.37)$$

with the distribution function $P(r) = P^{(g)}(r)$ and the density $p(r) = p^{(g)}(r)$ of the offspring of the previous generation g. These will again be approximations of type (5.25, 5.26). Therefore, $\overline{r}^{(g+1)}$, $s^{(g+1)}$, and $\gamma^{(g+1)}$ depend according to (5.19, 5.20) and (5.24) on $\overline{r}^{(g)}$, $s^{(g)}$, and $\gamma^{(g)}$.

With the integral (5.36), we have found an expression for the estimation of $\overline{r}^{(g+1)}$, $s^{(g+1)}$, and $\gamma^{(g+1)}$

$$\overline{r}^{(g+1)} = \mathrm{E}\{\overline{r}_{|R_m}\}, \qquad s^{(g+1)} = \sqrt{\mathrm{E}\{M_{2|R_m}\}}, \quad (5.38)$$

and

$$\gamma^{(g+1)} = \frac{1}{6} \frac{\mathrm{E}\{M_{3|R_m}\}}{\left(\mathrm{E}\{M_{2|R_m}\}\right)^{3/2}}, \quad (5.39)$$

however, in terms of μ-dimensional integrals. The calculation of (5.36) should be very difficult, if it is possible at all for any given μ. Therefore, it is reasonable to simplify the functions $f_i(R_1, \ldots, R_\mu)$ by appropriate approximations so that the integral will become tractable. The necessary derivations and approximations can be summarized in three independent parts. The first two will be performed in this section. The third step will follow in Sect. 5.2:

1. Simplification of $\overline{r}_{|R_m}$, $M_{2|R_m}$, and $M_{3|R_m}$. It will be the aim to substitute the explicit R_m dependency by the three random variables $\langle R \rangle$, $\langle (R - \langle R \rangle)^2 \rangle =: \langle (\Delta R)^2 \rangle$, and $\langle (R - \langle R \rangle)^3 \rangle =: \langle (\Delta R)^3 \rangle$. These approximations affect the sphere model, but not the hyperplane. In other words, the formulae are exact for $\sigma = \text{const}$ and $R \to \infty$. The derivation follows in Sect. 5.1.6.
2. Statistical approximation of $\overline{r}^{(g+1)}$, $s^{(g+1)}$, and $\gamma^{(g+1)}$. The aim is to express these quantities as functions of $\mathrm{E}\{\langle R \rangle\} = \overline{\langle R \rangle}$, $\overline{\langle (\Delta R)^2 \rangle}$, and $\overline{\langle (\Delta R)^3 \rangle}$. The corresponding expected value calculations are performed according to (5.36). This is the content of Sect. 5.1.7. This approximation is also exact for the hyperplane.

[5] This density can be constructed easily: the mth best descendant (of the previous generation) has the density $p(R_m)$. Using these densities, one obtains the product in the formula. All remaining $(\lambda - \mu)$ descendants must have $\tilde{R}$ values, each of which is larger than R_μ. This occurs with probability $\Pr(\tilde{R} > R_\mu) = 1 - P(R_\mu)$ for each single descendant. The factor $\lambda!/(\lambda - \mu)!$ counts again all possible constellations.

Nothing would become more tractable after these steps if one had to carry out the integration in (5.36). Therefore, in Sects. 5.1.8 and 5.1.9 integral expressions will be derived for $\overline{\langle(\Delta R)^2\rangle}$ and $\overline{\langle(\Delta R)^3\rangle}$. These will free us from the μ-fold integral (5.36).

3. Up to this point, the series expansion (5.25, 5.26) for the distribution of the offspring population $P(r)$, has not been used at all. This will be done in Sect. 5.2. The products of the densities $p(r)$ with the distribution functions $P(r)$ will be evaluated, yielding terms with powers of γ. We will make use of the linear approximation, i.e. neglect all γ^k with $k > 1$. This is admissible since the skewness of the offspring population can be assumed to be small.

5.1.6 The Simplification of $\overline{r}_{|R_m}$, $M_{2|R_m}$, and $M_{3|R_m}$

The simplification of (5.32, 5.34) and (5.35) is based on the assumption that the random variables R_m are concentrated around the mean value $\langle R\rangle$ (defined by (5.5)). Defining the deviation ΔR_m from the mean value $\langle R\rangle$ as

$$\Delta R_m := R_m - \langle R\rangle, \qquad \langle R\rangle = \frac{1}{\mu}\sum_{m=1}^{\mu} R_m, \tag{5.40}$$

one can express this assumption in another way: $|\Delta R_m| \ll \langle R\rangle$ should be satisfied in a *statistical mean*. This is the case if all the moments have this property, in other words

$$\mathrm{E}\left\{\left(\frac{|\Delta R_m|}{\langle R\rangle}\right)^k\right\} \ll 1 \quad \Longrightarrow \quad \frac{R_m}{\langle R\rangle} = \frac{\Delta R_m + \langle R\rangle}{\langle R\rangle} = 1 + \frac{\Delta R_m}{\langle R\rangle}. \tag{5.41}$$

Indeed, this condition is always fulfilled for sufficiently small mutation strengths σ: the average change in R per generation, ΔR, is given by φ; therefore, considering (5.8) one gets

$$\frac{\overline{\Delta R}}{\langle R\rangle} = \frac{\varphi^*}{N} = \mathcal{O}\left(\frac{1}{N}\right). \tag{5.42}$$

It is assumed here that φ^* is of order $\mathcal{O}(1)$ with respect to N. This is the case for $\varphi^* \gtrapprox 0$, as one knows from the N-dependent analysis of the $(1,\lambda)$-ES.

After these justifications, we can investigate $\overline{r}_{|R_m}$ closer. The square root in (5.32) will be expanded in a Taylor series. To this end, we write

$$\begin{aligned}\sqrt{R_m^2 + \sigma^2 N} &= \sqrt{(\langle R\rangle + \Delta R_m)^2 + \sigma^2 N} \\ &= \langle R\rangle\sqrt{1 + \frac{\sigma^2 N}{\langle R\rangle^2} + 2\left(\frac{\Delta R_m}{\langle R\rangle}\right) + \left(\frac{\Delta R_m}{\langle R\rangle}\right)^2}.\end{aligned}$$

The first terms of the Taylor expansion are

$$\sqrt{R_m^2+\sigma^2 N} = \langle R\rangle \left[\sqrt{1+\frac{\sigma^2 N}{\langle R\rangle^2}} + \frac{1}{\sqrt{1+\frac{\sigma^2 N}{\langle R\rangle^2}}}\left(\frac{\Delta R_m}{\langle R\rangle}\right) + \underbrace{\frac{1}{2}\frac{\frac{\sigma^2 N}{\langle R\rangle^2}}{\left(\sqrt{1+\frac{\sigma^2 N}{\langle R\rangle^2}}\right)^3}}_{=: h_1 \le 1/(\sqrt{3})^3}\left(\frac{\Delta R_m}{\langle R\rangle}\right)^2 - \mathcal{O}\left(\frac{\Delta R_m}{\langle R\rangle}\right)^3\right]. \tag{5.43}$$

Using this result, one gets for (5.32)

$$\overline{r}_{|R_m} = \frac{1}{\mu}\sum_{m=1}^{\mu}\sqrt{R_m^2+\sigma^2 N} = \langle R\rangle\sqrt{1+\frac{\sigma^2 N}{\langle R\rangle^2}} + \frac{\langle R\rangle}{\mu}\sum_{m=1}^{\mu}\mathcal{O}\left(\frac{\Delta R_m}{\langle R\rangle}\right)^2$$

$$\overline{r}_{|R_m} \approx \langle R\rangle\sqrt{1+\frac{\sigma^2 N}{\langle R\rangle^2}}. \tag{5.44}$$

Here, it was taken into account that the linear term of (5.43) vanishes because of (5.40), that is

$$\sum_{m=1}^{\mu}\Delta R_m = 0. \tag{5.45}$$

The simplification of $M_{2|R_m}$ in Eq. (5.34) requires two steps. We will treat $\tilde{\sigma}_m^2$ first. In order to shorten the notation, $x := x_0 + h = \langle R\rangle + \Delta R_m$ and $a := \sigma^2 N/2$ are substituted in (5.28). Thereafter, $\tilde{\sigma}_m^2/\sigma^2$ is expanded in a Taylor series at $x_0 = \langle R\rangle$

$$\frac{\tilde{\sigma}_m^2}{\sigma^2} = \frac{x^2+a}{x^2+2a} = 1 - \frac{a}{x^2+2a}$$

$$\frac{\tilde{\sigma}_m^2}{\sigma^2} = \frac{x_0^2+a}{x_0^2+2a} + \frac{2x_0 a}{(x_0^2+2a)^2}h - \frac{3ax_0^2-2a^2}{(x_0^2+2a)^3}h^2 + \cdots$$

$$\frac{\tilde{\sigma}_m^2}{\sigma^2} = \frac{1+\frac{\sigma^2 N}{2\langle R\rangle^2}}{1+\frac{\sigma^2 N}{\langle R\rangle^2}} + \frac{\frac{\sigma^2 N}{\langle R\rangle^2}}{\left(1+\frac{\sigma^2 N}{\langle R\rangle^2}\right)^2}\frac{\Delta R_m}{\langle R\rangle} - \underbrace{\frac{1}{2}\frac{3\frac{\sigma^2 N}{\langle R\rangle^2} - \left(\frac{\sigma^2 N}{\langle R\rangle^2}\right)^2}{\left(1+\frac{\sigma^2 N}{\langle R\rangle^2}\right)^3}}_{:= h_2(x)}\left(\frac{\Delta R_m}{\langle R\rangle}\right)^2 + \ldots . \tag{5.46}$$

The expression $h_2(x)$ underbraced in (5.46) with $x := \sigma^2 N/\langle R\rangle^2$ is bounded; its maximum/minimum is at $x_{1/2} = 4 \mp \sqrt{13}$. Hence, the last term in (5.46) is of order $\mathcal{O}(\Delta R_m/\langle R\rangle)^2$. Considering (5.45), the sum over $\tilde{\sigma}_m^2$ yields

$$\frac{1}{\mu}\sum_{m=1}^{\mu}\tilde{\sigma}_m^2 = \sigma^2\left[\frac{1+\frac{\sigma^2 N}{2\langle R\rangle^2}}{1+\frac{\sigma^2 N}{\langle R\rangle^2}} - \frac{h_2}{\mu}\sum_{m=1}^{\mu}\left(\frac{\Delta R_m}{\langle R\rangle}\right)^2 + \ldots\right]. \tag{5.47}$$

Due to (5.42), the error term, i.e. the sum in the brackets, vanishes with $\mathcal{O}(1/N^2)$. It will therefore be neglected in the following.

The second term to be simplified in the $M_{2|R_m}$ expression (5.34) is $\sqrt{R_m^2+\sigma^2 N} - \overline{r}_{|R_m}$. Using (5.43) and (5.44), one obtains

$$\sqrt{R_m^2+\sigma^2 N} - \overline{r}_{|R_m} = \frac{\Delta R_m}{\sqrt{1+\frac{\sigma^2 N}{\langle R\rangle^2}}} + h_1\left[\frac{(\Delta R_m)^2}{\langle R\rangle} - \frac{1}{\mu}\sum_{m'=1}^{\mu}\frac{(\Delta R_{m'})^2}{\langle R\rangle}\right] + \ldots. \tag{5.48}$$

The term in brackets in (5.48) is comparatively small: it is of order $\Delta R_m/\langle R\rangle$ since $|\Delta R_m|/\langle R\rangle \ll 1$ and therefore $(\Delta R_m)^2/\langle R\rangle \ll |\Delta R_m|$ considering (5.41, 5.42). Squaring (5.48) and summing over μ, taking (5.45) into account and the definition of h_1 (see (5.43)), one gets

$$\frac{1}{\mu}\sum_{m=1}^{\mu}\left(\sqrt{R_m^2+\sigma^2 N} - \overline{r}_{|R_m}\right)^2 = \frac{1}{\mu}\sum_{m=1}^{\mu}\frac{(\Delta R_m)^2}{1+\frac{\sigma^2 N}{\langle R\rangle^2}}\left[1+\frac{\frac{\sigma^2 N}{\langle R\rangle^2}}{1+\frac{\sigma^2 N}{\langle R\rangle^2}}\frac{\Delta R_m}{\langle R\rangle}\right] + \ldots. \tag{5.49}$$

Provided that (5.42) holds, the second term in the brackets is of order $\mathcal{O}(1/N)$. As a result, it can be neglected in the following. Now, $s_{|R_m}^2$ can be put together. Using (5.47) and (5.49), one gets $s_{|R_m}^2$ defined by (5.34) as

$$s_{|R_m}^2 = M_{2|R_m} = \sigma^2\frac{1+\frac{\sigma^2 N}{2\langle R\rangle^2}}{1+\frac{\sigma^2 N}{\langle R\rangle^2}} + \frac{1}{1+\frac{\sigma^2 N}{\langle R\rangle^2}}\frac{1}{\mu}\sum_{n=1}^{\mu}(\Delta R_n)^2 + \ldots. \tag{5.50}$$

Similarly, $M_{3|R_m}$ is obtained in two steps. The first term of (5.35) will be investigated first. For this purpose, (5.46) and (5.48) are used. Considering (5.45), one finds

$$\frac{3}{\mu}\sum_{m=1}^{\mu}\tilde{\sigma}_m^2\left(\sqrt{R_m^2+\sigma^2 N} - \overline{r}_{|R_m}\right) = 3\sigma^2\frac{\frac{\sigma^2 N}{2\langle R\rangle^2}}{\left(1+\frac{\sigma^2 N}{\langle R\rangle^2}\right)^{5/2}}\frac{1}{\mu}\sum_{m=1}^{\mu}\frac{(\Delta R_m)^2}{\langle R\rangle} + \ldots. \tag{5.51}$$

The second term is obtained as the sum of the third powers of (5.48)

$$\frac{1}{\mu}\sum_{m=1}^{\mu}\left(\sqrt{R_m^2+\sigma^2 N}-\overline{r}_{|R_m}\right)^3 = \frac{1}{\mu}\sum_{m=1}^{\mu}\frac{(\Delta R_m)^3}{\left(1+\frac{\sigma^2 N}{\langle R\rangle^2}\right)^{3/2}}$$
$$+\frac{3}{2}\frac{\frac{\sigma^2 N}{\langle R\rangle^2}}{\left(1+\frac{\sigma^2 N}{\langle R\rangle^2}\right)^{5/2}}\frac{1}{\mu}\sum_{m=1}^{\mu}\frac{1}{\langle R\rangle}\left[(\Delta R_m)^4-\frac{(\Delta R_m)^2}{\mu}\sum_{m'=1}^{\mu}(\Delta R_{m'})^2\right]+\ldots. \tag{5.52}$$

As can be shown easily, the sum over ΔR_m in the brackets is not negative. However, it is smaller than or equal to $\sum_m (\Delta R_m)^4/\langle R\rangle$. Therefore, the second line can be neglected compared to the first one. The second line is smaller, and is of order $\Delta R_m/\langle R\rangle$.

Now, $M_{3|R_m}$, Eq. (5.35), can be determined. Neglecting (5.51) and the second line of (5.52) one obtains

$$M_{3|R_m} = \frac{1}{\mu}\sum_{m=1}^{\mu}\frac{(\Delta R_m)^3}{\left(1+\frac{\sigma^2 N}{\langle R\rangle^2}\right)^{3/2}}. \tag{5.53}$$

5.1.7 The Statistical Approximation of $\overline{r}$, s, and γ

The following expected values must be determined for the statistical approximation of $\overline{r}$, s, and γ, according to (5.38) and (5.39):

$$\overline{r} = \mathrm{E}\left\{\sqrt{\langle R\rangle^2+\sigma^2 N}\right\}, \tag{5.54}$$

$$M_2 = \mathrm{E}\left\{\sigma^2\frac{\langle R\rangle^2+\frac{\sigma^2 N}{2}}{\langle R\rangle^2+\sigma^2 N}+\frac{1}{1+\frac{\sigma^2 N}{\langle R\rangle^2}}\frac{1}{\mu}\sum_{m=1}^{\mu}(\Delta R_m)^2\right\}, \tag{5.55}$$

$$M_3 = \mathrm{E}\left\{\frac{1}{\left(1+\frac{\sigma^2 N}{\langle R\rangle^2}\right)^{3/2}}\frac{1}{\mu}\sum_{m=1}^{\mu}(\Delta R_m)^3\right\}. \tag{5.56}$$

Since the R_m contain square roots and fractions, one cannot integrate (5.36) in an analytically closed form. Hence, a statistical approximation will be used here. This approximation assumes that the statistical fluctuations of $\langle R\rangle$ are small around the mean value $\overline{\langle R\rangle}$ as compared to the expected value $\overline{\langle R\rangle}$

$$\mathrm{E}\left\{\frac{|\langle R\rangle-\overline{\langle R\rangle}|}{\overline{\langle R\rangle}}\right\} \ll 1. \tag{5.57}$$

Geometrically, this assumption means that the statistical quantity "mean residual distance of the parental generation" $\langle R\rangle$ differs only slightly from its expected value $\overline{\langle R \rangle}$. This is again a valid assumption for the sphere model. For $\langle R \rangle \to \infty$, one obtains asymptotically *exact* formulae for the hyperplane.

Under the assumption (5.57), the functions of the expected values in (5.54, 5.55) and (5.56) can be expanded using a Taylor series. We start with the square root $\sqrt{\langle R \rangle^2 + \sigma^2 N}$

$$\sqrt{\langle R \rangle^2 + \sigma^2 N} = \sqrt{\left[\overline{\langle R \rangle} + \left(\langle R \rangle - \overline{\langle R \rangle}\right)\right]^2 + \sigma^2 N}$$

$$= \sqrt{\overline{\langle R \rangle}^2 + \sigma^2 N} + \overline{\langle R \rangle} \frac{\langle R \rangle - \overline{\langle R \rangle}}{\sqrt{\overline{\langle R \rangle}^2 + \sigma^2 N}} + \frac{\sigma^2 N}{2} \frac{\left(\langle R \rangle - \overline{\langle R \rangle}\right)^2}{\left(\overline{\langle R \rangle}^2 + \sigma^2 N\right)^{3/2}} + \ldots .$$

Evaluating the expected value, the linear term vanishes and one obtains

$$\overline{r} = \overline{\langle R \rangle} \left[\sqrt{1 + \frac{\sigma^2 N}{\overline{\langle R \rangle}^2}} + \frac{1}{2} \frac{\frac{\sigma^2 N}{\overline{\langle R \rangle}^2}}{\left(1 + \frac{\sigma^2 N}{\overline{\langle R \rangle}^2}\right)^{3/2}} \mathrm{E}\left\{ \frac{\left(\langle R \rangle - \overline{\langle R \rangle}\right)^2}{\overline{\langle R \rangle}^2} \right\} + \ldots \right].$$

The factor in the quadratic term is bounded (compare with the h_1 term in (5.43)). Hence, the quadratical term is of order $\mathcal{O}((\langle R \rangle - \overline{\langle R \rangle})/\overline{\langle R \rangle})^2$. It can be neglected because of (5.57). Considering the normalization (5.8) for σ, one obtains the result in terms of[6]

$$\overline{r} = \overline{\langle R \rangle}\, \zeta + \mathcal{R}_{\overline{r}} \qquad \text{with} \qquad \zeta := \sqrt{1 + \frac{\sigma^{*2}}{N}}. \tag{5.58}$$

The approximation of M_2, Eq. (5.55), has two terms. For the first one, the Taylor expansion at $\langle R \rangle = \overline{\langle R \rangle}$ gives

$$\sigma^2 \frac{\langle R \rangle^2 + \frac{\sigma^2 N}{2}}{\langle R \rangle^2 + \sigma^2 N} = \sigma^2 \left[1 - \frac{\sigma^2 N}{2} \frac{1}{\langle R \rangle^2 + \sigma^2 N} \right]$$

$$= \sigma^2 \left[\frac{\overline{\langle R \rangle}^2 + \frac{\sigma^2 N}{2}}{\overline{\langle R \rangle}^2 + \sigma^2 N} + \overline{\langle R \rangle} \sigma^2 N \frac{\langle R \rangle - \overline{\langle R \rangle}}{\left(\overline{\langle R \rangle}^2 + \sigma^2 N\right)^2} \right.$$

$$\left. - \frac{\sigma^2 N}{2} \frac{3\overline{\langle R \rangle}^2 - \sigma^2 N}{\left(\overline{\langle R \rangle}^2 + \sigma^2 N\right)^3} \left(\langle R \rangle - \overline{\langle R \rangle}\right)^2 + \ldots \right]. \tag{5.59}$$

Using (5.8), one obtains for the expected value

[6] In the following, the error term $\mathcal{R}_{\overline{r}}$ will not be explicitly stated, since already for $\overline{r}_{|R_m}$ the approximation (5.44) was used. A more satisfactory estimation of the error made by the approximation has not been done yet.

$$\mathrm{E}\left\{\sigma^2 \frac{\langle R\rangle^2 + \frac{\sigma^2 N}{2}}{\langle R\rangle^2 + \sigma^2 N}\right\}$$
$$= \sigma^2 \left[\frac{1+\frac{\sigma^{*2}}{2N}}{1+\frac{\sigma^{*2}}{N}} - \frac{1}{2}\frac{\frac{\sigma^{*2}}{N}\left(3-\frac{\sigma^{*2}}{N}\right)}{\left(1+\frac{\sigma^{*2}}{N}\right)^3} \mathrm{E}\left\{\frac{\left(\langle R\rangle - \overline{\langle R\rangle}\right)^2}{\overline{\langle R\rangle}^2}\right\} + \ldots\right]. \quad (5.60)$$

The factor in front of $\mathrm{E}\{\cdots\}$ in the braces is bounded. Consequently, the second term can be neglected under the assumption (5.57).

The second term of (5.55) should be handled more cautiously. It contains terms depending on R_m which are *not* caused by the sphere model. Hence, they should be conserved if possible. Such terms can be identified easily since they do not vanish for the limit $\langle R\rangle \to \infty$. In our case, this is the $\sum_m (\Delta R_m)^2$ term. Therefore, we will only expand the factor in front of the sum. It has the form $(1+\sigma^2 N/\langle R\rangle^2)^{-k/2}$, with $k=2$. For the general case, the Taylor expansion at $\langle R\rangle = \overline{\langle R\rangle}$ gives

$$\frac{1}{\left(1+\frac{\sigma^2 N}{\langle R\rangle^2}\right)^{k/2}} = \frac{1}{\left(1+\frac{\sigma^2 N}{\overline{\langle R\rangle}^2}\right)^{k/2}} \left[1 + k\frac{\frac{\sigma^2 N}{\overline{\langle R\rangle}^2}}{1+\frac{\sigma^2 N}{\overline{\langle R\rangle}^2}} \frac{\langle R\rangle - \overline{\langle R\rangle}}{\overline{\langle R\rangle}}\right.$$
$$\left. + \frac{k}{2}\frac{\frac{\sigma^2 N}{\overline{\langle R\rangle}^2}\left((k-1)\frac{\sigma^2 N}{\overline{\langle R\rangle}^2} - 3\right)}{\left(1+\frac{\sigma^2 N}{\overline{\langle R\rangle}^2}\right)^2} \left(\frac{\langle R\rangle - \overline{\langle R\rangle}}{\overline{\langle R\rangle}}\right)^2 + \ldots\right]. \quad (5.61)$$

Because of (5.57), the linear term can be assumed to be small yielding the expected value

$$\mathrm{E}\left\{\frac{1}{\left(1+\frac{\sigma^2 N}{\langle R\rangle^2}\right)^{k/2}} \frac{1}{\mu}\sum_{m=1}^{\mu} (\Delta R_m)^k\right\} = \frac{\overline{\langle(\Delta R)^k\rangle}}{\zeta^k} + \mathcal{R}_k. \quad (5.62)$$

Here the definition of ζ in (5.58) has been considered, and the abbreviation $\overline{\langle(\Delta R)^k\rangle}$ is introduced. It stands for

$$\overline{\langle(\Delta R)^k\rangle} := \frac{1}{\mu}\sum_{m=1}^{\mu} \mathrm{E}\big\{(\Delta R_m)^k\big\}. \quad (5.63)$$

The error term of (5.62) is not considered further. However, it vanishes for $\langle R\rangle \to \infty$ (hyperplane condition).

Now M_2, given by (5.55), can be assembled leading to $s = \sqrt{M_2}$. With (5.60, 5.62) and the definition of ζ in (5.58), one obtains

$$s = \frac{1}{\zeta}\sqrt{\sigma^2 \frac{1+\zeta^2}{2} + \overline{\langle(\Delta R)^2\rangle}} + \mathcal{R}_s. \quad (5.64)$$

$\mathcal{R}_s$ stands for the error term related to the sphere model. It will be ignored in the following calculations.

The approximation of γ starts with (5.39), $\gamma = M_3/6s^3$ (see (5.53) and (5.56)). Consequently, one gets, using (5.62) and (5.64),

$$\gamma = \frac{1}{6} \frac{\overline{\langle(\Delta R)^3\rangle}}{\left(\sqrt{\sigma^2 \frac{1+\zeta^2}{2} + \overline{\langle(\Delta R)^2\rangle}}\right)^3} \; + \; \mathcal{R}_\gamma. \tag{5.65}$$

Again, the error term $\mathcal{R}_\gamma$ is defined by the sphere model. Similar to $\mathcal{R}_{\overline{r}}$ in Eq. (5.58) and $\mathcal{R}_s$ in Eq. (5.64), it will be neglected in the following.

We have now expressed s and γ as functions of the expected values $\overline{\langle(\Delta R)^2\rangle}$ and $\overline{\langle(\Delta R)^3\rangle}$. These expected values should be determined by the μ-fold integral (5.36). However, there is an alternative way for the determination of the expected values without using (5.36). For $\overline{\langle(\Delta R)^2\rangle}$, the derivation follows in the next subsection. The case $\overline{\langle(\Delta R)^3\rangle}$ is treated in Sect. 5.1.9.

5.1.8 The Integral Expression of $\overline{\langle(\Delta R)^2\rangle}$

In order to calculate $\overline{\langle(\Delta R)^2\rangle}$, one needs to compute the expected value of the expression

$$\langle(\Delta R)^2\rangle = \frac{1}{\mu}\sum_{m=1}^{\mu}\left(R_m - \frac{1}{\mu}\sum_{m'=1}^{\mu} R_{m'}\right)^2, \tag{5.66}$$

considering (5.63) and (5.40). For this purpose, $\langle(\Delta R)^2\rangle$ is written in sums of products of *ordered* R_m. In other words, the aim of the rearrangement is to use only products $R_m \cdot R_k$ with $R_m \le R_k$ in the expression for $\langle(\Delta R)^2\rangle$. First of all, the square is performed in (5.66), yielding

$$\begin{aligned}\langle(\Delta R)^2\rangle &= \frac{1}{\mu}\sum_{m=1}^{\mu} R_m^2 - \left(\frac{1}{\mu}\sum_{m=1}^{\mu} R_m\right)^2 \\ &= \frac{1}{\mu}\sum_{m=1}^{\mu} R_m^2 - \frac{1}{\mu^2}\sum_{l=1}^{\mu}\sum_{k=1}^{\mu} R_k R_l.\end{aligned} \tag{5.67}$$

The first term in (5.67) already fulfills the ordering condition. However, the second one contains products $R_k R_l$ with $k > l$ and $R_k \not\le R_l$. Therefore, we continue with the transformations

$$\begin{aligned}\langle(\Delta R)^2\rangle &= \frac{\mu-1}{\mu}\frac{1}{\mu}\sum_{m=1}^{\mu} R_m^2 - \frac{1}{\mu^2}\sum_{l=1}^{\mu}\sum_{k\ne l} R_k R_l \\ \langle(\Delta R)^2\rangle &= \frac{\mu-1}{\mu}\frac{1}{\mu}\sum_{m=1}^{\mu} R_m^2 - \frac{1}{\mu^2}\sum_{l=2}^{\mu}\sum_{k=1}^{l-1} R_k R_l - \frac{1}{\mu^2}\sum_{k=2}^{\mu}\sum_{l=1}^{k-1} R_l R_k \\ \langle(\Delta R)^2\rangle &= \frac{\mu-1}{\mu}\frac{1}{\mu}\sum_{m=1}^{\mu} R_m^2 - \frac{2}{\mu^2}\sum_{l=2}^{\mu}\sum_{k=1}^{l-1} R_k R_l\,.\end{aligned} \tag{5.68}$$

In this way, we are now able to express $\langle(\Delta R)^2\rangle$ in terms of ordered products. These products obey the relation $R_k \le R_l$ for $k < l$. Consequently, the expected value of $\langle(\Delta R)^2\rangle$ can be written as

$$\overline{\langle(\Delta R)^2\rangle} = \frac{\mu-1}{\mu}\frac{1}{\mu}\sum_{m=1}^{\mu}\overline{R_m^2} - \frac{2}{\mu^2}\sum_{l=2}^{\mu}\sum_{k=1}^{l-1}\overline{R_k R_l}\,. \tag{5.69}$$

Now, one can continue with the integral expression of the product moments of the order statistics of R_m.

The second moment of the mth order statistic of the offspring distribution $P(r)$ is denoted by $\overline{R_m^2}$, and its density by $p_{m:\lambda}(r)$. In general, the ath moment reads

$$\overline{R_m^a} = \int_0^\infty r^a p_{m:\lambda}(r)\,\mathrm{d}r. \tag{5.70}$$

The density $p_{m:\lambda}(r)$ was already calculated. Using Eq. (5.10), one obtains

$$\frac{1}{\mu}\sum_{m=1}^{\mu}\overline{R_m^a}$$
$$= \frac{\lambda}{\mu}\int_0^\infty r^a p(r)\sum_{m=1}^{\mu}\frac{(\lambda-1)!}{(m-1)!\,(\lambda-m)!}\,[P(r)]^{m-1}[1-P(r)]^{\lambda-m}\mathrm{d}r. \tag{5.71}$$

Using the same (formal) transformations used to obtain Eq. (5.17) from Eq. (5.11), one gets

$$\frac{1}{\mu}\sum_{m=1}^{\mu}\overline{R_m^a} = (\lambda-\mu)\binom{\lambda}{\mu}\int_{t=0}^{t=\infty} p(t)\,[1-P(t)]^{\lambda-\mu-1}[P(t)]^{\mu-1}$$
$$\times \int_{r=0}^{r=t} r^a p(r)\,\mathrm{d}r\,\mathrm{d}t. \tag{5.72}$$

The calculation of the product moments $\overline{R_k R_l}$ over the ordered domain $r_k \le r_l$, or more generally

$$\overline{R_k^a R_l^b} = \int_{r_l=0}^{r_l=\infty}\int_{r_k=0}^{r_k=r_l} r_k^a\, r_l^b\, p_{k\,l:\lambda}(r_k, r_l)\,\mathrm{d}r_k\,\mathrm{d}r_l, \tag{5.73}$$

requires the knowledge of the joint density $p_{k\,l:\lambda}(r_k, r_l)$. The $p_{k\,l:\lambda}(r_k, r_l)$ density can be derived in a relatively easy way. Considering the "ordered event scale" of λ descendants $\tilde{R}_l$, where the $\tilde{R}_l$ are sorted by their magnitude, we have

$$\begin{array}{ccccc}
\underbrace{\tilde{R}_{1:\lambda}\cdots\tilde{R}_{k-1:\lambda}}_{(k-1)} & \underbrace{\le \tilde{R}_{k:\lambda}}_{(1)} & \underbrace{\le \tilde{R}_{k+1:\lambda}\cdots\tilde{R}_{l-1:\lambda}}_{(l-k-1)} & \underbrace{\le \tilde{R}_{l:\lambda}}_{(1)} & \underbrace{\le \tilde{R}_{l+1:\lambda}\cdots\tilde{R}_{\lambda:\lambda}}_{(\lambda-l)} \\
[\Pr(\tilde{R}\le r_k)]^{k-1} & p(r_k) & [\Pr(r_k\le\tilde{R}\le r_l)]^{l-k-1} & p(r_l) & [\Pr(\tilde{R}> r_l)]^{\lambda-l}.
\end{array} \tag{5.74}$$

The single terms in the product of probabilities and densities are immediately clear. Moreover, one can easily read the number of independent possible orderings. There are $\lambda!$ permutations in a total of λ descendants, $(k-1)!$ of them are in the subgroup $\tilde{R} \le r_k$, $(l-k-1)!$ in the subgroup $r_k \le \tilde{R} \le r_l$, and $(\lambda - l)!$ in the subgroup $\tilde{R} > r_l$. Therefore, the number ("No.") of independent orderings must fulfill $\lambda! = \text{No.} \times (k-1)!\,(l-k-1)!\,(\lambda-l)!$ Hence, one gets for the joint density

$$p_{k\,l:\lambda}(r_k, r_l) = \lambda!\,\frac{[P(r_k)]^{k-1}\,p(r_k)\,[P(r_l)-P(r_k)]^{l-k-1}\,p(r_l)\,[1-P(r_l)]^{\lambda-l}}{(k-1)!\,1!\,(l-k-1)!\,1!\,(\lambda-l)!}. \quad (5.75)$$

Substituting $u := r_k$, $v := r_l$ into (5.73), one obtains

$$\sum_{l=2}^{\mu}\sum_{k=1}^{l-1}\overline{R_k^a R_l^b} = \int_{v=0}^{v=\infty} v^b p(v) \int_{u=0}^{u=v} u^a p(u)\,\lambda! \sum_{l=2}^{\mu}\frac{[1-P(v)]^{\lambda-l}}{(\lambda-l)!} \times \sum_{k=1}^{l-1}\frac{[P(u)]^{k-1}[P(v)-P(u)]^{l-k-1}}{(k-1)!\,(l-k-1)!}\,\mathrm{d}u\,\mathrm{d}v. \quad (5.76)$$

The sum over k can be simplified using the binomial theorem

$$\begin{aligned}\sum_{k=1}^{l-1}\frac{[P(u)]^{k-1}[P(v)-P(u)]^{l-k-1}}{(k-1)!\,(l-k-1)!} &= \sum_{s=0}^{l-2}\frac{[P(u)]^{s}[P(v)-P(u)]^{l-2-s}}{s!\,(l-2-s)!} \\ &= \frac{1}{(l-2)!}\sum_{s=0}^{l-2}\frac{(l-2)!}{s!\,(l-2-s)!}\,[P(u)]^{s}[P(v)-P(u)]^{l-2-s} \\ &= \frac{1}{(l-2)!}\,[P(v)]^{l-2},\end{aligned} \quad (5.77)$$

consequently, the sum over l in (5.76) follows

$$\sum_{l=2}^{\mu}\frac{[P(v)]^{l-2}[1-P(v)]^{\lambda-l}}{(l-2)!\,(\lambda-l)!} = \sum_{m=1}^{\mu-1}\frac{[P(v)]^{m-1}[1-P(v)]^{\lambda-1-m}}{(m-1)!\,(\lambda-1-m)!}. \quad (5.78)$$

Comparing with (5.14) one finds that this sum can be expressed by an integral

$$\sum_{m=1}^{\mu-1}\frac{[P(v)]^{m-1}[1-P(v)]^{\lambda-1-m}}{(m-1)!\,(\lambda-1-m)!} = \frac{1}{(\lambda-\mu-1)!\,(\mu-2)!}\int_{x=0}^{x=1-P(v)} x^{\lambda-\mu-1}(1-x)^{\mu-2}\mathrm{d}x. \quad (5.79)$$

Therefore, one gets for (5.78)

$$\lambda!\sum_{l=2}^{\mu}\frac{[P(v)]^{l-2}[1-P(v)]^{\lambda-l}}{(l-2)!\,(\lambda-l)!} = \mu(\mu-1)(\lambda-\mu)\binom{\lambda}{\mu}\int_{x=0}^{x=1-P(v)} x^{\lambda-\mu-1}(1-x)^{\mu-2}\mathrm{d}x, \quad (5.80)$$

and consequently for (5.76)

$$\sum_{l=2}^{\mu}\sum_{k=1}^{l-1}\overline{R_k^a R_l^b} = \mu(\mu-1)(\lambda-\mu)\binom{\lambda}{\mu}\int_{v=0}^{v=\infty}\int_{x=0}^{x=1-P(v)} v^b p(v) \times \int_{u=0}^{u=v} u^a p(u)\,\mathrm{d}u\; x^{\lambda-\mu-1}(1-x)^{\mu-2}\mathrm{d}x\,\mathrm{d}v. \quad (5.81)$$

Substituting $v = r$ and using the same transformation as used for the derivation of (5.17) from (5.15), one obtains

$$\sum_{l=2}^{\mu}\sum_{k=1}^{l-1}\overline{R_k^a R_l^b} = \mu(\mu-1)(\lambda-\mu)\binom{\lambda}{\mu}\int_{t=0}^{t=\infty} p(t)[1-P(t)]^{\lambda-\mu-1}[P(t)]^{\mu-2} \times \int_{r=0}^{r=t} r^b p(r)\left(\int_{u=0}^{u=r} u^a p(u)\,\mathrm{d}u\right)\,\mathrm{d}r\,\mathrm{d}t. \quad (5.82)$$

The integral expression for (5.69) can be assembled now. With $a = 2$ in (5.72) and $a = b = 1$ in (5.82), one gets

$$\overline{\langle(\Delta R)^2\rangle} = \frac{\mu-1}{\mu}(\lambda-\mu)\binom{\lambda}{\mu}\left\{\int_{t=0}^{t=\infty} p(t)[1-P(t)]^{\lambda-\mu-1}[P(t)]^{\mu-1}\int_{r=0}^{r=t} r^2 p(r)\,\mathrm{d}r\,\mathrm{d}t - 2\int_{t=0}^{t=\infty} p(t)[1-P(t)]^{\lambda-\mu-1}[P(t)]^{\mu-2}\int_{r=0}^{r=t} r\,p(r)\left(\int_{u=0}^{u=r} u\,p(u)\,\mathrm{d}u\right)\mathrm{d}r\,\mathrm{d}t\right\}. \quad (5.83)$$

5.1.9 The Integral Expression of $\overline{\langle(\Delta R)^3\rangle}$

As in the case of $\overline{\langle(\Delta R)^2\rangle}$, the triple sum in the definition of $\langle(\Delta R)^3\rangle$

$$\langle(\Delta R)^3\rangle = \frac{1}{\mu}\sum_{m=1}^{\mu}(R_m - \langle R\rangle)^3$$

$$\langle(\Delta R)^3\rangle = \frac{1}{\mu}\sum_{m=1}^{\mu}\left(R_m^3 - 3R_m^2\langle R\rangle + 3R_m\langle R\rangle^2 - \langle R\rangle^3\right)$$

$$\langle(\Delta R)^3\rangle = \frac{1}{\mu}\sum_{m=1}^{\mu}R_m^3 - 3\left(\frac{1}{\mu}\sum_{m=1}^{\mu}R_m^2\right)\left(\frac{1}{\mu}\sum_{m=1}^{\mu}R_m\right) + 2\left(\frac{1}{\mu}\sum_{m=1}^{\mu}R_m\right)^3 \quad (5.84)$$

will be simplified. It will be reordered to contain only terms of the type $R_k \le R_l \le R_m$. The first sum in (5.84) already satisfies this. The other two shall be reordered. Without the factor $1/\mu$, the second term can be rewritten as

$$\left(\sum_{m=1}^{\mu} R_m^2\right)\left(\sum_{m=1}^{\mu} R_m\right) = \sum_{m=1}^{\mu} R_m^3 + \sum_{l=1}^{\mu}\sum_{k\neq l} R_k^2 R_l$$
$$= \sum_{m=1}^{\mu} R_m^3 + \sum_{l=2}^{\mu}\sum_{k=1}^{l-1} R_k^2 R_l + \sum_{k=2}^{\mu}\sum_{l=1}^{k-1} R_l R_k^2$$
$$= \sum_{m=1}^{\mu} R_m^3 + \sum_{l=2}^{\mu}\sum_{k=1}^{l-1} (R_k^2 R_l + R_k R_l^2) \qquad (5.85)$$

and for the third expression in (5.84)

$$\left(\sum_{m=1}^{\mu} R_m\right)^3 = \sum_{m=1}^{\mu}\sum_{l=1}^{\mu}\sum_{k=1}^{\mu} R_k R_l R_m$$
$$= \sum_{m=1}^{\mu} R_m^3 \;+\; 3\sum_{l=2}^{\mu}\sum_{k=1}^{l-1}(R_k^2 R_l + R_k R_l^2) \;+\; 6\sum_{m=3}^{\mu}\sum_{l=2}^{m-1}\sum_{k=1}^{l-1} R_k R_l R_m. \quad (5.86)$$

Combining all the terms, one obtains for $\overline{\langle(\Delta R)^3\rangle}$

$$\overline{\langle(\Delta R)^3\rangle} = \left(1 - \frac{3}{\mu} + \frac{2}{\mu^2}\right)\frac{1}{\mu}\sum_{m=1}^{\mu}\overline{R_m^3} - \left(3 - \frac{6}{\mu}\right)\frac{1}{\mu^2}\sum_{l=2}^{\mu}\sum_{k=1}^{l-1}\left(\overline{R_k^2 R_l} + \overline{R_k R_l^2}\right)$$
$$+ \frac{12}{\mu^3}\sum_{m=3}^{\mu}\sum_{l=2}^{m-1}\sum_{k=1}^{l-1}\overline{R_k R_l R_m}. \qquad (5.87)$$

The integral expressions for the first two terms are already given by (5.72) and (5.82). The third is represented by the triple integral

$$\sum_{m=3}^{\mu}\sum_{l=2}^{m-1}\sum_{k=1}^{l-1}\overline{R_k R_l R_m}$$
$$= \int_{w=0}^{w=\infty}\int_{v=0}^{v=w}\int_{u=0}^{u=v} uvw \sum_{m=3}^{\mu}\sum_{l=2}^{m-1}\sum_{k=1}^{l-1} p_{k\,l\,m:\lambda}(u,v,w)\,\mathrm{d}u\,\mathrm{d}v\,\mathrm{d}w. \quad (5.88)$$

The joint density $p_{k\,l\,m:\lambda}(u,v,w)$ with $u \le v \le w$, $k < l < m$ can be derived similar to (5.75)

$$p_{k\,l\,m:\lambda}(u,v,w) = \lambda!\,\frac{[P(u)]^{k-1}\;p(u)\;[P(v)-P(u)]^{l-k-1}\;p(v)}{(k-1)!\;1!\;(l-k-1)!\;1!}$$
$$\times\frac{[P(w)-P(v)]^{m-l-1}\;p(w)\;[1-P(w)]^{\lambda-m}}{(m-l-1)!\;1!\;(\lambda-m)!}. \qquad (5.89)$$

The sums in (5.88) can be simplified using the binomial theorem. The sum over k is used, given in (5.77). As a result, one gets

$$\sum_{m=3}^{\mu}\sum_{l=2}^{m-1}\sum_{k=1}^{l-1} p_{k\,l\,m:\lambda} = p(u)\,p(v)\,p(w)\,\lambda!\sum_{m=3}^{\mu}\frac{[1-P(w)]^{\lambda-m}}{(\lambda-m)!}$$

$$\times \sum_{l=2}^{m-1} \frac{[P(v)]^{l-2}[P(w) - P(v)]^{m-l-1}}{(m-l-1)!\,(l-2)!}$$

$$= p(u)\,p(v)\,p(w)\,\lambda! \sum_{m=3}^{\mu} \frac{[1-P(w)]^{\lambda-m}}{(\lambda-m)!\,(m-3)!}$$

$$\times \sum_{s=0}^{m-3} \frac{(m-3)!}{(m-3-s)!\,s!}[P(v)]^{s}[P(w)-P(v)]^{m-3-s}$$

$$= p(u)\,p(v)\,p(w)\,\lambda! \sum_{m=3}^{\mu} \frac{[1-P(w)]^{\lambda-m}[P(w)]^{m-3}}{(\lambda-m)!\,(m-3)!}$$

$$= p(u)\,p(v)\,p(w)\,\lambda! \sum_{m=1}^{\mu-2} \frac{[P(w)]^{m-1}[1-P(w)]^{\lambda-2-m}}{(\lambda-2-m)!\,(m-1)!}. \tag{5.90}$$

Similar to (5.14), the sum in (5.90) can again be expressed as an integral

$$\sum_{m=1}^{\mu-2} \frac{[P(w)]^{m-1}[1-P(w)]^{\lambda-2-m}}{(\lambda-2-m)!\,(m-1)!}$$

$$= \frac{1}{(\lambda-\mu-1)!\,(\mu-3)!} \int_{x=0}^{x=1-P(w)} x^{\lambda-\mu-1}(1-x)^{\mu-3}\,\mathrm{d}x, \tag{5.91}$$

and (5.90) becomes

$$\sum_{m=3}^{\mu}\sum_{l=2}^{m-1}\sum_{k=1}^{l-1} p_{k\,l\,m:\lambda} = \lambda! \frac{p(u)\,p(v)\,p(w)}{(\lambda-\mu-1)!\,(\mu-3)!} \int_{x=0}^{x=1-P(w)} x^{\lambda-\mu-1}(1-x)^{\mu-3}\,\mathrm{d}x.$$

Inserting this result into (5.88), one obtains

$$\sum_{m=3}^{\mu}\sum_{l=2}^{m-1}\sum_{k=1}^{l-1} \overline{R_k R_l R_m} = \frac{\lambda!}{(\lambda-\mu-1)!\,(\mu-3)!} \int_{w=0}^{w=\infty} w p(w)$$

$$\times \left(\int_{v=0}^{v=w} v p(v) \int_{u=0}^{u=v} u p(u)\,\mathrm{d}u\,\mathrm{d}v \right) \int_{0}^{1-P(w)} x^{\lambda-\mu-1}(1-x)^{\mu-3}\,\mathrm{d}x\,\mathrm{d}w.$$

Using the same method already applied to (5.15), yielding (5.17), the integration order $\int\int \mathrm{d}x\,\mathrm{d}w$ is exchanged with $\int\int \mathrm{d}w\,\mathrm{d}x$ and the substitution $x = 1 - P(t)$ is performed, resulting in

$$\sum_{m=3}^{\mu}\sum_{l=2}^{m-1}\sum_{k=1}^{l-1} \overline{R_k R_l R_m}$$

$$= \mu(\mu-1)(\mu-2)(\lambda-\mu)\binom{\lambda}{\mu} \int_{t=0}^{t=\infty} p(t)[1-P(t)]^{\lambda-\mu-1}[P(t)]^{\mu-3}$$

$$\times \int_{w=0}^{w=t} w p(w) \int_{v=0}^{v=w} v p(v) \int_{u=0}^{u=v} u p(u)\,\mathrm{d}u\,\mathrm{d}v\,\mathrm{d}w\,\mathrm{d}t. \tag{5.92}$$

Now one can construct the integral expression for $\overline{\langle(\Delta R)^3\rangle}$, Eq. (5.87). Using the intermediate results (5.72, 5.82) and (5.92) one obtains

$$\begin{aligned}
&\overline{\langle(\Delta R)^3\rangle} \\
&= (\lambda-\mu)\binom{\lambda}{\mu}\left\{\left(1-\frac{3}{\mu}+\frac{2}{\mu^2}\right)\int\limits_{t=0}^{t=\infty} p(t)[1-P(t)]^{\lambda-\mu-1}[P(t)]^{\mu-1}\right. \\
&\qquad\qquad\times\int\limits_{r=0}^{r=t} r^3 p(r)\,\mathrm{d}r\,\mathrm{d}t \\
&\quad -\left(3-\frac{6}{\mu}\right)\frac{\mu-1}{\mu}\int\limits_{t=0}^{t=\infty} p(t)[1-P(t)]^{\lambda-\mu-1}[P(t)]^{\mu-2} \\
&\qquad\qquad\times\int\limits_{r=0}^{r=t}\int\limits_{u=0}^{u=r}\left(ru^2+r^2u\right)\,p(u)\,p(r)\,\mathrm{d}u\,\mathrm{d}r\,\mathrm{d}t \\
&\quad +12\,\frac{\mu-1}{\mu}\,\frac{\mu-2}{\mu}\int\limits_{t=0}^{t=\infty} p(t)[1-P(t)]^{\lambda-\mu-1}[P(t)]^{\mu-3} \\
&\qquad\qquad\left.\times\int\limits_{w=0}^{w=t}\int\limits_{v=0}^{v=w}\int\limits_{u=0}^{u=v} wvu\,p(u)\,p(v)\,p(w)\,\mathrm{d}u\,\mathrm{d}v\,\mathrm{d}w\,\mathrm{d}t\right\}
\end{aligned}$$

and, finally,

$$\begin{aligned}
&\overline{\langle(\Delta R)^3\rangle} \\
&= \frac{\mu-1}{\mu}\,\frac{\mu-2}{\mu}(\lambda-\mu)\binom{\lambda}{\mu}\left\{\int\limits_{t=0}^{t=\infty} p(t)[1-P(t)]^{\lambda-\mu-1}[P(t)]^{\mu-1}\int\limits_{r=0}^{r=t} r^3 p(r)\,\mathrm{d}r\,\mathrm{d}t\right. \\
&\quad -3\int\limits_{t=0}^{t=\infty} p(t)[1-P(t)]^{\lambda-\mu-1}[P(t)]^{\mu-2}\int\limits_{r=0}^{r=t}\int\limits_{u=0}^{u=r}\left(ru^2+r^2u\right)\,p(u)\,p(r)\,\mathrm{d}u\,\mathrm{d}r\,\mathrm{d}t \\
&\quad +12\int\limits_{t=0}^{t=\infty} p(t)[1-P(t)]^{\lambda-\mu-1}[P(t)]^{\mu-3} \\
&\qquad\left.\times\int\limits_{w=0}^{w=t}\int\limits_{v=0}^{v=w}\int\limits_{u=0}^{u=v} wvu\,p(u)\,p(v)\,p(w)\,\mathrm{d}u\,\mathrm{d}v\,\mathrm{d}w\,\mathrm{d}t\right\}.
\end{aligned} \tag{5.93}$$

5.1.10 Approximation of the Stationary State – the Self-Consistent Method

At the end of this first section, a summary is needed to understand why these lengthy derivations were necessary and what is gained hereby.

The progress rate $\varphi_{\mu,\lambda}$ can be calculated if the offspring probability density $p(r)$ described in Sect. 5.1.3 is known. Therefore, the determination of $p(r)$ is a main aim of the chapter.

The density $p^{(g)}(r)$ at generation g is transformed by the (μ, λ)-ES algorithm to the density $p^{(g+1)}(r)$. This can be expressed formally by a (nonlinear) functional equation as

$$p^{(g+1)}(r) = \mathrm{F}_{(\mu,\lambda)}\left[p^{(g)}(r)\right]. \tag{5.94}$$

The concrete form of this functional equation is given by the expected value expression $p^{(g+1)}(r) = \mathrm{E}\{p(r\,|\,R_1, \ldots, R_\mu)\}$, with (5.36, 5.37), and (5.29). The $p^{(g)}(r) \mapsto p^{(g+1)}(r)$ mapping described hereby cannot be carried out in an analytically closed manner. Instead, an approximate approach must be chosen. In other words, the densities $p^{(g)}(r)$ should be approximated appropriately. An almost "natural" density approximation is achieved by the expansion of $p(r)$ in a series of Hermite polynomials (Sect. 5.1.4).

In the series expansion of $p(r)$, the symbols $\overline{r}$, s, and γ are used for the first three coefficients. These three coefficients as well as the others following can be calculated using the cumulants of the probability density $p(r)$. These cumulants are functions of the central moments of $p(r)$. From Sect. 5.1.5 to Sect. 5.1.9, the moments for the generation $g+1$ have been calculated for a given probability density $p^{(g)}(r)$. In this way, a system of functionals is obtained instead of the functional equation (5.94). This system is expressed with adequate accuracy by the Eqs. (5.56, 5.64) with (5.83), and (5.65) with (5.93) and (5.83)

$$\left.\begin{aligned} \overline{r}^{(g+1)} &= \mathrm{F}_{\overline{r}}\left[p^{(g)}(r)\right], \\ s^{(g+1)} &= \mathrm{F}_{s}\left[p^{(g)}(r)\right], \\ \gamma^{(g+1)} &= \mathrm{F}_{\gamma}\left[p^{(g)}(r)\right], \\ \vdots\quad & \quad\vdots \qquad \vdots \end{aligned}\right\}. \tag{5.95}$$

Introducing the approximation approach (5.25) in (5.95), one obtains a system of nonlinear equations. This system maps the coefficients $\overline{r}^{(g)}$, $s^{(g)}$, and $\gamma^{(g)}$, as well as all the others following, to the coefficients of the next generation $(g+1)$[7]

[7] Note, one obtains by the series expansion the functions $F(\cdot)$ from the functionals (5.95) $\mathrm{F}[\cdot]$.

$$\left.\begin{array}{lll} \overline{r}^{(g+1)} & = F_{\overline{r}}\left(\overline{r}^{(g)}, s^{(g)}, \gamma^{(g)}, \ldots\right), \\ s^{(g+1)} & = F_s\left(\overline{r}^{(g)}, s^{(g)}, \gamma^{(g)}, \ldots\right), \\ \gamma^{(g+1)} & = F_\gamma\left(\overline{r}^{(g)}, s^{(g)}, \gamma^{(g)}, \ldots\right), \\ \vdots & \vdots & \vdots \end{array}\right\}. \tag{5.96}$$

Here we are able to map the $p(r)$ evolution to the evolution of the coefficients.

The system of evolution equations obtained can be applied to the investigation of the (μ, λ)-ES dynamics. The question of the stationary $p(r)$ distribution, or the question of the behavior for large g ($g \to \infty$) becomes now the main question. It is obvious that a *static* stationary $p(r)$ density generally cannot exist, simply because at least the expected value $\overline{r}$ must usually change significantly from generation to generation. Otherwise, the population will stagnate in the parameter space $\mathcal{Y}$, as a result the ES would not be able to optimize. But one can expect that the higher moments of the distribution become stationary for sufficiently large g. Since exactly these moments determine the shape of the offspring population, this corresponds to the stability of the distribution. However, this requires that the higher moments are *not* functions of $\overline{r}$. This necessary condition is satisfied for the system of cumulants: the cumulants only depend on the central moments, they are *translation invariant.*

With the translation invariance of s, γ, etc., the system (5.96) for the determination of the stationary density reduces to

$$\left.\begin{array}{lll} s^{(g+1)} & = F_s\left(s^{(g)}, \gamma^{(g)}, \ldots\right), \\ \gamma^{(g+1)} & = F_\gamma\left(s^{(g)}, \gamma^{(g)}, \ldots\right), \\ \vdots & \vdots & \vdots \end{array}\right\}. \tag{5.97}$$

For $g \to \infty$ as the criterion for being stationary, if one postulates the existence of the limits for the higher moments and cumulants, $s = \lim_{g\to\infty} s^{(g)}$, $\gamma = \lim_{g\to\infty} \gamma^{(g)}$, etc., the self-consistency conditions are obtained as

$$s = s^{(g+1)} \stackrel{!}{=} s^{(g)}, \qquad \gamma = \gamma^{(g+1)} \stackrel{!}{=} \gamma^{(g)},$$

etc.[8] In this manner, one obtains a system of *self-consistent* equations having the stationary state as their solution

[8] While the existence of a stationary γ and of the higher cumulants (5.24) seems obvious on the sphere model (stability of the distribution shape), the stationarity of the population standard deviation s can only appear under special conditions, as e.g., constant mutation strength σ or the experimental setting considered in Sect. 5.3.1.2 below. The usual dynamical operation case with constant σ^* is covered, however, under the condition $N \to \infty$, because in that case $\varphi \to 0$, causing only a small change of the population on a moderate timescale leading to quasistationary conditions.

$$\left.\begin{array}{l} s \stackrel{!}{=} F_s(s, \gamma, \ldots), \\ \gamma \stackrel{!}{=} F_\gamma(s, \gamma, \ldots), \\ \vdots \ \vdots \qquad \vdots \end{array}\right\}. \tag{5.98}$$

These equations are obtained by setting $\overline{r} = 0$ in the $p(r)$ approximation (5.25) in (5.95)

$$p(r) = \frac{1}{\sqrt{2\pi}\, s} \exp\left[-\frac{1}{2}\left(\frac{r}{s}\right)^2\right]\left[1 + \gamma \mathrm{He}_3\left(\frac{r}{s}\right) + \ldots\right]. \tag{5.99}$$

Equation (5.99) must be inserted in the integral expression of s, (5.64), and that of γ, (5.65), respectively. The very long calculations related to these insertions are carried out in the next section.

It should be emphasized that the equations in (5.98) must be considered as a *necessary*, but not sufficient criterion for the existence of a stationary offspring population. Assuming that a solution $s = s_0$, $\gamma = \gamma_0, \ldots$ of (5.98) exists, then one should prove that this solution is also *stable*. That is, a stability analysis of the iterative map (5.97) should be carried out in the neighborhood of the state $s = s_0$, $\gamma = \gamma_0, \ldots$ Such an analysis can be omitted here: because of the linear approximation to be introduced in the next section, one obtains a system of equations. This system is linear in s^2 and γ, and has only one solution. Thus, the proof of the existence of a solution is trivial for this special case. A general proof of the existence for a stationary offspring distribution in the sense of the stationarity of higher-order moments remains as a task for future research.

5.2 On the Analysis in the Linear γ Approximation

5.2.1 The Linear γ Approximation

In order to derive a system of self-consistent equations (5.98), one has to calculate $s^{(g+1)}$ and $\gamma^{(g+1)}$ using the $p(r)$ approximation (5.99). Therefore, it is necessary to calculate the integral expression (5.83) for $\overline{\langle(\Delta R)^2\rangle}$ and (5.93) for $\overline{\langle(\Delta R)^3\rangle}$, respectively. As shown in (5.83) or (5.93), the integrands contain powers of the offspring distribution function $P(r)$. Consequently, the resulting self-consistent equations of (5.98) are *nonlinear* in γ. As a result, the analytical solution of (5.98) becomes a problem.

However, if one is interested in numerical values of s and γ only, this nonlinearity does not cause a problem. By the system (5.97), one even has a numerical iteration rule to calculate s and γ on a computer. Actually, this iteration rule has been used to compare the numerical error caused by the γ linearization, to be introduced next, with the nonlinear approach.

Since it is the aim to obtain analytical formulae for the progress rate, the nonlinear system of self-consistent equations must be linearized. This simplification is not as crucial as one might first suppose. The γ factors occur in $P(r)$

and $p(r)$ products in terms of γ^k powers. As already argued in Sect. 5.1.4, γ measures the skewness of the offspring distribution. Since this distribution is obtained from the parental distribution that has a relatively large skewness, with the addition of Gaussian mutations this distribution will have a small skewness.[9] The parental distribution is "smoothed" by the mutations. Hence, the powers of γ vanish very fast. The error made by ignoring them in the integrands of (5.17, 5.83) and (5.93) can be assumed to be small. This has been checked by numerical investigations on a computer algebra system (`Mathematica`). A much larger error is caused by the series cut-off after the γ term in (5.25) and (5.26).

The *linear approximation* will be used because of the above-mentioned arguments. In other words, all γ^k with $k > 1$ will be neglected in the integrands. The error terms $\mathcal{R}$ caused by that simplification are not considered further.

This section comprises four parts. In the next subsection, the formula will be derived for the progress rate $\varphi^*_{\mu,\lambda}$ in the linear γ approximation. However, this formula is not final. It still contains s and γ as unknowns. In Sect. 5.2.3, the self-consistent equation for s will be derived and in Sect. 5.2.4 for γ. The whole calculations of the self-consistent method culminate in Sect. 5.2.5: s and γ will be calculated *analytically*, and the $\varphi^*_{\mu,\lambda}$ formulae of different degree of approximation will be provided.

5.2.2 The Approximation of the $\varphi_{\mu,\lambda}$ Integral

The progress rate $\varphi_{\mu,\lambda}$ is given by the integral (5.17). Inserting the approximation (5.25) for $p(r)$ and (5.26) for $P(r)$, respectively, in (5.17), and using the standardization transformations

$$\frac{t-\overline{r}}{s} = x \qquad \text{and} \qquad \frac{r-\overline{r}}{s} = y \tag{5.100}$$

one obtains

$$\varphi_{\mu,\lambda} = \langle R\rangle - (\lambda-\mu)\binom{\lambda}{\mu}\int_{x=-\infty}^{x=\infty} p_z(x)[1-P_z(x)]^{\lambda-\mu-1}[P_z(x)]^{\mu-1} \times \int_{y=-\infty}^{y=x} (sy+\overline{r})p_z(y)\,\mathrm{d}y\,\mathrm{d}x, \tag{5.101}$$

with

$$p_z(z) = \frac{1}{\sqrt{2\pi}}\,\mathrm{e}^{-\frac{1}{2}z^2}\,\left[1 + \gamma\mathrm{He}_3(z) + \ldots\right] \tag{5.102}$$

and

[9] For the case $\mu = 2$, assuming the normal approximation (5.27), one even obtains (almost) exactly $\gamma = 0$. The reason is simple: the mixing of two normal distributions with (almost) the same standard deviations gives a symmetric probability density.

$$P_z(z) = \Phi(z) - \frac{\gamma}{\sqrt{2\pi}} \, \mathrm{e}^{-\frac{1}{2}z^2} \mathrm{He}_2(z) - \ldots . \tag{5.103}$$

The lower integration limits $x = y = -\overline{r}/s$ are shifted to $-\infty$. This is admissible since the largest part of the probability mass of $p_z(z)$ is concentrated around $x = 0$ because of standardization, and since $\overline{r} \gg s$ is valid.[10] The integration over y yields

$$\begin{aligned}
\int_{y=-\infty}^{y=x} (sy + \overline{r}) p_z(y) \, \mathrm{d}y &= \overline{r} P_z(x) + s \int_{-\infty}^{x} y p_z(y) \, \mathrm{d}y \\
&= \overline{r} P_z(x) + \frac{s}{\sqrt{2\pi}} \int_{-\infty}^{x} \mathrm{e}^{-\frac{1}{2}y^2} \, (y - 3\gamma y^2 + \gamma y^4) \, \mathrm{d}y \\
&= \overline{r} P_z(x) + \frac{s}{\sqrt{2\pi}} \left[I^{0,1}(x) - 3\gamma I^{0,2}(x) + \gamma I^{0,4}(x) \right] \\
&= \overline{r} P_z(x) - \frac{s}{\sqrt{2\pi}} \, \mathrm{e}^{-\frac{1}{2}x^2} (1 + \gamma x^3).
\end{aligned} \tag{5.104}$$

The formulae for $I^{0,\beta}$ (A.16) and (A.17) from Appendix A.2 are used in this derivation. Inserting (5.104) in (5.101), the first term of (5.104) yields an integral that can be simplified using (5.10)

$$\begin{aligned}
&(\lambda - \mu) \binom{\lambda}{\mu} \int_{x=-\infty}^{x=\infty} p_z(x) [1 - P_z(x)]^{\lambda-\mu-1} [P_z(x)]^{\mu} \mathrm{d}x \\
&= \frac{\lambda!}{((\mu+1)-1)! \, (\lambda - (\mu+1))!} \int_{-\infty}^{\infty} p_z(x) [P_z(x)]^{(\mu+1)-1} [1 - P_z(x)]^{\lambda-(\mu+1)} \mathrm{d}x \\
&= \int_{-\infty}^{\infty} p_{(\mu+1):\lambda}(x) \, \mathrm{d}x = 1.
\end{aligned} \tag{5.105}$$

Consequently, for the progress rate one gets the integral

$$\begin{aligned}
\varphi_{\mu,\lambda} = \langle R \rangle - \overline{r} + s \, (\lambda - \mu) \binom{\lambda}{\mu} \int_{x=-\infty}^{x=\infty} & p_z(x) [1 - P_z(x)]^{\lambda-\mu-1} [P_z(x)]^{\mu-1} \\
& \times \frac{\mathrm{e}^{-\frac{1}{2}x^2}}{\sqrt{2\pi}} \, (1 + \gamma x^3) \, \mathrm{d}x.
\end{aligned} \tag{5.106}$$

The integral in (5.106) is nonlinear in γ. As already justified in the introduction, Sect. 5.2.1, the terms with powers of γ^k, $k > 1$, will be neglected in the following. Using (5.103), one gets

$$\begin{aligned}
[P_z(x)]^{\mu-\alpha} &= \left[\Phi(x) - \frac{\gamma}{\sqrt{2\pi}} \, \mathrm{e}^{-\frac{1}{2}x^2} \mathrm{He}_2(x) \right]^{\mu-\alpha} \\
&= (\Phi(x))^{\mu-\alpha} - \frac{\gamma}{\sqrt{2\pi}} \, (\mu - \alpha) \, (\Phi(x))^{\mu-\alpha-1} \, \mathrm{e}^{-\frac{1}{2}x^2} \mathrm{He}_2(x) \\
&\quad + \mathcal{O}\left(\gamma^2\right).
\end{aligned} \tag{5.107}$$

---

[10] Compare also to the explanations for Eq. (5.31) on p. 151 and the discussion in Sect. 3.4.5, p. 111. For the hyperplane, the lower integration limit is $-\infty$ regardless.

Analogously, one obtains

$$[1-P_z(x)]^{\lambda-\mu-1} = (1-\Phi(x))^{\lambda-\mu-1} + \frac{\gamma}{\sqrt{2\pi}}(\lambda-\mu-1)(1-\Phi(x))^{\lambda-\mu-2}\,\mathrm{e}^{-\frac{1}{2}x^2}\mathrm{He}_2(x) + \mathcal{O}(\gamma^2). \quad (5.108)$$

The multiplication of $p_z(x)$ with (5.107) and (5.108) gives

$$\begin{aligned} p_z(x)[1-P_z(x)]^{\lambda-\mu-1}[P_z(x)]^{\mu-\alpha} &= \frac{1}{\sqrt{2\pi}}\mathrm{e}^{-\frac{1}{2}x^2}(1-\Phi(x))^{\lambda-\mu-1}(\Phi(x))^{\mu-\alpha} \\ &+ \frac{\gamma}{\sqrt{2\pi}}\mathrm{e}^{-\frac{1}{2}x^2}(1-\Phi(x))^{\lambda-\mu-1}(\Phi(x))^{\mu-\alpha}\mathrm{He}_3(x) \\ &+ \frac{\gamma}{2\pi}(\lambda-\mu-1)\mathrm{e}^{-x^2}(1-\Phi(x))^{\lambda-\mu-2}(\Phi(x))^{\mu-\alpha}\mathrm{He}_2(x) \\ &- \frac{\gamma}{2\pi}(\mu-\alpha)\mathrm{e}^{-x^2}(1-\Phi(x))^{\lambda-\mu-1}(\Phi(x))^{\mu-\alpha-1}\mathrm{He}_2(x) + \mathcal{O}(\gamma^2). \quad (5.109) \end{aligned}$$

Neglecting γ^k, now one can insert this in (5.106)

$$\begin{aligned} \varphi_{\mu,\lambda} = \langle R\rangle - \overline{r} + s(\lambda-\mu)\binom{\lambda}{\mu}\Bigg[&\frac{1}{2\pi}\int_{-\infty}^{\infty}\mathrm{e}^{-x^2}(1-\Phi(x))^{\lambda-\mu-1}(\Phi(x))^{\mu-1}\,\mathrm{d}x \\ &+ \gamma\frac{1}{2\pi}\int_{-\infty}^{\infty}\mathrm{e}^{-x^2}(1-\Phi(x))^{\lambda-\mu-1}(\Phi(x))^{\mu-1}(x^3-3x)\,\mathrm{d}x \\ &+ \gamma\frac{1}{2\pi}\int_{-\infty}^{\infty}\mathrm{e}^{-x^2}(1-\Phi(x))^{\lambda-\mu-1}(\Phi(x))^{\mu-1}x^3\,\mathrm{d}x \\ &+ \gamma\frac{(\lambda-\mu-1)}{(\sqrt{2\pi})^3}\int_{-\infty}^{\infty}\mathrm{e}^{-\frac{3}{2}x^2}(1-\Phi(x))^{\lambda-\mu-2}(\Phi(x))^{\mu-1}(x^2-1)\,\mathrm{d}x \\ &- \gamma\frac{(\mu-1)}{(\sqrt{2\pi})^3}\int_{-\infty}^{\infty}\mathrm{e}^{-\frac{3}{2}x^2}(1-\Phi(x))^{\lambda-\mu-1}(\Phi(x))^{\mu-2}(x^2-1)\,\mathrm{d}x\Bigg] \\ &+ \mathcal{R}. \quad (5.110) \end{aligned}$$

The influence of the higher-order γ terms is represented by the error term $\mathcal{R}$. Furthermore, the He_k polynomials according to (B.8) have been inserted. Using the integral identity (A.32) in Appendix A.3 for the case (A.34) and substituting $x=-t$, one obtains

$$\begin{aligned} \varphi_{\mu,\lambda} = \langle R\rangle - \overline{r} + s(\lambda-\mu)\binom{\lambda}{\mu}\Bigg[&\frac{1}{2\pi}\int_{-\infty}^{\infty}\mathrm{e}^{-t^2}(\Phi(t))^{\lambda-\mu-1}(1-\Phi(t))^{\mu-1}\,\mathrm{d}t \\ &- \gamma\frac{1}{2\pi}\int_{-\infty}^{\infty}t\,\mathrm{e}^{-t^2}(\Phi(t))^{\lambda-\mu-1}(1-\Phi(t))^{\mu-1}\,\mathrm{d}t\Bigg] + \mathcal{R}. \quad (5.111) \end{aligned}$$

The integrals in (5.111) cannot be solved analytically for large μ and λ. They can be interpreted as generalizations of the integrals (3.100) and (4.41). They will be called *generalized progress coefficients* $e_{\mu,\lambda}^{\alpha,\beta}$

$$e^{\alpha,\beta}_{\mu,\lambda} := \frac{\lambda-\mu}{\left(\sqrt{2\pi}\right)^{\alpha+1}} \binom{\lambda}{\mu} \int_{-\infty}^{\infty} t^{\beta} \, \mathrm{e}^{-\frac{\alpha+1}{2} t^2} \left(\Phi(t)\right)^{\lambda-\mu-1} \left(1-\Phi(t)\right)^{\mu-\alpha} \mathrm{d}t. \tag{5.112}$$

They contain the progress coefficients $c_{1,\lambda}$ and $d^{(k)}_{1,\lambda}$ as special cases, for instance

$$c_{1,\lambda} = e^{0,1}_{0,\lambda} = e^{1,0}_{1,\lambda}. \tag{5.113}$$

The second equality in (5.113) can be proven by partial integration.

Using the generalized progress coefficients, the progress rate (5.111) can be expressed as

$$\varphi_{\mu,\lambda} = \langle R \rangle - \overline{r} + s \left[e^{1,0}_{\mu,\lambda} - \gamma \, e^{1,1}_{\mu,\lambda} \right] + \mathcal{R}. \tag{5.114}$$

Substituting $\overline{r}$ by (5.58) and using the normalization (5.8), it follows that

$$\varphi^*_{\mu,\lambda} = N \left(1 - \sqrt{1 + \frac{\sigma^{*2}}{N}} \right) + \frac{N}{\langle R \rangle} \, s \left[e^{1,0}_{\mu,\lambda} - \gamma \, e^{1,1}_{\mu,\lambda} \right] + \mathcal{R}^*. \tag{5.115}$$

This progress rate formula still contains s and γ as unknown coefficients. However, one can already recognize the general form of φ^*. In particular, one gets the N-dependent progress rate formula (3.242) of the $(1, \lambda)$-ES as a special case of (5.115): if there is only one parent ($\mu = 1$), one has $\gamma = 0$. The standard deviation s of the offspring population is given by (5.28), i.e. $s = \tilde{\sigma}_1 = \tilde{\sigma}$ (see (3.246)). Considering (5.113) and $\langle R \rangle = R$ and the normalization (2.22), the equivalence to (3.242) is proven.

5.2.3 The Approximation of $s^{(g+1)}$

In order to carry out the self-consistent method, $s^{(g+1)} = F_s(s^{(g)}, \gamma^{(g)})$ in Eq. (5.98) should be established first. Using (5.64) and (5.83), this becomes

$$s^{(g+1)} = \frac{1}{\zeta} \sqrt{\sigma^2 \frac{1+\zeta^2}{2} + \frac{\mu-1}{\mu} \left[I_A\left(s^{(g)}, \gamma^{(g)}\right) - I_B\left(s^{(g)}, \gamma^{(g)}\right) \right]} + \mathcal{R}_s, \tag{5.116}$$

where I_A and I_B are given by

$$I_A := (\lambda-\mu) \binom{\lambda}{\mu} \int_{t=0}^{t=\infty} p(t) \left[1 - P(t)\right]^{\lambda-\mu-1} \left[P(t)\right]^{\mu-1} \times \int_{r=0}^{r=t} r^2 p(r) \, \mathrm{d}r \, \mathrm{d}t \tag{5.117}$$

and

$$I_B := 2(\lambda-\mu)\binom{\lambda}{\mu}\int_0^\infty p(t)\,[1-P(t)]^{\lambda-\mu-1}[P(t)]^{\mu-2} \times \int_{r=0}^{r=t} r\,p(r)\int_{u=0}^{u=r} u\,p(u)\,\mathrm{d}u\,\mathrm{d}r\,\mathrm{d}t. \tag{5.118}$$

The distribution $P(\cdot)$ and the density $p(\cdot)$ in (5.117) and (5.118) describe the offspring population at generation g. The density $p(\cdot)$ is given by (5.99)

$$p(r) = \frac{1}{\sqrt{2\pi}\,s^{(g)}}\exp\left[-\frac{1}{2}\left(\frac{r}{s^{(g)}}\right)^2\right]\left[1+\gamma^{(g)}\,\mathrm{He}_3\left(\frac{r}{s^{(g)}}\right)+\ldots\right]. \tag{5.119}$$

In order to simplify the notation, the generation index g will be omitted in the following calculations to be continued with the treatment of I_A.

5.2.3.1 The I_A Integral. Using $x = t/s$ and $y = r/s$, (5.117) is transformed to standardized integration variables. Analogous to (5.101), one obtains using (5.102) and (5.103)

$$I_A = s^2(\lambda-\mu)\binom{\lambda}{\mu}\int_{x=-\infty}^{x=\infty} p_z(x)\,[1-P_z(x)]^{\lambda-\mu-1}[P_z(x)]^{\mu-1} \times \int_{y=-\infty}^{y=x} y^2 p_z(y)\,\mathrm{d}y\,\mathrm{d}x. \tag{5.120}$$

The integration over y yields

$$\begin{aligned}\int_{y=-\infty}^{y=x} y^2 p_z(y)\,\mathrm{d}y &= \frac{1}{\sqrt{2\pi}}\int_{-\infty}^{x} \mathrm{e}^{-\frac{1}{2}y^2}(y^2-3\gamma y^3+\gamma y^5)\,\mathrm{d}y\\ &= \frac{1}{\sqrt{2\pi}}\left[I^{0,2}(x)-3\gamma I^{0,3}(x)+\gamma I^{0,5}(x)\right]\\ &= \Phi(x)-\frac{1}{\sqrt{2\pi}}\mathrm{e}^{-\frac{1}{2}x^2}\left[x-\gamma\,(6+3x^2-8-4x^2-x^4)\right]\\ &= \left(\Phi(x)-\frac{x}{\sqrt{2\pi}}\mathrm{e}^{-\frac{1}{2}x^2}\right)-\gamma\frac{1}{\sqrt{2\pi}}\mathrm{e}^{-\frac{1}{2}x^2}(2+x^2+x^4).\end{aligned} \tag{5.121}$$

The integrals (A.17) in Appendix A.2 were used in this step. Using (5.121) and (5.109), for $\alpha = 1$, Eq. (5.117) can be calculated further. In the following, the integration limits will be omitted as long as the integral is taken from $-\infty$ to ∞

$$\begin{aligned}I_A = s^2(\lambda-\mu)\binom{\lambda}{\mu}\Bigg[&\frac{1}{\sqrt{2\pi}}\int \mathrm{e}^{-\frac{1}{2}x^2}\,(1-\Phi(x))^{\lambda-\mu-1}\,(\Phi(x))^{\mu}\,\mathrm{d}x\\ &-\gamma\frac{1}{2\pi}\int \mathrm{e}^{-x^2}\,(1-\Phi(x))^{\lambda-\mu-1}\,(\Phi(x))^{\mu-1}\,x\,\mathrm{d}x\\ &-\gamma\frac{1}{2\pi}\int \mathrm{e}^{-x^2}\,(1-\Phi(x))^{\lambda-\mu-1}\,(\Phi(x))^{\mu-1}\,(2+x^2+x^4)\,\mathrm{d}x\end{aligned}$$

$$
\begin{aligned}
&+ \gamma \frac{1}{\sqrt{2\pi}} \int \mathrm{e}^{-\frac{1}{2}x^2} (1-\Phi(x))^{\lambda-\mu-1} (\Phi(x))^{\mu} (x^3 - 3x)\, \mathrm{d}x \\
&+ \gamma \frac{\lambda-\mu-1}{2\pi} \int \mathrm{e}^{-x^2} (1-\Phi(x))^{\lambda-\mu-2} (\Phi(x))^{\mu} (x^2 - 1)\, \mathrm{d}x \\
&- \gamma \frac{\mu-1}{2\pi} \int \mathrm{e}^{-x^2} (1-\Phi(x))^{\lambda-\mu-1} (\Phi(x))^{\mu-1} (x^2 - 1)\, \mathrm{d}x \\
&- \gamma \frac{1}{2\pi} \int \mathrm{e}^{-x^2} (1-\Phi(x))^{\lambda-\mu-1} (\Phi(x))^{\mu-1} (x^4 - 3x^2)\, \mathrm{d}x \\
&- \gamma \frac{\lambda-\mu-1}{\left(\sqrt{2\pi}\right)^3} \int \mathrm{e}^{-\frac{3}{2}x^2} (1-\Phi(x))^{\lambda-\mu-2} (\Phi(x))^{\mu-1} (x^3 - x)\, \mathrm{d}x \\
&+ \gamma \frac{\mu-1}{\left(\sqrt{2\pi}\right)^3} \int \mathrm{e}^{-\frac{3}{2}x^2} (1-\Phi(x))^{\lambda-\mu-1} (\Phi(x))^{\mu-2} (x^3 - x)\, \mathrm{d}x \Bigg] \\
&+ \mathcal{R}_A. \qquad (5.122)
\end{aligned}
$$

The first line of (5.122) gives s^2 (compare with Eq. (5.105)). The second line yields $s^2 e^{1,1}_{\mu,\lambda}$: one shall use the transformation $x = -t$, taking (C.8) into account, and comparing with (5.112). Furthermore, one combines the third and seventh line. The last two lines can be substituted using the integral identity (A.32), variant (A.37). The intermediate result reads

$$
\begin{aligned}
I_A = s^2 \left(1 + e^{1,1}_{\mu,\lambda}\right) &+ s^2 \gamma (\lambda-\mu) \binom{\lambda}{\mu} \Bigg[\\
&+ \frac{1}{2\pi} \int \mathrm{e}^{-x^2} (1-\Phi(x))^{\lambda-\mu-1} (\Phi(x))^{\mu-1} (2x^2 - 2x^4 - 2)\, \mathrm{d}x \\
&+ \frac{1}{2\pi} \int \mathrm{e}^{-x^2} (1-\Phi(x))^{\lambda-\mu-1} (\Phi(x))^{\mu-1} (1 - 5x^2 + 2x^4)\, \mathrm{d}x \\
&+ \frac{1}{\sqrt{2\pi}} \int \mathrm{e}^{-\frac{1}{2}x^2} (1-\Phi(x))^{\lambda-\mu-1} (\Phi(x))^{\mu} (x^3 - 3x)\, \mathrm{d}x \\
&+ \frac{\lambda-\mu-1}{2\pi} \int \mathrm{e}^{-x^2} (1-\Phi(x))^{\lambda-\mu-2} (\Phi(x))^{\mu} (x^2 - 1)\, \mathrm{d}x \\
&- \frac{\mu-1}{2\pi} \int \mathrm{e}^{-x^2} (1-\Phi(x))^{\lambda-\mu-1} (\Phi(x))^{\mu-1} (x^2 - 1)\, \mathrm{d}x \Bigg] \\
+ \mathcal{R}_A. & \qquad (5.123)
\end{aligned}
$$

Using the identity (A.32) with (A.33), the last two lines can be simplified

$$
\begin{aligned}
I_A = s^2 &\left(1 + e^{1,1}_{\mu,\lambda}\right) \\
+ s^2 \gamma (\lambda-\mu) \binom{\lambda}{\mu} &\Bigg[-2 \frac{1}{2\pi} \int \mathrm{e}^{-x^2} (1-\Phi(x))^{\lambda-\mu-1} (\Phi(x))^{\mu-1}\, \mathrm{d}x \\
&- 2 \frac{1}{2\pi} \int x^2 \mathrm{e}^{-x^2} (1-\Phi(x))^{\lambda-\mu-1} (\Phi(x))^{\mu-1}\, \mathrm{d}x \Bigg] + \mathcal{R}_A.
\end{aligned}
$$

With (5.112) and $x = -t$, one finally gets

$$I_A = s^2 \left[1 + e^{1,1}_{\mu,\lambda} - 2\gamma \left(e^{1,0}_{\mu,\lambda} + e^{1,2}_{\mu,\lambda}\right)\right] + \mathcal{R}_A. \tag{5.124}$$

5.2.3.2 The I_B Integral. Standardization transformations $x = t/s$, $y = r/s$, and $z = u/s$ are applied to the integral (5.118) that has the density given by (5.119). Considering (5.102) and (5.103), one obtains

$$I_B = 2s^2(\lambda - \mu)\binom{\lambda}{\mu} \int_{-\infty}^{\infty} p_z(x)[1 - P_z(x)]^{\lambda-\mu-1}[P_z(x)]^{\mu-2} \times \int_{y=-\infty}^{y=x} y\, p_z(y) \int_{z=-\infty}^{z=y} z\, p_z(z)\, \mathrm{d}z\, \mathrm{d}y\, \mathrm{d}x. \tag{5.125}$$

The integration over z is almost trivial. After making the necessary changes in (5.104), one obtains *mutatis mutandis*

$$\int_{z=-\infty}^{z=y} z\, p_z(z)\, \mathrm{d}z = -\frac{1}{\sqrt{2\pi}}\, \mathrm{e}^{-\frac{1}{2}y^2}(1 + \gamma y^3). \tag{5.126}$$

The integration $\int \mathrm{d}y$ is carried out by neglecting the γ^2 term

$$\begin{aligned}
&\int_{y=-\infty}^{y=x} y\, p_z(y) \int_{z=-\infty}^{z=y} z\, p_z(z)\, \mathrm{d}z\, \mathrm{d}y \\
&\quad = -\frac{1}{2\pi} \int_{y=-\infty}^{y=x} \mathrm{e}^{-y^2}\left[y - 3\gamma y^2 + 2\gamma y^4\right] \mathrm{d}y + \mathcal{O}(\gamma^2) \\
&\quad = -\frac{1}{2\pi}\left[I^{1,1}(x) + \gamma\left(-3I^{1,2}(x) + 2I^{1,4}(x)\right)\right] + \mathcal{O}(\gamma^2) \\
&\quad = \frac{1}{4\pi}\mathrm{e}^{-x^2}\left(1 + 2\gamma x^3\right) + \mathcal{O}(\gamma^2).
\end{aligned} \tag{5.127}$$

The integrals $I^{1,\beta}$ can be found in (A.19), Appendix A.2. Inserting (5.109) and (5.127) into (5.125), neglecting γ^k terms indicated by the error term $\mathcal{R}_B$, one gets

$$\begin{aligned}
I_B = s^2 (\lambda - \mu)\binom{\lambda}{\mu} \Bigg[&\frac{1}{\left(\sqrt{2\pi}\right)^3} \int \mathrm{e}^{-\frac{3}{2}x^2} (1 - \Phi(x))^{\lambda-\mu-1} (\Phi(x))^{\mu-2}\, \mathrm{d}x \\
&+ \gamma \frac{1}{\left(\sqrt{2\pi}\right)^3} \int \mathrm{e}^{-\frac{3}{2}x^2} (1 - \Phi(x))^{\lambda-\mu-1} (\Phi(x))^{\mu-2} (x^3 - 3x)\, \mathrm{d}x \\
&+ \gamma \frac{1}{\left(\sqrt{2\pi}\right)^3} \int \mathrm{e}^{-\frac{3}{2}x^2} (1 - \Phi(x))^{\lambda-\mu-1} (\Phi(x))^{\mu-2} (2x^3)\, \mathrm{d}x \\
&+ \gamma \frac{\lambda - \mu - 1}{\left(\sqrt{2\pi}\right)^4} \int \mathrm{e}^{-2x^2} (1 - \Phi(x))^{\lambda-\mu-2} (\Phi(x))^{\mu-2} (x^2 - 1)\, \mathrm{d}x \\
&- \gamma \frac{\mu - 2}{\left(\sqrt{2\pi}\right)^4} \int \mathrm{e}^{-2x^2} (1 - \Phi(x))^{\lambda-\mu-1} (\Phi(x))^{\mu-3} (x^2 - 1)\, \mathrm{d}x \Bigg] \\
&+ \mathcal{R}_B.
\end{aligned} \tag{5.128}$$

Using the integral identity (A.32) for $\alpha = 3$ with (A.35), one obtains

$$I_B = s^2 (\lambda - \mu) \binom{\lambda}{\mu} \left[\frac{1}{\left(\sqrt{2\pi}\right)^3} \int e^{-\frac{3}{2}x^2} (1 - \Phi(x))^{\lambda-\mu-1} (\Phi(x))^{\mu-2} \, dx \right.$$
$$\left. + 2\gamma \frac{1}{\left(\sqrt{2\pi}\right)^3} \int x \, e^{-\frac{3}{2}x^2} (1 - \Phi(x))^{\lambda-\mu-1} (\Phi(x))^{\mu-2} \, dx \right]$$
$$+ \mathcal{R}_B. \tag{5.129}$$

Substituting $x = -t$, considering the definition (5.112), one gets

$$I_B = s^2 \left(e^{2,0}_{\mu,\lambda} - 2\gamma \, e^{2,1}_{\mu,\lambda} \right) + \mathcal{R}_B. \tag{5.130}$$

5.2.3.3 Composing the $s^{(g+1)}$ Formula. The results in (5.124) and (5.130) will be inserted into Eq. (5.116). One obtains

$$s^{(g+1)} = \frac{1}{\zeta} \sqrt{\sigma^2 \frac{1+\zeta^2}{2} + s^2 \frac{\mu - 1}{\mu} \left[1 + e^{1,1}_{\mu,\lambda} - e^{2,0}_{\mu,\lambda} - 2\gamma \left(e^{1,0}_{\mu,\lambda} + e^{1,2}_{\mu,\lambda} - e^{2,1}_{\mu,\lambda} \right) \right]}$$
$$+ \mathcal{R}_{s'}. \tag{5.131}$$

All error terms emerging from the γ linearization and the sphere model approximation are collected in $\mathcal{R}_{s'}$. This error term will be neglected in the following.

5.2.4 The Approximation of $\gamma^{(g+1)}$

The skewness of the offspring distribution at generation $g + 1$ is defined by the formula (5.24). With (5.56, 5.62) and (5.63), this formula reads

$$\gamma^{(g+1)} = \frac{1}{6} \frac{M_3^{(g+1)}}{\left(s^{(g+1)}\right)^3} = \frac{1}{6} \frac{1}{\left(s^{(g+1)}\right)^3} \frac{\overline{\langle (\Delta R)^3 \rangle}}{\zeta^3} + \mathcal{R}_\gamma$$
$$\gamma^{(g+1)} = \frac{1}{6} \frac{1}{\left(s^{(g+1)}\right)^3} \frac{1}{\zeta^3} \frac{\mu - 1}{\mu} \frac{\mu - 2}{\mu} \left(I_C - 3I_D + 2I_E \right) + \mathcal{R}_\gamma. \tag{5.132}$$

According to (5.93), the integrals I_C, I_D, and I_E are defined as

$$I_C := (\lambda - \mu) \binom{\lambda}{\mu} \int_0^\infty p(t) [1 - P(t)]^{\lambda-\mu-1} [P(t)]^{\mu-1} \int_{r=0}^{r=t} r^3 p(r) \, dr \, dt, \tag{5.133}$$

$$I_D := (\lambda - \mu) \binom{\lambda}{\mu} \int_0^\infty p(t) [1 - P(t)]^{\lambda-\mu-1} [P(t)]^{\mu-2}$$
$$\times \int_{r=0}^{r=t} \int_{u=0}^{u=r} (r u^2 + r^2 u) p(u) p(r) \, du \, dr \, dt, \tag{5.134}$$

and

$$I_E := 6\,(\lambda-\mu)\binom{\lambda}{\mu}\int_0^{\infty} p(t)[1-P(t)]^{\lambda-\mu-1}[P(t)]^{\mu-3}$$
$$\times\int_{w=0}^{w=t}\int_{v=0}^{v=w}\int_{u=0}^{u=v} wvu\,p(u)p(v)p(w)\,\mathrm{d}u\,\mathrm{d}v\,\mathrm{d}w\,\mathrm{d}t. \quad (5.135)$$

5.2.4.1 The I_C Integral. Equation (5.133) differs from (5.117) by just having r^3 instead of r^2 in the integrand. Therefore, the derivation in Sect. 5.2.3.1 can be used as a template for I_C

$$I_C = s^3\,(\lambda-\mu)\binom{\lambda}{\mu}\int_{x=-\infty}^{x=\infty} p_z(x)\,[1-P_z(x)]^{\lambda-\mu-1}[P_z(x)]^{\mu-1}$$
$$\times\int_{y=-\infty}^{y=x} y^3 p_z(y)\,\mathrm{d}y\,\mathrm{d}x. \quad (5.136)$$

Because of (A.17), the integration with respect to y yields

$$\int_{-\infty}^{x} y^3 p_z(y)\,\mathrm{d}y = \frac{1}{\sqrt{2\pi}}\int_{-\infty}^{x} \mathrm{e}^{-\frac{1}{2}y^2}\left(y^3-3\gamma y^4+\gamma y^6\right)\mathrm{d}y$$
$$= \frac{1}{\sqrt{2\pi}}\left[I^{0,3}(x)-3\gamma I^{0,4}(x)+\gamma I^{0,6}(x)\right]$$
$$= \frac{1}{\sqrt{2\pi}}\left\{-\mathrm{e}^{-\frac{1}{2}x^2}(2+x^2)+\gamma\left[6\sqrt{2\pi}\,\Phi(x)-\mathrm{e}^{-\frac{1}{2}x^2}\left(6x+2x^3+x^5\right)\right]\right\}.$$

Inserting this into (5.136), considering (5.109), one obtains in the linear γ approximation

$$I_C = s^3(\lambda-\mu)\binom{\lambda}{\mu}\Bigg[-\frac{1}{2\pi}\int \mathrm{e}^{-x^2}\,(1-\Phi(x))^{\lambda-\mu-1}\,(\Phi(x))^{\mu-1}\,(2+x^2)\,\mathrm{d}x$$
$$+\gamma\frac{1}{2\pi}\int \mathrm{e}^{-x^2}\,(1-\Phi(x))^{\lambda-\mu-1}\,(\Phi(x))^{\mu-1}\,(6x+x^3-x^5)\,\mathrm{d}x$$
$$+\gamma\frac{6}{\sqrt{2\pi}}\int \mathrm{e}^{-\frac{1}{2}x^2}\,(1-\Phi(x))^{\lambda-\mu-1}\,(\Phi(x))^{\mu}\,\mathrm{d}x$$
$$+\gamma\frac{1}{2\pi}\int \mathrm{e}^{-x^2}\,(1-\Phi(x))^{\lambda-\mu-1}\,(\Phi(x))^{\mu-1}\,(-6x-2x^3-x^5)\,\mathrm{d}x$$
$$+\gamma\frac{\lambda-\mu-1}{\left(\sqrt{2\pi}\right)^3}\int \mathrm{e}^{-\frac{3}{2}x^2}\,(1-\Phi(x))^{\lambda-\mu-2}\,(\Phi(x))^{\mu-1}\,(2-x^2-x^4)\,\mathrm{d}x$$
$$-\gamma\frac{\mu-1}{\left(\sqrt{2\pi}\right)^3}\int \mathrm{e}^{-\frac{3}{2}x^2}\,(1-\Phi(x))^{\lambda-\mu-1}\,(\Phi(x))^{\mu-2}\,(2-x^2-x^4)\,\mathrm{d}x\Bigg]$$
$$+\mathcal{R}_C. \quad (5.137)$$

After the substitution $x=-t$, the first line of (5.137) can be expressed by the progress coefficients $e^{1,0}_{\mu,\lambda}$ and $e^{1,2}_{\mu,\lambda}$. The second and fourth line are joined together. Analogous to (5.105), the third line gives $6s^3\gamma$. The intermediate result reads

$$
\begin{aligned}
I_C = {} & s^3\left(-2e^{1,0}_{\mu,\lambda} - e^{1,2}_{\mu,\lambda} + 6\gamma\right) \; + \; s^3\gamma\,(\lambda-\mu)\binom{\lambda}{\mu}\Bigg[\\
& \frac{1}{2\pi}\int \mathrm{e}^{-x^2}\left(1-\Phi(x)\right)^{\lambda-\mu-1}\left(\Phi(x)\right)^{\mu-1}\left(-x^3-2x^5\right)\mathrm{d}x \\
& + \frac{\lambda-\mu-1}{\left(\sqrt{2\pi}\right)^3}\int \mathrm{e}^{-\frac{3}{2}x^2}\left(1-\Phi(x)\right)^{\lambda-\mu-2}\left(\Phi(x)\right)^{\mu-1}\left(2-x^2-x^4\right)\mathrm{d}x \\
& - \frac{\mu-1}{\left(\sqrt{2\pi}\right)^3}\int \mathrm{e}^{-\frac{3}{2}x^2}\left(1-\Phi(x)\right)^{\lambda-\mu-1}\left(\Phi(x)\right)^{\mu-2}\left(2-x^2-x^4\right)\mathrm{d}x\Bigg] \\
& + \mathcal{R}_C.
\end{aligned} \tag{5.138}
$$

Using the integral identity (A.32) with (A.39), this yields

$$
\begin{aligned}
I_C = {} & s^3\left(-2e^{1,0}_{\mu,\lambda} - e^{1,2}_{\mu,\lambda} + 6\gamma\right) \\
& + s^3\gamma\,(\lambda-\mu)\binom{\lambda}{\mu}\frac{1}{2\pi}\int \mathrm{e}^{-x^2}\left(1-\Phi(x)\right)^{\lambda-\mu-1}\left(\Phi(x)\right)^{\mu-1}\left(-6x-3x^3\right)\mathrm{d}x \\
& + \mathcal{R}_C.
\end{aligned} \tag{5.139}
$$

After the substitution $x=-t$ using (5.112), it follows that

$$
I_C = s^3\left[-2e^{1,0}_{\mu,\lambda} - e^{1,2}_{\mu,\lambda} + 3\gamma\left(2 + 2e^{1,1}_{\mu,\lambda} + e^{1,3}_{\mu,\lambda}\right)\right] \; + \; \mathcal{R}_C. \tag{5.140}
$$

5.2.4.2 The I_D Integral. Analogous to (5.118, 5.125), one gets for (5.134) after the standardization transformation

$$
\begin{aligned}
I_D = s^3\,(\lambda-\mu)\binom{\lambda}{\mu}\int_{x=-\infty}^{x=\infty} & p_z(x)[1-P_z(x)]^{\lambda-\mu-1}[P_z(x)]^{\mu-1} \\
& \times\left\{\int_{y=-\infty}^{y=x} y p_z(y)\int_{z=-\infty}^{z=y} z^2 p_z(z)\,\mathrm{d}z\,\mathrm{d}y\right. \\
& \qquad \left. + \int_{y=-\infty}^{y=x} y^2 p_z(y)\int_{z=-\infty}^{z=y} z p_z(z)\,\mathrm{d}z\,\mathrm{d}y\right\}\mathrm{d}x.
\end{aligned} \tag{5.141}
$$

The integrations over z in (5.141) were already carried out. Using (5.121) and (5.102), one gets for the first y integrand in the braces of (5.141)

$$
\begin{aligned}
& y p_z(y)\int_{-\infty}^{y} z^2 p_z(z)\,\mathrm{d}z \\
& = \frac{\mathrm{e}^{-\frac{1}{2}y^2}}{2\pi}\left[y+\gamma(y^4-3y^2)\right]\left[\left(\sqrt{2\pi}\,\Phi(y) - y\,\mathrm{e}^{-\frac{1}{2}y^2}\right) - \gamma\,\mathrm{e}^{-\frac{1}{2}y^2}\left(2+y^2+y^4\right)\right] \\
& = \frac{1}{\sqrt{2\pi}}\,y\mathrm{e}^{-\frac{1}{2}y^2}\Phi(y) - \frac{1}{2\pi}\,y^2\mathrm{e}^{-y^2} + \gamma\,\frac{1}{\sqrt{2\pi}}\left(y^4-3y^2\right)\mathrm{e}^{-\frac{1}{2}y^2}\Phi(y)
\end{aligned}
$$

$$+ \gamma \frac{1}{2\pi} e^{-y^2} \left(-2y + 2y^3 - 2y^5\right) + \mathcal{O}(\gamma^2) , \tag{5.142}$$

and for the second one, using (5.126),

$$\begin{aligned} y^2 p_z(y) \int_{-\infty}^{y} z p_z(z) \, dz \\ &= \frac{1}{2\pi} e^{-y^2} \left[y^2 + \gamma \left(y^5 - 3y^3\right)\right] \left(-1 - \gamma y^3\right) \\ &= -\frac{1}{2\pi} y^2 e^{-y^2} + \gamma \frac{1}{2\pi} e^{-y^2} \left(3y^3 - 2y^5\right) + \mathcal{O}(\gamma^2) . \end{aligned} \tag{5.143}$$

Therefore, the contents of the braces in (5.141) become

$$\begin{aligned} &\left\{ \int_{-\infty}^{x} \int_{-\infty}^{y} \left(yz^2 + y^2 z\right) p_z(z) \, p_z(y) \, dz \, dy \right\} \\ &= -\frac{1}{\pi} \int_{-\infty}^{x} y^2 e^{-y^2} \, dy + \frac{1}{\sqrt{2\pi}} \int_{-\infty}^{x} y e^{-\frac{1}{2} y^2} \Phi(y) \, dy \\ &\quad - \gamma \frac{1}{2\pi} \int_{-\infty}^{x} e^{-y^2} \left(2y - 5y^3 + 4y^5\right) dy \\ &\quad - \gamma \frac{1}{\sqrt{2\pi}} \int_{-\infty}^{x} \left(3y^2 - y^4\right) e^{-\frac{1}{2} y^2} \Phi(y) \, dy + \mathcal{O}(\gamma^2) . \end{aligned} \tag{5.144}$$

The integrals in (5.144) can be solved analytically. Using the formulae (A.19, A.24, A.25) and (A.27) in Appendix A.2, they become

$$\begin{aligned} \{(5.144)\} &= -\frac{1}{2} \frac{1}{\sqrt{\pi}} \Phi(\sqrt{2} x) + \frac{1}{2\pi} x e^{-x^2} \\ &\quad + \frac{1}{\sqrt{2}} \frac{1}{\sqrt{2\pi}} \Phi(\sqrt{2} x) - \frac{1}{\sqrt{2\pi}} e^{-\frac{1}{2} x^2} \Phi(x) \\ &\quad + \gamma \frac{1}{2\pi} e^{-x^2} \left(\frac{5}{2} + \frac{3}{2} x^2 + 2x^4\right) \\ &\quad + \gamma \frac{1}{\sqrt{2\pi}} \left[-\frac{3}{2} \sqrt{2\pi} \left(\Phi(x)\right)^2 + 3x e^{-\frac{1}{2} x^2} \Phi(x) + \frac{3}{2} \frac{1}{\sqrt{2\pi}} e^{-x^2}\right] \\ &\quad + \gamma \frac{1}{\sqrt{2\pi}} \left[\frac{3}{2} \sqrt{2\pi} \left(\Phi(x)\right)^2 + \left(-3x - x^3\right) e^{-\frac{1}{2} x^2} \Phi(x) \right. \\ &\qquad \left. + \frac{1}{\sqrt{2\pi}} \left(-2 - \frac{1}{2} x^2\right) e^{-x^2}\right] + \mathcal{O}(\gamma^2) \\ \{(5.144)\} &= \frac{1}{2\pi} x e^{-x^2} - \frac{1}{\sqrt{2\pi}} e^{-\frac{1}{2} x^2} \Phi(x) + \gamma \frac{1}{2\pi} e^{-x^2} \left(2 + x^2 + 2x^4\right) \\ &\quad - \gamma \frac{1}{\sqrt{2\pi}} x^3 e^{-\frac{1}{2} x^2} \Phi(x) + \mathcal{O}(\gamma^2) . \end{aligned} \tag{5.145}$$

Now we can continue with the calculation of (5.141). Inserting (5.145) for the braces in (5.141), taking (5.109) into account, one gets in a linear γ approximation

$$
\begin{aligned}
I_D = s^3(\lambda-\mu)\binom{\lambda}{\mu}\Bigg[& -\frac{1}{2\pi}\int \mathrm{e}^{-x^2}\left(1-\Phi(x)\right)^{\lambda-\mu-1}\left(\Phi(x)\right)^{\mu-1}\,\mathrm{d}x \\
& + \frac{1}{\left(\sqrt{2\pi}\right)^3}\int x\mathrm{e}^{-\frac{3}{2}x^2}\left(1-\Phi(x)\right)^{\lambda-\mu-1}\left(\Phi(x)\right)^{\mu-2}\,\mathrm{d}x\Bigg] \\
+ s^3\gamma(\lambda-\mu)\binom{\lambda}{\mu}\Bigg[& \\
& -\frac{1}{2\pi}\int \mathrm{e}^{-x^2}\left(1-\Phi(x)\right)^{\lambda-\mu-1}\left(\Phi(x)\right)^{\mu-1}\left(x^3-3x\right)\mathrm{d}x \\
& + \frac{1}{\left(\sqrt{2\pi}\right)^3}\int \mathrm{e}^{-\frac{3}{2}x^2}\left(1-\Phi(x)\right)^{\lambda-\mu-1}\left(\Phi(x)\right)^{\mu-2}\left(x^4-3x^2\right)\mathrm{d}x \\
& + \frac{1}{\left(\sqrt{2\pi}\right)^3}\int \mathrm{e}^{-\frac{3}{2}x^2}\left(1-\Phi(x)\right)^{\lambda-\mu-1}\left(\Phi(x)\right)^{\mu-2}\left(2+x^2+2x^4\right)\mathrm{d}x \\
& -\frac{1}{2\pi}\int \mathrm{e}^{-x^2}\left(1-\Phi(x)\right)^{\lambda-\mu-1}\left(\Phi(x)\right)^{\mu-1}x^3\,\mathrm{d}x \\
& + \frac{\lambda-\mu-1}{\left(\sqrt{2\pi}\right)^4}\int \mathrm{e}^{-2x^2}\left(1-\Phi(x)\right)^{\lambda-\mu-2}\left(\Phi(x)\right)^{\mu-2}\left(x^3-x\right)\mathrm{d}x \\
& - \frac{\mu-2}{\left(\sqrt{2\pi}\right)^4}\int \mathrm{e}^{-2x^2}\left(1-\Phi(x)\right)^{\lambda-\mu-1}\left(\Phi(x)\right)^{\mu-3}\left(x^3-x\right)\mathrm{d}x \\
& - \frac{\lambda-\mu-1}{\left(\sqrt{2\pi}\right)^3}\int \mathrm{e}^{-\frac{3}{2}x^2}\left(1-\Phi(x)\right)^{\lambda-\mu-2}\left(\Phi(x)\right)^{\mu-1}\left(x^2-1\right)\mathrm{d}x \\
& + \frac{\mu-2}{\left(\sqrt{2\pi}\right)^3}\int \mathrm{e}^{-\frac{3}{2}x^2}\left(1-\Phi(x)\right)^{\lambda-\mu-1}\left(\Phi(x)\right)^{\mu-2}\left(x^2-1\right)\mathrm{d}x\Bigg] \\
+ \mathcal{R}_D. & \qquad (5.146)
\end{aligned}
$$

The first two lines can be directly expressed by the progress coefficients (5.112). One obtains $-s^3 e^{1,0}_{\mu,\lambda}$ and $-s^3 e^{2,1}_{\mu,\lambda}$. The third and sixth line can be joined together as well as the fourth and fifth ones. The seventh and eighth lines are transformed using the integral identity (A.32), version (A.38). One gets as an intermediate result

$$
\begin{aligned}
I_D = s^3\left[-e^{1,0}_{\mu,\lambda} - e^{2,1}_{\mu,\lambda}\right] \;+\; s^3\gamma(\lambda-\mu)\binom{\lambda}{\mu}\Bigg[& \\
& -\frac{1}{2\pi}\int \mathrm{e}^{-x^2}\left(1-\Phi(x)\right)^{\lambda-\mu-1}\left(\Phi(x)\right)^{\mu-1}\left(2x^3-3x\right)\mathrm{d}x \\
& + \frac{1}{\left(\sqrt{2\pi}\right)^3}\int \mathrm{e}^{-\frac{3}{2}x^2}\left(1-\Phi(x)\right)^{\lambda-\mu-1}\left(\Phi(x)\right)^{\mu-2}\left(1+4x^2\right)\mathrm{d}x
\end{aligned}
$$

$$
\begin{aligned}
& -\frac{\lambda-\mu-1}{\left(\sqrt{2\pi}\right)^3} \int \mathrm{e}^{-\frac{3}{2}x^2} \left(1-\Phi(x)\right)^{\lambda-\mu-2} \left(\Phi(x)\right)^{\mu-1} \left(x^2-1\right) \mathrm{d}x \\
& +\frac{\mu-1}{\left(\sqrt{2\pi}\right)^3} \int \mathrm{e}^{-\frac{3}{2}x^2} \left(1-\Phi(x)\right)^{\lambda-\mu-1} \left(\Phi(x)\right)^{\mu-2} \left(x^2-1\right) \mathrm{d}x \\
& \left. -\frac{1}{\left(\sqrt{2\pi}\right)^3} \int \mathrm{e}^{-\frac{3}{2}x^2} \left(1-\Phi(x)\right)^{\lambda-\mu-1} \left(\Phi(x)\right)^{\mu-2} \left(x^2-1\right) \mathrm{d}x \right] \\
& + \mathcal{R}_D. \qquad (5.147)
\end{aligned}
$$

Joining the third and last line together and applying the integral identity (A.32) with (A.34), it follows that

$$
\begin{aligned}
I_D = {} & s^3 \left[-e^{1,0}_{\mu,\lambda} - e^{2,1}_{\mu,\lambda}\right] \\
& + s^3 \gamma(\lambda-\mu)\binom{\lambda}{\mu} \left[-\frac{1}{2\pi} \int x\,\mathrm{e}^{-x^2} \left(1-\Phi(x)\right)^{\lambda-\mu-1} \left(\Phi(x)\right)^{\mu-1} \mathrm{d}x \right. \\
& \left. + \frac{1}{\left(\sqrt{2\pi}\right)^3} \int \mathrm{e}^{-\frac{3}{2}x^2} \left(1-\Phi(x)\right)^{\lambda-\mu-1} \left(\Phi(x)\right)^{\mu-2} \left(2+3x^2\right) \mathrm{d}x \right] \\
& + \mathcal{R}_D. \qquad (5.148)
\end{aligned}
$$

The remaining integrals can be expressed by progress coefficients (5.112). One finally obtains

$$
I_D = -s^3 \left[e^{1,0}_{\mu,\lambda} + e^{2,1}_{\mu,\lambda} - \gamma \left(e^{1,1}_{\mu,\lambda} + 2e^{2,0}_{\mu,\lambda} + 3e^{2,2}_{\mu,\lambda} \right) \right] + \mathcal{R}_D. \qquad (5.149)
$$

5.2.4.3 The I_E Integral. For the I_E integral in (5.135), one obtains after the standardization transformation similar to the transformation from (5.118) to (5.125)

$$
\begin{aligned}
I_E = {} & 6s^3(\lambda-\mu)\binom{\lambda}{\mu} \int_{-\infty}^{\infty} p_z(x)[1-P_z(x)]^{\lambda-\mu-1}[P_z(x)]^{\mu-3} \\
& \times \int_{y=-\infty}^{y=x} y\,p_z(y) \int_{z=-\infty}^{z=y} z\,p_z(z) \int_{t=-\infty}^{t=z} t\,p_z(t)\,\mathrm{d}t\,\mathrm{d}z\,\mathrm{d}y\,\mathrm{d}x. \qquad (5.150)
\end{aligned}
$$

The inner integrations with respect to t and z were already carried out in a similar form, see Eq. (5.127). Considering (5.102) and (5.127) in the linear γ approximation, one obtains for the integral with respect to y

$$
\begin{aligned}
& \int_{y=-\infty}^{y=x} y\,p_z(y) \int_{z=-\infty}^{z=y} z\,p_z(z) \int_{t=-\infty}^{t=z} t\,p_z(t)\,\mathrm{d}t\,\mathrm{d}z\,\mathrm{d}y\,\mathrm{d}x \\
& = \frac{1}{2}\frac{1}{\left(\sqrt{2\pi}\right)^3} \int_{y=-\infty}^{y=x} \mathrm{e}^{-\frac{3}{2}y^2} \left(y - 3\gamma y^2 + 3\gamma y^4\right) \mathrm{d}y + \mathcal{O}(\gamma^2) \\
& = \frac{1}{2}\frac{1}{\left(\sqrt{2\pi}\right)^3} \left[I^{2,1}(x) - 3\gamma I^{2,2}(x) + 3\gamma I^{2,4}(x) \right] + \mathcal{O}(\gamma^2)
\end{aligned}
$$

$$= \frac{1}{2} \frac{1}{\left(\sqrt{2\pi}\right)^3} \left[-\frac{1}{3} \mathrm{e}^{-\frac{3}{2}x^2} + \gamma x \mathrm{e}^{-\frac{3}{2}x^2} - \gamma \mathrm{e}^{-\frac{3}{2}x^2} \left(x + x^3\right) \right] + \mathcal{O}(\gamma^2)$$

$$= -\frac{1}{6} \frac{1}{\left(\sqrt{2\pi}\right)^3} \mathrm{e}^{-\frac{3}{2}x^2} \left(1 + \gamma 3x^3\right) + \mathcal{O}(\gamma^2). \qquad (5.151)$$

The integration formulae (A.20) were used for this derivation. Using (5.109) and (5.151), one gets for the I_E integral

$$\begin{aligned} I_E = -s^3(\lambda - \mu)\binom{\lambda}{\mu} \Bigg[& \frac{1}{(2\pi)^2} \int \mathrm{e}^{-2x^2} (1 - \Phi(x))^{\lambda-\mu-1} (\Phi(x))^{\mu-3} \, \mathrm{d}x \\ & + \gamma \frac{1}{(2\pi)^2} \int \mathrm{e}^{-2x^2} (1 - \Phi(x))^{\lambda-\mu-1} (\Phi(x))^{\mu-3} \left(x^3 - 3x\right) \mathrm{d}x \\ & + \gamma \frac{1}{(2\pi)^2} \int \mathrm{e}^{-2x^2} (1 - \Phi(x))^{\lambda-\mu-1} (\Phi(x))^{\mu-3} \left(3x^3\right) \mathrm{d}x \\ & + \frac{\lambda - \mu - 1}{\left(\sqrt{2\pi}\right)^5} \int \mathrm{e}^{-\frac{5}{2}x^2} (1 - \Phi(x))^{\lambda-\mu-2} (\Phi(x))^{\mu-3} \left(x^2 - 1\right) \mathrm{d}x \\ & - \frac{\mu - 3}{\left(\sqrt{2\pi}\right)^5} \int \mathrm{e}^{-\frac{5}{2}x^2} (1 - \Phi(x))^{\lambda-\mu-1} (\Phi(x))^{\mu-4} \left(x^2 - 1\right) \mathrm{d}x \Bigg] \\ & + \mathcal{R}_E. \end{aligned} \qquad (5.152)$$

With (5.112), the first line gives $-s^3 e^{3,0}_{\mu,\lambda}$. The second and third lines are combined, the last two are transformed using the integral identity (A.32), using (A.36). One obtains

$$\begin{aligned} I_E = s^3 \Bigg[-e^{3,0}_{\mu,\lambda} - (\lambda - \mu)\binom{\lambda}{\mu} \frac{3\gamma}{(2\pi)^2} \int x \, \mathrm{e}^{-2x^2} & (1 - \Phi(x))^{\lambda-\mu-1} \\ & \times (\Phi(x))^{\mu-3} \, \mathrm{d}x \Bigg] + \mathcal{R}_E. \end{aligned} \qquad (5.153)$$

Finally, one gets with (5.112)

$$I_E = s^3 \left(-e^{3,0}_{\mu,\lambda} + 3\gamma e^{3,1}_{\mu,\lambda} \right) + \mathcal{R}_E. \qquad (5.154)$$

5.2.4.4 Composing the $\gamma^{(g+1)}$ Formula. The $\gamma^{(g+1)}$ formula (5.132) can be composed now. Using the terms (5.140, 5.149) and (5.154), one obtains

$$\begin{aligned} \gamma^{(g+1)} = \frac{1}{6} \frac{s^3}{\left(s^{(g+1)}\right)^3} \frac{1}{\zeta^3} \frac{\mu - 1}{\mu} \frac{\mu - 2}{\mu} & \Big[\left(e^{1,0}_{\mu,\lambda} - e^{1,2}_{\mu,\lambda} + 3e^{2,1}_{\mu,\lambda} - 2e^{3,0}_{\mu,\lambda} \right) \\ & + \gamma \left(6 + 3e^{1,1}_{\mu,\lambda} + 3e^{1,3}_{\mu,\lambda} - 6e^{2,0}_{\mu,\lambda} - 9e^{2,2}_{\mu,\lambda} + 6e^{3,1}_{\mu,\lambda} \right) \Big] \\ + \mathcal{R}_{\gamma'}. & \end{aligned} \qquad (5.155)$$

5.2.5 The Self-Consistent Method and the $\varphi^*_{\mu,\lambda}$ Formulae

After achieving the formulae (5.131) and (5.155), we are now able to derive the self-consistency equations (5.98), and to solve them stepwise. The self-consistency condition (see Sect. 5.1.10), requires the equality of $s^{(g+1)}$ and $s^{(g)}$, as well as of $\gamma^{(g+1)}$ and $\gamma^{(g)}$. According to this condition, we can introduce in (5.155) the stationary skewness $\gamma_{\mu,\lambda}$ and the stationary standard deviation $s_{\mu,\lambda}$. Using $s_{\mu,\lambda} := s^{(g+1)} \overset{!}{=} s^{(g)} \equiv s$ and $\gamma_{\mu,\lambda} := \gamma^{(g+1)} \overset{!}{=} \gamma^{(g)} \equiv \gamma$, one obtains

$$\gamma_{\mu,\lambda} = \frac{1}{\zeta^3} \frac{\mu-1}{\mu} \frac{\mu-2}{\mu} \frac{1}{6} \left[\left(e^{1,0}_{\mu,\lambda} - e^{1,2}_{\mu,\lambda} + 3e^{2,1}_{\mu,\lambda} - 2e^{3,0}_{\mu,\lambda} \right) + \gamma_{\mu,\lambda} \left(6 + 3e^{1,1}_{\mu,\lambda} + 3e^{1,3}_{\mu,\lambda} - 6e^{2,0}_{\mu,\lambda} - 9e^{2,2}_{\mu,\lambda} + 6e^{3,1}_{\mu,\lambda} \right) \right]. \qquad (5.156)$$

According to the linear γ approximation, $\mathcal{R}_\gamma = 0$ is assumed. As one can see, the self-consistency condition yields an equation linear in γ *independent* of the standard deviation s of the offspring population. Considering (5.58) for ζ, the solution reads

$$\boxed{\gamma_{\mu,\lambda} = \frac{e^{1,0}_{\mu,\lambda} - e^{1,2}_{\mu,\lambda} + 3e^{2,1}_{\mu,\lambda} - 2e^{3,0}_{\mu,\lambda}}{\frac{6\mu^2}{(\mu-1)(\mu-2)} \left(\sqrt{1 + \frac{\sigma^{*2}}{N}} \right)^3 - \left(6 + 3e^{1,1}_{\mu,\lambda} + 3e^{1,3}_{\mu,\lambda} - 6e^{2,0}_{\mu,\lambda} - 9e^{2,2}_{\mu,\lambda} + 6e^{3,1}_{\mu,\lambda} \right)}.} \qquad (5.157)$$

This $\gamma_{\mu,\lambda}$ formula *cannot* be used for the cases $\mu = 1,\ 2$. However, interpreting (5.156) as an iteration relation, it immediately follows that

$$\boxed{\gamma_{1,\lambda} = \gamma_{2,\lambda} = 0.} \qquad (5.158)$$

This should obviously be expected for this case, since the offspring are generated almost normally by $\mu = 1$ and $\mu = 2$ parents. Hence, the offspring population is (almost) symmetric.

With $\gamma_{\mu,\lambda} := \gamma^{(g+1)} \overset{!}{=} \gamma^{(g)} \equiv \gamma$ and $s_{\mu,\lambda} := s^{(g+1)} \overset{!}{=} s^{(g)} \equiv s$, the equation for the self-consistent determination of $s_{\mu,\lambda}$ follows from Eq. (5.131)

$$s_{\mu,\lambda} = \frac{1}{\zeta} \sqrt{\sigma^2 \frac{1+\zeta^2}{2} + s^2_{\mu,\lambda} \frac{\mu-1}{\mu} \left[1 + e^{1,1}_{\mu,\lambda} - e^{2,0}_{\mu,\lambda} - 2\gamma_{\mu,\lambda} \left(e^{1,0}_{\mu,\lambda} + e^{1,2}_{\mu,\lambda} - e^{2,1}_{\mu,\lambda} \right) \right]}. \qquad (5.159)$$

The error term $\mathcal{R}_s$ is neglected again. As one can see, (5.159) is linear in $s^2_{\mu,\lambda}$. Hence, this equation can also be solved easily. Considering the definition of ζ in (5.58), one obtains

$$\boxed{s_{\mu,\lambda} = \sigma \frac{\sqrt{1 + \frac{\sigma^{*2}}{2N}}}{\sqrt{1 + \frac{\sigma^{*2}}{N} - \frac{\mu-1}{\mu} \left[1 + e^{1,1}_{\mu,\lambda} - e^{2,0}_{\mu,\lambda} - 2\gamma_{\mu,\lambda} \left(e^{1,0}_{\mu,\lambda} + e^{1,2}_{\mu,\lambda} - e^{2,1}_{\mu,\lambda} \right) \right]}}.} \qquad (5.160)$$

As the last step of the long way to the $\varphi^*_{\mu,\lambda}$ formula, (5.160) is inserted into (5.115) and the normalization (5.8) is carried out. Finally, we obtain the *N-dependent formula for the progress rate* of the (μ, λ)-ES on the sphere model

$$\boxed{\begin{aligned}\varphi^*_{\mu,\lambda}(\sigma^*) &= N\left(1 - \sqrt{1 + \frac{\sigma^{*2}}{N}}\right) \\ &+ \sigma^* \frac{\sqrt{1 + \frac{\sigma^{*2}}{2N}}\left(e^{1,0}_{\mu,\lambda} - \gamma_{\mu,\lambda} e^{1,1}_{\mu,\lambda}\right)}{\sqrt{1 + \frac{\sigma^{*2}}{N} - \frac{\mu-1}{\mu}\left[1 + e^{1,1}_{\mu,\lambda} - e^{2,0}_{\mu,\lambda} - 2\gamma_{\mu,\lambda}\left(e^{1,0}_{\mu,\lambda} + e^{1,2}_{\mu,\lambda} - e^{2,1}_{\mu,\lambda}\right)\right]}} + \mathcal{R}_{\varphi^*}.\end{aligned}} \tag{5.161}$$

This relatively complicated formula depends on nine generalized progress coefficients, if one counts the different $e^{\alpha,\beta}_{\mu,\lambda}$ in (5.161) as well as in (5.157). If one substitutes in (5.161) $\gamma_{\mu,\lambda}$ by (5.157), five *independent* $e^{\alpha,\beta}_{\mu,\lambda}$ combinations still remain. For numerical evaluations, this is not a serious problem if a computer is used. The $e^{\alpha,\beta}_{\mu,\lambda}$ coefficients can be calculated easily by numerical integration of (5.112). The curves of Fig. 5.1 (p. 187) are obtained in this way. For the theoretical investigations, however, (5.161) is really unmanageable. Therefore, an approximate progress rate formula with a single progress coefficient will be derived in Sect. 5.3.2.

The $\varphi^*_{\mu,\lambda}$ formula in (5.161) is based on the approximation (5.25) of the offspring distribution $p(r)$ taking the γ term into account. Hence, it considers a certain asymmetry (skewness) of the offspring distribution. The simple special case $\gamma = 0$, i.e. the approximation of lower-order is also contained in (5.161). In this case, the offspring population is approximated (symmetrically) by a normal distribution with standard deviation s

$$\varphi^*_{\mu,\lambda} = e^{1,0}_{\mu,\lambda}\,\sigma^* \frac{\sqrt{1 + \frac{\sigma^{*2}}{2N}}}{\sqrt{1 + \frac{\sigma^{*2}}{N} - \frac{\mu-1}{\mu}\left(1 + e^{1,1}_{\mu,\lambda} - e^{2,0}_{\mu,\lambda}\right)}} - N\left(\sqrt{1 + \frac{\sigma^{*2}}{N}} - 1\right) + \ldots, \tag{5.162}$$

noting that[11]

$$s > \tilde{\sigma}, \qquad \tilde{\sigma} = \sigma \sqrt{\frac{1 + \frac{\sigma^{*2}}{2N}}{1 + \frac{\sigma^{*2}}{N}}}. \tag{5.163}$$

This is an immediate consequence of the (μ, λ)-ES algorithm: the r distribution of the parents in the parameter space $\mathcal{Y}$, has a standard deviation greater than zero. This r distribution is superimposed with the mutations having standard deviation $\tilde{\sigma}$. Therefore, $s > \tilde{\sigma}$ necessarily holds.

[11] Note that $\tilde{\sigma}$ is the standard deviation of the $\tilde{R}$ values generated by the mutations for a given parental R. Compare here Eqs. (3.246) and (5.28).

5.3 The Discussion of the (μ, λ)-ES

5.3.1 The Comparison with Experiments

In the derivation of the progress rate (5.161), several approximation assumptions have been made. These assumptions are related to the sphere model as well as to the approximation of the offspring distribution $p(r)$. The approximations used are based on plausibility considerations and on generalizations of results obtained from the $(1, \lambda)$-ES theory. An exact trace of the order of error made here has not been done. The derivations were very complicated. Therefore, no exact and at the same time numerically *relevant* statements can be made in general on the error term $\mathcal{R}_{\varphi^*}$.

In addition to the plausibility considerations, there are at least three further arguments showing the usefulness of the approximations:

1. The correspondence principle: the $\varphi^*_{\mu,\lambda}$ formula must become for $\mu = 1$ equal to the N-dependent $\varphi^*_{1,\lambda}$ formula (3.242). For this formula, the order of the error term (with respect to N) is known.
2. The sphere model $\varphi^*_{\mu,\lambda}$ must yield for $\langle R\rangle \to \infty$, $\sigma = \text{const} > 0$, the formula for the hyperplane (see Sect. 2.2.2). This will be investigated in Sect. 5.3.3.
3. The usefulness of a theoretical analysis has at least two aspects. Apart from the deeper understanding obtained by the analysis of the working principles of the algorithm, the *predictive* value should be emphasized here. Consequently, the experiment finally decides on the quality of the predictions, hence on the usability of the approximations made.

In this section, we are concerned with the experimental examination of the progress rate formula (5.161). There are two possibilities to carry out the ES experiments: (a) investigation of real ES runs; and (b) specially tailored ES simulations.

5.3.1.1 Data Extraction from ES Runs. In this case, the data of real ES runs are analyzed. The case of interest is the determination of φ^* values in a large interval of σ^*, and not only the ES working with maximal performance. Therefore, σ control mechanisms, such as the 1/5th rule or self-adaptation, should be switched off. In other words, the ES algorithm should run with *constant* σ. Since the mean radius $\langle R\rangle$ changes over the generations g, σ^* attains values in a certain interval.[12] Therefore, one can calculate the σ^* value for each generation g based on the measured $\langle R\rangle$ values. Similarly, one can calculate a random value $\tilde{\varphi}^*$ using the change in the $\langle R\rangle$ value from generation g to generation $(g+1)$. Since the actual change in $\langle R\rangle$ is of concern, which depends on σ^*, denoted as $\tilde{\varphi}^* = f(\sigma^*)$, the $\tilde{\varphi}^*(\sigma^*)$ value must be averaged accordingly (see Eq. (5.4)).

[12] This interval for σ^* is limited by the *stationary residual distance* $\langle R\rangle_\infty$ for large g, $g \to \infty$. For constant σ, the (μ, λ)-ES shows a behavior similar to the $(1, \lambda)$-ES. Compare the R_∞ formula in (3.182) for this purpose.

A problem emerges at this point. In order to calculate averages, many ES runs must be made. For a fixed generation number G of any of these runs there will be different σ^* values (stochastic process!). However, σ^* should be constant for the calculation of the average over $\tilde{\varphi}^*(\sigma^*)$. Therefore, calculating the $\varphi^*(\sigma^*)$ value from the ES runs requires the introduction of different σ^* classes. For these classes, the corresponding intervals must be chosen sufficiently small. The number of ES runs depends on the desired minimal number of σ^* values per class. In the classes, the averages are taken over $\tilde{\varphi}^*$ and σ^*.

Obviously, this method is very expensive as to the computational effort. Sufficiently fine class discrimination is desired, and the deviations of the mean values should be appropriately small. Therefore, this method should only be used if other possibilities of experimental φ^* determination are excluded. It is used in the investigation of the *dominant* multirecombination. For the (μ, λ)-ES, however, there exists a more elegant solution.

5.3.1.2 The ES Simulations for σ = const *and* σ^* = const. For the experimental investigation of the relation $\varphi^*_{\mu,\lambda} = f(\sigma^*)$, a technique will be used that provides $\varphi^*_{\mu,\lambda}$ for a given σ^*. If one additionally postulates $\sigma = \text{const}$, it follows because of (5.8) that $\langle R \rangle \stackrel{!}{=} \text{const} =: R^{(0)}$. In other words, $\langle R \rangle$ should be held constant from generation to generation. This can be realized easily. Starting from μ identical parents with the distance $R^{(0)} = \sigma N/\sigma^*$ to the optimum, one generation cycle of the (μ, λ)-ES is executed. As a result, one gets the parents of the next generation. The residual distances of these, the $R_m^{(1)}$ values, can be calculated. Therefore, one obtains $\langle R \rangle$ and $\tilde{\varphi} = R^{(0)} - \langle R \rangle$. Since $\langle R \rangle \neq R^{(0)}$, the parents generated must be *rescaled*, i.e. the new parent vectors $\mathbf{R}_m^{(1)}$ are multiplied by $R^{(0)}/\langle R \rangle$. The parents scaled in this way have again the mean residual distance $\langle R \rangle = R^{(0)}$ to the optimum, and the condition $\sigma^* = \text{const}$ is fulfilled. The rescaling of the parents is admissible since the actual placement of the parents in the parameter space $\mathcal{Y}$ is irrelevant because of the sphere symmetry. Only their residual distances count.

For the actual simulation, the ES algorithm should run for some generations G_1. After this period, stationary distribution of the parental population can be expected. Thereafter, the value $\tilde{\varphi}^{(g)} = R^{(0)} - \langle R \rangle^{(g)}$ is calculated over certain generations G_2. In this way, one obtains very accurate simulation results, assuming that the pseudo-random number generators are "sufficiently random."

The choice of G_1 is not critical, if one chooses G_2 sufficiently large, $G_2 \gg G_1$.[13] In the simulations, $G_1 = 100$ was chosen.

The simulation results obtained for the $N = 100$ dimensional sphere model are given here as an example. Figure 5.1 shows the $(\mu, 8)$-ES on the

[13] In principle, one can estimate the G_1-time accurately by numerically iterating Eqs. (5.97, 5.131) and (5.155). Alternatively, the transient behavior of s and γ can be described by differential equations approximately. The time constants emerging thereby can be used as an estimate for G_1.

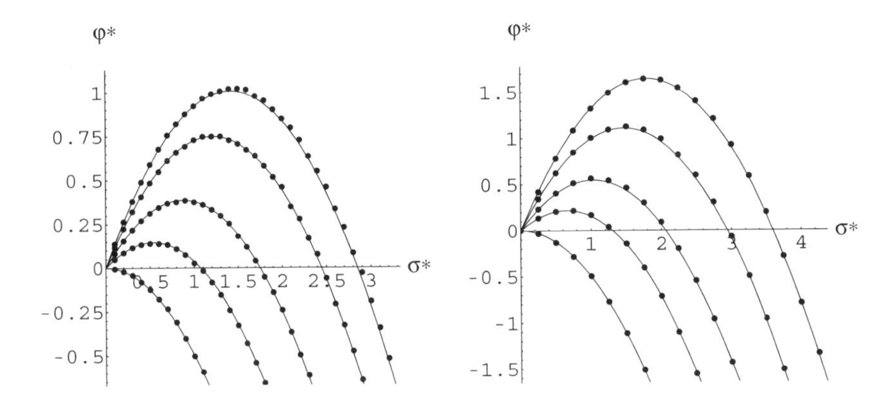

Fig. 5.1. The comparison of the theoretically computed progress rate (depicted as curves) and the simulation results (dots) for the (μ, λ)-ES on the sphere model. Left: the $(\mu, 8)$-ES with $\mu = 8,\ 6,\ 4,\ 2,\ 1$ (from the left curve to the right one). Right: the $(\mu, 80)$-ES with $\mu = 80,\ 60,\ 40,\ 20,\ 10$ (from the left curve to the right one)

left and the $(\mu, 80)$-ES on the right. For the strategies with $\lambda = 8$, each data point was obtained by averaging over $G_2 = 20,000$ generations. For $\lambda = 80$, it was $G_2 = 10,000$. The standard error of the empirical measurements are smaller than the diameter of the "dots."

One observes a good agreement between theory and experiments. For most of the cases, the limitation to s and γ in the $p(r)$ series (5.25) suffices.

From the curves in Fig. 5.1, one can infer some important properties of the (μ, λ)-ES. These will be given now without a formal proof. One has

$$\varphi^*_{\lambda,\lambda}(\sigma^*) \le 0. \tag{5.164}$$

That is, the ES algorithm cannot converge if it gives the same reproduction chances to all its offspring. If the selection pressure does not exist, as in the case $\mu = \lambda$ considered here, the population performs a *random walk* in the parameter space $\mathcal{Y}$. Since the progress rate is negative for $\sigma^* \neq 0$, the ES algorithm diverges, i.e. the distance of the population to the optimum increases in the statistical mean.

If one compares the progress rates for a constant $\sigma^* > 0$ and constant λ, one observes for the sphere model

$$\mu_1 > \mu_2 \qquad \Longrightarrow \qquad \varphi^*_{\mu_1,\lambda}(\sigma^*) < \varphi^*_{\mu_2,\lambda}(\sigma^*). \tag{5.165}$$

A decrease in the number of parents μ increases the progress rate for a constant number of descendants λ. The largest progress rate on the sphere model (without noise) is obtained for a single parent ($\mu = 1$), i.e. for the $(1, \lambda)$-ES.

5.3.2 Simplified $\varphi^*_{\mu,\lambda}$ Formulae and the Progress Coefficient $c_{\mu,\lambda}$

5.3.2.1 The Derivation of the $\varphi^*_{\mu,\lambda}$ Formulae. The progress rate formula (5.161) is not very handy. Therefore, a simplification will be introduced here. Instead of nine $e^{\alpha,\beta}_{\mu,\lambda}$ coefficients, a single progress coefficient will be derived for the (μ, λ)-ES. Particularly, the $(1, \lambda)$-ES analyzed in Chap. 3 described by Eq. (3.242) should appear as a special case of the progress rate formula for the (μ, λ)-ES. Considering (5.161) under this aspect, (5.161) can be rewritten equivalently as follows

$$\varphi^*_{\mu,\lambda}(\sigma^*) = N\left(1 - \sqrt{1 + \frac{\sigma^{*2}}{N}}\right) + \underbrace{\frac{\left(e^{1,0}_{\mu,\lambda} - \gamma_{\mu,\lambda}(\zeta)\, e^{1,1}_{\mu,\lambda}\right)}{\sqrt{1 - \frac{\mu-1}{\zeta^2\,\mu}\left[1 + e^{1,1}_{\mu,\lambda} - e^{2,0}_{\mu,\lambda} - 2\gamma_{\mu,\lambda}(\zeta)\left(e^{1,0}_{\mu,\lambda} + e^{1,2}_{\mu,\lambda} - e^{2,1}_{\mu,\lambda}\right)\right]}}}_{\stackrel{!}{=}\, c_{\mu,\lambda}} \sigma^* \sqrt{\frac{1 + \frac{\sigma^{*2}}{2N}}{1 + \frac{\sigma^{*2}}{N}}} + \mathcal{R}'_{\varphi^*}. \tag{5.166}$$

In the underbraced part, ζ is substituted according to Eq. (5.58). After this reordering, one gets a formula analogous to Eq. (3.242), if one *assumes* that the underbraced part of (5.166) is a constant factor

$$c_{\mu,\lambda} \stackrel{!}{\neq} f(\zeta). \tag{5.167}$$

For $\mu = 1$, this is fulfilled independent of $\zeta = \sqrt{1 + \sigma^{*2}/N}$. Trivially, one can see in Eq. (5.158) that $\gamma_{1,\lambda} = 0$ for this case, and (5.166) becomes (3.242) because of (5.113). For the case $\mu \geq 2$, the requirement (5.167) is fulfilled only approximately. In other words, one has to choose a ζ value. We choose $\zeta = 1$, which is equivalent to $N \to \infty$ for $\sigma^* = \text{const}$ and which is approximately fulfilled for $\sigma^{*2}/N \ll 1$. Under this approximation assumption one obtains the progress rate formula

$$\boxed{\varphi^*_{\mu,\lambda}(\sigma^*, N) = c_{\mu,\lambda}\sigma^* \sqrt{\frac{1 + \frac{\sigma^{*2}}{2N}}{1 + \frac{\sigma^{*2}}{N}}} - N\left(\sqrt{1 + \frac{\sigma^{*2}}{N}} - 1\right) + \mathcal{R}'_{\varphi^*}.} \tag{5.168}$$

Here, the $c_{\mu,\lambda}$ *progress coefficient* is given by the formula

$$c_{\mu,\lambda} = \frac{e^{1,0}_{\mu,\lambda} - \gamma_{\mu,\lambda}(1)\, e^{1,1}_{\mu,\lambda}}{\sqrt{1 - \frac{\mu-1}{\mu}\left[1 + e^{1,1}_{\mu,\lambda} - e^{2,0}_{\mu,\lambda} - 2\gamma_{\mu,\lambda}(1)\left(e^{1,0}_{\mu,\lambda} + e^{1,2}_{\mu,\lambda} - e^{2,1}_{\mu,\lambda}\right)\right]}} \tag{5.169}$$

with

$$\gamma_{\mu,\lambda}(1) = \frac{e^{1,0}_{\mu,\lambda} - e^{1,2}_{\mu,\lambda} + 3e^{2,1}_{\mu,\lambda} - 2e^{3,0}_{\mu,\lambda}}{\frac{6\mu^2}{(\mu-1)(\mu-2)} - \left(6 + 3e^{1,1}_{\mu,\lambda} + 3e^{1,3}_{\mu,\lambda} - 6e^{2,0}_{\mu,\lambda} - 9e^{2,2}_{\mu,\lambda} + 6e^{3,1}_{\mu,\lambda}\right)} \tag{5.170}$$

(see also Table 5.1). The order of the error term $\mathcal{R}'_{\varphi^*}$ in (5.168) has not been investigated yet. However, by a comparison with the error term of the N-dependent $(1, \lambda)$ formula (3.242), one can assume that $\mathcal{R}'_{\varphi^*} = \mathcal{O}(1/\sqrt{N})$ should be valid.

Table 5.1. The $c_{\mu,\lambda}$ progress coefficients. They are obtained by the numerical integration of $e^{\alpha,\beta}_{\mu,\lambda}$ in (5.169) and (5.170). Since the $c_{\mu,\lambda}$ values only change slightly, intermediate values can be obtained by interpolation.

μ \ λ	5	10	20	30	40	50	100	150	200	250
1	1.16	1.54	1.87	2.04	2.16	2.25	2.51	2.65	2.75	2.82
2	0.92	1.36	1.72	1.91	2.04	2.13	2.40	2.55	2.65	2.72
3	0.68	1.20	1.60	1.80	1.93	2.03	2.32	2.47	2.57	2.65
4	0.41	1.05	1.49	1.70	1.84	1.95	2.24	2.40	2.51	2.59
5	0.00	0.91	1.39	1.62	1.77	1.87	2.18	2.34	2.45	2.53
10	-	0.00	0.99	1.28	1.46	1.59	1.94	2.12	2.24	2.33
20	-	-	0.00	0.76	1.03	1.20	1.63	1.84	1.97	2.07
30	-	-	-	0.00	0.65	0.89	1.41	1.64	1.79	1.90
40	-	-	-	-	0.00	0.57	1.22	1.49	1.65	1.77
50	-	-	-	-	-	0.00	1.06	1.35	1.53	1.65
100	-	-	-	-	-	-	0.00	0.81	1.07	1.24

If one is only interested in the σ^* interval where the progress rate is positive or slightly negative, the formula in (5.168) can be expanded in a Taylor series, simplifying the square roots with respect to σ^{*2}/N. As a result, one obtains the asymptotic $(N \to \infty)$ $\varphi^*_{\mu,\lambda}$ formula

$$\boxed{\varphi^*_{\mu,\lambda}(\sigma^*) = c_{\mu,\lambda}\sigma^* - \frac{\sigma^{*2}}{2} + \dots .} \tag{5.171}$$

This formula gives a satisfactory approximation for most cases, Rechenberg (1994) also offered this formula. However, he did not give a derivation; he simply generalized the $\varphi^*_{1,\lambda}$ formula (3.101) by analogy and used $c_{\mu,\lambda}$ values obtained by ES simulations. A comparison of Rechenberg's $c_{\mu,\lambda}$ simulation results with those obtained from (5.169) and displayed in Table 5.1 shows good agreement. The reason for the remaining deviations of less than 2% will be discussed in Sect. 5.3.3.

5.3.2.2 EPP, the Properties of $c_{\mu,\lambda}$, and the Fitness Efficiency.

EPP. The structure of (5.171) agrees with the asymptotic progress rate formula of the $(1, \lambda)$-ES (3.101). Therefore, the arguments about the EPP on p. 73 are also valid for this case. As a result, no further reasoning is necessary.

Properties of the $c_{\mu,\lambda}$ coefficients. Considering the $c_{\mu,\lambda}$ coefficients in Table 5.1, one verifies the validity of the inequality (see also Fig. 5.3 on p. 193)

$$\lambda > \mu > 1 \qquad \Rightarrow \qquad 0 < c_{\mu,\lambda} < c_{1,\lambda}. \tag{5.172}$$

The correctness of this statement can be proven by a plausibility argument: in the $(1, \lambda)$-ES always the best offspring is chosen, whereas worse descendants also take part in the reproduction process of the (μ, λ)-ES. The progress rate of the (μ, λ)-ES is obtained by the mean value of the μ best descendants; therefore, it cannot be larger than the progress rate of the best descendant. Unfortunately, a rigorous proof for the relation (5.172) as well as the following similarly plausible inequality

$$\mu_1 > \mu_2 \qquad \Rightarrow \qquad c_{\mu_1,\lambda} < c_{\mu_2,\lambda} \tag{5.173}$$

is still pending. However,

$$c_{\lambda,\lambda} = 0 \tag{5.174}$$

follows immediately from (5.169) with $e^{1,0}_{\lambda,\lambda} = 0$ and $e^{1,1}_{\lambda,\lambda} = 0$.

The calculation of the asymptotic properties of $c_{\mu,\lambda}$ for $\lambda \to \infty$ is an open problem. Let us numerically investigate $c_{\mu,\lambda}$ as a function of the selection pressure ϑ, introduced in (2.8) on p. 27. One can immediately infer from Table 5.1 the conjecture that

$$c_{\vartheta\lambda,\lambda} \xrightarrow{\lambda\to\infty} c(\vartheta) \neq f(\lambda) \qquad (0 < \vartheta \leq 1) \tag{5.175}$$

should be valid for $\lambda \to \infty$. That is, for large λ the progress coefficient should depend on the selection pressure ϑ only. However, deriving an analytical $c(\vartheta)$ expression turns out to be a nontrivial problem. Probably, an asymptotically exact analytically closed expression does not exist, since there is no such expression for $p(r)$ using ϑ. That is $p(r, \vartheta)$ cannot be expressed by a finite set of cumulants.[14]

There is a certain chance of finding asymptotically exact formulae for the $\lambda \to \infty$ case with $\mu = \text{const} < \infty$. This is so, because an asymptotic stability of the shape of $p(r)$ should exists.

Fitness efficiency. The notion of fitness efficiency was introduced in Chap. 3. Its definition (3.104) for the $(1, \lambda)$-ES can be generalized to the (μ, λ)-ES case

$$\text{FITNESS EFFICIENCY:} \qquad \eta_{\mu,\lambda} := \frac{\max_{\sigma^*}\left[\varphi^*_{\mu,\lambda}(\sigma^*)\right]}{\lambda}. \tag{5.176}$$

Therefore, asymptotically ($N \to \infty$), one gets with Eq. (5.171)

$$\eta_{\mu,\lambda} = \frac{c^2_{\mu,\lambda}}{2\lambda}. \tag{5.177}$$

Figure 5.2 displays $\eta_{\mu,\lambda} = f_\mu(\lambda)$ as a function of λ with parameter μ. There is a $\lambda = \hat{\lambda}$ at any given μ for which η becomes maximal. However, the maximum is not very distinct for larger μ: small deviations from the $\hat{\lambda}$ value are not very critical. A collection of the fitness-optimal $(\mu, \hat{\lambda})$ strategies is given in Table 5.2. One recognizes that the $(1, \lambda)$-ES is the most fitness-efficient (μ, λ)-

[14] Recall that formula (5.169) is derived on the approximation (5.99) for $p(r)$.

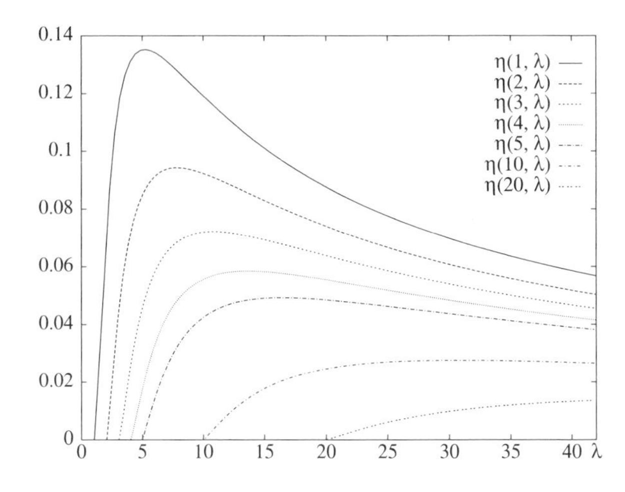

Fig. 5.2. The fitness efficiency η of the (μ, λ)-ES on the sphere model for different numbers of parents, $\mu = 1, 2, 3, 4, 5, 10$, and 20 (from top to bottom)

ES. The performance of the strategies with $\mu > 1$ are worse. This also follows from the property of the progress coefficients given in (5.172).

It is striking that for large μ obviously there is an optimal selection pressure $\vartheta = \hat{\vartheta}$. Its value is – in the scope of the $c_{\mu,\lambda}$ approximation used – at $\hat{\vartheta} \approx 0.35$ $(\ldots 0.36)$. The existence of such an optimum selection pressure for $\mu \to \infty$ immediately follows from (5.175) and (5.177)

$$\eta(\mu, \vartheta) = \eta_{\mu, \frac{\mu}{\vartheta}} = \frac{c(\vartheta)^2}{2\,\frac{\mu}{\vartheta}} = \frac{1}{2\mu}\,\vartheta c(\vartheta)^2. \tag{5.178}$$

In order to obtain a maximal η given a fixed μ, the $\vartheta c(\vartheta)^2$ term must be maximized. If one denotes the optimal ϑ value as $\hat{\vartheta}$, this relation reads

$$\hat{\eta}(\mu) = \frac{1}{2\mu}\,\hat{\vartheta} c(\hat{\vartheta})^2 \quad \text{with} \quad \hat{\vartheta} = \arg\max_{\vartheta}\left[\vartheta c(\vartheta)^2\right], \quad \frac{\hat{\vartheta} c(\hat{\vartheta})^2}{2} \approx 0.31. \tag{5.179}$$

Table 5.2. The fitness-efficient (μ, λ) strategies; note $\hat{\vartheta} := \mu/\hat{\lambda}$.

μ	1	2	3	4	5	6	7	8	9
$\hat{\lambda}$	5	8	11	14	17	19	22	25	28
$\hat{\eta}$	0.135	0.094	0.072	0.058	0.049	0.043	0.038	0.033	0.030
$\hat{\vartheta}$	0.200	0.250	0.273	0.286	0.294	0.316	0.318	0.320	0.321
μ	10	12	14	16	18	20	30	50	100
$\hat{\lambda}$	31	36	42	48	54	59	87	143	284
$\hat{\eta}$	0.028	0.024	0.020	0.018	0.016	0.015	0.010	0.006	0.003
$\hat{\vartheta}$	0.323	0.333	0.333	0.333	0.333	0.339	0.345	0.350	0.352

Thus, it is shown empirically (using Table 5.2) that there exists an optimal selection pressure $\hat{\vartheta}$ for large μ (μ, $\lambda \to \infty$). Provided that (5.175) is correct, this $\hat{\vartheta}$ is valid for *all* (μ, λ) strategies.

5.3.3 The (μ, λ)-ES on the Hyperplane

The progress coefficient $c_{\mu,\lambda}$ was introduced in Sect. 5.3.2.1 under the assumption that the condition $\zeta = 1$ is fulfilled approximately. With (5.58) and (5.8), this condition reads

$$1 = \sqrt{1 + \frac{\sigma^{*2}}{N}} = \sqrt{1 + \frac{\sigma^2 N}{\langle R \rangle^2}}. \tag{5.180}$$

As one can see, (5.180) is fulfilled exactly if the radius $\langle R \rangle$ goes to infinity for a given finite mutation strength $\sigma < \infty$. In other words, this means that the local curvature vanishes. As a result, one obtains the hyperplane.

The progress rate φ_{plane} on the hyperplane can be derived from the sphere model theory, Eq. (5.168). The related considerations were provided in Sect. 2.2.2. Equation (2.28) gives the relation between the progress rates on the sphere model and the hyperplane. For the calculation of φ_{plane}, the $\varphi^*_{\mu,\lambda}$ formula in (5.168) is differentiated at $\sigma^* = 0$. One obtains

$$\frac{\partial}{\partial \sigma^*} \varphi^*_{\mu,\lambda}(\sigma^*) \Big|_{\sigma^*=0} = c_{\mu,\lambda} \tag{5.181}$$

and consequently with (2.28)

$$\boxed{\text{HYPERPLANE:} \qquad \varphi_{\mu,\lambda}(\sigma) = c_{\mu,\lambda}\sigma.} \tag{5.182}$$

The formula (5.182) is exact if one knows the exact value of the progress coefficient $c_{\mu,\lambda}$. The $c_{\mu,\lambda}$ values given by (5.169) and Table 5.1 are based on the approximation of the offspring population. The quality of the approximations can be verified by simulations. The ES experiments on the hyperplane are used for this purpose.

The starting point for the experimental verification of the $c_{\mu,\lambda}$ values is Eq. (5.182). The case of $\sigma = 1$ yields $c_{\mu,\lambda} = \varphi_{\mu,\lambda}$. In other words, if the standard normal distribution $\mathcal{N}(0,1)$ is used for the generation of mutations, the value of $c_{\mu,\lambda}$ is directly obtained as the distance of the parental means of successive generations. However, the principal benefit is that these experiments can be carried out for the *one-dimensional* ES ($N = 1$), There is no dependency on N, and no overhead for calculating the fitness value. This leads to low computation costs for the determination of $c_{\mu,\lambda}$ values by simulations. Thus, a comparatively high accuracy can be obtained, provided that the pseudo-random number generators have a corresponding quality. The (μ, λ)-ES is realized in one x dimension as follows:

1. Start at $x = 0$, i.e. $x_m = 0$.
2. Generate λ descendants x_l by random selection from μ parents x_m and the addition of $\mathcal{N}(0,1)$ mutations.
3. Select the μ largest x_l values as the parents for the next generation and compute their mean $\langle x \rangle$.
4. Determine the distance $\Delta\langle x \rangle$ to the previous generation.
5. Go to 2.

After a number of generations G_1, the $\Delta\langle x \rangle$ values are recorded over G_2 generations. The mean value of the collected $\Delta\langle x \rangle$ is an estimation for $c_{\mu,\lambda} = \varphi_{\mu,\lambda}$.

In Fig. 5.3, the results of such simulations with $G_1 = 1,000$ and $G_2 = 1,000,000$ are displayed as dots. The curves are from the $c_{\mu,\lambda}$ values computed according to (5.169). The μ value is treated as a real-valued variable and λ as a constant.

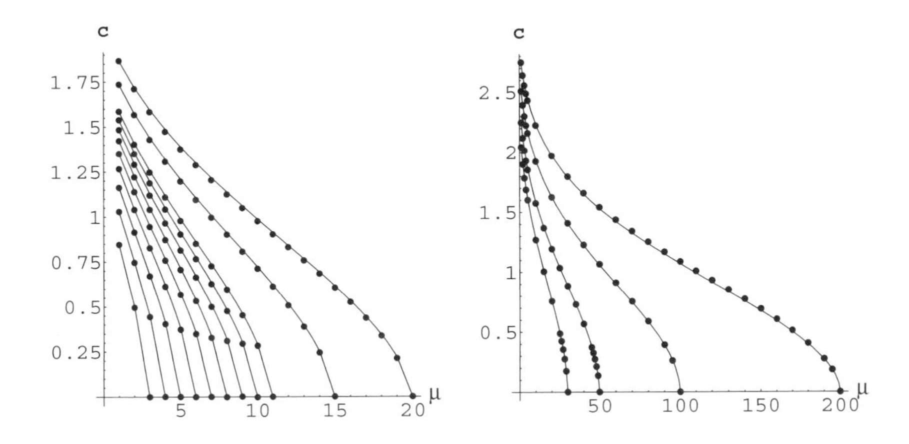

Fig. 5.3. Comparison of the theoretically calculated $c_{\mu,\lambda}$ coefficients (shown as curves) with the simulations on the hyperplane (dots). The figure on the left shows the $c_{\mu,\lambda}$ values for $\lambda = 3, 4, 5, 6, 7, 8, 9, 10, 11, 15, 20,$ and the right one shows the values for $\lambda = 30, 50, 100, 200$.

A good agreement between the theoretical predictions and the experiments is observed.The remaining deviations are smaller than 2%. They can definitely be attributed to the series cut-off (5.99) in the density approximation (5.25) of the offspring population. The linearization of the γ terms used in Sect. 5.2, however, plays a relatively inferior role. For the derivation of a $c_{\mu,\lambda}$ formula more accurate than (5.169), one has to consider a further coefficient γ_2 according to (5.25). In this work, only the skewness of the offspring distribution has been considered by γ. Taking γ_2 into account would correspond to the introduction of the excess of the distribution.

5.3.4 The Exploration Behavior of the (μ, λ)-ES

5.3.4.1 Evolution in the (r, γ_i) Picture. As already pointed out in Sect. 5.1.2, the (μ, λ)-ES algorithms show a distinct "exploration behavior" (search behavior) on the sphere model. This becomes especially clear if one describes the offspring $\tilde{\mathbf{R}}_l$ generated from the parental state[15] $\mathbf{R}_m$ in a *spherical coordinate system*. In this system, the states are described by the distance r to the origin, and by $N-1$ angles $\gamma_1, \dots, \gamma_{N-1}$, instead of the N-dimensional vector $(y_1, \dots, y_N)$ in the Cartesian system. A possible representation is (Miller, 1964)

$$\begin{gathered} y_j = r\left(\prod_{k=1}^{j-1} \sin\gamma_k\right)\cos\gamma_j, \qquad 1 \le j \le N-2, \\ y_{N-1} = r\left(\prod_{k=1}^{N-2} \sin\gamma_k\right)\cos\gamma_{N-1}, \quad y_N = r\left(\prod_{k=1}^{N-2} \sin\gamma_k\right)\sin\gamma_{N-1}. \end{gathered} \tag{5.183}$$

The angles are defined in the interval

$$0 \le \gamma_k < \pi, \quad 1 \le k \le N-2 \qquad \text{and} \qquad 0 \le \gamma_{N-1} < 2\pi. \tag{5.184}$$

This system is predestined for the analysis of the (μ, λ)-ES on the sphere model. Actually, this system has been used without mentioning it explicitly: because of the sphere-symmetry, the $N-1$ angle coordinates are *selectively neutral*. Therefore, only the coordinate r has been investigated.

The following question arises: what happens with the $N-1$ angle coordinates during the evolution process? Since these coordinate values are not subject to directed selection, they obviously perform a *random walk* in the angle coordinates of the descendants $\tilde{\mathbf{R}}_l$ and therefore also in the new parents $\mathbf{R}_m$. Thus, the parents $\mathbf{R}_m^{(g)}$, starting at the same "ancient parent" position $\mathbf{R}_m^{(0)} = \mathbf{R}_\star$, are distributed more or less erratically on the hypersphere surface. The dynamics of this propagation will be investigated using a simple model. For this purpose, the evolution of the angle $\alpha^{(g)}$ is considered

$$\alpha^{(g)} := \sphericalangle\left(\mathbf{R}_m^{(g)}, \mathbf{R}_\star\right) = \arccos\left(\frac{\mathbf{R}_\star^{\mathrm{T}}\mathbf{R}_m^{(g)}}{\|\mathbf{R}_\star\|\,\|\mathbf{R}_m^{(g)}\|}\right). \tag{5.185}$$

This angle is measured between the ancient parent $\mathbf{R}_\star$ and a parent $\mathbf{R}_m^{(g)}$ at generation g. Because of the statistical independence of the mutations and the selection neutrality of the angular components, the choice of the parent among the μ is irrelevant. Basically, the investigation of a $(1,1)$-ES like series of angular changes suffices starting from the initial angle $\alpha^{(0)} = 0$ to the angle $\alpha^{(g)}$, caused by a mutation-selection process. Hence, the actual evolution process in (5.185) can be modeled by a simple random walk in the $[0, 2\pi) \times [0, \pi]^{N-2}$ space (5.184).

[15] The optimum is placed at the origin for the sake of simplicity.

As a model of this process, the influence of a mutation $\mathbf{z}$ applied to the radius vector (parent) $\mathbf{R}$ is considered next. Because of the sphere symmetry $\mathbf{R} = (0, 0, \ldots, 0, R)^{\mathrm{T}}$ can be chosen w.l.o.g., i.e. the parent is on the Nth coordinate axis. In the spherical system, this is the point $r = R$, $\gamma_i = \pi/2$

$$\mathbf{R}_\star = \mathbf{R} = (0, 0, \ldots, R)^{\mathrm{T}} \qquad \Leftrightarrow \qquad r = R, \quad \gamma_i = \pi/2. \tag{5.186}$$

A mutation $\mathbf{z}$ generates a new state $\tilde{\mathbf{y}}$ according to

$$\tilde{\mathbf{y}} = \mathbf{R} + \mathbf{z} = (z_1, \ldots, z_{N-1}, R + z_N)^{\mathrm{T}} \quad \text{with} \quad z_i = \mathcal{N}(0, \sigma^2).$$

In spherical coordinates, this corresponds to a mutation of the coordinates r and γ_i by $\tilde{r} = R + \delta r$, $\tilde{\gamma}_i = \gamma_i + \delta\gamma_i$. Since the z_i mutations are small compared to R, their components δr and $\delta\gamma_i$ are also small. Hence, one can expand the trigonometric functions in (5.183) at $\gamma_i = \pi/2$ using a Taylor series

$$i = 1, \ldots, N-1: \quad \tilde{y}_i = z_i = (R + \delta r) \prod_{k=1}^{i-1} \left(1 - \frac{1}{2}(\delta\gamma_k)^2\right)(-\delta\gamma_i) \; + \; \ldots$$

$$i = N: \qquad \tilde{y}_N = R + z_N = (R + \delta r) \prod_{k=1}^{N-1} \left(1 - \frac{1}{2}(\delta\gamma_k)^2\right) \; + \; \ldots .$$

By cutting after the linear term one gets

$$\begin{aligned} \tilde{y}_i = z_i \qquad &= -R\,\delta\gamma_i \; + \; \ldots \quad \Rightarrow \quad \delta\gamma_i = -\frac{z_i}{R} \; + \; \ldots \\ \tilde{y}_N = R + z_N &= R + \delta r \; + \; \ldots \quad \Rightarrow \quad \delta r = z_N \; + \; \ldots . \end{aligned} \tag{5.187}$$

As one can see, the $z_i = \mathcal{N}(0, \sigma^2)$ mutations have in the spherical system the effect of a stochastic process of type

$$\tilde{\gamma}_i = \mathcal{N}\left(\gamma_i, \frac{\sigma^2}{R^2}\right) \; + \; \ldots \qquad \text{and} \qquad \tilde{r} = \mathcal{N}(R, \sigma^2) \; + \; \ldots . \tag{5.188}$$

Based on the statistical independence of the $\delta\gamma_i$ in the scope of the approximations used, it is sufficient to consider the random walk of a single component $\gamma = \gamma_i$[16].

5.3.4.2 The Random Walk in the Angular Space. The complete investigation of the dynamics of the random walk is relatively difficult, it will not be carried out here. The investigation of the special cases for a small number of generations g and for the stationary state $g \to \infty$ is completely sufficient in order to understand the functioning mechanism of the (μ, λ)-ES.

[16] The angle γ should not be mixed up with the series coefficients of the density $p(r)$ in Eq. (5.21).

The angular dynamics for a small number of generations g. Let us consider a single angular coordinate[16] $\gamma^{(g)}$ at generation g. Its probability density is described by $p(\gamma^{(g)})$. The angle $\gamma^{(g)}$ is transformed to the angle $\gamma^{(g+1)} = \tilde{\gamma}^{(g)}$ by the mutation $\mathbf{z}$ according to Eq. (5.188). Consequently, the density of $\gamma^{(g+1)}$ is given by the integral

$$p\Big(\gamma^{(g+1)}\Big) = \frac{1}{\sqrt{2\pi}} \frac{N}{\sigma^*} \int_{-\infty}^{\infty} \exp\left[-\frac{1}{2}\left(\frac{\gamma^{(g+1)} - \gamma^{(g)}}{\sigma^*/N}\right)^2\right] p\Big(\gamma^{(g)}\Big)\, \mathrm{d}\gamma^{(g)}. \tag{5.189}$$

The normalization (2.22), $\sigma/R = \sigma^*/N$, was already considered in this step.

For the sake of simplicity, the random walk is started at $\gamma^{(0)} = 0$. For the next generation, one obtains $\gamma^{(1)} = \mathcal{N}(0, \sigma^{*2}/N^2)$ according to Eq. (5.188), and after considering (5.189) it becomes clear that the $\gamma^{(g)}$ obey Gaussian probability densities of the type $\gamma^{(g)} = \mathcal{N}\big(0, (s^{(g)})^2\big)$, i.e.

$$p\Big(\gamma^{(g)}\Big) = \frac{1}{\sqrt{2\pi}\, s^{(g)}} \exp\left[-\frac{1}{2}\left(\frac{\gamma^{(g)}}{s^{(g)}}\right)^2\right]. \tag{5.190}$$

The standard deviation $s^{(g+1)}$ of the next generation can be calculated using (5.189). Inserting (5.190) into (5.189), substituting $z = \gamma^{(g)}/s^{(g)}$, one obtains the integral

$$p\Big(\gamma^{(g+1)}\Big) = \frac{1}{\sqrt{2\pi}} \frac{N}{\sigma^*} \frac{1}{\sqrt{2\pi}} \int_{-\infty}^{\infty} \mathrm{e}^{-\frac{1}{2}z^2} \exp\left[-\frac{1}{2}\left(\frac{-s^{(g)}}{\sigma^*/N} z + \frac{\gamma^{(g+1)}}{\sigma^*/N}\right)^2\right] \mathrm{d}z.$$

The solution to this integral is given by Eq. (A.6)

$$p\Big(\gamma^{(g+1)}\Big) = \frac{1}{\sqrt{2\pi}\, s^{(g+1)}} \exp\left[-\frac{1}{2}\left(\frac{\gamma^{(g+1)}}{s^{(g+1)}}\right)^2\right] \tag{5.191}$$

with

$$s^{(g+1)} = \sqrt{\big(s^{(g)}\big)^2 + (\sigma^*/N)^2}. \tag{5.192}$$

Because of $s^{(1)} = \sigma^*/N$ (see above), one obtains $s^{(2)} = \sqrt{2}\sigma^*/N$ and generally for $s^{(g)}$

$$s^{(g)} = \sqrt{g}\, \frac{\sigma^*}{N}. \tag{5.193}$$

Here we have obtained the typical $\sqrt{g}$ law of Gaussian diffusion processes.[17]

Because of the statistical independence of the γ_i angles, the joint density in the $(N-1)$-dimensional angular space is obtained from (5.190) as

[17] Such processes are also called *Wiener processes*, see, e.g., (Ahlbehrendt & Kempe, 1984).

$$p\Big(\gamma_1^{(g)}, \ldots, \gamma_{N-1}^{(g)}\Big) = \frac{1}{\left(\sqrt{2\pi}\, s^{(g)}\right)^{N-1}} \exp\left[-\frac{1}{2} \frac{\sum_{i=1}^{N-1} \left(\gamma_i^{(g)}\right)^2}{\left(s^{(g)}\right)^2}\right]. \quad (5.194)$$

Considering Eq. (5.190), or (5.194), it should be clear that the description of the random walk by (5.190) and (5.194) necessitates a periodical continuation of the definition interval in (5.184). However, since the following considerations are only valid for small angles anyway, this peculiarity is not considered in detail.

The actual aim of considering the random walk in the angular space was to calculate the dynamics of the angle $\alpha^{(g)}$, Eq. (5.185). This angle is composed nonlinearly by the $\gamma_i^{(g)}$ angles (see Eq. (5.201) further below). Considering an infinitesimal angle, the total angle is given as the Euclidean length of the angular components

$$\alpha^{(g)} = \sqrt{\sum_{i=1}^{N-1} \left(\gamma_i^{(g)}\right)^2 + \ldots}\,. \quad (5.195)$$

This is so, because both the Cartesian and the spherical coordinate systems have orthogonal coordinate lines (the metric tensor g_{ik} is diagonal; for its definition, see Eqs. (2.46) and (2.47)).

Because of (5.194), the angle $\alpha^{(g)}$ (5.195) has a χ distribution (see Sect. 3.2.1.1). For large N, the expected value $\overline{\alpha}^{(g)} = \mathrm{E}\left\{\alpha^{(g)}\right\}$ can be represented by the formula (3.51). One obtains

$$\overline{\alpha}^{(g)} \approx \sqrt{N}\, s^{(g)}, \quad (5.196)$$

and with (5.193)

$$\boxed{\overline{\alpha}^{(g)} \approx \sqrt{g}\, \frac{\sigma^*}{\sqrt{N}}.} \quad (5.197)$$

It should be explicitly emphasized that this formula is valid for small g. The Euclidean summation used provides only for small γ_i usable results for the total angle.

The asymptotic limit case for $g \to \infty$. For large g, the formula (5.197) cannot be used. However, one can in principle read the asymptotic distribution of the angular components of Eq. (5.190) by assuming the periodic continuation of the angular space: for sufficiently large $s^{(g)}$, $p\left(\gamma^{(g)}\right)$ changes over the angle interval $[0, \pi)$ only very little. In other words, one has a uniform distribution over the interval $\gamma \in [0, \pi)$.[18] For $\gamma \to \infty$, the probability density is constant at $p(\gamma) = 1/\pi$.

The probability density for the absolute value of an angle difference $\Delta\gamma$, $\Delta\gamma = \gamma - \gamma'$ can be investigated next. The question is: how large is the

[18] This is only valid for γ_i with $i = 1, \ldots, N-2$. For $i = N-1$, one has $\gamma \in [0, 2\pi)$.

expected absolute difference value of two random angles arbitrarily selected from $[0, \pi)$? For the following considerations, it suffices to determine the expected value $\overline{\Delta\gamma} = \mathrm{E}\{|\gamma - \gamma'|\}$. This can be determined by

$$\begin{aligned}
\overline{\Delta\gamma} &= \int_0^\pi \int_0^\pi |\gamma - \gamma'|\, p(\gamma)\, p(\gamma')\, \mathrm{d}\gamma\, \mathrm{d}\gamma' \\
&= \int_0^\pi \int_{\gamma=0}^{\gamma=\gamma'} (\gamma' - \gamma)\, p(\gamma)\, p(\gamma')\, \mathrm{d}\gamma\, \mathrm{d}\gamma' \\
&\quad + \int_0^\pi \int_{\gamma=\gamma'}^{\gamma=\pi} (\gamma - \gamma')\, p(\gamma)\, p(\gamma')\, \mathrm{d}\gamma\, \mathrm{d}\gamma'.
\end{aligned} \tag{5.198}$$

Inserting $p(\gamma) = p(\gamma') = 1/\pi$, the integration yields

$$\overline{\Delta\gamma} = \overline{\Delta\gamma_i} = \frac{\pi}{3}, \qquad 1 \le i \le N-2. \tag{5.199}$$

Analogously, one gets for the angular coordinate $\gamma_{N-1} \in [0, 2\pi)$

$$\overline{\Delta\gamma_{N-1}} = \frac{2}{3}\,\pi. \tag{5.200}$$

Using (5.199) and (5.200), the expected value of the angle $\alpha^{(g)}$ (5.185) can be determined approximately. Since only the y_N coordinate of the vector $\mathbf{R}_\star$ is nonzero, the scalar product reduces to this component. Using the spherical coordinates (5.183), one gets

$$\alpha^{(g)} = \arccos\left(\prod_{k=1}^{N-1} \sin \gamma_k^{(g)} \right). \tag{5.201}$$

The equality $r = R = \|\mathbf{R}^{(g)}\|$ for the sphere surface has been taken into account in this step.

The next step is the approximate calculation of the expected value of the stationary angle $\alpha_\infty := \mathrm{E}\{\alpha^{(\infty)}\}$. The angle $\alpha^{(\infty)} = f(\gamma_1^{(\infty)}, \ldots, \gamma_{N-1}^{(\infty)})$ is expanded at $\gamma_k^{(\infty)} = \overline{\gamma_k}^{(\infty)} = \gamma_k^{(0)} + \overline{\Delta\gamma_k}$ in a Taylor series for this purpose. The series is cut after the linear term, and the expected value is calculated. Considering (5.199) and (5.200) as well as $\gamma_k^{(0)} = \pi/2$, Eq. (5.186), one gets

$$\alpha_\infty = \arccos\left[\left(\prod_{k=1}^{N-2} \sin \frac{5}{6}\,\pi \right) \sin \frac{7}{6}\,\pi \right] + \mathcal{R}_\alpha. \tag{5.202}$$

The error term $\mathcal{R}_\alpha$ is a function of the central moments (and of the product moment, respectively) of γ_k. Neglecting these, the result is

$$\boxed{\alpha_\infty \approx \arccos\left[-\left(\frac{1}{2}\right)^{N-1} \right] \xrightarrow{N\to\infty} \frac{\pi}{2} \;\hat{=}\; 90^\circ.} \tag{5.203}$$

The descendants are distributed on the sphere surface such that on average the distance to the ancient parent is about 90°. As the consequence for the exploration behavior of the (μ, λ)-ES, one can conclude that this strategy does *not* "diffuse along the gradient path," as claimed in Rechenberg (1994, p. 75). If the individuals of the (μ, λ)-ES (and therefore also the individuals of the $(1, \lambda)$-ES) fluctuated around the gradient, i.e. around $\mathbf{R}_\star$, their angle α_∞ would not be so large.

5.3.4.3 Experimental Verification and Discussion. The predictions in (5.197) and (5.203) can be verified experimentally. As an example, the $(5, 30)$-ES with $N = 100$ is investigated for this purpose. In order to avoid side effects of the self-adaptation, the mutation strength is externally controlled such that σ^* is (almost) optimal, $\sigma^* = \hat{\sigma}^* = c_{5,30} = 1.62$ (see Table 5.1 on p. 189), by determining the actual residual distance $\langle R \rangle^{(g)}$. In other words, the mutation strength at generation g is obtained by $\sigma = \hat{\sigma}^* \langle R \rangle^{(g)} / N$. The offspring for generation $g + 1$ are generated by mutations with this standard deviation. The ES run starts at $\mathbf{y}^{(0)} = \mathbf{R}_\star = (0, 0, \ldots, R_0)$ with $R_0 = 100$. The angles $\alpha_m^{(g)}$ are calculated using (5.185). Thereafter, the mean value $\langle \alpha \rangle^{(g)} = \frac{1}{\mu} \sum_{m=1}^{\mu} \alpha_m^{(g)}$ is calculated. Alternatively, the angle of a single parent could be considered. The averaging here serves for the statistical smoothing. For the same reason, the result is averaged over L independent ES runs. Figure 5.4 shows the result with $L = 76$. As expected, the $\sqrt{g}$ law (5.197) has only a small interval of validity. For $g \lessapprox 20$ or $\alpha \lessapprox 45°$, the estimates are unusable.[19]

Eq. (5.197) fails for large α. Instead, the stationary state is predicted by Eq. (5.203). The transition to the stationary angle α_∞ for large g is shown in Fig. 5.4. As predicted, the stationary value is about 90°.

In Fig. 5.4, also the dynamics of $\langle R \rangle$ is shown. The dynamics is described by (2.103) with $R(g) = \langle R \rangle^{(g)}$. With (5.168) and Table 5.1 one gets $\varphi^*_{(5,30)}(\hat{\sigma}^*) = \varphi^*(1.62) \approx 1.29$ and with $g_0 = 0$, $R(g_0) = R_0 = 100$ it follows that

$$\langle R \rangle^{(g)} = R_0 \, \mathrm{e}^{-0.0129\, g}. \tag{5.204}$$

As one can see in Fig. 5.4, this theoretical curve (5.204) is in the confidence interval of the experiments. This is a further approval for the admissibility of the approximations used in this chapter.

5.3.4.4 Final Remarks on the Search Behavior of the (μ, λ)-ES. It is frequently said that evolutionary algorithms explore the search space "in a peculiar way." Therefore, they are believed to be predestined to localize global optima. Particularly, the genetic algorithms live with this "myth" (Goldberg, 1989). It is hard to assess how well this myth is based on provable reality, because a satisfactory theory on the exploration behavior of genetic algorithms is still missing.

[19] The statement $g \lessapprox 20$ is based on the example considered. As a general criterion, $\alpha \lessapprox 45°$, i.e. $\alpha \lessapprox \pi/4$, can be used. Thus, by using Eq. (5.197), a criterion for g can be derived.

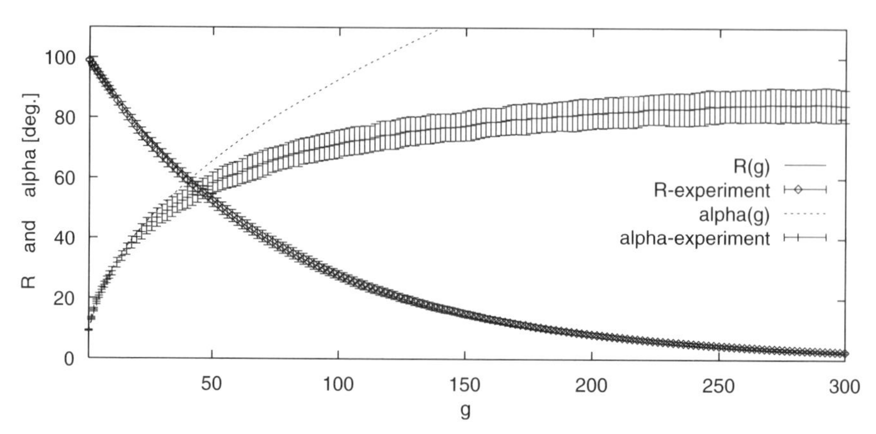

Fig. 5.4. A $(5, 30)$-ES experiment on the dynamics of the alpha(g) $:= \langle\alpha\rangle^{(g)}$ angle and the residual distance R(g) $:= \langle R\rangle^{(g)}$. The angles are measured in degrees. The experimental data are shown with "error bars," indicating the confidence interval for the mean values. The fluctuation interval shows the standard error, i.e. the standard deviation of the mean over L runs. Note that only every second measured value is plotted after generation $g = 10$ for better visibility.

The relations are a bit better for the field of evolution strategies. At least, the principal question of how the mutative changes are placed in the search space $\mathcal{Y}$, in other words, the principal question of the search behavior, is understood to a certain extent. It is in relation to the evolutionary progress principle (EPP) introduced in Sect. 1.4.2.1. Further evidence for the EPP was shown in Sect. 3.2.1.3 on p. 68 and in Sect. 3.2.3.1 on p. 73. According to this principle, the effect of a mutation $\mathbf{z}$ can be decomposed in two components; one in the progress direction toward optimum and in a vector $\mathbf{h}$ perpendicular to it. The former is of order σ; however, even if the mutation was successful and surpassed the parental fitness value, the descendant produced will be placed a distance $||\mathbf{h}|| \approx \sigma\sqrt{N}$ sideways from the line between the parent and the optimum. In other words, the descendant is pushed away from this gradient line. Therefore, the search continues more in the directions other than in the gradient direction. This does not necessarily imply that the μ parents are distributed uniformly on the hypersphere. The parental distribution may be temporarily collapsing – similar to the genetic drift phenomenon. This is also observed when the $\frac{\mu(\mu-1)}{2}$ angles $\beta^{(g)}_{m\,m'}$ between the μ parents $\mathbf{R}^{(g)}_m$ and $\mathbf{R}^{(g)}_{m'}$ for a fixed generation g are investigated over several ES runs. For example, Fig. 5.5 shows this for the $(5, 30)$-ES experiment discussed above. In Fig. 5.5, the respective maximum $\hat{\beta}^{(g)}$ of the $\beta^{(g)}_{m\,m'}$ angle

$$\hat{\beta}^{(g)} := \max_{m,m'}\left[\beta^{(g)}_{m\,m'}\right] \quad \text{with} \quad \beta^{(g)}_{m\,m'} := \arccos\left(\frac{\mathbf{R}^{(g)\mathrm{T}}_m \mathbf{R}^{(g)}_{m'}}{||\mathbf{R}^{(g)}_m||\,||\mathbf{R}^{(g)}_{m'}||}\right) \tag{5.205}$$

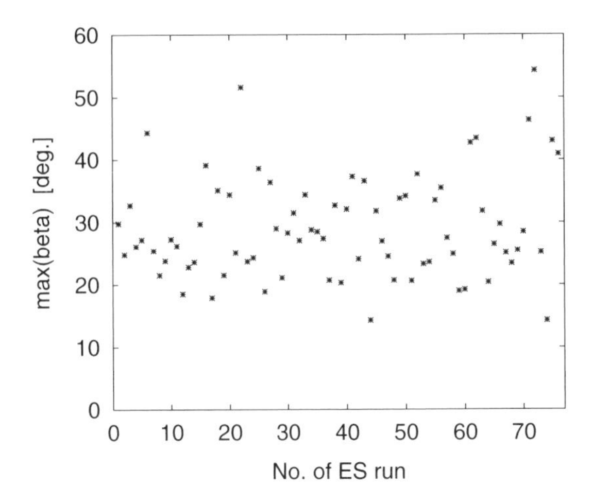

Fig. 5.5. The maximal angle max(beta) $:= \hat{\beta}^{(g)}$ between the parents $\mathbf{R}_m^{(g)}$ at generation $g = 100$ is shown over $L = 76$ runs of the $(5, 30)$-ES.

is displayed over L ES runs. One can see that the population can be spread as well as collapsed. Consequently, one can state that the (μ, λ)-ES is a search algorithm that may be signified as *explorative*.[20,21]

As we have seen in the analysis of the sphere model, the (μ, λ)-ES "favors" the states perpendicular to the gradient direction. The behavior can be called "favoring" because the ratio of the lengths of the components in the direction toward the optimum and perpendicular to it is 1 to $\sqrt{N}$. In general, this is also true for fitness landscapes that are not spherically shaped. To make an analogy, the behavior of the (μ, λ)-ES is reminiscent of a hill-walker: it avoids making steps which require too much exertion. This is in contrast to the principle of following the steepest descent/ascent, realized in gradient strategies. The search paths of the (μ, λ)-ES are rather similar to the serpentines which can be found in mountains.

To summarize, the picture of the "gradient diffusion" is not correct for the sphere model. Therefore, its general applicability to other fitness landscapes seems questionable. A detailed investigation of the (μ, λ)-ES on landscapes with nonspherical curvature remains to be done. From the viewpoint of the theoretical analysis, it would be helpful to investigate random walk models in generally curved spaces. Investigations in this field should be in the program of future research.

[20] The property of being "explorative" is used here intuitively. A mathematical definition to quantify "explorativeness" remains to be developed.

[21] Further discussions on this topic can be found in Beyer (1998).

6. The $(\mu/\mu, \lambda)$ Strategies – or Why "Sex" May be Good

Two special cases of multirecombination will be investigated – the intermediate (μ/μ_I) recombination and the dominant (μ/μ_D) recombination. The first section deals mainly with the $(\mu/\mu_I, \lambda)$-ES. The N-dependent φ will be derived for the sphere model. Thereafter, the reasons for the higher performance of the recombinative strategies are investigated. This higher performance compared to the non-recombinative strategies will be explained by the genetic repair hypothesis (GR). The second part of the chapter is devoted to the $(\mu/\mu_D, \lambda)$-ES. In contrast to the theory of the $(\mu/\mu_I, \lambda)$-ES, the analysis of the $(\mu/\mu_D, \lambda)$-ES is still in its infancy. A simple model with surrogate mutations will be introduced in Sect. 6.2. This model describes the fundamental aspects of dominant recombination for $N \to \infty$. The dominant recombination generates offspring populations reminiscent of a species. Therefore, the MISR principle (mutation-induced speciation by recombination) is postulated. The asymptotic behavior of the recombinative strategies will be investigated in Sect. 6.3. It will be shown that the $(\mu/\mu, \lambda)$-ES attains in the asymptotic limit case a progress rate which is λ times larger than that of the $(1+1)$-ES.

6.1 The Intermediate $(\mu/\mu_I, \lambda)$-ES

"Sexuality" – in the sense of recombination of genetic information – is an almost universal mechanism in nature. It can be observed even in the simplest life forms, e.g. bacteria (see Birge, 1994). Therefore, one may claim that recombination had emerged already in the early phases of evolution history. It should have some benefits for the organisms that applied it. Most probably, it still serves these benefits.

The recombination in evolutionary algorithms can be emulated in the sense of an analogy. Therefore one expects that the performance of the ES improves when recombination is applied. In genetic algorithms, this idea is massively propagated; however, recombination techniques were also integrated into the ES algorithms relatively early in ES history (see Rechenberg, 1973, pp. 83–86; Schwefel, 1974).

Recombinative ES is described by the $(\mu/\rho, \lambda)$ notation proposed by Rechenberg (1978). It was introduced here in Sect. 1.2.1. The symbol ρ stands for the number of parents involved in the procreation of a *single* offspring. The case $\rho > 2$ is called multirecombination.

Two special recombination types will be investigated here. The difference between the *intermediate* and *dominant* cases gave rise to a refined $(\mu/\rho \overset{+}{,} \lambda)$ notation (Beyer, 1995b)

$$\begin{aligned}&\text{INTERMEDIATE RECOMBINATION: } (\mu/\rho_I \overset{+}{,} \lambda)\\&\text{DOMINANT RECOMBINATION: } \quad (\mu/\rho_D \overset{+}{,} \lambda)\end{aligned} \tag{6.1}$$

These recombination types have been shown to increase the performance in numerous ES experiments and in practical applications. They are both used successfully at the level of object parameters (parameter space $\mathcal{Y}$), as well as of strategy parameters ($\mathcal{S}$ space).

The analysis will be carried out only in the parameter space $\mathcal{Y}$. For the investigation in the strategy parameter space, a theory of self-adaptation is required. A framework for such a theory will be developed in Chap. 7, but without recombination. The theoretical analysis of recombination in the strategy parameter space remains as a task for the future.

6.1.1 Foundations of the $(\mu/\mu, \lambda)$ Theory

The $(\mu/\mu_I, \lambda)$-ES in the real-valued parameter space $\mathcal{Y}$ is considered here. However, some of the approaches, definitions, and results can be directly adapted to the $(\mu/\mu_D, \lambda)$-ES. Therefore, the notation $(\mu/\mu, \lambda)$ was used in the title.

6.1.1.1 The $(\mu/\mu_I, \lambda)$ Algorithm. The analysis of recombinative strategies may seem difficult. For example, one may expect that the related theory should be more complicated than that for the nonrecombinative (μ, λ)-ES in Chap. 5. In general, this is also true; however, there is an exceptional case: the $(\mu/\mu_I, \lambda)$-ES.

Intermediate recombination means computing the center of mass of ρ parents that are randomly chosen among μ parents (see Sect. 1.2.2.3, p. 13). As an exception, in the case $\rho = \mu$ all μ parents take part equally weighted. That is, *per definitionem* there is *no* random sampling. Therefore, all of these parents $\mathbf{y}_m^{(g)}$ of generation g are involved in the calculation of the arithmetic mean which gives the center of mass – also called centroid – of the population. This centroid $\langle \mathbf{y} \rangle^{(g)}$ is computed using

$$\text{CENTER OF MASS:} \qquad \langle \mathbf{y} \rangle^{(g)} = \frac{1}{\mu} \sum_{m=1}^{\mu} \mathbf{y}_m^{(g)}. \tag{6.2}$$

The offspring $\tilde{\mathbf{y}}_l$ are generated from this parental centroid $\langle \mathbf{y} \rangle$ by adding normally distributed isotropic mutations $\mathbf{z}$

$$\mathbf{z} := \sigma \left(\mathcal{N}(0,1), \dots, \mathcal{N}(0,1) \right)^{\mathrm{T}}. \tag{6.3}$$

At generation $g+1$, the descendants are obtained by

$$\tilde{\mathbf{y}}_l^{(g+1)} := \langle \mathbf{y} \rangle^{(g)} + \mathbf{z}_l. \tag{6.4}$$

The parents $\mathbf{y}_m^{(g)}$ form the center of mass as in (6.2). These parents are obtained by the (μ, λ) selection applied to the descendants $\tilde{\mathbf{y}}_l^{(g)}$, one has (see Sect. 5.1.2).

$$(\mu, \lambda) \text{ SELECTION:} \qquad \mathbf{y}_m^{(g)} := \tilde{\mathbf{y}}_{m;\lambda}^{(g)}. \tag{6.5}$$

In summary, the $(\mu/\mu_I, \lambda)$-ES algorithm can be written compactly as

$$\boxed{\underline{(\mu/\mu_I, \lambda)\text{-ES:}} \qquad \forall\, l = 1, \ldots, \lambda: \quad \tilde{\mathbf{y}}_l^{(g+1)} := \frac{1}{\mu} \sum_{m=1}^{\mu} \tilde{\mathbf{y}}_{m;\lambda}^{(g)} + \mathbf{z}_l.} \tag{6.6}$$

6.1.1.2 The Definition of the Progress Rate $\varphi_{\mu/\mu,\lambda}$. Different possibilities for the measurement of the progress in the parameter space $\mathcal{Y}$ were introduced in Sect. 2.1.2. Since one already computes the center of mass in Algorithm (6.6), the progress rate measure with collective evaluation is the appropriate measure. That is, the progress rate is defined by the formula (2.13).

As in the analysis of the (μ, λ)-ES, the optimum $\hat{\mathbf{y}}$ is w.l.o.g. placed in the coordinate origin. This assumption simplifies the derivations: the center of mass vector $\langle \mathbf{y} \rangle$ of the parents is at the same time the radius vector. Using (6.2), it follows that

$$\langle \mathbf{y} \rangle^{(g)} = \frac{1}{\mu} \sum_{m=1}^{\mu} \mathbf{y}_m^{(g)} =: \mathbf{R}. \tag{6.7}$$

The length of $\mathbf{R}$ is the residual distance R of the center of mass of the parental population at generation g, $R = \|\mathbf{R}\| = \|\langle \mathbf{y} \rangle^{(g)}\|$. Analogously, for the parents of generation $g+1$ one has

$$\langle \mathbf{y} \rangle^{(g+1)} = \frac{1}{\mu} \sum_{m=1}^{\mu} \mathbf{y}_m^{(g+1)} = \frac{1}{\mu} \sum_{m=1}^{\mu} \tilde{\mathbf{y}}_{m;\lambda}^{(g+1)} =: \tilde{\mathbf{R}}. \tag{6.8}$$

Here, it was taken into account that the parents of the respective generation are constituted by the μ best descendants $\tilde{\mathbf{y}}_l$ according to (6.5). Inserting (6.7) and (6.8) in the definition of the progress rate (2.13), one obtains

$$\varphi_{\mu/\mu,\lambda} = \mathrm{E}\Big\{ \|\langle \mathbf{y} \rangle^{(g)}\| - \|\langle \mathbf{y} \rangle^{(g+1)}\| \Big\} = \mathrm{E}\Big\{ R - \|\langle \mathbf{y} \rangle^{(g+1)}\| \Big\}. \tag{6.9}$$

According to (6.8), the center of mass $\langle \mathbf{y} \rangle^{(g+1)}$ is obtained from the first μ vectors $\tilde{\mathbf{y}}_{m;\lambda}^{(g+1)}$. Taking (6.4) and (6.7) into account, it follows that

$$\tilde{\mathbf{y}}_{m;\lambda} = \Big(\langle \mathbf{y} \rangle^{(g)} + \mathbf{z} \Big)_{m;\lambda} =: \mathbf{R} + \mathbf{z}_{m;\lambda}, \tag{6.10}$$

in other words the mth best descendant is generated by the "mth best mutation," $\mathbf{z}_{m;\lambda}$.

Because of the rotation symmetry, the mutations can be decomposed in a component x in the direction toward the optimum and in a component vector $\mathbf{h}$ perpendicular to it. This decomposition was already introduced and used in Chap. 3 (cf. Eq. (3.8)). The symbols and notation here are adopted from there, their geometrical meaning can be extracted from Fig. 3.1, p. 53. One gets

$$\tilde{\mathbf{y}}_{m;\lambda} = R\mathbf{e}_R - x_{m;\lambda}\mathbf{e}_R + \mathbf{h}_{m;\lambda} \quad \text{with} \quad \mathbf{e}_R := \frac{\mathbf{R}}{||\mathbf{R}||}, \quad \mathbf{e}_R^{\mathrm{T}}\mathbf{h}_{m;\lambda} = 0. \tag{6.11}$$

The symbol $x_{m;\lambda}$ denotes the component of $\mathbf{z}_{m;\lambda}$ in $-\mathbf{R}$ direction and $\mathbf{h}_{m;\lambda}$ the vector component perpendicular to it. Hence, for the center of mass (6.8), it follows that

$$\langle \mathbf{y} \rangle^{(g+1)} = (R - \langle x \rangle)\, \mathbf{e}_R \; + \; \langle \mathbf{h} \rangle, \tag{6.12}$$

where the mean values are introduced as

$$\langle x \rangle := \frac{1}{\mu} \sum_{m=1}^{\mu} x_{m;\lambda} \qquad \text{and} \qquad \langle \mathbf{h} \rangle := \frac{1}{\mu} \sum_{m=1}^{\mu} \mathbf{h}_{m;\lambda}. \tag{6.13}$$

Substituting (6.12) into (6.9), the progress rate becomes

$$\varphi_{\mu/\mu,\lambda} = \mathrm{E}\{\Delta R\} = \mathrm{E}\{R - \tilde{R}\} = \mathrm{E}\left\{ R - \sqrt{(R - \langle x \rangle)^2 + \langle \mathbf{h} \rangle^2} \right\}. \tag{6.14}$$

Consequently, the progress rate is an expected value expression of the random variables $\langle x \rangle$ and $\langle \mathbf{h} \rangle^2$.

Similar to the other ES algorithms analyzed on the sphere model, it is reasonable to introduce normalized quantities. The normalization (2.22) is used as a model. However, instead of the local radius R, the length of the center of mass vector $R = ||\mathbf{R}|| = ||\langle \mathbf{y} \rangle^{(g)}||$ is used. Hence, the normalization reads

$$\text{NORMALIZATION:} \qquad \varphi^* := \varphi \frac{N}{R} \quad \text{and} \quad \sigma^* := \sigma \frac{N}{R}. \tag{6.15}$$

6.1.1.3 The Statistical Approximation of $\varphi_{\mu/\mu,\lambda}$. For the approximate calculation of (6.14) it is assumed that the deviations of the quantities $\langle x \rangle$ and $\langle \mathbf{h} \rangle^2$ are sufficiently small around their mean values $\overline{\langle x \rangle}$ and $\overline{\langle \mathbf{h} \rangle^2}$. Under this assumption, the distance function occurring in (6.14) $\Delta R = f(\langle x \rangle, \langle \mathbf{h} \rangle^2)$

$$\Delta R(\langle x \rangle, \langle \mathbf{h} \rangle^2) = R - \sqrt{(R - \langle x \rangle)^2 + \langle \mathbf{h} \rangle^2} \tag{6.16}$$

can be expanded into a Taylor series at $\langle x \rangle = \overline{\langle x \rangle}$ and $\langle \mathbf{h} \rangle^2 = \overline{\langle \mathbf{h} \rangle^2}$. Including the quadratic terms one obtains

$$
\begin{aligned}
&\Delta R\big(\langle x\rangle, \langle \mathbf{h}\rangle^2\big) = \\
&R - \sqrt{\left(R - \overline{\langle x\rangle}\right)^2 + \overline{\langle \mathbf{h}\rangle^2}} \\
&+ \frac{R - \overline{\langle x\rangle}}{\sqrt{\left(R - \overline{\langle x\rangle}\right)^2 + \overline{\langle \mathbf{h}\rangle^2}}} \left(\langle x\rangle - \overline{\langle x\rangle}\right) - \frac{1}{2} \frac{\left(\langle \mathbf{h}\rangle^2 - \overline{\langle \mathbf{h}\rangle^2}\right)}{\sqrt{\left(R - \overline{\langle x\rangle}\right)^2 + \overline{\langle \mathbf{h}\rangle^2}}} \\
&- \frac{1}{2} \frac{\overline{\langle \mathbf{h}\rangle^2}}{\left(\sqrt{\left(R - \overline{\langle x\rangle}\right)^2 + \overline{\langle \mathbf{h}\rangle^2}}\right)^3} \left(\langle x\rangle - \overline{\langle x\rangle}\right)^2 \\
&+ \frac{1}{8} \frac{\left(\langle \mathbf{h}\rangle^2 - \overline{\langle \mathbf{h}\rangle^2}\right)^2}{\left(\sqrt{\left(R - \overline{\langle x\rangle}\right)^2 + \overline{\langle \mathbf{h}\rangle^2}}\right)^3} \\
&- \frac{2}{4} \frac{R - \overline{\langle x\rangle}}{\left(\sqrt{\left(R - \overline{\langle x\rangle}\right)^2 + \overline{\langle \mathbf{h}\rangle^2}}\right)^3} \left(\langle x\rangle - \overline{\langle x\rangle}\right) \left(\langle \mathbf{h}\rangle^2 - \overline{\langle \mathbf{h}\rangle^2}\right) + \ldots . \qquad (6.17)
\end{aligned}
$$

If one takes the expected value of (6.17), the two linear terms in the second line vanish

$$
\begin{aligned}
&\overline{\Delta R(\langle x\rangle, \langle \mathbf{h}\rangle^2)} = \\
&R - \sqrt{\left(R - \overline{\langle x\rangle}\right)^2 + \overline{\langle \mathbf{h}\rangle^2}} \\
&- \frac{1}{2} \frac{\overline{\langle \mathbf{h}\rangle^2}}{\left(\sqrt{\left(R - \overline{\langle x\rangle}\right)^2 + \overline{\langle \mathbf{h}\rangle^2}}\right)^3} \overline{\left(\langle x\rangle - \overline{\langle x\rangle}\right)^2} \\
&+ \frac{1}{8} \frac{\overline{\left(\langle \mathbf{h}\rangle^2 - \overline{\langle \mathbf{h}\rangle^2}\right)^2}}{\left(\sqrt{\left(R - \overline{\langle x\rangle}\right)^2 + \overline{\langle \mathbf{h}\rangle^2}}\right)^3} \\
&- \frac{1}{2} \frac{R - \overline{\langle x\rangle}}{\left(\sqrt{\left(R - \overline{\langle x\rangle}\right)^2 + \overline{\langle \mathbf{h}\rangle^2}}\right)^3} \overline{\left(\langle x\rangle - \overline{\langle x\rangle}\right) \left(\langle \mathbf{h}\rangle^2 - \overline{\langle \mathbf{h}\rangle^2}\right)} + \ldots . \qquad (6.18)
\end{aligned}
$$

The expected values in (6.18) will be discussed and approximated next. First, the expected value $\overline{\langle x\rangle}$ will be calculated in Sect. 6.1.2.1. Its quantity can be approximated using (6.13). The $x_{m;\lambda}$ variates are the $\mathbf{e}_R$-projections of the mutation vectors $\mathbf{z}$. Due to the isotropy of the mutations, these projections fluctuate according to $\mathcal{N}(0, \sigma^2)$. Hence, one has the order relation $\overline{\langle x \rangle} = \mathcal{O}(\sigma)$. For the same reason, the variance of $\langle x \rangle$ is of order $\mathrm{D}^2\{\langle x \rangle\} = \overline{(\langle x \rangle - \overline{\langle x \rangle})^2} = \mathcal{O}(\sigma^2)$. The length square of the averaged $\mathbf{h}$ vectors will be derived in Sect. 6.1.2.2. As in the $(1, \lambda)$-ES case (see Eq. (3.220)) one obtains for this quantity $\overline{\langle \mathbf{h} \rangle^2} = \mathcal{O}(N\sigma^2)$. The variance of $\langle \mathbf{h} \rangle^2$ is at most of order[1] $\mathrm{D}^2\{\langle \mathbf{h} \rangle^2\} = \overline{(\langle \mathbf{h} \rangle^2 - \overline{\langle \mathbf{h} \rangle^2})^2} = \mathcal{O}(N^2\sigma^4)$.

The covariance term $\overline{(\langle x \rangle - \overline{\langle x \rangle})(\langle \mathbf{h} \rangle^2 - \overline{\langle \mathbf{h} \rangle^2})}$ gives (at most) a magnitude of $\mathcal{O}(N\sigma^3)$.

After these considerations and the introduction of the normalization (6.15), the error term in (6.18) can be estimated as $\mathcal{O}(1/\sqrt{N})$. Using (6.14) and (6.18), the resulting normalized progress rate $\varphi^*_{\mu/\mu,\lambda}$ reads

$$\varphi^*_{\mu/\mu,\lambda} = N\left(1 - \frac{1}{R}\sqrt{\left(R - \overline{\langle x \rangle}\right)^2 + \overline{\langle \mathbf{h} \rangle^2}}\right) + \mathcal{O}\left(\frac{1}{\sqrt{N}}\right). \tag{6.19}$$

The expected value of the square of the mean value of the $\mathbf{h}$ vectors occurring in (6.19) can be rewritten as the mean value of the expected length squares of the $\mathbf{h}$ vectors. Using (6.13), one gets step by step

$$\begin{aligned}
\overline{\langle \mathbf{h} \rangle^2} &= \mathrm{E}\left\{\left(\frac{1}{\mu}\sum_{m=1}^{\mu} \mathbf{h}_{m;\lambda}\right)^2\right\} \\
&= \frac{1}{\mu^2}\mathrm{E}\left\{\sum_{m=1}^{\mu}\sum_{k=1}^{\mu} \left(\mathbf{h}_{k;\lambda}^{\mathrm{T}}\, \mathbf{h}_{m;\lambda}\right)\right\} \\
&= \frac{1}{\mu^2}\mathrm{E}\left\{\sum_{m=1}^{\mu} \left(\mathbf{h}_{m;\lambda}^{\mathrm{T}}\, \mathbf{h}_{m;\lambda}\right)\right\} + \frac{1}{\mu^2}\mathrm{E}\left\{\sum_{m=1}^{\mu}\sum_{k\neq m} \left(\mathbf{h}_{k;\lambda}^{\mathrm{T}}\, \mathbf{h}_{m;\lambda}\right)\right\} \\
&= \frac{1}{\mu^2}\sum_{m=1}^{\mu} \mathrm{E}\{\mathbf{h}_{m;\lambda}^2\} + \frac{1}{\mu^2}\sum_{m=1}^{\mu}\sum_{k\neq m} \mathrm{E}\{\mathbf{h}_{k;\lambda}^{\mathrm{T}}\, \mathbf{h}_{m;\lambda}\} \\
&= \frac{1}{\mu^2}\sum_{m=1}^{\mu} \overline{\mathbf{h}_{m;\lambda}^2} + \frac{1}{\mu^2}\sum_{m=1}^{\mu}\sum_{k\neq m} \overline{\mathbf{h}_{k;\lambda}^{\mathrm{T}}\, \mathbf{h}_{m;\lambda}}.
\end{aligned} \tag{6.20}$$

It is *very important* to note that the expected value of the scalar product $\mathbf{h}_{k;\lambda}^{\mathrm{T}}\, \mathbf{h}_{m;\lambda}$ vanishes for $k \neq m$

$$k \neq m \quad \Longrightarrow \quad \mathrm{E}\left\{\mathbf{h}_{k;\lambda}^{\mathrm{T}}\, \mathbf{h}_{m;\lambda}\right\} = 0. \tag{6.21}$$

The reason for that is the statistical independence of the $\mathbf{h}_{k;\lambda}$ vectors. Since these vectors are perpendicular to the $\mathbf{R}$ direction, they are – at the same

[1] A more accurate estimation yields asymptotically $\mathrm{D}^2\{\langle \mathbf{h} \rangle^2\} \sim 2\frac{N}{\mu^2}\sigma^4$ (see also Eq. (3.222)).

length – selectively neutral because of the symmetry of the sphere model. Thus, different **h** vectors are statistically independent of each other. Furthermore, because of the isotropy of the mutations applied, all possible directions for **h** are equally probable. Consequently, the expected value of the **h** vectors is zero. For fixed k and m $(k \neq m)$, one has

$$\mathrm{E}\left\{\mathbf{h}_{k;\lambda}^{\mathrm{T}}\,\mathbf{h}_{m;\lambda}\right\} = \mathrm{E}\left\{\mathbf{h}_{k;\lambda}^{\mathrm{T}}\right\}\,\mathrm{E}\left\{\mathbf{h}_{m;\lambda}\right\} = 0. \tag{6.22}$$

As a result, only the first term remains in (6.20)

$$\overline{\langle \mathbf{h} \rangle^2} = \frac{1}{\mu^2}\sum_{m=1}^{\mu}\overline{\mathbf{h}_{m;\lambda}^2} = \frac{1}{\mu}\overline{\langle \mathbf{h}^2 \rangle} \qquad \text{with} \qquad \overline{\langle \mathbf{h}^2 \rangle} := \frac{1}{\mu}\sum_{m=1}^{\mu}\overline{\mathbf{h}_{m;\lambda}^2}. \tag{6.23}$$

Inserting this result into (6.19), one obtains

$$\boxed{\varphi^*_{\mu/\mu,\lambda} = N\left[1 - \sqrt{\left(1 - \frac{\overline{\langle x \rangle}}{R}\right)^2 + \frac{\overline{\langle \mathbf{h}^2 \rangle}}{\mu R^2}}\right] + \mathcal{O}\left(\frac{1}{\sqrt{N}}\right).} \tag{6.24}$$

Equation (6.24) is the starting point for several approximations. Knowing $\overline{\langle x \rangle}$ and $\overline{\langle \mathbf{h}^2 \rangle}$, one can obtain the final $\varphi^*_{\mu/\mu,\lambda}$ formula. For the case of the $(\mu/\mu_I, \lambda)$-ES, $\overline{\langle x \rangle}$ and $\overline{\langle \mathbf{h}^2 \rangle}$ will be calculated in the following section. At this point, note that the quadratic term $(\overline{\langle x \rangle}/R)^2$ of (6.24) can usually be neglected in Taylor expansions because of

$$\frac{\overline{\langle x \rangle}}{R} = \mathcal{O}\left(\frac{1}{N}\right) \qquad \text{and} \qquad \frac{\overline{\langle \mathbf{h}^2 \rangle}}{\mu R^2} = \mathcal{O}\left(\frac{1}{N}\right). \tag{6.25}$$

If one rewrites (6.24) as

$$\varphi^*_{\mu/\mu,\lambda} = N\left(1 - \sqrt{1 + \frac{\overline{\langle \mathbf{h}^2 \rangle}}{\mu R^2}}\sqrt{1 - 2\frac{\frac{\overline{\langle x \rangle}}{R}\left(1 - \frac{1}{2}\frac{\overline{\langle x \rangle}}{R}\right)}{1 + \frac{\overline{\langle \mathbf{h}^2 \rangle}}{\mu R^2}}}\right) + \mathcal{O}\left(\frac{1}{\sqrt{N}}\right), \tag{6.26}$$

and expands the second square root of (6.26) into a Taylor series (see Appendix B.1), one obtains

$$\varphi^*_{\mu/\mu,\lambda} = N\left(1 - \sqrt{1 + \frac{\overline{\langle \mathbf{h}^2 \rangle}}{\mu R^2}}\right) + \frac{N\frac{\overline{\langle x \rangle}}{R}}{\sqrt{1 + \frac{\overline{\langle \mathbf{h}^2 \rangle}}{\mu R^2}}} - \frac{\frac{N}{2}\left(\frac{\overline{\langle x \rangle}}{R}\right)^2}{\sqrt{1 + \frac{\overline{\langle \mathbf{h}^2 \rangle}}{\mu R^2}}} + \ldots$$

and therefore

$$\varphi^*_{\mu/\mu,\lambda} = N\left(1 - \sqrt{1 + \frac{\overline{\langle \mathbf{h}^2 \rangle}}{\mu R^2}}\right) + \frac{N\frac{\overline{\langle x \rangle}}{R}}{\sqrt{1 + \frac{\overline{\langle \mathbf{h}^2 \rangle}}{\mu R^2}}} + \mathcal{O}\left(\frac{1}{\sqrt{N}}\right). \tag{6.27}$$

The concrete form of the progress rate formula still depends on the expected values of $\overline{\langle x \rangle}$ and $\overline{\langle \mathbf{h}^2 \rangle}$. Hence, it depends on the strategy used. In the following, these expected values will be calculated for the $(\mu/\mu_I, \lambda)$-ES.

6.1.2 The Calculation of $\varphi_{\mu/\mu_I,\lambda}$

6.1.2.1 The Derivation of the Expected Value $\overline{\langle x \rangle}$. As one can see in (6.13), the expected value of $\langle x \rangle$ is obtained from the mean value of the expected values of the first μ $x_{m;\lambda}$ variables

$$\overline{\langle x \rangle} = \frac{1}{\mu} \sum_{m=1}^{\mu} \overline{x_{m;\lambda}} = \frac{1}{\mu} \sum_{m=1}^{\mu} \int_{-\infty}^{\infty} x p_{m;\lambda}(x) \, \mathrm{d}x. \tag{6.28}$$

The probability density of the x value of the mth best descendant is denoted by $p_{m;\lambda}(x)$. Note that $p_{m;\lambda}(x)$ is *not* a usual order statistic. This distribution depends indirectly on the order of the offspring with respect to its fitness (or residual distance) $\tilde{R}_l = \|\tilde{\mathbf{y}}_l\|$. Obviously, this case is a generalization of order statistics. Such statistics will be called *induced order statistics* in this book. The respective notation "$m;\lambda$" was already introduced in Sect. 1.2.2.1.

The determination of the $p_{m;\lambda}(x)$ density. The probability density $p_{m;\lambda}(x)$ of the induced order statistic $x_{m;\lambda}$ will be derived in the following. All λ descendants $\tilde{\mathbf{y}}_l$ are generated by isotropic, i.e. rotation-invariant mutations $\mathbf{z}$. Therefore, the projection of $\mathbf{z}$ to any arbitrarily chosen direction in the parameter space is normally distributed with $\mathcal{N}(0, \sigma^2)$. Consequently, one has for the x components of the $\mathbf{z}$ mutations, i.e. for the projection in the direction toward the optimum,

$$p(x) = \frac{1}{\sqrt{2\pi}\,\sigma} \exp\left(-\frac{1}{2} \left(\frac{x}{\sigma} \right)^2 \right). \tag{6.29}$$

We randomly pick an individual $\tilde{\mathbf{y}}$ from the λ descendants. The individual's x-component has the density $p(x)$. The residual distance (or fitness value) of this individual may be denoted by $r := \|\tilde{\mathbf{y}}\|$. For a given x, this quantity is a conditional random variable. Its probability density is denoted by $p(r|x)$, it will be specified further below. To be the mth best, this individual with the specific (x, r) combination must be on the mth place with respect to its length r in the $\tilde{R}_l$ ordering of the descendants

$$\tilde{R}_{1:\lambda} \le \tilde{R}_{2:\lambda} \le \ldots \le \tilde{R}_{m:\lambda} \le \ldots \le \tilde{R}_{\mu:\lambda}. \tag{6.30}$$

In other words, $m - 1$ of the descendants should have $\tilde{R} \le r$, and the other remaining $(\lambda - m)$ descendants should have $\tilde{R} > r$. The probability of $\tilde{R} \le r$ is given by the distribution function $P(r) = \Pr(\tilde{R} \le r)$. For the complement, one has $\Pr(\tilde{R} > r) = 1 - \Pr(\tilde{R} \le r) = 1 - P(r)$. Considering the number of independent constellations, as occurs in the case of the usual $m : \lambda$ order statistic (cf. p. 146), the resulting joint density reads

$$p(x) \frac{\lambda!}{(m-1)!\,(\lambda-m)!}\, p(r|x) \left[\Pr(\tilde{R} \le r)\right]^{m-1} \left[\Pr(\tilde{R} > r)\right]^{\lambda-m}. \qquad (6.31)$$

After integrating over all possible r values, the *probability density of the induced* $x_{m;\lambda}$ *order statistic* is obtained

$$p_{m;\lambda}(x) = p(x) \frac{\lambda!}{(m-1)!\,(\lambda-m)!} \int_{r=0}^{\infty} p(r|x)\, [P(r)]^{m-1}\, [1-P(r)]^{\lambda-m}\, \mathrm{d}r. \qquad (6.32)$$

The distribution $P(r) = \Pr(\tilde{R} \le r) = \Pr(\tilde{R}^2 \le r^2)$ occurring in this expression was already determined in Sect. 3.4.2, Eq. (3.227). Only the derivation of $p(r|x)$ remains to be done. It will be based on the considerations made at the beginning of Sect. 3.4.2, p. 105. The density $p(r|x)$ can be obtained by differentiation of the conditional distribution $P(r|x) = \Pr(\tilde{R} \le r \,|\, x) = \Pr(\tilde{R}^2 \le r^2 \,|\, x)$. The random variable $\tilde{R}^2$ is the square of the residual distance length of a descendant before selection. According to the decomposition (3.7, 3.8) one has $\tilde{R}^2 = (R-x)^2 + \mathbf{h}^2 \le r^2$. As in the case of Eq. (3.217), the x^2 term can be neglected because of its smallness compared to $\mathbf{h}^2$ and R^2. Resolving for $\mathbf{h}^2$ one obtains

$$R^2 - 2Rx + \mathbf{h}^2 \le r^2 \qquad \Longrightarrow \qquad \mathbf{h}^2 \le r^2 - R^2 + 2Rx. \qquad (6.33)$$

The distribution of the square of the length $\mathbf{h}^2 =: v_2$ has already been derived in Sect. 3.4.2. With Eq. (3.224) one gets

$$\Pr(\tilde{R} \le r|x) = \int_{v_2=0}^{v_2=r^2-R^2+2Rx} p_{h^2}(v_2)\, \mathrm{d}v_2 = \Phi\left(\frac{r^2 - R^2 - N\sigma^2 + 2Rx}{\sqrt{2N\sigma^2}}\right)$$

and consequently with $p(r|x) = \frac{\mathrm{d}}{\mathrm{d}r} P(r|x) = \frac{\mathrm{d}}{\mathrm{d}r} \Pr(\tilde{R} \le r \,|\, x)$

$$p(r|x) = \frac{\mathrm{d}}{\mathrm{d}r} P(r|x) = \frac{2r}{\sqrt{2\pi}\,\sqrt{2N\sigma^2}} \exp\left[-\frac{1}{2}\left(\frac{r^2 - R^2 - N\sigma^2 + 2Rx}{\sqrt{2N\sigma^2}}\right)^2\right]. \qquad (6.34)$$

The determination of $\overline{\langle x \rangle}$. Now, we can continue with the calculation of $\overline{\langle x \rangle}$. Inserting (6.32) into (6.28), and exchanging the integration order, one obtains

$$\overline{\langle x \rangle} = \frac{\lambda!}{\mu} \int_{r=0}^{\infty} \left(\int_{-\infty}^{\infty} x p(x) p(r|x)\, \mathrm{d}x \right) \sum_{m=1}^{\mu} \frac{[P(r)]^{m-1}[1-P(r)]^{\lambda-m}}{(m-1)!\,(\lambda-m)!}\, \mathrm{d}r. \qquad (6.35)$$

Considering (6.29) and (6.34) and substituting $t = x/\sigma$, the integral in the parentheses of (6.35) yields

$$\frac{2r}{\sqrt{2\pi}\,\sqrt{2N}\sigma} \frac{1}{\sqrt{2\pi}} \int_{-\infty}^{\infty} t\, \mathrm{e}^{-\frac{1}{2}t^2} \exp\left[-\frac{1}{2}\left(\frac{2R}{\sqrt{2N}\sigma}\, t + \frac{r^2 - R^2 - N\sigma^2}{\sqrt{2N\sigma^2}}\right)^2\right] \mathrm{d}t.$$

The obtained integral can be solved using Eq. (A.8)

$$\int_{-\infty}^{\infty} x p(x) p(r|x) \, dx = \frac{-1}{\sqrt{2\pi}} \frac{\sigma}{\sqrt{1+\frac{\sigma^2 N}{2R^2}}} \frac{2r}{2R\sigma\sqrt{1+\frac{\sigma^2 N}{2R^2}}} \frac{r^2 - R^2 - N\sigma^2}{2R\sigma\sqrt{1+\frac{\sigma^2 N}{2R^2}}} \times \exp\left[-\frac{1}{2}\left(\frac{r^2 - R^2 - N\sigma^2}{2R\sigma\sqrt{1+\frac{\sigma^2 N}{2R^2}}}\right)^2\right]. \tag{6.36}$$

Inserting this result back into (6.35), using (3.227) for $P(r)$, and applying the substitution

$$y = \frac{r^2 - R^2 - N\sigma^2}{2R\sigma\sqrt{1+\frac{N\sigma^2}{2R^2}}}, \tag{6.37}$$

one obtains

$$\overline{\langle x \rangle} = -\frac{\sigma}{\sqrt{1+\frac{N\sigma^2}{2R^2}}} \frac{1}{\sqrt{2\pi}} \frac{\lambda!}{\mu} \int_{y=-\infty}^{y=\infty} y \mathrm{e}^{-\frac{1}{2}y^2} \sum_{m=1}^{\mu} \frac{(\Phi(y))^{m-1} (1-\Phi(y))^{\lambda-m}}{(m-1)!\,(\lambda-m)!} \, dy. \tag{6.38}$$

The lower integration limit is extended – as usual – to $-\infty$ (cf. the discussion below Eq. (3.238), p. 108). The sum occurring in (6.38) can be expressed using an integral. This was shown in Sect. 5.1.3; using Eq. (5.14) and $P = \Phi(y)$, it follows that

$$\overline{\langle x \rangle} = -\frac{\sigma}{\sqrt{1+\frac{N\sigma^2}{2R^2}}} \frac{\lambda-\mu}{\sqrt{2\pi}} \binom{\lambda}{\mu} \times \int_{y=-\infty}^{y=\infty} \int_{x=0}^{x=1-\Phi(y)} y \mathrm{e}^{-\frac{1}{2}y^2} x^{\lambda-\mu-1} (1-x)^{\mu-1} \, dx \, dy. \tag{6.39}$$

Similar to the procedure from (5.15) and (5.16), by exchanging the integration order and using the substitution $x = 1 - \Phi(t)$ thereafter, one gets

$$\overline{\langle x \rangle} = \frac{-\sigma}{\sqrt{1+\frac{N\sigma^2}{2R^2}}} \frac{\lambda-\mu}{\sqrt{2\pi}} \binom{\lambda}{\mu} \int_{t=-\infty}^{t=\infty} \frac{\mathrm{e}^{-\frac{1}{2}t^2}}{\sqrt{2\pi}} (1-\Phi(t))^{\lambda-\mu-1} (\Phi(t))^{\mu-1} \times \int_{y=-\infty}^{y=t} y \, \mathrm{e}^{-\frac{1}{2}y^2} \, dy \, dt. \tag{6.40}$$

The inner integral over y yields $-\exp(-t^2/2)$. After that, the substitution $t \to -t$ is applied and the symmetry relation $\Phi(-t) = 1 - \Phi(t)$ is used. The remaining integral is compared with the definition of *generalized progress coefficient* $e_{\mu,\lambda}^{\alpha,\beta}$ given in (5.112). One recognizes the $e_{\mu,\lambda}^{1,0}$ coefficient. Therefore, the result for $\overline{\langle x \rangle}$ reads

$$\overline{\langle x \rangle} = \frac{\sigma}{\sqrt{1+\frac{N\sigma^2}{2R^2}}} e_{\mu,\lambda}^{1,0}. \tag{6.41}$$

6.1.2.2 The Derivation of the Expected Value $\langle\overline{\mathbf{h}^2}\rangle$. The expected value $\langle\overline{\mathbf{h}^2}\rangle$ is defined by Eq. (6.23)

$$\langle\overline{\mathbf{h}^2}\rangle = \frac{1}{\mu}\sum_{m=1}^{\mu}\overline{\mathbf{h}^2_{m;\lambda}} = \frac{1}{\mu}\sum_{m=1}^{\mu}\int_{v_2=0}^{\infty} v_2 p_{m;\lambda}(v_2)\,\mathrm{d}v_2. \tag{6.42}$$

The random variable "square of the length $\mathbf{h}$" is symbolized by $v_2 := \mathbf{h}^2$. The variate v_2 is again the *induced order statistic* of the mth best trial. Therefore, its density $p_{m;\lambda}(v_2)$ can be obtained using the same considerations as for the $p_{m;\lambda}(x)$ density in Sect. 6.1.2.1. Hence, Eq. (6.32) can be used *mutatis mutandis*

$$p_{m;\lambda}(v_2) = p_{h^2}(v_2)\,\frac{\lambda!}{(m-1)!\,(\lambda-m)!} \times \int_{r=0}^{\infty} p(r|v_2)[P(r)]^{m-1}[1-P(r)]^{\lambda-m}\,\mathrm{d}r. \tag{6.43}$$

Here, the density of the $\mathbf{h}^2$ value of a single descendant was denoted by $p(v_2) = p_{h^2}(v_2)$. It was already calculated in Sect. 3.4.3, Eq. (3.224). The conditional density $p(r|v_2)$ is obtained from $p(r|v_2) = p(r|\mathbf{h}^2) = \frac{\mathrm{d}}{\mathrm{d}r}P(r|\mathbf{h}^2)$ with $P(r|\mathbf{h}^2) = \Pr(\tilde{R} \le r|\mathbf{h}^2)$. To this end, $\Pr(\tilde{R} \le r|\mathbf{h}^2) = \Pr(\tilde{R}^2 \le r^2|\mathbf{h}^2)$ must be calculated. Reordering the first inequality of (6.33) for x, it follows that

$$R^2 - 2Rx + \mathbf{h}^2 \le r^2 \qquad \Longrightarrow \qquad x \ge \frac{\mathbf{h}^2 + R^2 - r^2}{2R}. \tag{6.44}$$

Hence, the statement $\tilde{R}^2 \le r^2$ is equivalent to the inequality for x. The density of the random variable x is given by Eq. (6.29). Thus, the distribution $P(r|\mathbf{h}^2) = \Pr(\tilde{R}^2 \le r^2|\mathbf{h}^2)$ reads

$$\begin{aligned} P(r|\mathbf{h}^2) &= \Pr(\tilde{R} \le r|\mathbf{h}^2) = \Pr(\tilde{R}^2 \le r^2|\mathbf{h}^2) = \Pr\left(x \ge \frac{\mathbf{h}^2 + R^2 - r^2}{2R}\right) \\ &= \frac{1}{\sqrt{2\pi}\,\sigma}\int_{x=\frac{\mathbf{h}^2+R^2-r^2}{2R}}^{\infty} \exp\left(-\frac{1}{2}\left(\frac{x}{\sigma}\right)^2\right)\mathrm{d}x \\ &= 1 - \Phi\left(\frac{\mathbf{h}^2 + R^2 - r^2}{2R\sigma}\right), \end{aligned} \tag{6.45}$$

and the conditional density is obtained by differentiation

$$p(r|v_2) = \frac{r}{\sqrt{2\pi}\,R\sigma}\exp\left[-\frac{1}{2}\left(\frac{v_2 + R^2 - r^2}{2R\sigma}\right)^2\right]. \tag{6.46}$$

Now, the calculation of $\langle\overline{\mathbf{h}^2}\rangle$ can be resumed. To this end, (6.43) is inserted into Eq. (6.42)

$$\langle \overline{\mathbf{h}^2} \rangle = \frac{\lambda!}{\mu} \int_{r=0}^{\infty} \left(\int_{v_2=0}^{\infty} v_2 p_{h^2}(v_2) p(r|v_2)\, \mathrm{d}v_2 \right) \times \sum_{m=1}^{\mu} \frac{[P(r)]^{m-1} [1-P(r)]^{\lambda-m}}{(m-1)!\,(\lambda-m)!}\, \mathrm{d}r. \tag{6.47}$$

Using (6.46) and (3.224), the integral for v_2 in the parentheses of (6.47) becomes

$$(\cdot) = \frac{r}{\sqrt{2\pi}\, R\sigma} \frac{1}{\sqrt{2\pi}\sqrt{2N}\sigma^2} \times \int_{v_2=0}^{v_2=\infty} v_2 \exp\left[-\frac{1}{2} \left(\frac{v_2 - \sigma^2 N}{\sqrt{2N}\sigma^2} \right)^2 - \frac{1}{2} \left(\frac{v_2 + R^2 - r^2}{2R\sigma} \right)^2 \right] \mathrm{d}v_2. \tag{6.48}$$

Using the substitution $t = (v_2 - \sigma^2 N)/\sqrt{2N}\sigma^2$ one gets

$$(\cdot) = \frac{r\sigma^2 N}{\sqrt{2\pi}\,\sigma R} \frac{1}{\sqrt{2\pi}} \int_{-\infty}^{\infty} \mathrm{e}^{-\frac{1}{2}t^2} \exp\left[-\frac{1}{2} \left(\frac{\sqrt{2N}\,\sigma}{2R} t + \frac{R^2 + \sigma^2 N - r^2}{2R\sigma} \right)^2 \right] \mathrm{d}t$$

$$+ \frac{r\sigma^2 \sqrt{2N}}{\sqrt{2\pi}\,\sigma R} \frac{1}{\sqrt{2\pi}} \int_{-\infty}^{\infty} t\,\mathrm{e}^{-\frac{1}{2}t^2} \exp\left[-\frac{1}{2} \left(\frac{\sqrt{2N}\,\sigma}{2R} t + \frac{R^2 + \sigma^2 N - r^2}{2R\sigma} \right)^2 \right] \mathrm{d}t.$$

The lower integration limit was extended to $-\infty$. The solutions to both integrals can be found in Appendix A.1, Eqs. (A.6) and (A.8), respectively. One obtains

$$\int_{v_2=0}^{\infty} v_2 p_{h^2}(v_2) p(r|v_2)\, \mathrm{d}v_2 = \frac{\sigma^2 N}{\sqrt{2\pi}} \frac{r}{\sigma R \sqrt{1 + \frac{N\sigma^2}{2R^2}}} \left(1 - \frac{\sigma}{R\sqrt{1 + \frac{N\sigma^2}{2R^2}}} \frac{R^2 + \sigma^2 N - r^2}{2R\sigma\sqrt{1 + \frac{N\sigma^2}{2R^2}}} \right) \times \exp\left[-\frac{1}{2} \left(\frac{R^2 + \sigma^2 N - r^2}{2R\sigma\sqrt{1 + \frac{N\sigma^2}{2R^2}}} \right)^2 \right]. \tag{6.49}$$

Inserting this result into (6.47), considering Eq. (3.227) for $P(r)$, applying the substitution in (6.37), and extending the lower integration limit to $-\infty$, one obtains for $\langle \overline{\mathbf{h}^2} \rangle$

$$\langle \overline{\mathbf{h}^2} \rangle = \frac{\sigma^2 N}{\sqrt{2\pi}} \frac{\lambda!}{\mu} \int_{y=-\infty}^{y=\infty} \left(1 + \frac{\sigma}{R\sqrt{1 + \frac{N\sigma^2}{2R^2}}} y \right) \mathrm{e}^{-\frac{1}{2}y^2} \times \sum_{m=1}^{\mu} \frac{(\Phi(y))^{m-1} (1 - \Phi(y))^{\lambda-m}}{(m-1)!\,(\lambda-m)!}\, \mathrm{d}y. \tag{6.50}$$

This integral has two parts, as can be seen in the large parentheses. The first one, associated with "1" of the parentheses, yields $\sigma^2 N$. This follows immediately from the density formula $p_{m:\lambda}(y)$ of the $m:\lambda$ order statistic: trivially, $\int_{-\infty}^{\infty} p_{m:\lambda}(y)\,\mathrm{d}y = 1$ is valid for it (cf. Eq. (5.10) for $P(y) = \Phi(y)$). The remaining part of (6.50) is similar to the integral in (6.38), and the result in (6.41) can be adopted directly by taking the sign into account. Finally, one obtains the result

$$\langle \overline{\mathbf{h}^2} \rangle = \sigma^2 N \left[1 - \frac{\sigma}{R\sqrt{1 + \frac{N\sigma^2}{2R^2}}}\, e^{1,0}_{\mu,\lambda} \right]. \tag{6.51}$$

6.1.2.3 The $(\mu/\mu_I, \lambda)$ Progress Rate. The $\varphi^*_{\mu/\mu,\lambda}$ formula for intermediate recombination can be derived now. Inserting the partial results of (6.41) and (6.51) into (6.24), and applying the normalization (6.15), one obtains

$$\varphi^*_{\mu/\mu_I,\lambda} = N \left[1 - \sqrt{\left(1 - e^{1,0}_{\mu,\lambda} \frac{\frac{\sigma^*}{N}}{\sqrt{1 + \frac{\sigma^{*2}}{2N}}} \right)^2 + \frac{\sigma^{*2}}{\mu N} - e^{1,0}_{\mu,\lambda} \frac{\frac{\sigma^*}{N}\frac{\sigma^{*2}}{\mu N}}{\sqrt{1 + \frac{\sigma^{*2}}{2N}}}} \right] + \mathcal{O}\!\left(\frac{1}{\sqrt{N}}\right). \tag{6.52}$$

In order to compare (6.52) with the formulae obtained for the $(1,\lambda)$- and the (μ,λ)-ES, i.e. (3.242) and (5.168), (6.52) is simplified using a Taylor expansion. To this end, (6.52) is reordered as

$$\varphi^*_{\mu/\mu_I,\lambda} = \mathcal{O}\!\left(\frac{1}{\sqrt{N}}\right) + N - N\sqrt{1 + \frac{\sigma^{*2}}{\mu N}} \times \sqrt{1 - 2e^{1,0}_{\mu,\lambda} \frac{\sigma^*}{N} \left(\frac{1 + \frac{\sigma^{*2}}{2\mu N}}{\sqrt{1 + \frac{\sigma^{*2}}{2N}} \left(1 + \frac{\sigma^{*2}}{\mu N}\right)} - \frac{1}{2}\frac{\sigma^*}{N} \frac{e^{1,0}_{\mu,\lambda}}{\left(1 + \frac{\sigma^{*2}}{2N}\right)\left(1 + \frac{\sigma^{*2}}{\mu N}\right)} \right)}.$$

The second square root of (6.53) will be expanded using a Taylor series (see Appendix B.1) and cut after the linear term

$$\varphi^*_{\mu/\mu_I,\lambda} = N \left(1 - \sqrt{1 + \frac{\sigma^{*2}}{\mu N}} \right) + e^{1,0}_{\mu,\lambda} \sigma^* \frac{1 + \frac{\sigma^{*2}}{2\mu N}}{\sqrt{1 + \frac{\sigma^{*2}}{2N}}\sqrt{1 + \frac{\sigma^{*2}}{\mu N}}} - \frac{1}{2} e^{1,0}_{\mu,\lambda} \frac{\sigma^{*2}}{N} \frac{e^{1,0}_{\mu,\lambda}}{\left(1 + \frac{\sigma^{*2}}{2N}\right)\sqrt{1 + \frac{\sigma^{*2}}{\mu N}}} + \ldots. \tag{6.53}$$

Neglecting the terms of order $1/N$, the *N-dependent progress rate* formula of the $(\mu/\mu_I, \lambda)$-ES is obtained

$$\varphi^*_{\mu/\mu_I,\lambda}(\sigma^*) = c_{\mu/\mu,\lambda}\,\sigma^* \frac{1+\frac{\sigma^{*2}}{2\mu N}}{\sqrt{1+\frac{\sigma^{*2}}{2N}}\sqrt{1+\frac{\sigma^{*2}}{\mu N}}} - N\left(\sqrt{1+\frac{\sigma^{*2}}{\mu N}}-1\right) + \mathcal{O}\left(\frac{1}{\sqrt{N}}\right). \tag{6.54}$$

The *progress coefficient* $c_{\mu/\mu,\lambda}$ was introduced here

$$c_{\mu/\mu,\lambda} := e^{1,0}_{\mu,\lambda}. \tag{6.55}$$

The $c_{\mu/\mu,\lambda}$ coefficient can be calculated relatively easy by numerical integration of the *general progress integral* $e^{1,0}_{\mu,\lambda}$, Eq. (5.112). Table 6.1 shows a collection of coefficients obtained in this way. The asymptotic properties and approximations of $c_{\mu/\mu,\lambda}$ will be discussed in Sect. 6.3.1.

Table 6.1. A collection of $c_{\mu/\mu,\lambda}$ progress coefficients

μ \ λ	10	20	30	40	50	100	150	200	300
1	1.539	1.867	2.043	2.161	2.249	2.508	2.649	2.746	2.878
2	1.270	1.638	1.829	1.957	2.052	2.328	2.478	2.580	2.718
3	1.065	1.469	1.674	1.810	1.911	2.201	2.357	2.463	2.607
4	0.893	1.332	1.550	1.694	1.799	2.101	2.263	2.372	2.520
5	0.739	1.214	1.446	1.596	1.705	2.018	2.185	2.297	2.449
10	0.000	0.768	1.061	1.242	1.372	1.730	1.916	2.040	2.206
20	-	0.000	0.530	0.782	0.950	1.386	1.601	1.742	1.928
30	-	-	0.000	0.414	0.634	1.149	1.390	1.545	1.746
40	-	-	-	0.000	0.343	0.958	1.225	1.393	1.608
50	-	-	-	-	0.000	0.792	1.085	1.265	1.494
100	-	-	-	-	-	0.000	0.542	0.795	1.088

The progress rate formula (6.54) contains as a special case the progress rate of the $(1, \lambda)$-ES, Eq. (3.242). This can be shown by setting $\mu = 1$ and considering $c_{1/1,\lambda} = e^{1,0}_{1,\lambda} = c_{1,\lambda}$ according to (5.113).

6.1.3 The Discussion of the $(\mu/\mu_I, \lambda)$ Theory

6.1.3.1 The Comparison with Experiments and the Case $N \to \infty$.

The experimental verification of the N-dependent progress rate formula in (6.54) can be performed by *"one-generation experiments"* of the $(\mu/\mu_I, \lambda)$-ES in the same manner as in Sect. 3.4.4. For this purpose, the mutations are generated from a fixed parent $\mathbf{y}_p$ with distance R to the optimum. These λ mutations $\mathbf{z}$ are generated with a constant σ according to Eq. (6.4). The center of mass $\langle\tilde{\mathbf{y}}\rangle$ of the best μ descendants $\tilde{\mathbf{y}}_{m;\lambda}$ is determined. The distance

$\tilde{R} = \|\langle\tilde{\mathbf{y}}\rangle\|$ of this center of mass to the optimum is calculated. The difference $\Delta R := R - \tilde{R}$ is averaged over G independent generation trials. Finally, the normalization in (6.15) is applied.

The $(8/8_I, 30)$-ES with $N = 30$ and $N = 200$ is used as an example. Figure 6.1 shows the results obtained for $G = 10,000$ "one-generation experiments" per data point. The progress coefficient is $c_{8/8,30} = 1.196$ (computed

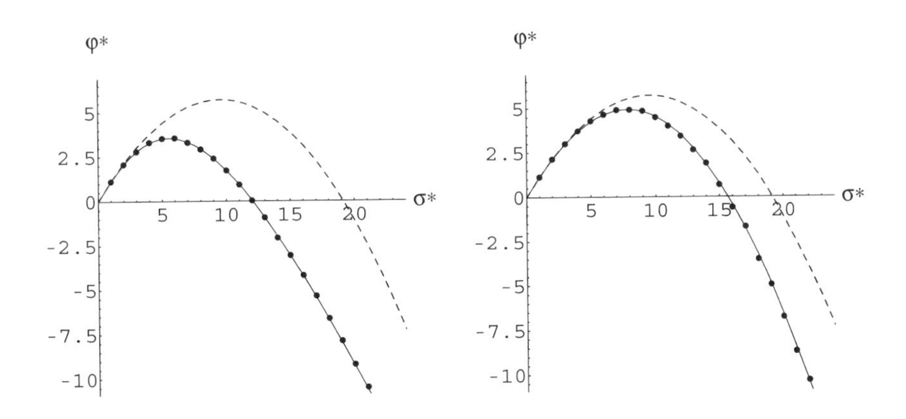

Fig. 6.1. The comparison of the $\varphi^*_{\mu/\mu_I,\lambda}$ formula with experiments. The solid curve represents Eq. (6.54), the dashed one Eq. (6.56). The figure on the left shows the $(8/8_I, 30)$-ES with $N = 30$, the one on the right shows the same ES algorithm for the case $N = 200$.

by numerical integration). One observes a very good agreement of theory and experiment. This accordance is also valid for large values of σ^* ($\sigma^* > N$, although not shown here).

Figure 6.1 also contains a dashed curve. This curve is obtained by the *asymptotically ($N \to \infty$) exact* progress rate formula

$$\varphi^*_{\mu/\mu_I,\lambda} = c_{\mu/\mu,\lambda}\sigma^* - \frac{\sigma^{*2}}{2\mu} + \dots . \tag{6.56}$$

This formula is obtained by Taylor expansion of (6.54) with the condition

$$\frac{\sigma^{*2}}{N} \ll 1. \tag{6.57}$$

The formula (6.56) was offered by Rechenberg (1994). It deviates considerably from the progress rate observed. Therefore, it is only conditionally applicable for practical purposes (see, particularly, Sect. 6.1.3.4, p. 226). In spite of that, Eq. (6.56) is in principle useful since it is simple and analytically manageable; it will be used for the discussion of the EPP. It also serves as the starting point for the investigations of the $\mu, \lambda \to \infty$ asymptotics of the $(\mu/\mu, \lambda)$-ES (see Sect. 6.3).

6.1.3.2 On the Benefit of Recombination – or Why "Sex" May be Good. A lot of speculations on the benefit of recombination in evolutionary algorithms have been offered. In the GA community, the application of recombination in terms of the *crossing-over* is almost a "holy duty" (Goldberg, 1989). If one examines the theoretical basis of the recombination dogma of the GA, one finds almost only the *schema theorem* of Holland (1975) and the *building block hypothesis* (BBH) of Goldberg (1989). However, this consideration results in confusion. The actual question should be: can one attain by recombination a performance improvement beyond simple algorithms with mutation and selection only? The schema theorem as well as the BBH cannot resolve this question. One might construct many numerical experiments to show the advantages of one strategy over another one; however, this is not a reliable justification method, as shown, e.g., in the work of Fogel and Stayton (1994). Even if one could succeed in presenting a simulation indicating the advantages of recombination, the actual questions remain unanswered: where does the advantage of recombination come from, and *why* is it advantageous in this case?

The question of performance increase as well as the question of the working mechanisms of recombination are elucidated in the ES field, at least to some satisfactory extent. In the following, the performance increase of the $(\mu/\mu_I, \lambda)$-ES over the (μ, λ)-ES will be shown first by an example. The ultimate causes of the performance increase are worked out next. A more general proof for the asymptotic limit case follows in Sect. 6.3.

A comparison of the $(\mu/\mu_I, \lambda)$*-ES with the* (μ, λ)*-ES.* Before comparing the performance of different strategies, one has to define how a "fair" comparison should be performed. From the author's point of view, three principal requirements should be met:

1. *Problem equivalence*: The fitness functions and the initial conditions must be the same[2] and should *not* be "tailored" for one of the algorithms. This condition may be satisfied best by using *simple* fitness functions.
2. *Resource equivalence*: As to the ES, this chiefly means using the same λ, provided that the fitness evaluation for an individual consumes much more computation time than the "recombination overhead."
3. *Maximal performance*: Each algorithm should be enabled to run at its maximal possible performance.

For a fair comparison of the $(\mu/\mu_I, \lambda)$-ES with the (μ, λ)-ES, we do not have too many choices as to the first two requirements. The fitness function is the sphere model; the resource equivalence is guaranteed by the choice of the same λ. The progress rate φ is used as the progress measure, in other words, the *local* decrease in the residual distance per generation. Because of

[2] Although this condition seems trivial, one can construct some tricky cases: even though the initial conditions may appear fair, they may still contain the hidden bias of an algorithm (Fogel & Beyer, 1996).

Point 1 and Point 2, R and N are to be chosen the same for both algorithms. Hence, using the conversion $\varphi = \varphi^* R/N$, the maximal performance values based on the normalized progress rate $\hat{\varphi}^* = \max_{\sigma^*} \varphi^*$ can be compared as well as the fitness efficiency $\eta = \hat{\varphi}^*/\lambda$ (see Eq. (5.176)).

As an example, both algorithms will be investigated for $\lambda = 50$ and $N = 100$. In Fig. 6.2, the maximal performance $\hat{\varphi}^*$ of the $(\mu/\mu_I, 50)$-ES is compared with that of the $(\mu, 50)$-ES. The curves are obtained by maximiz-

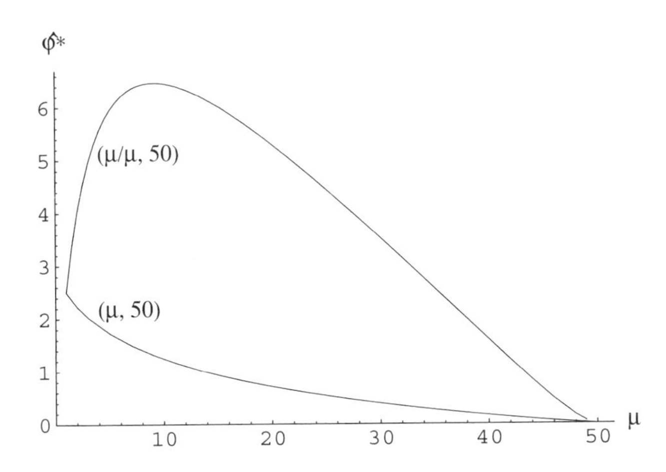

Fig. 6.2. Comparison of the maximal performance of the $(\mu/\mu_I, 50)$-ES with the $(\mu, 50)$-ES

ing the formulae (5.161) and (6.54) with respect to σ^*

$$\hat{\varphi}^*(\mu) = \max_{\sigma^*}[\varphi^*(\sigma^*, \mu, \lambda, N)]\Big|_{\substack{\lambda=50\\N=100}} . \tag{6.58}$$

The comparison shows the performance advantage of the recombinant strategies over the simple (μ, λ)-ES. The former algorithm uses mutation, selection, and recombination operators, whereas the latter one uses only mutation and selection. The $(1, \lambda)$-ES is obviously the most fitness-efficient (μ, λ)-ES (see p. 190). Even this best case is outperformed by the $(\mu/\mu_I, \lambda)$-ES in a wide μ interval, and the recombinant algorithm attains clearly larger performance. If one compares both strategy types for the same μ, the recombinant one is always better (with the exception of the case $\mu = \lambda$).

It is obvious that the $(\mu/\mu_I, \lambda)$-ES attains its maximal $\hat{\varphi}^*(\mu)$ *not* for $\mu = 1$. Instead, there is a *μ-optimal strategy*. This observation will be investigated further in Sect. 6.1.3.4.

The remaining question is how the performance advantage is attained by the recombination, i.e. its ultimate reason. This will be extracted next from the EPP.

The EPP and the genetic repair principle (GR). In order to identify the main difference between the performances $\varphi^*_{\mu,\lambda}$ and $\varphi^*_{\mu/\mu,\lambda}$, it suffices to investigate the asymptotic case $N \to \infty$. If one compares the respective Eqs. (5.171) and (6.56)

$$\varphi^*_{\mu,\lambda}(\sigma^*) = c_{\mu,\lambda}\sigma^* - \frac{\sigma^{*2}}{2} \quad \Bigg| \quad \varphi^*_{\mu/\mu_I,\lambda}(\sigma^*) = c_{\mu/\mu,\lambda}\sigma^* - \frac{\sigma^{*2}}{2\mu}, \qquad (6.59)$$

there is a striking similarity between these two formulae.

The *gain parts* of the EPP are given by the linear σ^* term and the corresponding progress coefficients $c_{\mu,\lambda}$ and $c_{\mu/\mu,\lambda}$. If one compares these quantities using Table 5.1, p. 189, and Table 6.1, p. 216, one realizes[3]

$$\boxed{c_{\mu,\lambda} \geq c_{\mu/\mu,\lambda}.} \qquad (6.60)$$

In other words, if the same mutation strength σ^* is used, the gain part of the recombinant strategies is *even smaller* than that of the (μ, λ)-ES.

The other term, the *loss part* in (6.59), is the rather interesting one. For the recombinative $(\mu/\mu, \lambda)$-ES, it is by a factor of $1/\mu$ *smaller* than in the nonrecombinant (μ, λ)-ES. As a result, the $(\mu/\mu, \lambda)$-ES can run at (much) larger mutation strengths than the (μ, λ)-ES. Actually, the maximum performance $\hat{\varphi}^*$ is attained at different $\hat{\sigma}^*$ values (for $N \to \infty$)

$$\hat{\sigma}^*_{\mu,\lambda} = \arg\max_{\sigma^*}\left[\varphi^*_{\mu,\lambda}(\sigma^*)\right] = c_{\mu,\lambda} \qquad (6.61a)$$

$$\hat{\sigma}^*_{\mu/\mu_I,\lambda} = \arg\max_{\sigma^*}\left[\varphi^*_{\mu/\mu_I,\lambda}(\sigma^*)\right] = \mu c_{\mu/\mu,\lambda}. \qquad (6.61b)$$

This yields up to μ times larger $\hat{\sigma}^*$ values for the $(\mu/\mu, \lambda)$-ES.

Obviously, the loss term $\sigma^{*2}/2\mu$ has something to do with the center of mass composition, i.e. the averaging over μ parents in the parameter space. If the loss term is traced back in the analytical derivation, one arrives via (6.54) and (6.51) at (6.24). The loss term is connected with the $\mathbf{h}$ portions of the mutations $\mathbf{z}$.[4] More exactly, it is connected with the length square of the resulting $\langle\mathbf{h}\rangle$ vector. The fundamental effect of (intermediate) recombination is the reduction of the length square $\langle\mathbf{h}\rangle^2$ of the averaged $\mathbf{h}$ vectors to a $1/\mu$th of $\langle\mathbf{h}^2\rangle$, as can be seen in Eq. (6.23).

The $\mathbf{h}$ components of the mutations $\mathbf{z}$ can be interpreted as the "harmful" parts of the mutations, since they cause an increase in the residual distance (and decrease in the fitness). The recombination reduces the influence of these parts, or to put it another way, it partly compensates them. Hence, the effect of recombination can be termed *statistical error correction.* Identifying the components of the object parameter vectors as "genes," the recombination causes something like a (partial) repair of the genes, that were changed by mutation (in the sense of a *black-box analogy*). Hence, the notion "genetic repair" or *genetic repair* (GR), introduced by the author (Beyer, 1995b), seems

[3] A rigorous mathematical proof of this relation is still pending, since the coefficients $c_{\mu,\lambda}$ and $c_{\mu/\mu,\lambda}$ are of the same order of magnitude.

[4] The mutation $\mathbf{z}$ can always be decomposed in the form $\mathbf{z} = -x\mathbf{e}_R + \mathbf{h}$ (cf. (6.11) as well as (3.6, 3.7)).

to be appropriate. It is conjectured that this is a fundamental functioning principle of evolutionary algorithms:

genetic repair principle (GR)

Intermediate Recombination	=	Statistic Error Correction	=	Principle of Genetic Repair (GR)

The mechanism of GR can be further clarified graphically. Figure 6.3 shows a two-dimensional section of the parameter space $\mathcal{Y}$ with two curves of constant fitness $F(\mathbf{y}) = a = \text{const}$ and $F(\mathbf{y}) = b = \text{const}$ assuming b to be "fitter" than a. Let us consider a $(4/4_I, \lambda)$-ES. The λ mutations are generated

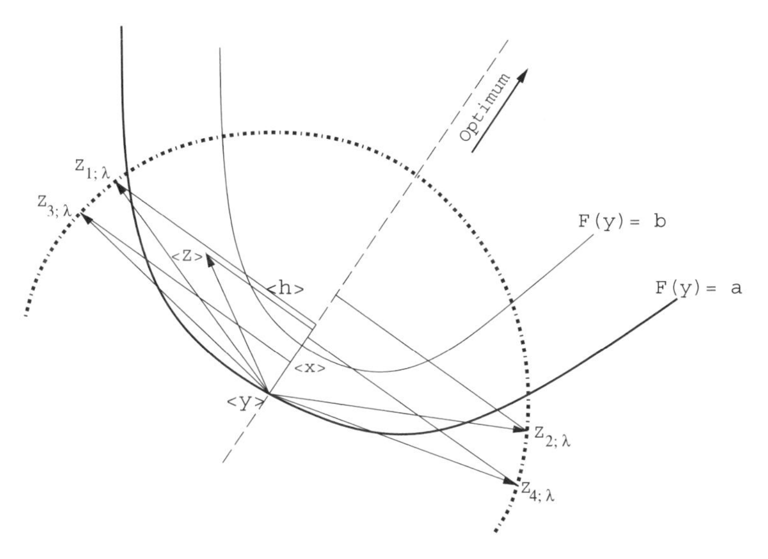

Fig. 6.3. Visualizing the *genetic-repair* effect for the intermediate recombination.

from the state $\langle\mathbf{y}\rangle$ according to (6.3, 6.4), and the four best individuals are selected. The corresponding mutations $\mathbf{z}_{1;\lambda}$ up to $\mathbf{z}_{4;\lambda}$ are displayed in the figure.[5] Because of the selection, these mutation vectors have the tendency to be directed rather toward the optimum than in the opposite direction. Each of these (selected) $\mathbf{z}$ vectors are formally decomposed in two directions. The component in the direction toward the optimum is denoted by x, the one perpendicular to it by $\mathbf{h}$. Intermediate recombination means computing the center of mass, i.e. all components of the $\mathbf{z}$ vectors are to be averaged.

[5] The $\mathbf{z}$ mutations can only be shown schematically: the length of the mutation vectors is in the first approximation $\sqrt{N}\sigma$, i.e. they lie approximately on a hyperspherical shell, but not in the two-dimensional subspace (as shown here).

For the $\mathbf{h}$ components, this gives a vector $\langle\mathbf{h}\rangle$, with a length which is by a factor about $1/\sqrt{4} = 0.5$ smaller than the mean expected length of a single mutation (see Eq. (6.23)).[6]

The error correction effect (GR) can be very impressive. For example, in Fig. 6.3, *none* of the four best descendants has a higher fitness than the parent $\langle\mathbf{y}\rangle$. However, the recombination result $\langle\mathbf{y}\rangle + \langle\mathbf{z}\rangle$ yields an essential fitness improvement with respect to $\langle\mathbf{y}\rangle$. Exactly this is expected from a genetic repair mechanism.

It is important to realize that the effect of recombination lies in *compensating the statistical errors.* The effect is *not a "(re)-combination of good traits"*, also known as the *building block hypothesis.* The "good" properties of the recombination partners $\mathbf{z}_{m;\lambda}$ are their $x_{m;\lambda}$ components. These components are averaged according to (6.13). Hence, it is clear that the recombination result $\langle x\rangle$ *cannot be better* (i.e. larger) than the largest x value, $\hat{x} = \max[x_{m;\lambda}]$ of the μ recombination partners. In other words, the recombination dampens both "especially good" and "especially bad" traits of the parents. It makes them similar and extracts or conserves the *common* properties of the μ recombination partners. In summary, this gives reason to the general GR hypothesis:

GR hypothesis

> *The benefit of (intermediate) recombination lies in genetic repair (GR). The effect of the recombination is the extraction of similarities.*

The GR principle (see p. 221) was only proven for the intermediate (μ/μ) recombination on the sphere model. Therefore, it was signified here as "hypothesis." The proof for other recombination types – with the exception of the dominant (μ/μ_D) recombination – is still an open problem. This question is especially interesting in the research area of genetic algorithms, where the *building block hypothesis* (BBH) (Goldberg, 1989) is still the (only) established but *unproven* explanation for the benefit of recombination. Whether the recombination gives any (unique) benefit at all is further investigated in Sect. 6.2.2.2.

6.1.3.3 System Conditions of Recombination. The strategies with recombination do not necessarily have performance advantages over the simpler (μ, λ) strategies with mutation and selection only. The benefit of recombination rather depends on several conditions. These conditions are specified by the algorithm as well as by the fitness landscape, in short, by the whole EA system. That is why they are called here *system conditions of recombination*:

[6] Note that $\langle\mathbf{z}\rangle$ *cannot* be obtained in Fig. 6.3 by simple (two-dimensional) geometrical addition of the four $\mathbf{z}$ vectors, since only the two-dimensional cross-section can be shown here.

1. local "convexity" of the fitness landscape (or $P_s(\sigma) < 0.5$);
2. statistical independence of the mutants to be recombined;
3. a sufficiently large mutation strength σ^*;
4. an appropriate choice of ρ, $1 < \rho \leq \mu < \lambda$ (see also Sect. 6.1.3.4).

1. The local "convexity" of the fitness landscape. As can be seen in Fig. 6.3, GR makes sense in curved fitness landscapes only. For instance, it is not advantageous for the hyperplane. The respective progress rate φ is obtained by applying (2.28) to (6.56)

$$\text{HYPERPLANE:} \qquad \varphi_{\mu/\mu_I,\lambda}(\sigma) = c_{\mu/\mu,\lambda}\sigma. \tag{6.62}$$

This result can be compared with the (μ, λ) formula (5.182). Because of (6.60), the $(\mu/\mu_I, \lambda)$-ES attains at the same mutation strength σ a worse performance than the (μ, λ)-ES. Since the choice of σ is arbitrary here (no EPP loss term), this does not play a role in the comparison. Similarly, the mutation strength can be increased (theoretically) arbitrarily in concave fitness landscapes, resulting in the same argument on comparison.

The concept of "convexity" can be used for characterizing smooth fitness landscapes. The immediate question is whether the recombination is advantageous in convex landscapes in general. Moreover, can this concept obtained for the sphere model be generalized to discrete and combinatorial optimization problems, e.g., even for the optimization of bitstrings? Having a generalized notion of convexity, could it be used as a characterization in the sense of: "advantageous for recombination?" As a matter of fact, the fitness landscape cannot be analyzed independent of the genetic operators acting on it. Moreover, if mutations are of the *nonisotropic* kind, the hyperplane can "behave" as if it were a curved fitness landscape. Perhaps the *success probability* P_s could serve as a characterization here, as proposed in Beyer (1997). However, its usefulness remains to be proven in each particular case.

The success probability $P_s(\sigma)$ (see Sect. 3.1.1) stands for the probability that the fitness of the parental state $\mathbf{y}$ is improved by the mutation $\mathbf{z}$ (i.e. the effect of *one* mutation is considered). From a geometrical point of view it seems plausible that one has $P_s < 0.5$ if the local curvature is convex in the neighborhood of the parental state. For the hyperplane case, one has exactly $P_s = 0.5$ for isotropic mutations. In contrast to the convex case, one observes $P_s > 0.5$ for the concave curvature. It shall be conjectured here that recombination will increase the performance only in landscapes with $P_s(\sigma) < 0.5$. In landscapes with $P_s(\sigma) \geq 0.5$ there is no need to apply recombination: obviously, the performance can be simply increased by raising the mutation strength σ (or the mutation rate p_m).

2. Statistical independence of the mutants to be recombined. The performance-increasing effect of recombination is coupled with the decrease in the EPP loss term by the factor $1/\mu$. As shown in Sect. 6.1.3.2, the averaging over the selected $\mathbf{h}$ vectors is responsible for that. The mean length of the vector

$\langle \mathbf{h} \rangle$ is approximately $1/\sqrt{\mu}$th of the length of the vector $\mathbf{h}$, as can be read in Eqs. (6.23) or (6.51).

The condition for the validity of (6.23) is the (approximate) *statistical independence* of the $\mathbf{h}$ vectors, as used in Condition (6.21) for the derivation of (6.23). The GR mechanism will fail if this condition is violated. For example, it makes no sense to recombine two identical or two linearly dependent vectors. The statistical error correction works only (in the sense of increasing the performance) if those components of $\mathbf{z}$, which are supposed to be corrected, are *not* correlated.

In other words, this can be expressed using the formulation of the GR hypothesis on p. 222: the intermediate recombination extracts correlations – or similarities – from the mutants. The *genetic repair* is based on the genetic variety of the parents to be recombined. A lack of variety will result in an increased probability of genetic defects. Exactly this is observed in biology, too.

There seem to be deep analogies between the recombination in ES and in biology. These will be discussed further in connection with the dominant recombination. At this point, I just remark that the phenomenon of *inbreeding* is relatively seldom observed in free-living animal populations. Actually, for animals as well as for human beings, it would be much simpler to mate with relatives. However, there exists something like an *incest taboo*, called this by Bischof (1985), widely prohibiting the mating of relatives. If mating and reproduction succeeds in spite of that, the descendants often suffer from hereditary diseases. What one observes in such a case is just the occurrence of common (i.e. correlated) negative traits and common deficits.

3. A sufficiently large mutation strength σ^*. The optimal mutation strengths of the (μ, λ)-ES and of the $(\mu/\mu_I, \lambda)$-ES are compared in (6.61a), (6.61b). One needs almost μ times larger mutation strength values for the maximal performance of the intermediate recombination compared to the (μ, λ)-ES. In other words, if the mutation strength is chosen too small, then the $(\mu/\mu_I, \lambda)$-ES shows no performance gain compared to the (μ, λ)-ES.

The performance gain of the $(\mu/\mu_I, \lambda)$-ES emerges as a result of the increased mutation strength σ (or mutation rate p_m, respectively): larger mutations $\mathbf{z}$ possess both larger $x_{m;\lambda}$ components (performance gain) and larger $\mathbf{h}$ components (performance loss). The effect of the latter is diminished by the GR mechanism of the recombination. However, the "common part" of the selected mutations, their large $x_{m;\lambda}$ components, remain more or less unchanged. As a result, increased progress rate φ values are observed.

6.1.3.4 The $(\mu/\rho_I, \lambda)$-ES and the Optimal μ Choice.

$(\mu/\rho_I, \lambda)$-strategies. The theoretical analysis of the intermediate recombination was performed for the $(\mu/\mu_I, \lambda)$-ES. This case was chosen in order to keep the analysis as simple as possible. The analysis of the $(\mu/\rho_I, \lambda)$-ES with $1 < \rho < \mu$ would be more difficult, e.g., than that of the (μ, λ)-ES.

Regardless of the practical difficulties in the determination of the progress rate for the case $\rho \neq \mu$, one can obtain general statements on the behavior of such strategies by plausibility considerations. The effect of the (μ/ρ_I) recombination on the EPP terms will be discussed for this purpose.

By decomposing the selected mutations $\mathbf{z}$ in x and $\mathbf{h}$ according to Eq. (6.11), one obtains for the average value $\langle x\rangle_\rho$ analogous to (6.13)

$$\langle x\rangle_\rho := \frac{1}{\rho}\sum_{m=1}^{\rho} x_{m;\lambda}. \tag{6.63}$$

Analogous to (6.13) and (6.23), one obtains the loss term $\langle\mathbf{h}\rangle_\rho$

$$\overline{\langle\mathbf{h}\rangle_\rho^2} = \frac{1}{\rho^2}\sum_{m=1}^{\rho}\overline{\mathbf{h}_{m;\lambda}^2} = \frac{1}{\rho}\langle\overline{\mathbf{h}^2}\rangle_\rho \qquad \text{with} \qquad \langle\overline{\mathbf{h}^2}\rangle_\rho := \frac{1}{\rho}\sum_{m=1}^{\rho}\overline{\mathbf{h}_{m;\lambda}^2}. \tag{6.64}$$

For the sake of simplicity, the ρ mates to be selected randomly from the μ parents are assumed to be the ρ best parents. Therefore, the performance obtained in a real random selection should be worse, since it does not exclude the self-fertilization ($\approx$ incest) and the mating of worse individuals.

According to the EPP, for a high progress rate the $\langle x\rangle_\rho$ value should be as large as possible and conversely $\langle\overline{\mathbf{h}^2}\rangle_\rho$ as small as possible. Since these mean values depend on ρ, one can influence the gain and loss terms by changing ρ. However, the resulting effects are diametrally opposed.

In order to attain an optimal GR, and therefore a minimal loss term, ρ should be as large as possible. The extreme case is $\rho = \lambda$; however, this implies $\mu = \lambda$, resulting in $\mathrm{E}\{\langle x\rangle_\lambda\} \equiv 0$. The gain term is zero for $\rho = \lambda$, since one has to integrate over all λ $\mathcal{N}(0, \sigma^2)$ normally distributed x components.

In order to maximize the gain term, one should choose $\rho = \mu = 1$, since the selection of the individual with the largest x value would be favored statistically. However, GR does not work any more (only one parent).

Consequently, one has to *compromise* between the gain maximization and loss minimization. This compromise consists of the decision for *maximal* genetic repair at a given μ, $1 < \mu < \lambda$. That is, one has to choose $\rho = \mu$, where the centroid is obtained deterministically by using all μ parents as described by (6.2).

Hence, after choosing $\rho = \mu$, the remaining problem concerns the determination of the optimal μ for a given λ.

The optimal μ choice. The choice of the strategy parameters μ and λ for the $(\mu/\mu_I, \lambda)$-ES depends on different factors. However, if either μ or λ is fixed, the other parameter depends on the optimality condition for the maximal progress rate.

The value of λ is usually kept fixed, since this quantity gives the number of fitness evaluations to be performed per generation. If a parallel computer has ϖ processors, one will choose

$$\lambda = k\,\varpi, \qquad k \in \mathbb{N}. \tag{6.65}$$

Therefore, λ is determined in a natural manner. The remaining question is the corresponding optimal μ value $\hat{\mu}$.

Let us recall Fig. 6.2 on p. 219 to get acquainted with the problem. This figure displays the maximal progress rate values for the $(\mu/\mu_I, \lambda)$-ES with $\rho = \mu$, $\lambda = 50$. For increasing μ, the performance increases first, caused by the reduction of the EPP loss term. At the same time, however, the quantity of the EPP gain term also decreases, as a result of the decrease in the selection pressure of the (μ, λ) truncation selection. If μ is increased further, the decrease of the selection pressure cannot be compensated for by the increase of the GR effect: the maximum is reached and the performance starts to decrease.

The μ value for the maximal performance may be denoted by $\hat{\mu}$. In order to determine the optimal strategy, one has to solve the nonlinear optimization problem

$$\hat{\mu}(\lambda, N) = \arg\max_{\mu}\left[\max_{\sigma^*}\left[\varphi^*_{\mu/\mu_I,\lambda}(\sigma^*, N)\right]\right] \tag{6.66}$$

for given values of λ and N, using (6.54). The solution can be obtained by numerical methods, considering μ as a real-valued parameter. Figure 6.4 shows the plots of the function $\hat{\mu}(\lambda, N)$ having parameter N. For the case $N \to \infty$, the asymptotically exact formula (6.56) was used instead of (6.54).

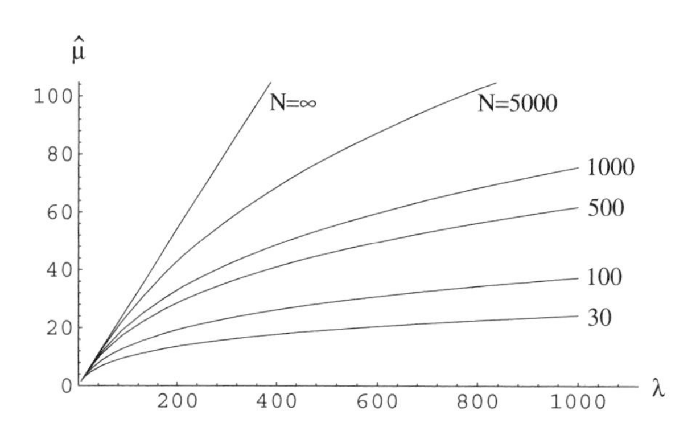

Fig. 6.4. The choice of optimal μ depending on the number of descendants λ for different numbers of variables N

One observes a relatively strong dependence of the optimal μ value on the dimension of the parameter space. For $N \to \infty$ and sufficiently large λ values (see Table 6.2), one observes a linear increase with a slope of approximately 0.27. For finite N, however, one obtains curves with sublinear increase. The optimal mutation strengths $\hat{\sigma}^*$ corresponding to $\hat{\mu}$ and the resulting progress rates $\hat{\varphi}^*$

$$\hat{\sigma}^*(\lambda, N) = \arg\max_{\sigma^*}\left[\varphi^*_{\hat{\mu}/\hat{\mu}_I,\lambda}(\sigma^*, N)\right], \tag{6.67a}$$

$$\hat{\varphi}^*(\lambda, N) = \max_{\sigma^*}\left[\varphi^*_{\hat{\mu}/\hat{\mu}_I,\lambda}(\sigma^*, N)\right] \tag{6.67b}$$

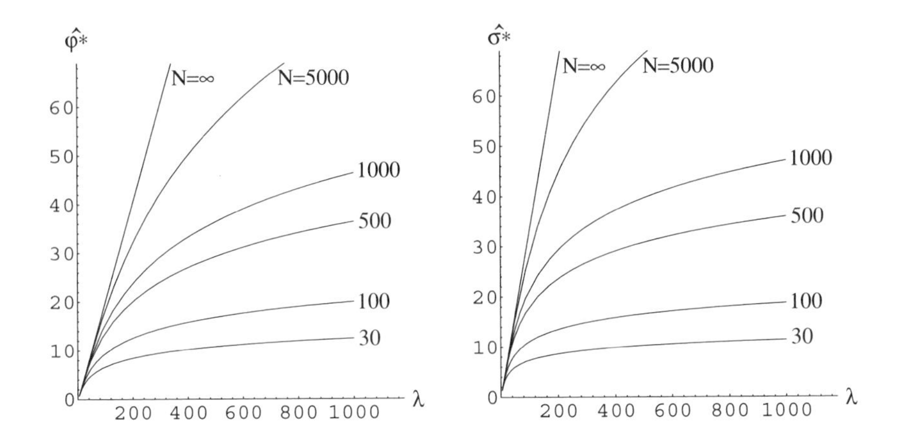

Fig. 6.5. The maximal progress rate (left) and the corresponding optimal mutation strength σ^* (right) of the $\mu = \hat{\mu}$-optimal $(\mu/\mu_I, \lambda)$-ES

are displayed in Fig. 6.5. For the asymptotic limit case $N \to \infty$, the maximal performance scales linearly with the number of descendants λ. This remarkable property will be proven in Sect. 6.3. Unfortunately, this law is not valid for finite N. However, for $N < \infty$, one also obtains considerable progress rates.

Figures 6.4 and 6.5 are not very suitable for practical purposes, Table 6.2 should be used instead. This table contains the optimal μ values, rounded to integer values

$$\hat{\mu}(\lambda, N) = \left\lfloor \frac{1}{2} + \arg\max_{\mu} \left[\max_{\sigma^*} \left[\varphi^*_{\mu/\mu_I, \lambda}(\sigma^*, N) \right] \right] \right\rfloor. \tag{6.68}$$

Using the $\hat{\mu}$ values obtained in this way, the values for $\hat{\sigma}^*$ and $\hat{\varphi}^*$ are obtained by applying (6.67a) and (6.67b).

Fitness efficiency. The concept of fitness efficiency was introduced in Sect. 3.2.3.1. It can also be used in the analysis of the $(\mu/\mu_I, \lambda)$-ES (see also Sect. 5.3.2.2, p. 190). By generalizing Definition (3.104), we define

FITNESS EFFICIENCY:

$$\eta_{\mu/\mu_I, \lambda}(\mu, \lambda, N) := \frac{\max_{\sigma^*} \left[\varphi^*_{\mu/\mu_I, \lambda}(\sigma^*, N) \right]}{\lambda}. \tag{6.69}$$

This function depends on μ, λ, and N. Its asymptotic properties will be investigated in Sect. 6.3. At this point, the scaling behavior of the strategy $\mu = \hat{\mu}$ with the optimal performance will be discussed briefly. If one substitutes $\mu = \hat{\mu} = f(\lambda, N)$ in (6.69) according to (6.68), one obtains a measure that will be called

Table 6.2. The optimal $(\mu/\mu_I, \lambda)$-ES

	$\hat{\mu}$				$\hat{\sigma}^*$				$\hat{\varphi}^*$			
λ \ N	30	100	1000	∞	30	100	1000	∞	30	100	1000	∞
10	3	3	3	3	2.788	3.035	3.177	3.196	1.529	1.638	1.695	1.703
20	4	5	5	6	4.016	5.172	5.936	6.663	2.805	3.261	3.629	3.699
30	5	7	8	8	4.826	6.763	9.055	9.568	3.692	4.545	5.503	5.721
40	6	8	10	11	5.451	7.697	11.38	13.05	4.370	5.597	7.246	7.746
50	7	9	12	14	5.966	8.468	13.49	16.54	4.916	6.469	8.865	9.766
60	8	10	14	16	6.407	9.131	15.41	19.42	5.369	7.214	10.37	11.79
70	8	11	16	19	6.562	9.717	17.16	22.91	5.754	7.864	11.76	13.81
80	9	12	18	22	6.928	10.24	18.77	26.40	6.095	8.440	13.06	15.84
90	10	13	19	25	7.258	10.72	19.86	29.88	6.394	8.955	14.26	17.86
100	10	14	21	27	7.359	11.17	21.26	32.77	6.665	9.420	15.39	19.89
150	12	17	28	41	8.126	12.57	26.09	49.60	7.706	11.25	20.10	30.01
200	14	19	33	54	8.731	13.46	29.29	65.83	8.444	12.58	23.75	40.13
250	15	21	38	68	9.084	14.21	31.96	82.67	9.016	13.61	26.71	50.25
300	16	23	42	81	9.391	14.86	34.01	98.90	9.481	14.46	29.19	60.37
350	17	25	45	95	9.666	15.45	35.59	115.7	9.872	15.17	31.33	70.50
400	18	26	49	108	9.917	15.81	37.25	132.0	10.21	15.79	33.21	80.62
450	19	27	52	122	10.15	16.15	38.55	148.8	10.51	16.33	34.88	90.74
500	19	29	55	135	10.23	16.62	39.75	165.0	10.77	16.82	36.39	100.9
600	21	31	60	162	10.63	17.18	41.71	198.1	11.22	17.66	39.01	121.1
700	22	33	64	189	10.88	17.69	43.26	231.2	11.61	18.37	41.24	141.4
800	23	34	68	216	11.10	18.01	44.68	264.2	11.93	18.98	43.18	161.6
900	23	36	72	243	11.18	18.44	45.98	297.3	12.22	19.52	44.90	181.8
1000	24	37	76	270	11.37	18.71	47.19	330.3	12.48	19.99	46.43	202.1
1500	28	43	90	406	12.10	19.99	51.35	496.3	13.45	21.81	52.33	303.3
2000	30	47	101	541	12.50	20.81	54.26	661.6	14.13	23.09	56.50	404.5

μ-OPTIMAL FITNESS EFFICIENCY:

$$\eta_{\hat{\mu}/\hat{\mu}_I,\lambda}(\lambda, N) := \frac{\max_{\sigma^*}\left[\varphi^*_{\hat{\mu}/\hat{\mu}_I,\lambda}(\sigma^*, N)\right]}{\lambda}. \tag{6.70}$$

Table 6.3 gives the μ-optimal fitness efficiency of some $(\mu/\mu_I, \lambda)$-ES algorithms. These values are obtained directly from Table 6.2. One observes that

Table 6.3. The $\eta_{\hat{\mu}/\hat{\mu}_I,\lambda}$-fitness efficiency

λ	$N = 30$	$N = 100$	$N = 1000$	$N = \infty$
10	0.153	0.164	0.169	0.170
20	0.140	0.163	0.181	0.185
30	0.123	0.152	0.183	0.191
40	0.109	0.140	0.181	0.194
50	0.098	0.129	0.177	0.195
100	0.066	0.094	0.154	0.199
1000	0.012	0.020	0.046	0.202

the $\eta_{\hat{\mu}/\hat{\mu}_I,\lambda}$-fitness efficiency shows as a function of λ a different behavior depending on N. For moderate N values, $\eta_{\hat{\mu}/\hat{\mu}_I,\lambda}$ is even for small λ a monotonically decreasing function. For larger values of N, such as $N = 1000$, there exists a weak maximum. For $N = \infty$, one gets a monotonic increase with an asymptotic limit for $\lambda \to \infty$ that will be calculated in Sect. 6.3.3.2.

Since the fitness efficiency decreases monotonically for moderate values of N, it does not make any sense to choose k in Eq. (6.65) larger than that for the $(\mu/\mu_I, \lambda)$-ES running on a parallel computer with ϖ processors. Therefore, one should use $\lambda = \varpi$.

6.1.3.5 The Exploration Behavior of the $(\mu/\mu_I, \lambda)$-ES.

The dynamics of $\alpha^{(g)}$ *and* $R(g)$. In the investigation of the exploration behavior of the (μ, λ)-ES on the sphere model, it was shown that the parents perform a *random walk* on the hypersphere surface. This result is in contrast to the "gradient diffusion picture" sometimes invoked to explain the exploration behavior of the ES (Rechenberg, 1994). The random walk property is principally also valid for the $(\mu/\mu_I, \lambda)$-ES, it will be investigated for the parental centroid $\langle \mathbf{y} \rangle$.

Analogous to (5.185), we define the angle α between the centroid $\langle \mathbf{y} \rangle_\star$ at generation 0 and the centroid $\langle \mathbf{y} \rangle^{(g)}$ at generation g

$$\alpha^{(g)} := \sphericalangle\left(\langle \mathbf{y} \rangle^{(g)}, \langle \mathbf{y} \rangle_\star\right) = \arccos\left(\frac{\langle \mathbf{y} \rangle_\star^{\mathrm{T}} \langle \mathbf{y} \rangle^{(g)}}{\|\langle \mathbf{y} \rangle_\star\| \, \|\langle \mathbf{y} \rangle^{(g)}\|}\right). \tag{6.71}$$

The framework developed in Sect. 5.3.4 can be applied directly to the $(\mu/\mu_I, \lambda)$-ES case. In particular, the considerations on the limit angle α_∞ for a sufficiently large number of generations g are valid (cf. Eq. (5.203)).

The case for a small number of generations g, described by Eq. (5.197), however, must be modified by substituting the normalized mutation strength in (5.197) by $\sigma^*/\sqrt{\mu}$, since every component of the parental centroid is obtained by averaging over μ (approximately) normally distributed components of the mutation vectors $\mathbf{z}$. Every single component has the standard deviation σ, and a sum over μ such random quantities has the standard deviation $\sqrt{\mu}\,\sigma$. Consequently, the standard deviation of the average is obtained (dividing by μ) as $\sigma/\sqrt{\mu}$. Therefore, instead of (5.197), one gets

$$\boxed{\overline{\alpha}^{(g)} \approx \sqrt{g}\,\frac{\sigma^*}{\sqrt{\mu N}},} \qquad \text{for} \quad \overline{\alpha}^{(g)} \lessapprox \pi/4. \tag{6.72}$$

The theoretical estimates should be verified by experiments. In the following, the results obtained will be discussed in the experimental framework introduced in Sect. 5.3.4.3, adapted for the case of intermediate recombination. For the $(5/5_I, 30)$-ES with $N = 100$, one finds in Table 6.1 $c_{5/5,30} = 1.446$ and by numerical maximization of (6.54) $\hat{\sigma}^* = 5.897$ and $\hat{\varphi}^* = 4.456$. The ES algorithm is operated with the optimal $\sigma^* = \hat{\sigma}^*$ value. The results are shown in Fig. 6.6.

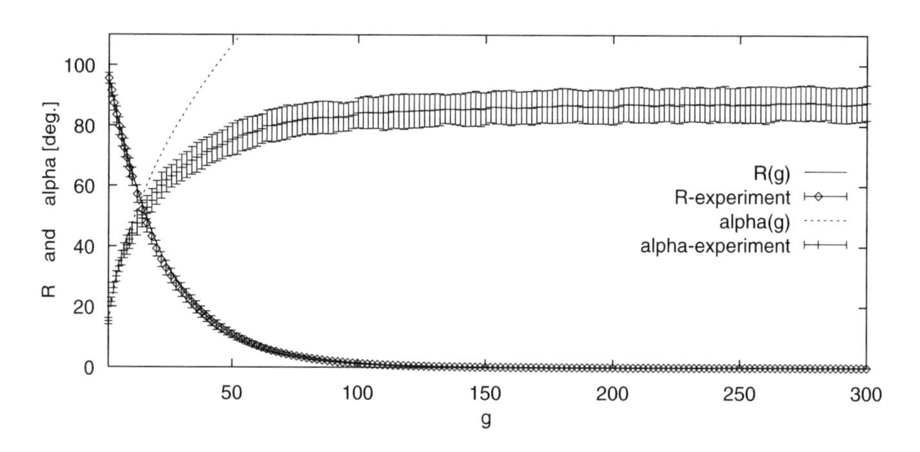

Fig. 6.6. A $(5/5_I, 30)$-ES experiment on the dynamics of the alpha$(g) := \overline{\alpha}^{(g)}$ angle and of the residual distance R$(g) := \|\langle\mathbf{y}\rangle^{(g)}\|$. The angular quantities are given in degrees. The experimental data are displayed using "error plots," indicating the deviation region of the average value ($\pm$ standard deviation) of L runs. Note that for $g \geq 10$ only every second measured value is displayed for better visibility.

These results can be compared with the results in Fig. 5.4, p. 200, obtained for the $(5, 30)$-ES experiment. One observes that the tendency in the angular dynamics is the same. However, caused by the larger optimal mutation strength for the $(5/5_I, 30)$-ES, the parental centroid departs much faster from the initial parent than in the $(5, 30)$-ES case.

Additionally, the R dynamics are shown in Fig. 6.6. The initial residual distance of the parental centroid was chosen as $R(0) = R_0 = 100$. The magnitude of this residual distance over generations can be computed using (2.103)

$$R(g) = R_0\, \mathrm{e}^{-0.04456\, g}. \tag{6.73}$$

A good agreement between the theory and experiment can be observed.

The β angles. In addition to the dynamics of the α angle, the behavior of the ES in a *single* generation is also of interest. For this purpose, the β angles were introduced in Sect. 5.3.4.4, Eq. (5.205). These angles are a measure for the exploration behavior of the population within one generation.

In the (μ, λ)-ES case, the distribution of β has a large variance, as can be seen in Fig. 5.5, p. 201, for the maximal β values. In contrast, the angles of the $(\mu/\mu_I, \lambda)$-ES are relatively constant (see Fig. 6.7). The average angle

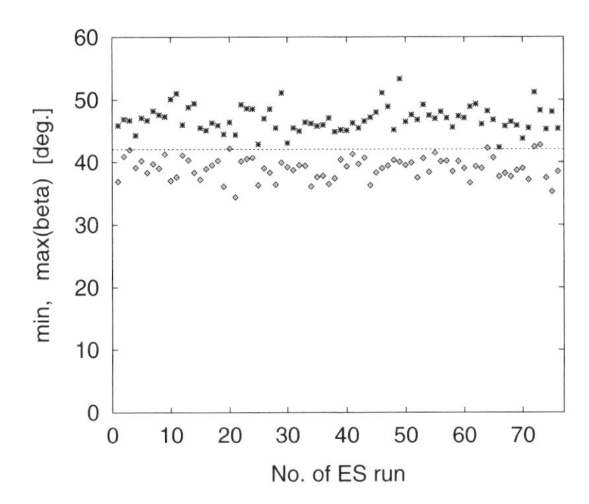

Fig. 6.7. The spread of the β angles: the minimum min(beta) ($\diamond$) and the maximum max(beta) ($\ast$) of the angles between the parents of a generation ($g = 100$ was chosen) are averaged over $L = 76$ independent runs of a $(5/5_I, 30)$-ES. The dashed horizontal line $\beta = \overline{\beta} = 42.1°$ is the expected value of β as predicted by (6.76).

$\overline{\beta}$ can be calculated approximately. Since all offspring $\tilde{\mathbf{y}}_l$ of a generation are generated from the same recombinant $\langle \mathbf{y} \rangle$, one has for two parents $\mathbf{y}$ and $\mathbf{y}'$ of the new generation

$$\mathbf{y} = (R + z_1, z_2, \ldots, z_N)^{\mathrm{T}} \qquad \text{and} \qquad \mathbf{y}' = (R + z'_1, z'_2, \ldots, z'_N)^{\mathrm{T}}. \tag{6.74}$$

The z_i and z'_i are $\mathcal{N}(0, \sigma^2)$ normally distributed components of the mutation vectors $\mathbf{z}$ and $\mathbf{z}'$. It was assumed w.l.o.g. $\langle \mathbf{y} \rangle = (R, 0, \ldots, 0)^{\mathrm{T}}$. The angle between these two parents is obtained by the scalar product

$$\begin{aligned} \beta &= \arccos\left(\frac{\mathbf{y}^{\mathrm{T}}\mathbf{y}'}{\|\mathbf{y}\|\,\|\mathbf{y}'\|} \right) \\ &= \arccos\left(\frac{R^2 + R(z_1 + z'_1) + \mathbf{z}^{\mathrm{T}}\mathbf{z}'}{\sqrt{(R^2 + 2Rz_1 + \|\mathbf{z}\|^2)\,(R^2 + 2Rz'_1 + \|\mathbf{z}'\|^2)}} \right). \end{aligned} \tag{6.75}$$

For the approximate calculation of the expected value $\overline{\beta}$, (6.75) is expanded in a Taylor series. Considering the statistical independence of z_i and z_i', as well as the expected values $\overline{z_i} = 0$, $\overline{z_i z_i'} = 0$, $\overline{\mathbf{z}^{\mathrm{T}}\mathbf{z}'} = 0$, and $\overline{||\mathbf{z}||^2} \approx N\sigma^2$, the first approximation reads

$$\overline{\beta} \approx \arccos\left(\frac{R^2}{\sqrt{(R^2 + N\sigma^2)\,(R^2 + N\sigma^2)}}\right) = \arccos\left(\frac{1}{1 + \frac{\sigma^{*2}}{N}}\right). \tag{6.76}$$

This formula gives for the $(5/5_I, 30)$-ES experiment with $\sigma^* = \hat{\sigma}^* = 5.897$ an angle $\overline{\beta} \approx 42.1°$. A glance at Fig. 6.7 validates that (6.76) describes the reality quite well.

6.2 The Dominant $(\mu/\mu_D, \lambda)$-ES

6.2.1 A First Approach to the Analysis of the $(\mu/\mu_D, \lambda)$-ES

6.2.1.1 The $(\mu/\mu_D, \lambda)$-ES Algorithm and the Definition of $\varphi_{\mu/\mu_D,\lambda}$.

The algorithm. The term "dominant recombination" is used for expressing the *exclusive* and independent heredity of (homologous) genes from the pool of the μ parents. Every ith component of the descendant vector $\tilde{\mathbf{y}}_l$ is obtained by randomly selecting one of the μ ith components of the parents $\mathbf{y}_m$, and the additive application of the $\mathcal{N}(0, \sigma^2)$ mutation thereafter. As in the (μ/μ_I)-ES case, the parents are obtained by (μ, λ) truncation selection of the descendants of the previous generation; therefore, Eq. (6.5) is valid.

If the $\mathbf{y}$ vectors are expressed by an orthonormal basis $\{\mathbf{e}_i\}$

$$\mathbf{y} = \sum_{i=1}^{N} y_i\, \mathbf{e}_i, \tag{6.77}$$

the $(\mu/\mu_D, \lambda)$-ES algorithm can be expressed considering (6.5) in a compact form

$$\underline{(\mu/\mu_D, \lambda)\text{-ES:}} \quad \forall\, l = 1, \ldots, \lambda: \quad \begin{cases} m_i = \texttt{Random}\{1, \ldots, \mu\} \\ \tilde{\mathbf{y}}_l^{(g+1)} := \displaystyle\sum_{i=1}^{N} \left(\mathbf{e}_i^{\mathrm{T}} \tilde{\mathbf{y}}_{m_i;\lambda}^{(g)}\right) \mathbf{e}_i \,+\, \mathbf{z}_l. \end{cases} \tag{6.78}$$

The vector $\mathbf{z}$ is defined by (6.3). The notation $\texttt{Random}\{1, \ldots, \mu\}$ was described in footnote 2 on p. 144.

The progress rate. Because of comparability reasons and the close relation of the $(\mu/\mu_D, \lambda)$-ES with the $(\mu/\mu_I, \lambda)$-ES, the progress rate definition (6.9) will be used

$$\varphi_{\mu/\mu_D,\lambda} = \mathrm{E}\Big\{ \|\langle \mathbf{y}\rangle^{(g)}\| - \|\langle \mathbf{y}\rangle^{(g+1)}\| \Big\} \quad \text{with} \quad \langle \mathbf{y}\rangle = \frac{1}{\mu}\sum_{m=1}^{\mu} \tilde{\mathbf{y}}_{m;\lambda}. \tag{6.79}$$

The parental centers of mass occurring in (6.79) are not needed in Algorithm (6.78). They are only needed for the performance measurement. Therefore, they must be calculated in the ES simulations explicitly.

6.2.1.2 A Simple Model for the Analysis of the $(\mu/\mu_D, \lambda)$-ES. The performance analysis of the $(\mu/\mu_D, \lambda)$-ES is significantly more difficult to carry out than the $(\mu/\mu_I, \lambda)$-ES, since Algorithm (6.78) does not explicitly use the parental centroid $\langle \mathbf{y}\rangle$ (6.79) as the initial point for the mutations $\mathbf{z}$. As in the (μ, λ)-ES case, there is a distributed population of μ individuals. A rigorous analysis requires the self-consistent determination of the parental distribution and/or of the offspring distribution in the parameter space $\mathcal{Y}$.

In the (μ, λ)-ES case on the sphere model, the analysis has been reduced to the self-consistent determination of the r-distribution of the descendants (see Sect. 5.1.2), because the angular components in the (r, γ_i) representation (see Sect. 5.3.4.1) of the mutations $\mathbf{z}$ did not play a role as to the fitness (selectively neutral). Unfortunately, this does *not* hold for the $(\mu/\mu_D, \lambda)$-ES. By the dominant recombination, the angular components are "mixed together," which must be described appropriately. It is still an open problem how and in which representation this mixing process should be modeled in the easiest way.

In the analysis of the $(\mu/\mu_I, \lambda)$-ES, the mutations were decomposed in the Cartesian representation in a part, x, in the direction toward the optimum, and in another one, denoted by $\mathbf{h}$, perpendicular to it. This decomposition has gained preference over the angular representation (r, γ_i), because the Cartesian representation has the important advantage that the GR effect can be extracted in a very simple way. Furthermore, the centroid composition can be carried out in the Cartesian coordinates much easier than in the spherical system. Therefore, we will analyze the $(\mu/\mu_D, \lambda)$-ES also in the Cartesian system.

As noted above, the parental distribution of the $\mathbf{y}$ vectors and/or of the offspring must be known for the analysis of the $(\mu/\mu_D, \lambda)$-ES. This distribution evolves as a result of the process selection $\to$ recombination $\to$ mutation. From the mathematical point of view, there is no reason to consider recombination as "something special" and to investigate it separately from the mutation process. Instead, the effects of both operators can be described by a *common* probability distribution. However, this means that the net effect of both operators can formally be created by a special kind of mutation, which would exactly have the resulting distribution of both operators.[7] These special mutations will be called ***surrogate mutations***, denoted by $\mathbf{S}$. They are applied to the *imaginary* parental centroid $\langle \mathbf{y} \rangle$. Formally, analogous to the $(\mu/\mu_I, \lambda)$-ES, Eq. (6.4), the descendants are given by

$$\tilde{\mathbf{y}}_l^{(g+1)} := \langle \mathbf{y} \rangle^{(g)} + \mathbf{S}_l. \tag{6.80}$$

The recombination is "hidden" by the introduction of surrogate mutations in the mutation distribution $p^{(g)}(\mathbf{S})$. As an advantage of this approach, certain results obtained in the analysis of the $(\mu/\mu_I, \lambda)$-ES can be directly adapted to the $(\mu/\mu_D, \lambda)$-ES. Particularly, the decomposition of the *selected surrogate mutations* in a component in the optimum direction and in a direction perpendicular to it can be applied. As a result, the EPP is applicable, and the formulae in (6.24, 6.27) yield the progress rate also for the case of the dominant $(\mu/\mu_D, \lambda)$-ES.

As attractive as it seems, this method of surrogate mutations bears a crucial intermediate step: the *self-consistent* determination of the probability density $p^{(g)}(\mathbf{S})$ of the surrogate mutations. The calculation of this density in any desired approximation accuracy must remain as a task for future research. Here, an *asymptotic model* of surrogate mutations will be developed which describes the probability density for $N \to \infty$ exactly.

The principal idea for the asymptotic model of surrogate mutations follows from the scaling property of the mutations, which is valid for all strategies in the sphere model: the ratio of the selection-dependent traveled distance x in the optimum direction to the length of the $\mathbf{h}$ vector is of the order $\mathcal{O}(1/\sqrt{N})$. Consequently, for $N \to \infty$, the influence of the selection is negligible, and one has $\varphi \propto 1/N \to 0$. In other words, for sufficiently large N, the selection can be considered as a *negligible perturbation*; it affects the form of the offspring distribution only slightly. *Hence, in the asymptotic limit case, the selection can be neglected in the self-consistent determination of the offspring distribution.* That is, the selection will be considered in this model only for the calculation of the progress rate φ^* given the offspring distribution by the surrogate model.

The neglection of the selection for the determination of the offspring distribution is the crucial assumption in this model. This has a particular consequence for the distribution of surrogate mutations: since all components of the descendant $\tilde{\mathbf{y}}_l$ are recombined independently of each other, the probability distributions of the components $\mathbf{e}_i^{\mathrm{T}}\mathbf{S}$ of the surrogate mutation vector $\mathbf{S}$ are all the same. Therefore, it suffices to consider only the probability density $p(s)$ of a single component $s := \mathbf{e}_i^{\mathrm{T}}\mathbf{S}$ of the surrogate mutations.

In the following it will be assumed that the components of the surrogate mutations are approximately normally distributed: if one starts with an ar-

[7] A formal proof of the correctness of this assertion has been given by Fogel and Ghozeil (1997) for discrete search spaces.

bitrary parent and applies the *physical* mutation[8] $\mathbf{z}$, the result is a normally distributed random variate. The recombination process is a sampling process from the normally distributed population. Therefore, the components are approximately normally distributed after recombination, and so forth. Consequently, the density of any component $s = \mathbf{e}_i^{\mathrm{T}}\mathbf{S}$ of the surrogate mutations reads

$$p(s) = \frac{1}{\sqrt{2\pi}\,\sigma_s} \exp\left(-\frac{1}{2}\left(\frac{s}{\sigma_s}\right)^2\right). \tag{6.81}$$

The only unknown in (6.81) is the mutation strength σ_s of the *isotropic surrogate mutations*. The mutation strength depends on the strength σ of the physical mutations and the number of parents taking part in the recombination, $\sigma_s = \sigma_s(\mu, \sigma)$. It will be calculated in the next section. If the strength of the surrogate is known, then the progress rate φ^* can be calculated according to Eq. (6.27). This second step will follow in Sect. 6.2.1.4.

6.2.1.3 The Isotropic Surrogate Mutations. As already explained, the investigation of the statistic for a single component of the (selected) surrogate vectors $\mathbf{S}_{m;\lambda}$ is sufficient. We will symbolize this component by $s_m^{(g)}$. The superscript g stands for the generation counter, and the subscript m for the corresponding parent. According to the concept of surrogate mutations, the descendants are generated on top of the parental centroid $\langle \mathbf{y} \rangle$ (see (6.80)). If one starts for the sake of simplicity at the state $\mathbf{y}_m^{(0)} = \mathbf{0}$, the process "recombination → (physical) mutation" can be described by the evolution of the parental centroid $\langle \mathbf{S} \rangle^{(g)}$ of the surrogate mutations. For a single component $\langle s \rangle^{(g)}$, one has

$$\langle s \rangle^{(g)} = \frac{1}{\mu} \sum_{m=1}^{\mu} s_m^{(g)}. \tag{6.82}$$

The generation of the surrogate $s^{(g+1)}$ for the $(g+1)$th generation will be considered next. The intermediate state s' is formed first by recombination. The resulting s' population is called "recombinant." The recombinant is just a sample from a (approximate) normal population, thus, it must be (approximately) normally distributed with a density $\mathcal{N}(0, (\sigma_{\mathrm{r}}^{(g)})^2)$. However, the standard deviation $\sigma_{\mathrm{r}}^{(g)}$ of the recombinant is not known yet. After this step, the physical $\mathcal{N}(s', \sigma^2)$ mutation follows. That is, from the mathematical point of view one gets the surrogate $s^{(g+1)}$ as the sum of statistically independent, normally distributed random variables. Hence the standard deviation $\sigma_s^{(g+1)}$ of this random variable is obtained as the square root of the sum of both variances

[8] It is important to remark that one has to distinguish between the surrogate mutation $\mathbf{S}$ and the actual mutation $\mathbf{z}$. To guarantee unambiguity, we will denote the $\mathcal{N}(\mathbf{0}, \sigma^2\mathbf{E})$ normally distributed $\mathbf{z}$ mutations by the term "physical."

$$\sigma_s^{(g+1)} = \sqrt{\left(\sigma_\mathrm{r}^{(g)}\right)^2 + \sigma^2}. \tag{6.83}$$

The recombinative standard deviation $\sigma_\mathrm{r}^{(g)}$ can be interpreted as an (statistical) *estimate* of the deviation parameter of the parental $s_m^{(g)}$ around its centroid (average value) $\langle s\rangle^{(g)}$

$$\sigma_\mathrm{r}^{(g)} = \sqrt{\mathrm{E}\left\{\left(s_m^{(g)} - \langle s\rangle^{(g)}\right)^2\right\}}. \tag{6.84}$$

This interpretation is compelling, since *per definitionem* the surrogate mutations are applied on top of the parental centroid. It is very important to note that $\langle s\rangle^{(g)}$, defined by (6.82), is itself a random variate.

Before starting with the calculation of $\sigma_\mathrm{r}^{(g)}$, the statistical properties of the random variables $s_m^{(g)}$ are recalled. The expected value of the surrogate is chosen w.l.o.g.

$$\mathrm{E}\left\{s_m^{(g)}\right\} = 0. \tag{6.85}$$

The surrogates are not correlated. Therefore, considering (6.85), one has

$$\mathrm{E}\left\{s_k^{(g)}\, s_l^{(g)}\right\} = \begin{cases} 0, & k \neq l \\ \left(\sigma_s^{(g)}\right)^2, & k = l. \end{cases} \tag{6.86}$$

The expected value of the surrogate can be determined now for a given parent $s^{(g)} = s_m^{(g)}$. One obtains stepwise

$$\begin{aligned}
\mathrm{E}\left\{\left(s_m^{(g)} - \langle s\rangle^{(g)}\right)^2\right\} &= \mathrm{E}\left\{\left(s_m^{(g)} - \frac{1}{\mu}\sum_{k=1}^{\mu} s_k^{(g)}\right)^2\right\} \\
&= \mathrm{E}\left\{\left(\left(1-\frac{1}{\mu}\right) s_m^{(g)} - \frac{1}{\mu}\sum_{k\neq m} s_k^{(g)}\right)^2\right\} \\
&= \mathrm{E}\left\{\left(1-\frac{1}{\mu}\right)^2 \left(s_m^{(g)}\right)^2 - 2\left(1-\frac{1}{\mu}\right)\frac{1}{\mu}\sum_{k\neq m} s_m^{(g)}\, s_k^{(g)} \right. \\
&\qquad\qquad \left. + \frac{1}{\mu^2}\sum_{k\neq m}\sum_{l\neq m} s_k^{(g)}\, s_l^{(g)}\right\} \\
&= \left(1-\frac{1}{\mu}\right)^2 \mathrm{E}\left\{s_m^{(g)}\, s_m^{(g)}\right\} + \frac{1}{\mu^2}\sum_{k\neq m}\sum_{l\neq m}\mathrm{E}\left\{s_k^{(g)}\, s_l^{(g)}\right\} \\
&= \left(1-\frac{1}{\mu}\right)^2 \mathrm{E}\left\{\left(s_m^{(g)}\right)^2\right\} + \frac{\mu-1}{\mu^2}\mathrm{E}\left\{\left(s_m^{(g)}\right)^2\right\}
\end{aligned}$$

$$\mathrm{E}\left\{\left(s_m^{(g)} - \langle s\rangle^{(g)}\right)^2\right\} = \frac{\mu-1}{\mu}\,\mathrm{E}\left\{\left(s_m^{(g)}\right)^2\right\} = \frac{\mu-1}{\mu}\left(\sigma_s^{(g)}\right)^2. \tag{6.87}$$

Therefore the iterative schema[9] follows

$$\sigma_s^{(g+1)} = \sqrt{\frac{\mu-1}{\mu}\left(\sigma_s^{(g)}\right)^2 + \sigma^2}. \tag{6.88}$$

After recursive insertion, this formula reads

$$\sigma_s^{(g+1)} =$$

$$\sigma\sqrt{\left[1 + \frac{\mu-1}{\mu} + \ldots + \left(\frac{\mu-1}{\mu}\right)^{g-1} + \left(\frac{\mu-1}{\mu}\right)^{g}\right] + \left(\frac{\mu-1}{\mu}\right)^{g+1}\left(\frac{\sigma_s^{(0)}}{\sigma}\right)^2}. \tag{6.89}$$

The sum in the brackets of (6.89) is a finite geometric series; thus one obtains

$$\sigma_s^{(g+1)} = \sigma\sqrt{\frac{1-\left(\frac{\mu-1}{\mu}\right)^{g+1}}{1-\left(\frac{\mu-1}{\mu}\right)} + \left(\frac{\mu-1}{\mu}\right)^{g+1}\left(\frac{\sigma_s^{(0)}}{\sigma}\right)^2}. \tag{6.90}$$

Using this formula, the dynamic behavior of the recombination-mutation process can be investigated. Especially, for the steady-state case $g \to \infty$ one gets a remarkable result[10]

$$\sigma_s := \lim_{g\to\infty} \sigma_s^{(g)} = \sigma\sqrt{\mu}. \tag{6.91}$$

Obviously, the strength of the initial surrogate mutations (at $g = 0$) is totally irrelevant. If one chooses $\sigma_s^{(0)} = 0$, i.e., all individuals are initialized in a single point, the distribution inflates successively to the value $\sigma\sqrt{\mu}$. Conversely, if the individuals are initialized too far from each other, the population shrinks. One may say that the recombination-mutation process gradually *forgets* the distribution of the predecessors (fading memory). The discussion of this interesting phenomenon will be continued in Sect. 6.2.2.2.

6.2.1.4 The Progress Rate $\varphi^*_{\mu/\mu_D,\lambda}$. If the strength of the surrogate mutations is known, the $\varphi^*_{\mu/\mu,\lambda}$ formula (6.27) can be applied. To this end, one has to determine $\overline{\langle \mathbf{h}^2\rangle}$ and $\overline{\langle x\rangle}$. These are obtained for the case of dominant recombination not from the physical mutations with the strength σ, but with the surrogate strength $\sigma_s = \sqrt{\mu}\,\sigma$.

[9] It is interesting to note that this schema can be obtained without any further derivation from the theory of the (μ, λ)-ES: if one inserts $\zeta = 1$ into Eq. (5.159) and switches the selection off by substituting $\lambda = \mu$, one gets the same result.

[10] Instead of the derivation provided here for the stationary standard derivation, one could also solve Eq. (6.88) by using the self-consistent method equating $\sigma_s^{(g+1)} = \sigma_s^{(g)} = \sigma_s$.

Since the actual influence of the selection process on the distribution of the length of $\mathbf{h}$ is not known, the simplest approximation of $\langle\overline{\mathbf{h}^2}\rangle$ is obtained by neglecting the influence of the selection on $\mathbf{h}$. In other words, the $\mathbf{h}$ components of the surrogate mutations $\mathbf{S}$ are assumed to be approximately $\mathcal{N}(0, \sigma_s^2)$ normally distributed. Consequently, the expected value of the length squared reads $\overline{\mathbf{h}^2} \approx N\sigma_s^2$ (cf. (3.220)). Hence, using (6.91), one gets

$$\langle\overline{\mathbf{h}^2}\rangle \approx N\sigma_s^2 = \mu N \sigma^2. \tag{6.92}$$

This formula can be obtained directly from (6.51), by substituting σ by σ_s and by "switching off" the selection by the choice $\lambda = \mu$.

Because of the surrogate model, also the x components of the $\mathbf{S}$ vectors are asymptotically $\mathcal{N}(0, \sigma_s^2)$ distributed. According to (6.13), the average $\langle x \rangle$ must be taken over the μ best x values $x_{m;\lambda}$, i.e. one has $x_{m;\lambda} = x_{\lambda+1-m:\lambda}$. The density $p_{m;\lambda}(x) = p_{\lambda+1-m:\lambda}(x)$ for the mth best trial is given by Eq. (5.10), with $P(x) = \Phi(x/\sigma_s)$. The technical aspects of the expected value calculation $\overline{\langle x \rangle}$

$$\overline{\langle x \rangle} = \frac{1}{\mu} \sum_{m=1}^{\mu} \overline{x_{m;\lambda}} = \frac{1}{\mu} \sum_{m=1}^{\mu} \int_{-\infty}^{\infty} x \, p_{\lambda+1-m:\lambda}(x) \, \mathrm{d}x \tag{6.93}$$

are not repeated here, since similar derivations were carried out in Chap. 5 and in Sect. 6.1.2.1, p. 211f. The result of Sect. 6.1.2.1, Eq. (6.41), can be adopted directly by considering the $R \to \infty$ case[11] and substituting the mutation strength σ by the surrogate mutation strength σ_s

$$\langle \overline{x} \rangle = \sigma_s \, e_{\mu,\lambda}^{1,0} = \sigma_s \, c_{\mu/\mu,\lambda} = \sqrt{\mu} \, \sigma \, c_{\mu/\mu,\lambda}. \tag{6.94}$$

Finally, after inserting (6.92) and (6.94) in (6.27) and applying the normalization (6.15), the *progress rate* of the $(\mu/\mu_D, \lambda)$-ES follows

$$\boxed{\varphi^*_{\mu/\mu_D,\lambda}(\sigma^*, N) = c_{\mu/\mu,\lambda} \, \sigma^* \frac{\sqrt{\mu}}{\sqrt{1 + \frac{\sigma^{*2}}{N}}} - N\left(\sqrt{1 + \frac{\sigma^{*2}}{N}} - 1\right) + \ldots .} \tag{6.95}$$

A first comparison with the intermediate recombination, Eq. (6.54), shows that the EPP loss term does not have the factor $1/\mu$. However, the gain term has an additional $\sqrt{\mu}$ factor. Consequently, the expected maximal performance of both strategy types should be of the same order. A comparison based on experiments for the case of the $(8/8, 30)$-ES (Figs. 6.1 and 6.8) reveals certain performance advantages of the intermediate variant over the dominant one. This can be attributed to the more perfect GR mechanism of the intermediate variant (see the relevant discussion in Sect. 6.2.2.2).

---

[11] The derivation of (6.41) was related to the sphere model. However, the calculation of the expected value in (6.93) is done for the one-dimensional distribution of the $x_{m;\lambda}$ statistics. The latter one is equivalent to the hyperplane case obtained from the sphere model for $R \to \infty$.

6.2.2 The Discussion of the $(\mu/\mu_D, \lambda)$-ES

6.2.2.1 The Comparison with Experiments and the Case $N \to \infty$. The $\varphi^*_{\mu/\mu_D,\lambda}$ formula will be compared with ES simulation results using the ES algorithm with $\mu = 8$ and $\lambda = 30$. These parameters were already used for the same purpose in Sect. 6.1.3.1 for intermediate recombination. However, the experimental setup of the simulation for the $(8/8_D, 30)$-ES differs slightly from that for the $(8/8_I, 30)$-ES. "One-generation experiments" are not possible for the $(\mu/\mu_D, \lambda)$-ES since Algorithm (6.78) does not calculate the parental centroid. However, this centroid is necessary for the calculation of the progress rate. For the case considered here, one has to deal with a real population exhibiting a varying probability density over the ES run. Therefore, starting from an initial parental distribution, one has to wait at least a certain number of generations $g = G_1$ – the transient time g_t. After this transient phase, the stationary state is reached and the measurements can be started.

The transient time can be estimated using Eq. (6.90). If all parents are initialized at the same state in the search space, the $\sigma_s^{(0)}$ term in (6.90) vanishes, and one obtains for the relative error $f_t := (\sigma_s - \sigma_s^{(g_t)})/\sigma_s$ using (6.90, 6.91)

$$f_t = 1 - \sqrt{1 - \left(\frac{\mu - 1}{\mu}\right)^{g_t}} \approx \frac{1}{2}\left(\frac{\mu - 1}{\mu}\right)^{g_t}. \tag{6.96}$$

Here, a Taylor expansion of the square root has been applied. The transient time is obtained after resolving for g_t

$$g_t \approx \frac{\ln 2f_t}{\ln \frac{\mu-1}{\mu}}. \tag{6.97}$$

For the $(8/8_D, 30)$-ES, the relative error of 1% is reached after a transient time of $g_t \approx 30$. Therefore, $G_1 = 30$ is chosen for the simulation.

Due to the angular dependency of the dominant recombination, the simulation method using population rescaling, as explained in Sect. 5.3.1.2, causes distorted results. Therefore, the data extraction method for φ in Sect. 5.3.1.1 was used in the following ES simulations. The results of the $(8/8_D, 30)$-ES for $N = 30$ and $N = 200$ are shown in Fig. 6.8. They are obtained from $L = 100$ independent ES runs for $N = 30$, and from $L = 200$ runs for $N = 200$.

As expected, there are considerable deviations between the real $\varphi^*_{\mu/\mu_D,\lambda}$ values and the predictions of Eq. (6.95). The basic assumption of the *isotropic* surrogate model is only asymptotically fulfilled for $N \to \infty$. Therefore, usable predictions are to be expected if the length $\sqrt{\langle \mathbf{h}^2 \rangle}$ is much larger than $\overline{\langle x \rangle}$, in other words if

$$\sqrt{\mu/N} \ll 1 \tag{6.98}$$

is valid.

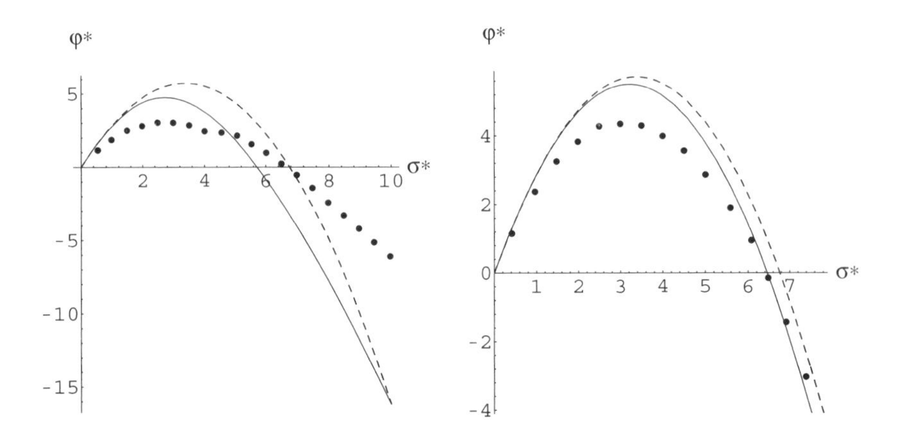

Fig. 6.8. Comparison of the $\varphi^*_{\mu/\mu_D,\lambda}$ formula with experiments. The solid curves are obtained by Eq. (6.95), the dashed by Eq. (6.99). The left part shows the $(8/8_D, 30)$-ES with $N = 30$. The right part shows the same strategy for the case $N = 200$.

If one is mainly interested in the interval of small mutation strengths σ^*, Eq. (6.98) can be expanded into a Taylor series, assuming the condition $\sigma^{*2}/N \ll 1$. Thus, one obtains the N-*in*dependent asymptotic progress rate formula[12]

$$\varphi^*_{\mu/\mu_D,\lambda} = \sqrt{\mu}\, c_{\mu/\mu,\lambda}\, \sigma^* - \frac{\sigma^{*2}}{2} + \dots, \qquad \text{for } \frac{\sigma^{*2}}{N} \ll 1. \tag{6.99}$$

It can also be found in Rechenberg (1994) who derived it from his model of "Thales recombination." In Fig. 6.8, this formula is shown as a dashed curve. In the interval shown for smaller values of σ^*, the quality of the formulae (6.95) and (6.99) are comparable. The essence of these plots is to show that both yield the *correct order of magnitude* of the $\varphi^*_{\mu/\mu_D,\lambda}$ function. The derivation of more accurate formulae remains as a challenge for future research.

6.2.2.2 The MISR Principle, the Genetic Drift, and the GR Hypothesis. The evidence for the principal applicability of the surrogate mutation model on the analysis of dominant recombination was given in the previous point. The next question is on the fundamental principles that are responsible for the functioning of the ES. A remarkable result obtained from the comparison of intermediate and dominant recombination is that they exhibit the same performance properties for the asymptotic $N \to \infty$ limit, if the strength of *physical* mutations is chosen appropriately (see footnote 12).

[12] This formula can be directly obtained from the N-*in*dependent progress rate formula of the $(\mu/\mu_I, \lambda)$-ES, just by substituting the physical mutation strength σ^* with the normalized strength of surrogate mutations (6.91), $\sigma^*_s = \sqrt{\mu}\,\sigma^*$, in (6.56). Therefore, both strategies show the same performance for $N \to \infty$.

Therefore, a deeper relation can be presumed and will be investigated in detail below.

MISR and genetic drift. The basic idea of the surrogate mutation model was to substitute the effect of both recombination and mutation acting on the parent $\mathbf{y}_m$ by a surrogate mutation $\mathbf{S}$. This surrogate is applied to the respective (virtual) parental centroid $\langle\mathbf{y}\rangle$. From the mathematical point of view, this substitution is admissible if the offspring distribution generated by the surrogate and by the original algorithm are the same, for a given parental distribution.

The distribution of the surrogate mutations was determined in 6.2.1.3 by neglecting the selection. This assumption is admissible for $N \to \infty$. The result in Eq. (6.91) of Sect. 6.2.1.3 is remarkable because the algorithm consisting of dominant recombination *and* mutation generates a *stationary* offspring distribution with a *finite* standard deviation. Although *no* environmental effects (fitness-based selection) are present, the "genetic information" does *not* melt in time, as one could expect intuitively. Rather, the process consisting of recombination *and* mutation produces descendants showing a cohesion. This cohesion resembles the properties of a *species*: the descendants are concentrated around a (virtual) center. This population center may be identified with the notion of the *wild-type* known from population biology (see Strickberger, 1988).

It is really important to make it clear that the process of "species formation" does not require fitness-related selection! By dominant recombination, the "physical mutations" with strength σ are transformed in "genetic variations" of strength $\sigma_s = \sqrt{\mu}\,\sigma$. The stationary offspring distribution is therefore a result of a process that may be called *"mutation-induced speciation by recombination"*, or, in short, MISR

MISR principle

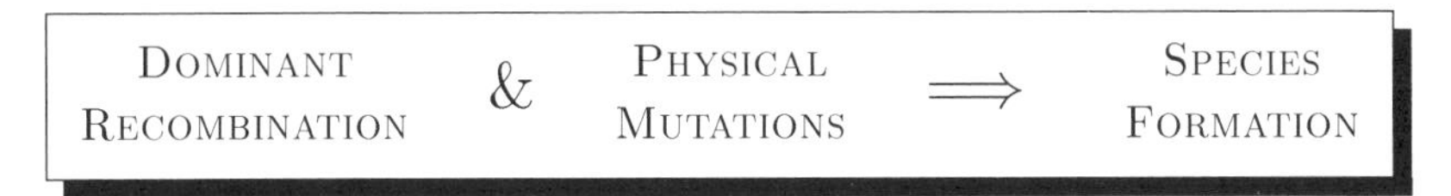

Note that the operators "dominant recombination" and "mutation" alone are not able to generate a stable offspring distribution. If only mutation were applied, one would observe some kind of $\lambda \times (1,1)$-ES, since no selection is present. As a result, the individuals perform a *random walk* in the parameter space $\mathcal{Y}$. Consequently, the standard deviation of the offspring distribution will increase according to the $\sqrt{g}$-law.[13]

If only recombination is applied instead, the "genetic drift" phenomenon emerges: the initial distribution of the population reduces continuously in its

[13] This is a Gaussian diffusion process. It can be handled after the scheme in Sect. 5.3.4.2.

deviation, and finally collapses at a point ("gene fixing"). The location of this point in the parameter space $\mathcal{Y}$ is random, since the selection is switched off and the population centroid performs a *random walk*.

The drift phenomenon can be investigated using Eq. (6.90) quantitatively. For $\sigma = 0$ (no mutations) one obtains

$$\sigma_s^{(g)} = \sigma_s^{(0)} \left(\frac{\mu - 1}{\mu} \right)^{g/2} . \tag{6.100}$$

The population contracts exponentially. If one considers sufficiently large parental populations, the μ term can be simplified by using the binomial theorem and can be cut after the linear term, $((\mu-1)/\mu)^{g/2} = 1 - g/2\mu + \dots$ After resolving for g, one gets

$$g \approx 2\mu \left(1 - \frac{\sigma_s^{(g)}}{\sigma_s^{(0)}} \right) . \tag{6.101}$$

As one can see, the *fixation time* $(\sigma_s^{(g)} \to 0)$ is approximately proportional to the population size μ. At this point, too, there are parallels to evolution biology. Results in population genetics show that the fixation time of an allele has the same order of magnitude as the *effective population size*, which is approximately the number of individuals taking part in the reproduction (Christiansen & Feldman, 1986). This proportionality was also observed in the field of genetic algorithms (Asoh & Mühlenbein, 1994). Unfortunately, mutation was not considered in the latter work. Therefore, the proposed MISR principle remains a hypothesis in the GA field.[14]

Dominant recombination and GR. The ultimate reason for the high progress rates of the ES with intermediate recombination lies in the possibility of the application of larger mutation strengths than in the (μ, λ)-ES case. The effect of harmful components of these mutations are dampened by the GR mechanism. Genetic repair proceeds in the algorithm by the calculation of the parental centroid, i.e. the mean values of the N components of μ object parameter vectors $\mathbf{y}_m$ are determined. Genetic repair and "taking the mean" are, for the $(\mu/\mu_I, \lambda)$-ES algorithm (6.6), synonymous, since the centroid composition is executed *explicitly* over the *selected* parents.

In contrast, the ES algorithm with dominant recombination (6.78) does not contain centroid composition over the parents. Instead, each component of the descendant vector is obtained by the random choice out of the μ

[14] There is an interesting work of Goldberg and Segrest (1987). It analyzes GAs by assuming a similar scenario using numerical Markov chain simulations. However, that work had a different focus so that the MISR effect was not noticed. Meanwhile, the MISR effect has also been analyzed in binary GAs (Beyer, 1999). However, unlike the dominant recombination in ES, it appeared that this effect is due to the random sampling and not a result of the (standard) crossover in GAs.

parental values. From probability theory's point of view, this is just a sampling process. By the dominant sampling, the actual distribution of the μ parents is *estimated* statistically. Since this distribution can be characterized by its moments, the dominant sampling process implicitly contains the estimation of the mean value $\langle s \rangle$ and therefore the center of the parental $\mathbf{y}_m$ vectors.

Mutation with the distribution $\mathcal{N}(\mathbf{0}, \sigma^2 \mathbf{E})$ is applied after the dominant recombination. It changes the standard deviation of the resulting distribution $p(\mathbf{S})$ of the surrogate mutations according to (6.83), but not its first moment. In other words, the dominant recombination statistically estimates the center of the parents, and thereby the center of the surrogate mutations. Since building the center and GR are synonymous, the effect of dominant recombination can be regarded as an *implicit* GR. This important relation is graphically described in Fig. 6.9.

As a consequence of the GR hypothesis on p. 222, the GR effect causes the extraction of similarities (common features). If one arbitrarily picks out one of the μ elements in an urn, the elements with higher frequency (or density of occurrence) in the urn will be selected with a higher probability. If one defines "similarity" as a property that is observed at many individuals, the dominant recombination consequently extracts these "similarities."

The extraction of similarities or the implicit GR through dominant recombination are statistical phenomena with statistical deviations. Thus, the GR mechanism does not operate perfectly. As a consequence, the performance in real applications ($N < \infty$) is in its statistical expectation less than that obtained by the explicit centroid composition through intermediate recombination. If the sampling (the dominant recombination) were to be done based on the same (i.e. g = const) parental distribution instead, the exact centroid would be better and better approximated as the number of samples increased. However, since the $\mathbf{y}_m$ population of the real algorithm moves in the parameter space $\mathcal{Y}$, *directed by selection*, the distribution to be approximated changes over the generations.

The progress rate φ is a measure for the directed change of the parental centroid. Hence, the relation of φ to the geometric dispersion $\sqrt{N}\,\sigma_s$ of the population[15] is a measure for the expected change of the distribution. By inserting the value for $\sigma_s = \sqrt{\mu}\,\sigma$ that maximizes (6.99), one concludes that the change is of order $\mathcal{O}(1/\sqrt{N})$. Consequently, dominant recombination attains for $N \to \infty$ the same GR quality, which is obtained by the intermediate recombination through explicit centroid composition.

GR, MISR, and genetic algorithms – a speculative discussion. We will complete our findings on the GR effect for dominant recombination with a hypothesis on a general working principle of evolutionary algorithms:

[15] The expected length l of the surrogate mutations is used here in order to measure the dispersion of the population, cf. Eq. (3.51).

GR hypothesis of dominant recombination

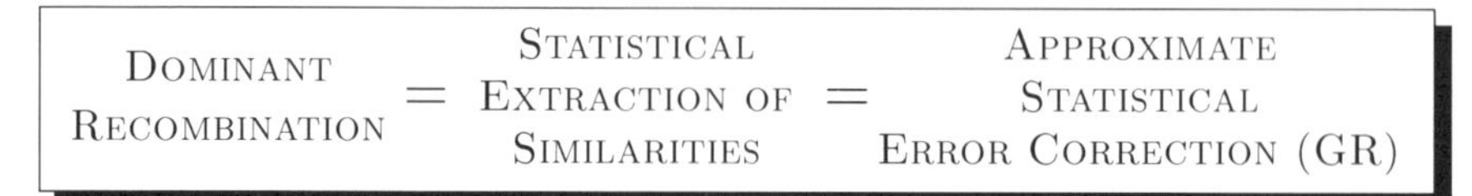

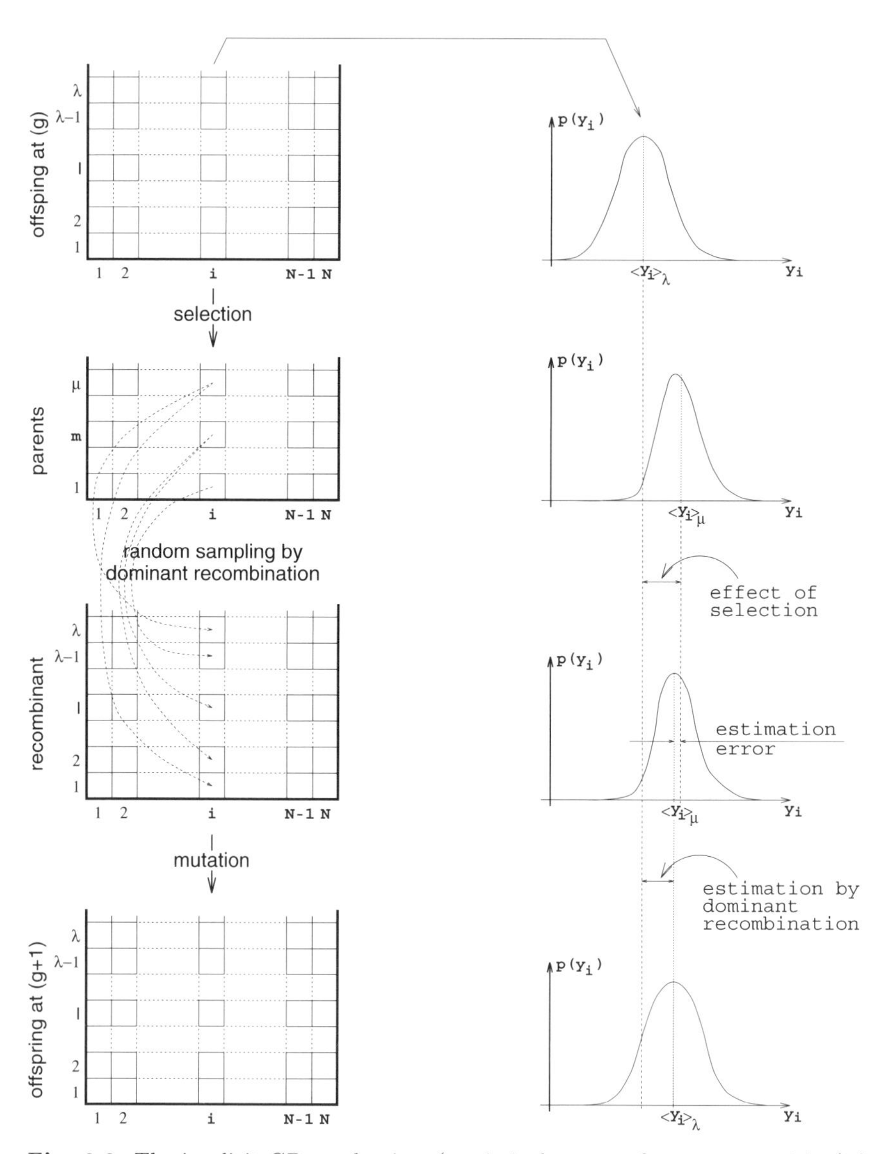

Fig. 6.9. The implicit GR mechanism (statistical center of mass composition) for the dominant recombination. The statistical parameters of the parental distribution are estimated by the "dominant sampling." In this manner, the "similarities" (common features) of the parents are transferred to the descendants.

The emphasis on "hypothesis" is appropriate here, since the GR principle is not rigorously proven yet, plausibility arguments have been presented instead. Moreover, whether or not a significant GR effect can occur in genetic algorithms is also an open problem. For the time being, the GR hypothesis is only an alternative explanation model for the working mechanism of GA (Beyer, 1995a) which is in contraposition to the *building block hypothesis* (BBH). One should note the principal difference between the BBH and GR hypothesis. Goldberg (1989, p. 41) writes:

> Just as a child creates magnificent fortresses through the arrangement of simple blocks, so does a genetic algorithm seek near optimal performance through the juxtaposition of short, low-order, high-performance schemata, or building blocks.

The explanatory approach of the BBH by combining good building blocks is on the surface very plausible. It emerges from the "engineer's approach" to obtain technical solutions by combining components. However, it overlooks the fact that such a technical approach by an engineer *always* presupposes problem-specific knowledge. In contrast, the GA (as well as the ES) recombines randomly. In general, it does not take advantage of problem specific knowledge.[16]

The idea of the BBH can also be considered on the platform of evolution biology. Possibly it may be desirable (or maybe not) to combine good properties of different species. Unfortunately (or luckily), the recombination does not function between different species. Reproduction is only (almost exclusively) possible within the boundaries of the species. Actually, a species is defined just by this very observation. From the viewpoint of the GR hypothesis, this is obvious. Given individuals from different species, the differences between the recombination partners are simply too large; therefore, the result of such a recombination corresponds to a surrogate mutation with a very large EPP loss part. Conversely, the similarities between these individuals are too small, since the parents were exactly chosen because of their *different* (but desired) properties, because of their different building blocks. Obviously, this cannot work!

An alternative explanation for the usefulness of recombination can be offered by the combination of the GR and the MISR hypotheses. Through the dominant recombination, the physical mutations (bit-flips) are transformed in genetic variations. The resulting structure is similar to a species: it drifts in the binary parameter space, guided by the selection. Up to this point it is totally open whether a GA can accomplish something like genetic repair. It

[16] Actually, such a problem-specific knowledge may be available. In other words, the partial solutions and the building blocks may be known *a priori*. In such a case, it is favorable to combine these building blocks by a deterministic algorithm, and not randomly. Typical examples are the separable fitness functions $F(y_1, \dots, y_N) = \sum_{i=1}^{N} f_i(y_i)$; these can be optimized by N separate *one-dimensional* search steps.

can absolutely be possible that the only effect of recombination is the increase of the genetic variety. By consequence, one could renounce recombination totally, and use an increased mutation rate p_m instead, obtaining the same effect. However, it seems more likely that, depending on the local properties of the fitness landscape, different recombination operators perform GR – or at least some kind of similarity extraction – with different quality.

There are many indications that, e.g., uniform crossover (Syswerda, 1989) shows better performance than the one-point crossover. From the viewpoint of the GR principle in dominant recombination, this is obviously understandable. The one-point crossover is at least in the ES domain the worst imaginable way to estimate the centroid of a population. On the other hand, from the GR point of view, it would be better to increase the number of parents taking part in the recombination from $\mu = 2$ to $\mu > 2$ (Beyer, 1994b, 1995b). Investigations done by Eiben at al. (1994) point in this direction. Therefore, the application of multirecombination seems to be also advantageous for genetic algorithms.[17] Unfortunately, these investigations were done without an appropriate theoretical background such that some important data are not available. Therefore, one cannot decide, based on the work of Eiben et al., whether the observed performance gain is a consequence of a GR-MISR synergism, or only a result of an increased variation. Actually, such a variation could be attained simply by increasing the mutation rate. An interesting field opens at this point for future research.

6.3 The Asymptotic Properties of the $(\mu/\mu, \lambda)$-ES

This section investigates the asymptotic properties of the multirecombinative $(\mu/\mu, \lambda)$-ES for $N \to \infty$ and $\lambda \to \infty$ $(\lambda/N \ll 1)$. For this purpose, first the $c_{\mu/\mu,\lambda}$ progress coefficient will be asymptotically expanded. Based on this result, the asymptotic progress law of the $(\mu/\mu, \lambda)$-ES will be derived. This result will be applied to the calculation of fitness efficiency and to the determination of the optimal selection strength $\hat{\vartheta}$. At the end, it will be shown that the $(\mu/\mu, \lambda)$-ES can attain a performance λ times larger than that of the $(1+1)$-ES. The results will be used for the investigation of the ES dynamics and for the estimation of the ES time complexity.

[17] The idea of multirecombination as well as the uniform crossover can be traced back to the early 1960s (Bremermann et al., 1966). However, Bremermann did not find a significantly conclusive advantage of these recombination techniques.

6.3.1 The $c_{\mu/\mu,\lambda}$ Coefficient

Because of (5.112), the progress coefficient $c_{\mu/\mu,\lambda}$, Eq. (6.55), has an elegant integral expression (Beyer, 1994b)[18]

$$c_{\mu/\mu,\lambda} = \frac{\lambda-\mu}{2\pi}\binom{\lambda}{\mu}\int_{-\infty}^{\infty} \mathrm{e}^{-t^2}\left(\Phi(t)\right)^{\lambda-\mu-1}\left(1-\Phi(t)\right)^{\mu-1}\,\mathrm{d}t. \tag{6.102}$$

This integral expression is appropriate for theoretical investigations as well as numerical calculation of concrete values. Beside expression (6.102) there is a variety of alternative $c_{\mu/\mu,\lambda}$ expressions consisting of sums as well as double sums over $\Phi(t)$-integrals (e.g. the one in footnote 18). It is important to keep in mind that the computation time increases linearly with μ for a simple integral sum and it is proportional to μ^2 for double sums. In contrast to these, the computational costs for (6.102) are almost independent of μ.

6.3.1.1 The Asymptotic Expansion of the $c_{\mu/\mu,\lambda}$ Coefficient. For the investigation of the asymptotic behavior of $c_{\mu/\mu,\lambda}$, (6.102) is transformed by the substitution $z = 1-\Phi(t)$ yielding the integral

$$c_{\mu/\mu,\lambda} = \frac{\lambda}{\mu}\frac{1}{\sqrt{2\pi}}\frac{1}{\mathrm{B}(\lambda-\mu,\mu)}\int_0^1 \exp\left[-\frac{1}{2}\left(\Phi^{-1}(1-z)\right)^2\right] \times z^{\mu-1}(1-z)^{\lambda-\mu-1}\,\mathrm{d}z. \tag{6.103}$$

As an intermediate step, we have expressed the binomial coefficient in (6.102) by factorial terms of λ and μ, yielding $(\lambda-1)!/((\lambda-\mu-1)!\,(\mu-1)!)$. Thereafter, this quotient was expressed by the *complete beta function* $\mathrm{B}(\lambda-\mu,\mu)$, it holds that (Abramowitz & Stegun, 1984, p. 79)

$$\mathrm{B}(\lambda-\mu,\mu) = \frac{(\lambda-\mu-1)!\,(\mu-1)!}{(\lambda-1)!} = \frac{\Gamma(\lambda-\mu)\,\Gamma(\mu)}{\Gamma(\lambda)} = \int_0^1 z^{\mu-1}(1-z)^{\lambda-\mu-1}\,\mathrm{d}z. \tag{6.104}$$

The term $z^{\mu-1}(1-z)^{\lambda-\mu-1}$, occurring in both (6.103) and (6.104), exhibits a maximum which gets sharper for increasing λ and $\lambda-\mu$. Hence, it suggests the expansion of the integrand of (6.103) at this maximum $z=\hat{z}$. One gets for $\hat{z}$

[18] Instead of this formula, Rechenberg (1994, p. 146) proposes for $c_{\mu/\mu,\lambda}$ a double sum over $c_{1,\lambda-k}$ progress coefficients

$$c_{\mu/\mu,\lambda} = \frac{1}{\mu}\sum_{k=0}^{\mu-1}\sum_{i=k}^{\mu-1}(-1)^{i-k}\frac{\lambda-i}{\lambda-k}\binom{\lambda}{i}\binom{i}{k}c_{1,\lambda-k}$$

and reports considerable numerical difficulties in practice. Therefore, it is recommended to use (6.102) instead for numerical calculations, because of its numerical stability. A proof of the equivalence of these expressions can be given, but will be omitted here.

$$\hat{z} := \arg\max_{z} \left[z^{\mu-1}(1-z)^{\lambda-\mu-1} \right] = \frac{\mu-1}{\lambda-2}. \tag{6.105}$$

If one formally expands the remaining part of the integrand in (6.103), i.e.

$$f(z) := \exp\left[-\frac{1}{2} \left(\Phi^{-1}(1-z) \right)^2 \right], \tag{6.106}$$

in a Taylor series at $z = \hat{z}$, one gets for (6.103)

$$c_{\mu/\mu,\lambda} = \frac{\lambda}{\mu} \frac{1}{\sqrt{2\pi}} \sum_{m=0}^{\infty} \frac{1}{m!} \left. \frac{\mathrm{d}^m f(z)}{\mathrm{d}z^m} \right|_{z=\hat{z}} \times \frac{1}{\mathrm{B}(\lambda-\mu,\mu)} \int_0^1 (z-\hat{z})^m z^{\mu-1}(1-z)^{\lambda-\mu-1} \mathrm{d}z. \tag{6.107}$$

The integral in (6.107) is a special case of the *hyper-geometric function* $\mathrm{F}(a, b; c; z)$ with the factor $1/\mathrm{B}(\lambda - \mu, \mu)$ (Abramowitz & Stegun, 1984, p. 215). It holds that

$$\mathrm{F}\left(-m, \mu; \lambda; \frac{1}{\hat{z}} \right) = \frac{1}{(-\hat{z})^m} \frac{1}{\mathrm{B}(\lambda-\mu,\mu)} \int_0^1 (z-\hat{z})^m z^{\mu-1}(1-z)^{\lambda-\mu-1} \mathrm{d}z. \tag{6.108}$$

Therefore, the series expansion of the progress coefficient $c_{\mu/\mu,\lambda}$ reads

$$c_{\mu/\mu,\lambda} = \frac{\lambda}{\mu} \frac{1}{\sqrt{2\pi}} \sum_{m=0}^{\infty} \frac{(-\hat{z})^m}{m!} \left. \frac{\mathrm{d}^m f(z)}{\mathrm{d}z^m} \right|_{z=\hat{z}} \mathrm{F}\left(-m, \mu; \lambda; \frac{1}{\hat{z}} \right). \tag{6.109}$$

In the following, it suffices to consider only the terms $m = 0$ and $m = 1$ of the series (6.109). As one can easily show by direct integration, it holds that

$$\mathrm{F}\left(0, \mu; \lambda; \frac{1}{\hat{z}} \right) = 1 \qquad \text{and} \qquad -\hat{z}\,\mathrm{F}\left(-1, \mu; \lambda; \frac{1}{\hat{z}} \right) = \frac{\mu}{\lambda} - \hat{z}. \tag{6.110}$$

Considering (C.13) and (6.105), the first derivative of $f(z)$, Eq. (6.106), at $z = \hat{z}$ gives

$$\left. \frac{\mathrm{d}f}{\mathrm{d}z} \right|_{z=\hat{z}} = \sqrt{2\pi}\, \Phi^{-1}\left(1 - \frac{\mu-1}{\lambda-2} \right). \tag{6.111}$$

After the insertion of the intermediate results (6.105, 6.106, 6.110, 6.111) into (6.109) and substitution of the μ/λ quotient by the *selection strength* $\vartheta = \mu/\lambda$ (see Definition (2.8)) one obtains

$$c_{\mu/\mu,\lambda} = \frac{1}{\vartheta} \frac{1}{\sqrt{2\pi}} \left\{ \exp\left[-\frac{1}{2} \left[\Phi^{-1}\left(1 - \vartheta \left(\frac{1-1/\mu}{1-2/\lambda} \right) \right) \right]^2 \right] + \frac{\sqrt{2\pi}}{\lambda} \left(\frac{1-2\vartheta}{1-2/\lambda} \right) \Phi^{-1}\left(1 - \vartheta \left(\frac{1-1/\mu}{1-2/\lambda} \right) \right) + \ldots \right\}. \tag{6.112}$$

This formula can serve for sufficiently large λ values ($\lambda \gtrsim 200$) and moderate selection strengths $\vartheta = \mu/\lambda$ ($0.03 \lesssim \vartheta \lesssim 0.97$) as an approximation formula for the progress coefficient $c_{\mu/\mu,\lambda}$.[19] However, its usefulness lies mainly in the approach to the *asymptotically exact* $c_{\mu/\mu,\lambda}$ formula. This formula is obtained for the limit case λ, $\mu \to \infty$ under the condition $\vartheta \neq 0$

$$\boxed{c_{\mu/\mu,\lambda} \simeq \frac{1}{\vartheta}\frac{1}{\sqrt{2\pi}} \exp\left[-\frac{1}{2}\left(\Phi^{-1}(1-\vartheta)\right)^2\right] \quad \text{with} \quad \vartheta = \frac{\mu}{\lambda}.} \tag{6.113}$$

Interestingly, this formula is also valid for $\mu = \lambda$ and for $\mu = 1$. The first case can be verified immediately by inserting $\mu = \lambda$ ($c_{\lambda/\lambda,\lambda} \equiv 0$). The validity for $\mu = 1$ can be shown by an alternative derivation of $c_{1/1,\lambda} = c_{1,\lambda}$ (see Eq. (D.20) in Appendix D.1).

In Eq. (6.113), it is remarkable that the progress coefficient $c_{\mu/\mu,\lambda}$ depends only on the selection strength $\vartheta = \mu/\lambda$ in the asymptotic limit case. A similar behavior was observed for the $c_{\mu,\lambda}$ progress coefficients (5.175) of the (μ, λ)-ES; however, no proof exists for conjecture (5.175).

Eq. (6.113) can be used for $\lambda \gtrsim 1\,000$ as a numerical approximation formula. However, the value of $\vartheta = \mu/\lambda$ should not be too near to zero. Figure 6.10 shows its approximation quality for some λ values.

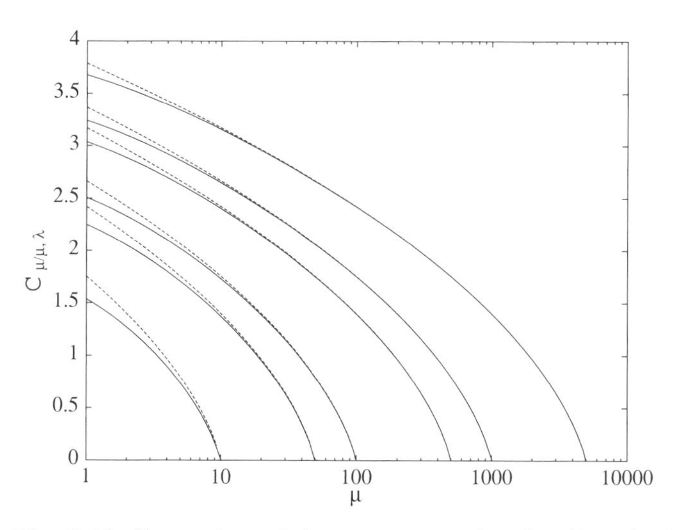

Fig. 6.10. Comparison of the $c_{\mu/\mu,\lambda}$ approximation (6.113) with the exact values obtained by numerical integration of (6.102). The cases $\lambda = 10$, 50, 100, 500, 1000, and 5000 are depicted (from bottom to top). The approximation (dashed line) overestimates the exact $c_{\mu/\mu,\lambda}$ values.

[19] One can derive better $c_{\mu/\mu,\lambda}$ expansions for which the integral vanishes for the linear term of the Taylor expansion of $f(z)$. The principal approach to such expansions will be exemplified in Appendix D.1 for the $c_{1,\lambda}$ progress coefficient.

6.3.1.2 The Asymptotic Order of the $c_{\mu/\mu,\lambda}$ Coefficients. Although the asymptotic exact formula in (6.113) allows for simpler investigations than the integral expression in (6.102), one can even go a step further and ask for the principal tendency of the $c_{\mu/\mu,\lambda}$ coefficient, i.e. the μ-, λ order.

For the derivation of the μ-, λ order, (6.113) is solved for ϑ in the brackets

$$\vartheta \simeq 1 - \Phi\left(\sqrt{-\ln\left(2\pi\vartheta^2 c^2_{\mu/\mu,\lambda}\right)}\right). \tag{6.114}$$

The product $\vartheta c_{\mu/\mu,\lambda}$ goes asymptotically to zero for $\vartheta \to 0$. This can be verified easily using Eq. (6.113). Therefore, the argument in the square root in (6.114) goes to ∞. Consequently, one has $\Phi(\cdots) \to 1$. Therefore, $\Phi(t)$ can be expanded asymptotically. Using (3.70), one obtains

$$\begin{aligned} c_{\mu/\mu,\lambda} &\simeq \sqrt{-\ln\left(2\pi\vartheta^2 c^2_{\mu/\mu,\lambda}\right)} \\ c^2_{\mu/\mu,\lambda} &\simeq -\ln(2\pi) - 2\ln\vartheta - 2\ln c_{\mu/\mu,\lambda}. \end{aligned} \tag{6.115}$$

For sufficiently large $c_{\mu/\mu,\lambda}$, the logarithm of $c_{\mu/\mu,\lambda}$ can be neglected as compared to its square, and one gets

$$c_{\mu/\mu,\lambda} \sim \sqrt{-2\ln\vartheta}, \tag{6.116}$$

or, as an order relation,

$$\boxed{c_{\mu/\mu,\lambda} = \mathcal{O}\left(\sqrt{\ln\frac{\lambda}{\mu}}\right).} \tag{6.117}$$

As a special case, the order relation for the progress coefficient $c_{1,\lambda}$ is included in (6.117). This special case was already used in Chap. 3, Eq. (3.114).

6.3.2 The Asymptotic Progress Law

The advantage of asymptotic formulae is based on their clearness and analytical manageability. These allow to make statements on the general behavior of the algorithms. One of these derivations, the $N \to \infty$ asymptotic of the progress rate formula was already carried out. Using (6.56) and (6.99), we have

$$\varphi^*_{\mu/\mu_I,\lambda}(\sigma^*) = c_{\mu/\mu,\lambda}\sigma^* - \frac{\sigma^{*2}}{2\mu} + \ldots \tag{6.118a}$$

$$\varphi^*_{\mu/\mu_D,\lambda}(\sigma^*) = \sqrt{\mu}\, c_{\mu/\mu,\lambda}\sigma^* - \frac{\sigma^{*2}}{2} + \ldots. \tag{6.118b}$$

Its applicability domain for $N < \infty$ is limited by the conditions

$$\sigma^{*2} \ll N \qquad \text{and} \qquad \mu^2 \ll N. \tag{6.119}$$

However, because of their analytical simplicity, they are perfectly suited for the investigation of the asymptotic performance properties of the $(\mu/\mu, \lambda)$-ES.

Particularly, one can analytically calculate the maximal obtainable (normalized) progress rate $\hat{\varphi}^* = \varphi^*(\hat{\sigma}^*) = \max_{\sigma^*}[\varphi^*(\sigma^*)]$. From (6.118a) one obtains

$$\hat{\varphi}^*_{\mu/\mu_I,\lambda} = \mu \frac{c^2_{\mu/\mu,\lambda}}{2} \quad \text{for} \quad \hat{\sigma}^*_{\mu/\mu_I,\lambda} = \mu c_{\mu/\mu,\lambda} \tag{6.120}$$

and from (6.118b) it follows that

$$\hat{\varphi}^*_{\mu/\mu_D,\lambda} = \mu \frac{c^2_{\mu/\mu,\lambda}}{2} \quad \text{for} \quad \hat{\sigma}^*_{\mu/\mu_D,\lambda} = \sqrt{\mu}\, c_{\mu/\mu,\lambda}. \tag{6.121}$$

As already mentioned and justified in Sect. 6.2.2.2, both recombination types show in the asymptotic limit case $N \to \infty$ the same performance

$$N \to \infty: \qquad \hat{\varphi}^*_{\mu/\mu_I,\lambda} = \hat{\varphi}^*_{\mu/\mu_D,\lambda} = \hat{\varphi}^*_{\mu/\mu,\lambda} = \mu \frac{c^2_{\mu/\mu,\lambda}}{2}. \tag{6.122}$$

Therefore, the maximal performance is a function of μ and λ. As a result, the scaling behavior should be investigated. Inserting (6.116) into (6.122), one gets

$$\hat{\varphi}^*_{\mu/\mu,\lambda} \sim \mu \ln \frac{\lambda}{\mu}. \tag{6.123}$$

In other words, for the maximal performance, one obtains

ASYMPTOTIC PROGRESS LAW OF THE $(\mu/\mu, \lambda)$-ES:

$$\boxed{\hat{\varphi}^*_{\mu/\mu,\lambda} = \mathcal{O}\left(\mu \ln \frac{\lambda}{\mu}\right).} \tag{6.124}$$

As one can see, the performance of the $(\mu/\mu, \lambda)$-ES increases logarithmically with λ. The influence of μ can be decomposed into two parts: a linear increasing term $\mu \ln \lambda$ and a nonlinear loss term $-\mu \ln \mu$. For small values of μ (λ = const), the linear term dominates. Consequently, an increase of μ results first in an increase of the performance $\hat{\varphi}^*(\mu, \lambda)$. This increase is reduced by the nonlinear loss term $-\mu \ln \mu$ if μ is increased further. Finally, this yields a decrease in the performance. For $\mu \to \lambda$, the performance goes to zero.

From this discussion, it becomes evident that there must exist an optimal μ/λ ratio for which the $(\mu/\mu, \lambda)$-ES should work with maximal efficiency. However, the question as to the optimal $\hat{\vartheta} = \mu/\lambda$ cannot be answered satisfactorily by maximizing (6.123), simply because the expression (6.116) was used for the progress coefficient $c_{\mu/\mu,\lambda}$ occurring in (6.123). This expression was obtained by neglecting some terms in Eq. (6.115). If one ignores this fact, and maximizes formula (6.123) with respect to μ, one obtains $\lambda/\mu = \mathrm{e} \approx 2.718$, and for the selection strength $\vartheta \approx 0.368$. Surprisingly, this result does not deviate much from the precise value $\hat{\vartheta} \approx 0.270$, which will be determined in the next section.

6.3.3 Fitness Efficiency and the Optimal ϑ Choice

6.3.3.1 Asymptotic Fitness Efficiency $\eta_{\mu/\mu,\lambda}$ of the $(\mu/\mu, \lambda)$-ES. The concept of fitness efficiency can be adopted directly from the N-dependent case (6.69). Taking (6.122) into account, one obtains for $N \to \infty$

$$\eta_{\mu/\mu,\lambda} = \frac{\mu}{\lambda} \frac{c_{\mu/\mu,\lambda}^2}{2}. \tag{6.125}$$

For large values of μ and λ, $c_{\mu/\mu,\lambda}$ is substituted by the asymptotically exact expression (6.113). In this way, one obtains a formula for η depending only on the selection strength ϑ

$$\eta_{\mu/\mu,\lambda}(\vartheta) = \frac{1}{4\pi} \frac{1}{\vartheta} \exp\left[- \left(\Phi^{-1}(1-\vartheta) \right)^2 \right]. \tag{6.126}$$

Figure 6.11 shows this function.

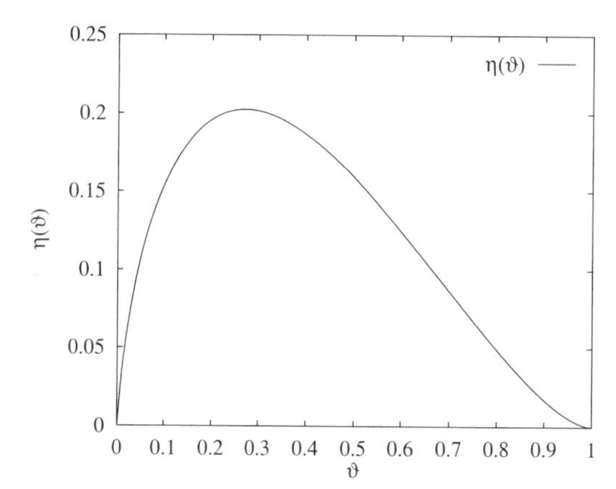

Fig. 6.11. The asymptotic fitness efficiency $\eta_{\mu/\mu,\lambda}$ as a function of the selection strength $\vartheta = \mu/\lambda$

The maximum $\hat{\eta} = \eta(\hat{\vartheta}) = \max_{\vartheta}[\eta(\vartheta)]$ can be located easily. Unfortunately, the maximum cannot be determined analytically. The numerical maximization of (6.126) gives $\hat{\eta} \approx 0.202$ at $\hat{\vartheta} \approx 0.270$. This result is astonishing since these numerical values are well known. They emerge for the $(1+1)$-ES on the sphere model as the maximal value of the normalized progress rate $\hat{\varphi}^*_{1+1} \approx 0.202$ (see Sect. 3.2.1.3, p. 69) and as the optimal success probability $P_{\text{opt}} \approx 0.27$ (see Sect. 3.3.3.3, p. 98). As astonishing as this result may be, it is *not* a coincidence. There is a deeper relation between the $(1+1)$-ES and the multirecombinative $(\mu/\mu, \lambda)$-ES.

6.3.3.2 The Relation to the $(1+1)$-ES. The derivation of the relation proceeds in two steps. The extreme value condition is applied first to the η formula (6.126). Thereafter, a similar procedure is carried out for the φ^*_{1+1} formula. The desired relation is obtained after the comparison.

The location of the maximum of η may be denoted by $\hat{\vartheta}$. For $\hat{\vartheta}$, one has $\left.\frac{\mathrm{d}\eta}{\mathrm{d}\vartheta}\right|_{\vartheta=\hat{\vartheta}} = 0$. Applying this to (6.126), one obtains

$$\frac{\mathrm{d}\eta}{\mathrm{d}\vartheta} \stackrel{!}{=} 0 \quad \Longrightarrow \quad 1 = 2\hat{\vartheta}\sqrt{2\pi}\,\Phi^{-1}(1-\hat{\vartheta})\,\exp\left[\frac{1}{2}\left(\Phi^{-1}(1-\hat{\vartheta})\right)^2\right]. \tag{6.127}$$

The substitution $x := \Phi^{-1}(1-\hat{\vartheta})$, in other words

$$\hat{\vartheta} = 1 - \Phi(x), \tag{6.128}$$

gives the nonlinear equation

$$0 = \frac{1}{\sqrt{2\pi}}\exp\left(-\frac{1}{2}x^2\right) - 2x\,[1-\Phi(x)]\,. \tag{6.129}$$

For determining the location of the maximum $\hat{\vartheta}$ explicitly, this nonlinear equation should be solved numerically; however, this is not necessary. It suffices to assume that the solution of (6.129) for $\hat{\vartheta}$ can be obtained using (6.128). Hence, the maximum $\hat{\eta}$ can be determined by the insertion in (6.126). The maximum $\hat{\eta}$ can also be presented using an alternative formula: such a formula is obtained if the exponential term in (6.126) is expressed for $\vartheta = \hat{\vartheta}$ by the exponential term in (6.127)[20]

$$\exp\left[-\frac{1}{2}\left(\Phi^{-1}(1-\hat{\vartheta})\right)^2\right] = 2\hat{\vartheta}\sqrt{2\pi}\,\Phi^{-1}(1-\hat{\vartheta})$$

$$\Longrightarrow \quad \hat{\eta}_{\mu/\mu,\lambda} = 2\left[\Phi^{-1}(1-\hat{\vartheta})\right]^2\hat{\vartheta}. \tag{6.130}$$

In the second step, we formulate the condition for the maximal progress rate (3.83) of the (1 + 1)-ES. The function $\varphi^*_{1+1}(\sigma^*)$ must be differentiated with respect to σ^* for this purpose, one obtains

$$\frac{\mathrm{d}\varphi^*}{\mathrm{d}\sigma^*} \stackrel{!}{=} 0 \quad \Longrightarrow \quad 0 = \frac{1}{\sqrt{2\pi}}\exp\left(-\frac{1}{8}\hat{\sigma}^{*2}\right) - \hat{\sigma}^*\left[1-\Phi\left(\frac{\hat{\sigma}^*}{2}\right)\right]. \tag{6.131}$$

The mutation strength that maximizes φ^*_{1+1} is denoted by $\hat{\sigma}^*$.

Comparing (6.131) with (6.129), one finds that both equations are equivalent: one just has to substitute $x = \frac{\hat{\sigma}^*}{2}$ into (6.129). Hence, using (6.128), the following relation results

$$\hat{\vartheta} = 1 - \Phi\left(\frac{\hat{\sigma}^*}{2}\right). \tag{6.132}$$

This relation will be used in the derivation later.

For a given $\hat{\sigma}^*$, one can calculate $\hat{\varphi}^*_{1+1}$ using Eq. (3.83). Alternatively, $\hat{\varphi}^*_{1+1}$ can also be expressed using (6.131), if one substitutes the exponential term occurring in (3.83)

[20] Note that Eq. (6.130) is only valid for $\vartheta = \hat{\vartheta}$.

$$\frac{1}{\sqrt{2\pi}} \exp\left(-\frac{1}{8}\hat{\sigma}^{*2}\right) = \hat{\sigma}^* \left[1 - \Phi\left(\frac{\hat{\sigma}^*}{2}\right)\right]$$
$$\Longrightarrow \quad \hat{\varphi}^*_{1+1} = \frac{1}{2}\hat{\sigma}^{*2}\left[1 - \Phi\left(\frac{\hat{\sigma}^*}{2}\right)\right]. \qquad (6.133)$$

Considering (6.132), it follows that[21]

$$\hat{\varphi}^*_{1+1} = 2\left[\Phi^{-1}(1-\hat{\vartheta})\right]^2 \hat{\vartheta}. \qquad (6.134)$$

Now we can compare (6.134) with (6.130) and obtain the remarkable result

$$\boxed{\hat{\eta}_{\mu/\mu,\lambda} = \hat{\varphi}^*_{1+1} = \eta_{1+1}.} \qquad (6.135)$$

That is, provided that $N \to \infty$ and $\lambda \to \infty$ such that $N \gg \lambda$ $(\lambda/N \to 0)$ one can state:

In the asymptotic limit, the maximal fitness efficiency of the $(\mu/\mu, \lambda)$-ES and of the $(1+1)$-ES are equal to each other.

Besides this more or less formal relation between the $(\mu/\mu, \lambda)$-ES and the $(1+1)$-ES, there is a further interesting relation. From the $(1+1)$-ES theory on the sphere model, Sect. 3.2.1.3, one knows that the *success probability* P_{s1+1} is given by Eq. (3.87), $P_s(\sigma^*) = 1 - \Phi(\sigma^*/2)$. If one inserts the optimal normalized mutation strength $\sigma^* = \hat{\sigma}^*$ into this expression, one obtains the optimal success probability $P_{opt} = P_{s1+1}(\hat{\sigma}^*)$. A comparison with (6.132) shows that this quantity is at the same time the optimal selection strength $\hat{\vartheta}$ of the $(\mu/\mu, \lambda)$-ES

$$\boxed{P_{opt} = \hat{\vartheta} = 1 - \Phi\left(\frac{\hat{\sigma}^*}{2}\right),} \qquad P_{opt} \approx 0.27027. \qquad (6.136)$$

For maximal performance of the $(1+1)$-ES on the sphere model, σ should be controlled in such a way that in the statistical mean every $1/P_{opt} \approx 1/0.27 \approx 3.7$th descendant will substitute its parent. In other words, a (one) parent has to produce on average 3.7 descendants in order to be substituted in an optimal manner (as far as the progress rate is concerned). Accordingly, the μ parents of the $(\mu/\mu, \lambda)$-ES must produce $\hat{\lambda} = \mu/\hat{\vartheta} \approx 3.7\mu$ offspring in order to perform optimally.

It seems that the $(1+1)$-ES has obtained by the $(\mu/\mu, \lambda)$-ES its natural extension into the domain of multimembered evolution strategies. It should be noted that the extension cannot be accomplished by a simple upscaling of

[21] Note that this $\hat{\varphi}^*_{1+1}$ formula must not be used for values $\vartheta \neq \hat{\vartheta}$ (cf. also footnote 20).

the $(1+1)$-ES in terms of the $(\lambda+\lambda)$-ES[22] or the $(1+\lambda)$-ES, as one might have guessed. The reason is that $\eta_{1+\lambda}$ is a monotonically decreasing function of λ (see Fig. 3.8 on p. 78). For these strategies, an increase of the number of descendants causes a sublinear increase in the maximal possible progress rate. However, the multirecombinative $(\mu/\mu, \lambda)$-ES has a linear scaling behavior in the asymptotic limit

$$\hat{\varphi}^*_{\mu/\mu,\lambda} = \lambda \hat{\eta}_{\mu/\mu,\lambda} \approx \lambda \times 0.20245. \qquad (6.137)$$

In other words, it attains a progress rate value λ times larger than that of the $(1+1)$-ES.

6.3.4 The Dynamics and the Time Complexity of the $(\mu/\mu, \lambda)$-ES

The notion of ES time complexity was introduced in Sect. 2.4.2 on p. 49. In the considerations, it is assumed that the ES algorithm operates near to its theoretical performance maximum, by using appropriate methods. These methods should adapt the mutation strength σ during the evolution process by a suitable σ control technique. These "controlling techniques," such as σ-self-adaptation (see Chap. 7), Meta-ES, 1/5th rule, etc., are not considered for the time being. Generally, one expects that these control mechanisms guarantee a $\varphi^* \approx \text{const} \approx \hat{\varphi}^* > 0$. Hence, it is sufficient to assume that

$$\varphi^*(\sigma^*) = \varepsilon\, \varphi^*(\hat{\sigma}^*) = \varepsilon\, \hat{\varphi}^* \qquad \text{with} \qquad 0 < \varepsilon \le 1 \qquad (6.138)$$

is fulfilled. Under this condition, the ES algorithms show (locally) linear convergence order, and the number of generations G for a desired relative distance change toward the optimum is given by Eq. (2.105) on p. 50. Considering the definition of the fitness efficiency and using (6.138), Eq. (2.105) yields

$$G = \ln\left(\frac{R(g_0)}{R(g)}\right) \frac{N}{\varphi^*(\sigma^*)} = \ln\left(\frac{R(g_0)}{R(g)}\right) \frac{N}{\varepsilon\, \hat{\varphi}^*} = \ln\left(\frac{R(g_0)}{R(g)}\right) \frac{N}{\varepsilon\, \lambda \eta}. \qquad (6.139)$$

In the asymptotic limit, $\eta = \eta_{\mu/\mu,\lambda}$ is a function of the selection strength ϑ, having its minimum at $\vartheta = 0$ and $\vartheta = 1$. Therefore, for $0 < \vartheta < 1$, there is always a lower bound for the fitness efficiency denoted by $\check{\eta}$ independent of λ, with $0 < \check{\eta} \le \eta_{\mu/\mu,\lambda}$.[23] Hence, the order of the ES time complexity becomes

$$(\mu/\mu, \lambda)\text{-ES TIME COMPLEXITY:} \qquad G_{\mu/\mu,\lambda} = \mathcal{O}\left(\frac{N}{\lambda}\right). \qquad (6.140)$$

[22] The analytical derivation of the progress rate of the $(\lambda+\lambda)$-ES and therefore also of the fitness efficiency is still an unsolved problem. However, it seems very plausible that for $\lambda > 1$ the inequality $\varphi^*_{\lambda+\lambda}(\sigma^*) < \varphi^*_{1+\lambda}(\sigma^*)$ should hold.

[23] Naturally, from the pragmatic point of view, one would not choose ϑ arbitrarily, as is done here. Instead, one would choose $\vartheta = \hat{\vartheta}$, so that one has $\eta = \hat{\eta}_{\mu/\mu,\lambda}$ and therefore the most fitness-efficient strategy in the application.

The ES time complexity of the $(\mu/\mu, \lambda)$-ES is linear in N and inversely proportional to the population size λ. This result can be compared with that of the $(1, \lambda)$-ES, Eq. (3.116). The $(1, \lambda)$-ES marks at the same time the lower limit for the time complexity of the (μ, λ)-ES. The principal difference can be recognized easily: for the nonrecombinative ES, the time complexity improves only with the logarithm of the population size.

Because of the scaling behavior (6.140), the $(\mu/\mu, \lambda)$-ES is really predestined for parallel computers. However, the performance gain obtained when compared to the serial computers is significant only if the computation time for the fitness evaluation and for the parts of the ES algorithm that can be parallelized (mainly the mutations) are sufficiently large compared to the communication time.

The ES time complexity G gives the number of generations necessary for a given relative distance change toward the optimum. In addition to that, one can alternatively consider the number of fitness evaluations ν^F

$$\text{FITNESS EVALUATIONS:} \qquad \nu^F = \lambda G. \tag{6.141}$$

The value for this measure is obtained for the $(1, \lambda)$-ES using (3.116) and for the $(\mu/\mu, \lambda)$-ES using (6.140)

$$\nu^F_{1,\lambda} = \mathcal{O}\left(\frac{\lambda}{\ln \lambda} N\right) \qquad \text{and} \qquad \nu^F_{\mu/\mu,\lambda} = \mathcal{O}(N). \tag{6.142}$$

As one can see, the number of necessary fitness evaluations for the multirecombinative case is of the same order as the $(1+1)$-ES. However, the order of ν^F increases sublinearly in λ for the $(1, \lambda)$-ES. In summary, it is to be stated that one cannot decrease the number of necessary fitness evaluations by using multiparent ES as compared to the $(1 + 1)$-ES. This statement holds for the local behavior of the ES on the sphere model. The relations may change, particularly if the fitness evaluation is disturbed by noise. However, a theory for the (μ, λ)-ES and the $(\mu/\mu, \lambda)$-ES operating with noisy fitness data has not been available up until now.[24]

---

[24] Recently, first results on the noisy $(\mu/\mu, \lambda)$-ES have been reported in Arnold and Beyer (2001).

7. The $(1, \lambda)$-σ-Self-Adaptation

This chapter is devoted to the analysis of the self-adaptation (SA) of the mutation strength σ on algorithms similar to the $(1, \lambda)$-ES. The basic $(1, \lambda)$-σSA algorithm is introduced and explained in the first section. The second section extends the theoretical framework to enable the analysis of the σSA. This framework includes the microscopic part, i.e. the analysis of the "driving forces" of the σSA, as well as the evolutionary dynamics. In the third section, the microscopic measures progress rate φ and self-adaptation response (SAR) ψ are determined and approximated for the case of the sphere model. Using these results, the dynamics of the self-adaptive $(1, \lambda)$-ES will be calculated in the fourth and and fifth sections. The essential aspects of the dynamics can be investigated by neglecting the statistical fluctuations. The most important results of the analysis are the scaling rule for the learning parameters and the prediction of the adaptation time. In the fifth section, these results are validated by considering the fluctuations: it will be shown that the $(1, \lambda)$-σSA-ES attains a linear convergence order if the learning parameters are chosen appropriately. The sixth section concludes this chapter and discusses the applicability of the results obtained to ES algorithms with $\mu > 1$.

7.1 Introduction

7.1.1 Concepts of σ-Control

The performance of the ES algorithms depends crucially on the value of the strategy parameter "mutation strength σ." This can be observed for *any* performance measure, i.e. both on the progress rate φ and on the quality gain $\overline{Q}$. The *dynamic* adaptation of the σ value to the local topology of the fitness landscape is a peculiarity of the ES. This peculiarity can already be found in the "oldest variant," the $(1+1)$-ES, in terms of the $1/5$th rule (see Sect. 3.3.3.3).[1,2] The $1/5$th rule presents more or less a rigid control concept. In addition to that rule, there are other methods for the adaptation of the

[1] This is a fundamental difference to the standard binary GA. The strategy parameters are in general kept constant over time in the standard GA.

[2] There are also real-coded GA variants which keep track of the local topology of the fitness landscape. Such GA can exhibit a performance behavior which is similar to ES behavior. See Deb and Beyer (1999) for further information.

mutation strength and of other (endogenous) strategy parameters. They can be classified in deterministic and evolutive approaches.

The deterministic methods have an operating mechanism similar to the 1/5th rule: they analyze the statistical features of the selected mutations in order to change the strategy parameters toward the optimum value. For instance, one can consider the isotropic normal mutations with strength σ and measure the distance that mutation vectors $\mathbf{z}$ without selection (*random walk*) would travel over G generations. This distance is expected to have the value $L = \mathrm{E}\{\|\tilde{\mathbf{y}}^{(g+G)} - \tilde{\mathbf{y}}^{(g)}\|\} \approx \sqrt{G\,N}\,\sigma$ (see Eq. 3.51). However, if the strategy parameter σ is chosen too large, the actual measured distance $L_s = \mathrm{E}\{\|\mathbf{y}_s^{(g+G)} - \mathbf{y}_s^{(g)}\|\}$ will differ significantly from L because of *selection*. Hence, the relation of the lengths L and L_s can be used for controlling the mutation strength. This idea of cumulative path-length control is used in the so-called cumulative step-size adaptation (CSA) rule by Ostermeier et al. (1995). Similarly, it is possible to detect the favored directions of the evolution in the parameter space and to perform a covariance matrix adaptation (CMA) as proposed by Hansen and Ostermeier (1996; 1997).

In contrast to these deterministic inference approaches, one can alternatively attempt to apply evolution itself to the adjustment of the optimal strategy parameter values. There are two established methods for that purpose: *self-adaptation* (Rechenberg, 1973; Schwefel, 1974) and *Meta-ES* (see, e.g., Herdy, 1992).

The concept of Meta-ES seems to be simple at first glance: one considers the whole ES algorithm optimizing in the parameter space $\mathcal{Y}$ as a function of the strategy parameters. Using this function, one generates a corresponding fitness function which is treated as the objective function in an appropriate optimization problem. Naturally, the ES algorithm can be used for this purpose. Thus one obtains hierarchically nested evolution strategies (Rechenberg, 1994). Their general hierarchical structure can be expressed compactly by the notation

$$\ldots \left\{\mu''/\rho'' \overset{+}{,} \lambda'' \left[\mu'/\rho' \overset{+}{,} \lambda' \left(\mu/\rho \overset{+}{,} \lambda\right)^{\gamma}\right]^{\gamma'}\right\} \ldots . \tag{7.1}$$

Some problems and questions immediately arise from this formal description, when trying to implement these strategies for practical applications: the parameters with the prime sign (e.g., μ', ρ', and λ') and the isolation period γ must be chosen appropriately taking the choice of the other exogenous strategy parameters into account. Moreover, it is not quite clear how to choose the fitness function for the optimization of the strategy parameters. The problem gets even more complex when one recalls the following observation: the quality gain and the actual progress rate have in general different optimal mutation strengths (see, e.g., Sect. 4.3.3). The quality gain can be measured directly; however, the actual progress toward the optimum is a "hidden measure." In principle, this fact also does not change if λ' populations are independently evolved over γ generations. A certain problem-specific knowledge is always required for getting benefits from the strategy nesting.

Meta-ES aims at the adaptation of the strategy parameters in the next higher hierarchy. Alternatively, *self-adaptation* as defined here adjusts the endogenous strategy parameters at the same hierarchical level. That is, the endogenous strategy parameters and object parameters are optimized at the same level and on the fly. Therefore, the evolution of the strategy parameters is coupled very strongly with the evolution of the object parameters. The adaptation process is by default local (in time). In other words, the local quality gain decides on the survival of the strategy parameters. This is not necessarily desirable; however, it significantly simplifies the theoretical analysis, since the (stochastic) dynamics of the system is fully determined by the system state of the offspring population at generation g. Without this condition, one obtains strategies with parameters for individual life spans (Schwefel & Rudolph, 1995). Although, performance advantages of such extended strategies have not been reported up to now, the theoretical analysis of those strategies should be much more difficult than the analysis of comma strategies.

7.1.2 The σ-Self-Adaptation

The principal idea (Rechenberg, 1973, pp. 132–137) for the σ-self-adaptation (σSA) is the *individual* coupling of evolvable strategy parameters with the object parameters.[3] In other words, every individual $\mathfrak{a} := (\mathbf{y}, \mathbf{s}, F(\mathbf{y}))$ (see Eq. (1.3)) has its own set of strategy parameters. If a descendant $\tilde{\mathfrak{a}}_l$ is selected because of its fitness $\tilde{F}_l := F(\tilde{\mathbf{y}}_l)$, not only its object parameter vector $\tilde{\mathbf{y}}_l$ survives, but also its corresponding strategy parameter set $\tilde{\mathbf{s}}_l$. Since the strategy parameters influence the statistical properties of mutation vectors $\mathbf{z}$ and since these vectors are used to generate the object parameter vector $\tilde{\mathbf{y}}_l$, one can expect (or hope) that the individuals "learn" the respective optimal strategy parameters (e.g., the mutation strength σ) during the progression of the evolution. Here, optimality is meant with respect to the quality gain, since only the fitness improvement can be evaluated by the algorithm.

The main elements of a self-adaptive ES were already included in the basic algorithm of the $(\mu/\rho \overset{+}{,} \lambda)$-ES on p. 6. However, the question of how the strategy parameters should be mutated (line 8) must be discussed now. The $(1, \lambda)$-σSA-ES with isotropic Gaussian mutations will be investigated in detail in the following. Therefore, the question reduces to the manner in which the mutation strength σ should be mutated. The answer is: *multiplicatively*, in contrast to the additive mutations of object parameters (see, e.g., Eq. (1.9)). This choice is justified by the following plausibility considerations for the sphere model.

---

[3] Note that there are also exogenous strategy parameters. Such parameters remain constant during the evolution process. For example, the number λ of the descendants is usually kept constant.

The *aim* of self-adaptation is the adjustment of the mutation strength σ in order to obtain maximal quality gain. This statement has on the sphere model (approximately) the same meaning as the maximization of the progress rate. This can be seen after comparing (4.108) with (3.101). If the fitness landscape can be approximated locally by a sphere model, one can define normalized mutation strength σ^* and progress rate φ^* (compare Eqs. (2.22), (2.92), and (2.93)) by using the local curvature radius $R = R_{\langle\varkappa\rangle}$ (see Sect. 2.3)

$$\sigma^* := \sigma \frac{N}{R} \qquad \text{and} \qquad \varphi^* := \varphi \frac{N}{R}. \tag{7.2}$$

Let us assume that the ES operates at its optimal mutation strength $\sigma^{(g)} = \hat{\sigma}$ at generation g which maximizes the quality gain, then there exists an optimal normalized mutation strength $\sigma^* = \hat{\sigma}^*$ according to (7.2). If the σSA-algorithm works correctly, the quantity $\hat{\sigma}^*$ should be realized by appropriate "adjustment" of σ at generation $g+1$. In other words, for each generation

$$\hat{\sigma}^* = \sigma^{(g)} \frac{N}{R^{(g)}} = \sigma^{(g+1)} \frac{N}{R^{(g+1)}} \tag{7.3}$$

should be fulfilled. Since the expected value of the local radius changes from generation g to $g+1$ according to (see the definition of φ)

$$\varphi = R^{(g)} - R^{(g+1)}, \tag{7.4}$$

one obtains using (7.3) and (7.2)

$$\sigma^{(g+1)} = \sigma^{(g)} \frac{R^{(g+1)}}{R^{(g)}} = \sigma^{(g)} \left(1 - \frac{\varphi^*(\hat{\sigma}^*)}{N}\right). \tag{7.5}$$

Assuming that one has an optimally adjusted mutation strength at generation g, the optimal mutation strength at generation $g+1$ is obtained by *multiplying* $\sigma^{(g)}$ by a constant factor.

In practice, the sphere model condition is usually not fulfilled. Neither can one assume that the mutation strength is optimal at generation g. Therefore, the controlling rule (7.5) cannot be applied directly. However, (7.5) reflects the discovery that the mutation strength should be changed multiplicatively. This guarantees a certain scaling invariance of the mutation operator: the absolute change of the mutation strength does not play a role, only the *relative* change counts. For this reason, the mutation of mutation strength is realized by the multiplication of the parental $\sigma^{(g)}$ values with random numbers ξ

$$\tilde{\sigma}_l := \xi_l \sigma^{(g)}, \qquad l = 1 \ldots \lambda. \tag{7.6}$$

The statistical properties necessary for ξ will be investigated further below. Naturally, not all arbitrary distributions $p(\xi)$ are appropriate for that purpose. However, one can immediately conclude from (7.5) that

$$\mathrm{E}\{\xi\} \approx 1 \tag{7.7}$$

should be fulfilled roughly, since $|\varphi^*(\hat{\sigma}^*)| \le \max_{\sigma^*}[\varphi^*(\sigma^*)] \ll N$ is valid for φ^* in general, as long as σ^* is sufficiently small.

7.1.3 The $(1, \lambda)$-σSA Algorithm

Using these results, the general algorithm (Algorithm 2, p. 6) can be substantiated for the special case of the $(1, \lambda)$-ES (Algorithm 4).

Procedure $(1,\lambda)$-σSA-ES;	line
Begin	1
$g := 0;$	2
initialize$\left(\mathbf{y}^{(0)}, \sigma^{(0)}\right)$	3
Repeat	4
For $l := 1$ **To** λ **Do Begin**	5
$\tilde{\sigma}_l := \xi \sigma^{(g)};$	6
$\tilde{\mathbf{y}}_l := \mathbf{y}^{(g)} + \tilde{\sigma}_l \left(\mathcal{N}(0,1), \ldots, \mathcal{N}(0,1)\right)^{\mathrm{T}};$	7
$\tilde{F}_l := F(\tilde{\mathbf{y}}_l)$	8
End;	9
$l_p :=$ selection$_{1;\lambda}\left(\tilde{F}_1, \tilde{F}_2, \ldots, \tilde{F}_\lambda\right);$	10
$\sigma^{(g+1)} := \tilde{\sigma}_{l_p};$	11
$\mathbf{y}^{(g+1)} := \tilde{\mathbf{y}}_{l_p};$	12
$g := g + 1;$	13
Until stop_criterion	14
End	15

Algorithm 4. The $(1, \lambda)$-ES with σ-self-adaptation

In SA Algorithm 4, it is important to note that the strategy parameter is mutated *first*. Therefore, in line 6, the parental mutation strength σ is mutated according to Eq. (7.6) (see Sect. 7.1.4 for the realization). The new mutation strength obtained in this way is used in line 7 as the standard deviation for the mutation operator in the procreation of the lth descendant: its object parameter vector $\tilde{\mathbf{y}}_l$ is generated from the parental object parameter vector $\mathbf{y}^{(g)}$. The $(1, \lambda)$-selection operator follows in line 10. It uses only the fitness values of the descendants and returns the number l_p of the descendant with the best fitness. This descendant becomes the parent of the next generation.

7.1.4 Operators for the Mutation of the Mutation Strength

The Condition (7.7) on the random variable ξ is not very restrictive. Nevertheless, only a few probability densities have been applied for ξ. One can differentiate in general between continuous and discrete types of distributions for the mutation operator.

Probably the best known mutation operator of the continuous type is the *log-normal operator* (Schwefel, 1974). The random number ξ is obtained by an exponential transformation of a $\mathcal{N}(0, \tau^2)$ normally distributed random number

$$\boxed{\text{LOG-NORMAL OPERATOR:} \qquad \tilde{\sigma} := \xi\sigma, \quad \xi := \mathrm{e}^{\tau\mathcal{N}(0,1)}.} \tag{7.8}$$

The symbol τ is called the *learning parameter*. It is an *exogenous* strategy parameter. The correct choice of this parameter influences the adaptation behavior of the σSA. The influence of τ on the ES-dynamics will be investigated in this chapter in detail. The choice of the learning parameter τ is usually in the interval $0 < \tau \lessapprox 1$.

By applying the mutation operator to a given parental mutation strength σ, an offspring mutation strength $\tilde{\sigma}$ is generated which itself is a random variate. If the mutation rule (7.8) is used, this variate is log-normally distributed. This property is also reflected in the name of the operator. The probability density of $\tilde{\sigma}$ is in this case

$$p_\sigma(\tilde{\sigma}) = \frac{1}{\sqrt{2\pi}\tau} \frac{1}{\tilde{\sigma}} \exp\left[-\frac{1}{2}\left(\frac{\ln(\tilde{\sigma}/\sigma)}{\tau}\right)^2\right] =: p_{\ln}(\tilde{\sigma}). \tag{7.9}$$

In a special variant of *evolutionary programming*, called meta-EP (Fogel, 1992), the following mutation operator is used

$$\boxed{\text{META-EP-OPERATOR:} \qquad \tilde{\sigma} := \xi\sigma, \quad \xi := 1 + \tau\mathcal{N}(0,1).} \tag{7.10}$$

As result, given a parental σ the descendants $\tilde{\sigma}$ are normally distributed. The density reads

$$p_\sigma(\tilde{\sigma}) = \frac{1}{\sqrt{2\pi}\tau} \frac{1}{\sigma} \exp\left[-\frac{1}{2}\left(\frac{\tilde{\sigma}/\sigma - 1}{\tau}\right)^2\right] =: p_{\mathcal{N}}(\tilde{\sigma}). \tag{7.11}$$

Interestingly, this mutation rule of the meta-EP can be considered as a limit case of the log-normal rule – at least for small learning parameters τ. In this case, the exponential function of (7.8) can be expanded in a Taylor series. By cutting after the linear term, one obtains (7.10). Therefore, both mutation operators should yield a similar self-adaptation dynamics for small τ.

The simplest mutation operator of discrete type generates σ-descendants using a symmetric two-point distribution according to[4]

$$p_\sigma(\tilde{\sigma}) = \frac{1}{2}\,[\delta(\tilde{\sigma} - \alpha\sigma) \,+\, \delta(\tilde{\sigma} - \sigma/\alpha)] =: p_{\mathrm{II}}(\tilde{\sigma}), \tag{7.12}$$

$$\alpha = 1 + \beta, \qquad \beta > 0. \tag{7.13}$$

The parameter β is an exogenous strategy parameter. It is similar to the learning parameter τ of the log-normal rule (7.8) and also is of the same

[4] Dirac's delta function is used here. See p. XII for its definition.

order of magnitude. An implementation of this two-point operator can be found in Rechenberg (1994, p. 47), using (0, 1]-uniformly distributed random numbers u(0, 1]

$$\textsc{Two-Point Operator:} \quad \tilde{\sigma} := \begin{cases} \sigma\,(1+\beta), & \text{if} \quad \mathrm{u}(0,1] \le 1/2 \\ \sigma/(1+\beta), & \text{if} \quad \mathrm{u}(0,1] > 1/2. \end{cases} \tag{7.14}$$

From the theoretical point of view, the symmetric two-point distribution (7.13) can be generalized (Beyer, 1995d, 1996)

$$\begin{aligned} p_\sigma(\tilde{\sigma}) &= (1-p_-)\,\delta(\tilde{\sigma} - (1+\beta_+)\sigma) \; + \; p_-\delta(\tilde{\sigma} - (1-\beta_-)\sigma) \\ &=: p_\beta(\tilde{\sigma}). \end{aligned} \tag{7.15}$$

The density proposed has three exogenous strategy parameters

$$\beta_+ > 0, \qquad \beta_- > 0, \qquad \text{and} \qquad 0 < p_- < 1. \tag{7.16}$$

The aim is to influence the dynamic behavior of self-adaptation more strongly. A detailed experimental and/or theoretical investigation of this distribution has not been done yet. However, the analysis tools to be developed in this chapter can principally be used without any change for the case (7.15).

In addition to the (discrete) two-point operator, one can also construct generalizations toward K-point operators. For instance, the symmetric 3-point variant (Muth, 1982) increases the parental σ with a probability $p_- = 1/3$ by a factor of α or decreases with $p_- = 1/3$ by a factor of $1/\alpha$. The remaining descendants receive the parental σ. Instead of the density (7.12), the function

$$p_\sigma(\tilde{\sigma}) = \frac{1}{3}\;[\delta(\tilde{\sigma} - \alpha\sigma) \; + \; \delta(\tilde{\sigma} - \sigma) \; + \; \delta(\tilde{\sigma} - \sigma/\alpha)] =: p_{\text{III}}(\tilde{\sigma}) \tag{7.17}$$

is to be used in the theoretical analysis.

In the following analysis, the microscopic aspects of the $(1,\lambda)$-σSA-ES will be investigated for the log-normal operator and the two-point operator. A very remarkable result of the analysis will be that the effect of both operators is the same in the microscopic picture if the learning parameters τ and α are chosen appropriately. As a consequence, the essentials of their ES-dynamics should also be similar.

7.2 Theoretical Framework for the Analysis of the σSA

Theoretical investigations on the functioning mechanisms of the $(1,\lambda)$-σ-self-adaptation were carried out first for the case of the log-normal operator (7.8) by Schwefel (1977) and for the two-point operator (7.14) by Scheel (1985).

Both of these works reflect certain aspects of self-adaptation correctly.[5] However, a *systematic* approach is missing in both of them. Therefore, one cannot evaluate those results – mostly obtained by arguing on the plausibility – on a unified basis.

An extension of the theoretical framework is necessary for the analysis of the σ-self-adaptation: the evolution process of this algorithm does not occur in the parameter space $\mathcal{Y}$ of the object parameters $\mathbf{y}$ only. Instead, it is closely coupled with the evolution of the strategy parameters $\mathbf{s}$ in the space $\mathcal{S}$. Moreover, it will be shown in the σSA case that the evolution in the space $\mathcal{Y}$ will be *governed* by the evolution in the space of strategy parameter σ.

If one considers the ES algorithm as a *dynamical system*, the analysis can be divided into two parts. The first one models the dynamics of the evolution process: the *relevant* state variables must be extracted and the equations of motion must be derived. Using these, the "macroscopic behavior" of the ES over generations can be determined. The second part asks for the *cause* of the evolution, i.e. for the *driving forces* in the equations of motion. This is the "microscopic" aspect of the analysis.

In this section, the stochastic equations of motion are formulated first. Thereafter, the driving microscopic forces in them are extracted. At this point, the transition probability densities must be determined. These densities describe the stochastic system behavior on the microscopic scale (i.e. from generation g to $g+1$). These are complicated functions and integrals; therefore, their system dynamics are impossible to solve analytically. By applying some asymptotically exact simplifications, the most important aspects of the σSA-ES can be approximated. This leads us to the central quantities called *progress rate* and *self-adaptation response* (SAR). Their approximate determination will be performed in Sect. 7.3. Based on these approximations, the ES dynamics will be determined in Sects. 7.4 and 7.5.

7.2.1 The Evolutionary Dynamics of the $(1, \lambda)$-σSA-ES

By assuming the local sphere model approximation, the state of the $(1, \lambda)$-ES algorithm is uniquely defined by determining the local curvature radius and parental mutation strength ς at generation g. Note that the symbols $(\varsigma, r)^{\mathrm{T}}$ are used instead of $(\sigma, R)^{\mathrm{T}}$ to underline the stochastic nature of the evolutionary dynamics. The symbols $(\sigma, R)^{\mathrm{T}}$ will be reserved in the following to signify the *mean value dynamics*, and the random variables ς and r will describe the evolution in the "space of probabilities."

One can interpret the ES dynamics as a stochastic mapping or in general as a Markovian process that maps the random variables $(\varsigma^{(g)}, r^{(g)})^{\mathrm{T}}$ at generation g to the random variables $(\varsigma^{(g+1)}, r^{(g+1)})^{\mathrm{T}}$ at generation $g+1$

[5] For example, Schwefel (1974) found the $\tau \propto 1/\sqrt{N}$ scaling law by using plausibility arguments. Scheel (1985, p. 65) was able to show that the theoretically attainable maximal progress rate decreases with the increase of the learning parameter α.

$$\begin{bmatrix} \varsigma^{(g)} \\ r^{(g)} \end{bmatrix} \quad \mapsto \quad \begin{bmatrix} \varsigma^{(g+1)} \\ r^{(g+1)} \end{bmatrix}. \tag{7.18}$$

The dynamics of the random variables $(\varsigma, r)^{\mathrm{T}}$ is described by so-called Chapman-Kolmogorov equations (see, e.g., Fisz, 1971). However, these integral equations will not be formulated here explicitly: beyond the formal formulation of these equations, no useful information can be obtained from them as long as the transition densities are unknown or not approximated appropriately. For example, it is not even possible to specify an analytically closed expression for the transition density of ς. Therefore, a "step-by-step approach" is preferred here. The aim is to simplify the evolution equations by extracting the important properties of the σSA process successively. At the end, it will be possible to treat the results analytically. The Chapman-Kolmogorov equations will be established and solved approximately in Sect. 7.5.

7.2.1.1 The r Evolution. One can denote the parental state of the ES system at generation g as $(\varsigma^{(g)}, r^{(g)})^{\mathrm{T}}$. The transition to a new parental r state at generation $g+1$ can be described by the evolution equation

$$r^{(g+1)} = r^{(g)} - \varphi(\varsigma^{(g)}, r^{(g)}) + \epsilon_R(\varsigma^{(g)}, r^{(g)}). \tag{7.19}$$

The progress rate φ describes the expected change in r conditional on $(\varsigma^{(g)}, r^{(g)})^{\mathrm{T}}$

$$\varphi(\varsigma^{(g)}, r^{(g)}) := \mathrm{E}\left\{ r^{(g)} - r^{(g+1)} \mid \varsigma^{(g)}, r^{(g)} \right\}, \tag{7.20}$$

and ϵ_R is a fluctuation term with the necessary condition

$$\mathrm{E}\left\{ \epsilon_R(\varsigma^{(g)}, r^{(g)}) \right\} = 0. \tag{7.21}$$

The term ϵ_R reflects the stochastic part of the r evolution. The density of the ϵ_R fluctuations can be determined using the conditional r transition density $p_{1;\lambda}(r)$, which will be derived below (see also Sect. 7.2.2.2). However, for the considerations in Sect. 7.5, an approximation is used: the quantity ϵ_R is approximated by a Gaussian random variate

$$\epsilon_R = D_\varphi(\varsigma^{(g)}, r^{(g)})\, \mathcal{N}(0,1) + \ldots. \tag{7.22}$$

Using (7.19) and (7.21), the conditional variance $D_\varphi^2 = \mathrm{E}\{\epsilon_R^2 \,|\, \varsigma^{(g)}, r^{(g)}\}$ becomes

$$D_\varphi^2 = \mathrm{E}\left\{ \left[-(r^{(g)} - r^{(g+1)}) + \varphi \right]^2 \right\}$$

$$D_\varphi^2 = \mathrm{E}\left\{ (r^{(g)} - r^{(g+1)})^2 - 2\,(r^{(g)} - r^{g+1})\,\varphi + \varphi^2 \right\}$$

$$D_\varphi^2 = \mathrm{E}\left\{ (r^{(g)} - r^{(g+1)})^2 \mid \varsigma^{(g)}, r^{(g)} \right\} - \left(\varphi(\varsigma^{(g)}, r^{(g)}) \right)^2. \tag{7.23}$$

The approximation assumption (7.22) is not very restrictive, since we are mainly interested in the mean value dynamics $R(g) := \overline{r^{(g)}}$. According to

Condition (7.21), it is independent of the ϵ_R fluctuations in the first order. The influence of ϵ_R back to $R(g)$ is only via the ς dynamics.

A main aim of the "microscopic" part of the theory is the determination of the expected values in (7.20) and in (7.23). After introducing the definition of *higher-order progress rates*[6] $\varphi^{(k)}$

$$\varphi^{(k)}(\varsigma^{(g)}, r^{(g)}) := \mathrm{E}\left\{(r^{(g)} - r^{(g+1)})^k \mid \varsigma^{(g)}, r^{(g)}\right\}$$
$$= \int_{r=0}^{\infty} (r^{(g)} - r)^k \, p_{1;\lambda}(r \mid \varsigma^{(g)}, r^{(g)}) \, \mathrm{d}r, \qquad (7.24)$$

(7.19) and (7.23) can be expressed in terms of

$$r^{(g+1)} = r^{(g)} - \varphi^{(1)}(\varsigma^{(g)}, r^{(g)}) + \epsilon_R(\varsigma^{(g)}, r^{(g)}) \qquad (7.25)$$

and

$$D_\varphi^2 = \varphi^{(2)}(\varsigma^{(g)}, r^{(g)}) - \left(\varphi(\varsigma^{(g)}, r^{(g)})\right)^2. \qquad (7.26)$$

The transition density $p_{1;\lambda}(r \mid \varsigma^{(g)}, r^{(g)})$ will be derived in Sect. 7.2.2.2. Later on it will be shown that $p_{1;\lambda}$ can also be determined for normalized quantities, similar to (7.2). In order to generalize (7.2), normalized quantities are introduced.

$$\begin{aligned} \varphi^{*(g)} &:= \varphi^{(g)} \frac{N}{r^{(g)}}, \qquad & D_\varphi^* &:= D_\varphi \frac{N}{r^{(g)}} \\ \varsigma^{*(g)} &:= \varsigma^{(g)} \frac{N}{r^{(g)}}, \qquad & \epsilon_R^* &:= \epsilon_R \frac{N}{r^{(g)}}. \end{aligned} \qquad (7.27)$$

If one applies the normalization to the evolution equation (7.19), one obtains

$$\boxed{r^{(g+1)} = r^{(g)} \left(1 - \frac{1}{N} \varphi^*(\varsigma^{(g)})\right) + \frac{r^{(g)}}{N} D_\varphi^*(\varsigma^{(g)}) \, \mathcal{N}(0,1).} \qquad (7.28)$$

This equation is the starting point for the derivations on the R dynamics in Sects. 7.4 and 7.5.

7.2.1.2 The ς Evolution. The ς evolution can be expressed by the decomposition[7]

$$\varsigma^{(g+1)} := \varsigma^{(g)} + \varsigma^{(g)} \, \psi(\varsigma^{(g)}, r^{(g)}) + \epsilon_\sigma(\varsigma^{(g)}, r^{(g)}). \qquad (7.29)$$

As in the case of the r evolution in Eq. (7.19), it is required that the ϵ_σ term represents the stochastic part with the expected value zero

[6] The exponent "(k)" should not be mixed up with the generation counter "(g)." For $k = 1$, one obtains the usual progress rate. To simplify the notation we agree on $\varphi := \varphi^{(1)}$.

[7] This decomposition reflects that the mutation of the σ parameter is performed multiplicatively.

$$\mathrm{E}\left\{\epsilon_\sigma \mid \varsigma^{(g)}, r^{(g)}\right\} = 0. \tag{7.30}$$

Therefore, the function[8] ψ presents the expected relative change in ς

$$\psi(\varsigma^{(g)}, r^{(g)}) := \mathrm{E}\left\{ \frac{\varsigma^{(g+1)} - \varsigma^{(g)}}{\varsigma^{(g)}} \;\Big|\; \varsigma^{(g)}, r^{(g)} \right\}. \tag{7.31}$$

We will call this function the *self-adaptation response*, or in short "SAR." The generalization $\psi^{(k)}$ of SAR functions of higher-order reads

$$\psi^{(k)}(\varsigma^{(g)}, r^{(g)}) := \int_{\varsigma=0}^{\infty} \left(\frac{\varsigma - \varsigma^{(g)}}{\varsigma^{(g)}} \right)^k p_{1;\lambda}(\varsigma \mid \varsigma^{(g)}, r^{(g)})\, \mathrm{d}\varsigma. \tag{7.32}$$

The transition density $p_{1;\lambda}$ is used in the integral (7.32). It will be derived in Sect. 7.2.2.3. It is interesting and important to notice that the SAR functions are invariant with respect to the normalization (7.27). As will be shown in Sect. 7.2.2.4, $\psi^{(k)}$ depends only on the normalized mutation strength $\varsigma^{*(g)}$

$$\psi^{(k)}(\varsigma^{(g)}, r^{(g)}) = \psi^{(k)}(\varsigma^{*(g)}, N). \tag{7.33}$$

As in the ϵ_R case, the fluctuation term ϵ_σ will be approximated by a Gaussian distribution

$$\epsilon_\sigma = D_\sigma(\varsigma^{(g)}, r^{(g)})\, \mathcal{N}(0,1) \;+\; \ldots . \tag{7.34}$$

One obtains the variance $D_\sigma^2 = \mathrm{E}\left\{\epsilon_\sigma^2 \mid \varsigma^{(g)}, r^{(g)}\right\}$ considering (7.29)

$$D_\sigma^2 = \mathrm{E}\left\{ \left[(\varsigma^{(g+1)} - \varsigma^{(g)}) - \varsigma^{(g)}\psi^{(1)} \right]^2 \right\}$$

$$D_\sigma^2 = \mathrm{E}\left\{ (\varsigma^{(g+1)} - \varsigma^{(g)})^2 - 2\,(\varsigma^{(g+1)} - \varsigma^{(g)})\,\varsigma^{(g)}\psi^{(1)} + (\varsigma^{(g)})^2(\psi^{(1)})^2 \right\}$$

$$D_\sigma^2 = (\varsigma^{(g)})^2\, \mathrm{E}\left\{ \left(\frac{\varsigma^{(g+1)} - \varsigma^{(g)}}{\varsigma^{(g)}} \right)^2 \right\} \;-\; (\varsigma^{(g)})^2(\psi^{(1)})^2. \tag{7.35}$$

Hence the standard deviation D_σ follows taking Eq. (7.32) into account and $\psi = \psi^{(1)}$

$$D_\sigma = \varsigma^{(g)} D_\psi \qquad \text{with} \qquad D_\psi := \sqrt{\psi^{(2)}(\varsigma^{*(g)}) - (\psi(\varsigma^{*(g)}))^2}. \tag{7.36}$$

Equation (7.29) describes the evolution of the mutation strength ς. However, using the normalized mutation strength ς^* is more suitable for the following considerations. The results obtained in this manner will be independent of the r state.

The next step is to obtain the equation of motion of the normalized mutation strength. To this end, the left-hand side of (7.29) is expressed according

[8] Note that the symbol δ was used in Beyer (1995d, 1996), instead of ψ. However, here the symbol ψ will be used, firstly because of its "lexicographic neighborhood" to φ, and secondly to avoid confusion with Dirac's delta function.

to the normalization equations in (7.27) by $\varsigma^{*(g+1)}$. Thereafter, $r^{(g+1)}$ is substituted by (7.28). Thus, one gets for the left-hand side of (7.29)

$$\varsigma^{(g+1)} = \varsigma^{*(g+1)} \frac{r^{(g+1)}}{N} = \varsigma^{*(g+1)} \frac{r^{(g)}}{N} \left[\left(1 - \frac{\varphi^*}{N}\right) + \frac{D_\varphi^*}{N} \mathcal{N}(0,1) \right]. \quad (7.37)$$

For the right-hand side of (7.29), considering (7.34, 7.36) and the normalization (7.27), one obtains

$$\varsigma^{(g+1)} = \varsigma^{*(g)} \frac{r^{(g)}}{N} \left[(1+\psi) + D_\psi \mathcal{N}(0,1) \right]. \quad (7.38)$$

After equating (7.37) and (7.38) and reordering terms for $\varsigma^{*(g+1)}$, the equation of the ς^* dynamics reads[9]

$$\boxed{\varsigma^{*(g+1)} = \varsigma^{*(g)} \frac{1 + \psi(\varsigma^{*(g)}, N) + D_\psi(\varsigma^{*(g)}, N) \mathcal{N}_\psi(0,1)}{1 - \frac{1}{N}\varphi^*(\varsigma^{*(g)}, N) + \frac{1}{N} D_\varphi^*(\varsigma^{*(g)}, N) \mathcal{N}_\varphi(0,1)}.} \quad (7.39)$$

This is a very remarkable result. It shows that the evolution of ς^* is *independent* of the r evolution. Equation (7.39) is decoupled from Eq. (7.28). In contrast, the evolution of r (7.28) is *controlled* by the evolution of ς^* (7.39). As a result, the investigation of the dynamics of self-adaptation is simplified considerably: the dynamics of ς^* can be treated independently.

Assuming that $|\varphi^*|/N$ and D_φ^*/N are small compared to 1, Eq. (7.39) can be simplified. This condition is usually satisfied if ς^* is sufficiently small, i.e. if $\varphi^*(\varsigma^*) \gtrapprox 0$ is valid. Whereas the case of larger ς^* values can only be observed for a few generations: if ς^* is initially chosen too large, it results in a large negative φ^* value. Hence, the denominator of (7.39) becomes large accordingly. As a result, ς^* is reduced in a few generations to values fulfilling the condition on smallness. A more precise discussion will follow in Sect. 7.4.2.4.

For sufficiently small $|\varphi^*|/N$, the denominator of (7.39) can be expanded in a Taylor series

$$\varsigma^{*(g+1)} = \varsigma^{*(g)} \left[1 + \psi(\varsigma^{*(g)}) + D_\psi \mathcal{N}_\psi\right] \left(1 + \frac{\varphi^*(\varsigma^{*(g)})}{N} - \frac{D_\varphi^*}{N} \mathcal{N}_\varphi\right) + \ldots . \quad (7.40)$$

In Sects. 7.4 and 7.5, Eqs. (7.39) and (7.40) will be investigated further.

7.2.2 The Microscopic Aspects

This subsection aims at the determination of the transition densities of the stochastic mappings $r^{(g)} \mapsto r^{(g+1)}$ and $\varsigma^{(g)} \mapsto \varsigma^{(g+1)}$. Furthermore, the integral expressions for the moments $\varphi^{(k)}$ and $\psi^{(k)}$ are to be derived. The r-density $p_{1;1}(r)$ of a single descendant must be derived first for this purpose.

[9] In order to differentiate between the two independent stochastic processes in (7.39), the indices ψ and φ are used for the random variables $\mathcal{N}$.

7.2.2.1 The Density $p_{1;1}(r)$ of a Single Descendant. The way of generating a single descendant according to Algorithm 4 should be recalled here. The ES system is at the parental state $(\varsigma^{(g)}, r^{(g)})^{\mathrm{T}}$. In line 6, $\varsigma^{(g)}$ is mutated first. This occurs with the conditional probability density $p_\sigma(\varsigma \mid \varsigma^{(g)})$, which depends on the σ mutation operator chosen. Some variants of the mutation operator were given in (7.9, 7.11, 7.12, 7.15, 7.17).[10] The ς value obtained serves as the mutation strength for the mutation of object parameters in line 7. This mutation changes the parental state $r^{(g)}$ into the new offspring state r. The conditional probability density $p_r(r \mid \varsigma, r^{(g)})$ for the new r state was already derived in Sect. 3.4. We will use the normal approximation of Sect. 3.4.5, Eqs. (3.245), (3.246), with the symbols of this chapter; the density reads

$$p_r(r \mid \varsigma, r^{(g)}) = \frac{1}{\sqrt{2\pi}\,\tilde{\varsigma}(\varsigma)} \exp\left[-\frac{1}{2}\left(\frac{r - \sqrt{(r^{(g)})^2 + \varsigma^2 N}}{\tilde{\varsigma}(\varsigma)}\right)^2\right] \tag{7.41}$$

with

$$\tilde{\varsigma}(\varsigma) := \varsigma \sqrt{\frac{(r^{(g)})^2 + \frac{\varsigma^2 N}{2}}{(r^{(g)})^2 + \varsigma^2 N}}. \tag{7.42}$$

Eq. (7.41) is a density function with parameter ς which appears with conditional probability $p_\sigma(\varsigma \mid \varsigma^{(g)})\,\mathrm{d}\varsigma$. After integrating over all possible ς values, one obtains the conditional probability density of state r of a single descendant

$$p_{1;1}(r \mid \varsigma^{(g)}, r^{(g)}) = \int_{\varsigma=0}^{\infty} p_r(r \mid \varsigma, r^{(g)})\, p_\sigma(\varsigma \mid \varsigma^{(g)})\,\mathrm{d}\varsigma. \tag{7.43}$$

The conditional distribution function $P_{1;1}(r) = P_{1;1}(r \,|\, \varsigma^{(g)}, r^{(g)})$ can be obtained by integrating over r

$$P_{1;1}(r) := \Pr\Big(\|\tilde{\mathbf{y}}_l\| < r \mid \varsigma^{(g)}, r^{(g)}\Big) = \int_{\tilde{r}=0}^{\tilde{r}=r} p_{1;1}(\tilde{r} \mid \varsigma^{(g)}, r^{(g)})\,\mathrm{d}\tilde{r}. \tag{7.44}$$

Using (7.41) and (7.43), one gets for (7.44)

$$\begin{aligned} P_{1;1}(r) &= P_{1;1}(r \,|\, \varsigma^{(g)}, r^{(g)}) \\ &= \int_{\varsigma=0}^{\infty} \Phi\left(\frac{r - \sqrt{(r^{(g)})^2 + \varsigma^2 N}}{\tilde{\varsigma}(\varsigma)}\right) p_\sigma(\varsigma \mid \varsigma^{(g)})\,\mathrm{d}\varsigma. \end{aligned} \tag{7.45}$$

For the integration over r it was assumed – as usual – that the lower integration limit can be changed such that $\Phi\Big(-\sqrt{(r^{(g)})^2 + \varsigma^2 N}\Big/\tilde{\varsigma}(\varsigma)\Big) \to 0$.

Equation (7.45) can also be expressed using the normalized ς variable $\varsigma^* := \varsigma N / r^{(g)}$ (7.27). One obtains after the substitution

[10] One has to set $\tilde{\sigma} = \tilde{\varsigma}$ and $\sigma = \varsigma^{(g)}$ in this step.

$$P_{1;1}(r) = P_{1;1}\left(r \,|\, \varsigma^{*(g)}, r^{(g)}\right) = \int_{\varsigma^*=0}^{\infty} \Phi\left(N \frac{r/r^{(g)} - \sqrt{1+\varsigma^{*2}/N}}{\tilde{\varsigma}^*(\varsigma^*)}\right) p_\sigma^*(\varsigma^* \mid \varsigma^{*(g)})\, \mathrm{d}\varsigma^*. \qquad (7.46)$$

The quantity $\tilde{\varsigma}^*$ is obtained using (7.42)

$$\tilde{\varsigma}^*(\varsigma^*) = \varsigma^* \sqrt{\frac{1+\varsigma^{*2}/2N}{1+\varsigma^{*2}/N}}. \qquad (7.47)$$

The density $p_\sigma^*(\varsigma^* \mid \varsigma^{*(g)})$ in (7.46) reads for the cases (7.9, 7.11, 7.12, 7.15) and (7.17) in Sect. 7.1.4

$$p_{\mathrm{ln}}^*(\varsigma^*) := \frac{1}{\sqrt{2\pi}\,\tau} \frac{1}{\varsigma^*} \exp\left[-\frac{1}{2}\left(\frac{\ln(\varsigma^*/\varsigma^{*(g)})}{\tau}\right)^2\right], \qquad (7.48)$$

$$p_{\mathcal{N}}^*(\varsigma^*) := \frac{1}{\sqrt{2\pi}\,\tau} \frac{1}{\varsigma^{*(g)}} \exp\left[-\frac{1}{2}\left(\frac{\varsigma^*/\varsigma^{*(g)} - 1}{\tau}\right)^2\right], \qquad (7.49)$$

$$p_{\mathrm{II}}^*(\varsigma^*) := \frac{1}{2}\left[\delta\left(\varsigma^* - \alpha\varsigma^{*(g)}\right) + \delta\left(\varsigma^* - \varsigma^{*(g)}/\alpha\right)\right], \qquad (7.50)$$

$$p_\beta^*(\varsigma^*) := (1-p_-)\ \delta\left(\varsigma^* - (1+\beta_+)\varsigma^{*(g)}\right) + p_-\ \delta\left(\varsigma^* - (1-\beta_-)\varsigma^{*(g)}\right), \qquad (7.51)$$

$$p_{\mathrm{III}}^*(\varsigma^*) := \frac{1}{3}\left[\delta\left(\varsigma^* - \alpha\varsigma^{*(g)}\right) + \delta\left(\varsigma^* - \varsigma^{*(g)}\right) + \delta\left(\varsigma^* - \varsigma^{*(g)}/\alpha\right)\right]. \qquad (7.52)$$

7.2.2.2 The Transition Density $p_{1;\lambda}(r)$. The density $p_{1;1}(r)$ of the descendants has been determined in the previous section. Based on that result, the transition density $p_{1;\lambda}(r) := p_{1;\lambda}(r \,|\, \varsigma^{(g)}, r^{(g)})$ required in (7.24) can be expressed now. The derivation has already been offered in a similar manner in Chap. 3. As a result, (3.216) can be taken here immediately

$$p_{1;\lambda}(r) := p_{1;\lambda}(r \mid \varsigma^{(g)}, r^{(g)}) = \lambda\, p_{1;1}(r)\, [1 - P_{1;1}(r)]^{\lambda-1}. \qquad (7.53)$$

However, $p_{1;1}(r)$ and $P_{1;1}(r)$ are given by (7.43) and (7.45, 7.46) as integrals, respectively. These integrals can be evaluated analytically in exceptional cases only. Therefore, there is in general no analytically closed expression for the transition density (7.53). Exceptions are the discrete σ mutation operators. Therefore, it is possible to specify the transition densities $p_{1;\lambda}(r)$ for the cases (7.50, 7.51, 7.52). The simplest case is the symmetric two-point operator (7.14). The corresponding density $p_\sigma^*(\varsigma^* \mid \varsigma^{*(g)}) = p_{\mathrm{II}}^*(\varsigma^*)$ has been given by (7.50). A simple calculation yields

$$
\begin{aligned}
&p_{1;\lambda}(r \mid \varsigma^{*(g)}, r^{(g)}) \\
&= \frac{\lambda}{2} \frac{1}{\sqrt{2\pi}} \left\{ \frac{1}{\tilde{\varsigma}^*(\alpha\varsigma^{*(g)})} \exp\left[-\frac{1}{2} \left(\frac{r/r^{(g)} - \sqrt{1 + (\alpha\varsigma^{*(g)})^2/N}}{\tilde{\varsigma}^*(\alpha\varsigma^{*(g)})/N} \right)^2 \right] \right. \\
&\qquad \left. + \frac{1}{\tilde{\varsigma}^*(\varsigma^{*(g)}/\alpha)} \exp\left[-\frac{1}{2} \left(\frac{r/r^{(g)} - \sqrt{1 + (\varsigma^{*(g)})^2/\alpha^2 N}}{\tilde{\varsigma}^*(\varsigma^{*(g)}/\alpha)/N} \right)^2 \right] \right\} \\
&\quad \times \left[1 - \frac{1}{2} \Phi\left(\frac{r/r^{(g)} - \sqrt{1 + (\alpha\varsigma^{*(g)})^2/N}}{\tilde{\varsigma}^*(\alpha\varsigma^{*(g)})/N} \right) \right. \\
&\qquad \left. - \frac{1}{2} \Phi\left(\frac{r/r^{(g)} - \sqrt{1 + (\varsigma^{*(g)})^2/\alpha^2 N}}{\tilde{\varsigma}^*(\varsigma^{*(g)}/\alpha)/N} \right) \right]^{\lambda-1} .
\end{aligned} \tag{7.54}
$$

7.2.2.3 The Transition Density $p_{1;\lambda}(\varsigma)$. $p_{1;\lambda}(\varsigma) := p_{1;\lambda}(\varsigma \mid \varsigma^{(g)}, r^{(g)})$ is the density of the ς value of the best descendant, i.e. $p_{1;\lambda}(\varsigma) = p(\varsigma_{1;\lambda})$. It is important to note that $\varsigma_{1;\lambda}$ is not obtained by direct selection from the set of the $\tilde{\varsigma}_l$ values of the λ descendants. Instead, it is selected indirectly: it is the ς value of the individual with the best fitness $\tilde{F}_{1;\lambda}$ that is selected among the λ descendants (see lines 10 and 11 in Algorithm 4, p. 261). This is an *induced* $1;\lambda$ order statistic (see also Sect. 6.1.2.1).

For the derivation of the $p_{1;\lambda}(\varsigma)$ formula, we consider a random sample ς from the conditional distribution density $p_\sigma(\varsigma \mid \varsigma^{(g)})$ according to line 6 in Algorithm 4. In order to survive among λ descendants, the individual generated at line 7 must have the best r value. This is the case if its r value (residual distance!) is smaller than the smallest r value of the remaining $(\lambda-1)$ trials. The latter case will be denoted by $r_{1;\lambda-1}$; and the corresponding probability will be symbolized by $\Pr[r < r_{1;\lambda-1} \mid \varsigma]$. Therefore, the density of ς of the best descendant reads

$$
p_{1;\lambda}(\varsigma) = p_\sigma(\varsigma \mid \varsigma^{(g)}) \Pr[r < r_{1;\lambda-1} \mid \varsigma] . \tag{7.55}
$$

The probability $\Pr[r < r_{1;\lambda-1} \mid \varsigma]$ can be derived as follows. If r denotes the residual distance of an offspring generated by a mutation conditional to a given mutation strength ς, its density $p_r(r \mid \varsigma, r^{(g)})$ is given by Eq. (7.41). We consider the case that the offspring with this r value is the best among all λ descendants $\tilde{\mathbf{y}}$. In order to fulfill this condition, all remaining $(\lambda-1)$ descendants must have larger residual distances $\|\tilde{\mathbf{y}}\|$. In other words, the condition $r < \|\tilde{\mathbf{y}}\|$ must hold. For a single individual, this is the case with probability $\Pr(r < \|\tilde{\mathbf{y}}\|) = 1 - \Pr(\|\tilde{\mathbf{y}}\| \le r)$. Since all $(\lambda-1)$ descendants are created independently, the condition $r < \|\tilde{\mathbf{y}}\|$ must hold for all of them. Consequently, the probability that the individual with r is the best one in all $(\lambda-1)$ comparisons becomes $[1 - \Pr(\|\tilde{\mathbf{y}}\| \le r)]^{\lambda-1}$. Finally, one obtains the probability density that a given r value is the best

$$
p_r(r \mid \varsigma, r^{(g)}) \; [1 - \Pr(\|\tilde{\mathbf{y}}\| \le r)]^{\lambda-1} . \tag{7.56}
$$

This density is valid for a single sequence of events. However, there are λ different exclusive possibilities (event sequences) for the best among λ descendants. Therefore, this formula must be multiplied by the factor λ. Integrating the formula obtained over all possible r values and considering Eq. (7.44), the result reads

$$\Pr\left[r < r_{1;\lambda-1} \mid \varsigma\right] = \lambda \int_{r=0}^{\infty} p_r(r \mid \varsigma, r^{(g)}) \left[1 - P_{1;1}(r \mid \varsigma^{(g)}, r^{(g)})\right]^{\lambda-1} \mathrm{d}r.$$

If this result is inserted into (7.55), one finally obtains the desired transition density

$$p_{1;\lambda}(\varsigma \mid \varsigma^{(g)}, r^{(g)}) = \lambda p_\sigma(\varsigma \mid \varsigma^{(g)}) \int_{r=0}^{\infty} p_r(r \mid \varsigma, r^{(g)}) \left[1 - P_{1;1}(r \mid \varsigma^{(g)}, r^{(g)})\right]^{\lambda-1} \mathrm{d}r. \quad (7.57)$$

As one can see, it is in general not possible to obtain analytical expressions for the transition density $p_{1;\lambda}(\varsigma)$. The integrand in (7.57) is too complicated.

7.2.2.4 The $\varphi^{(k)}$ and $\psi^{(k)}$-SAR Functions. An approach based on approximations is necessary because of enormous difficulties in the analytical calculation of the transition densities. Such an approach should follow the same scheme that has been used in Chap. 4 (see also Appendix B.3.2). That is, the density is expressed by a series of Hermite polynomials and the series coefficients are obtained from the moments (or cumulants) of the transition densities $p_{1;\lambda}(r)$ and $p_{1;\lambda}(\varsigma)$.

As already stated in Sects. 7.2.1.1 and 7.2.1.2, we will limit the approximation in the following to the first two moments. In other words, the transition densities will be approximated by Gaussian distributions. However, in this subsection one still can consider the general case and express the kth order progress functions by integrals.

From the definition of the progress rate (7.24), one gets using (7.53) the integral

$$\varphi^{(k)}(\varsigma^{(g)}, r^{(g)}) = \lambda \int_{r=0}^{\infty} (r^{(g)} - r)^k p_{1;1}(r \mid \varsigma^{(g)}, r^{(g)}) \left[1 - P_{1;1}(r \mid \varsigma^{(g)}, r^{(g)})\right]^{\lambda-1} \mathrm{d}r.$$

Because of (7.46), normalized quantities can be used. One first obtains

$$\varphi^{(k)}(\varsigma^{*(g)}, r^{(g)}) = \int_{r=0}^{\infty} (r^{(g)} - r)^k \left(-\frac{\mathrm{d}}{\mathrm{d}r}\right) \left[1 - P_{1;1}(r \mid \varsigma^{*(g)}, r^{(g)})\right]^{\lambda} \mathrm{d}r$$

and after the substitution $r := r^{(g)} t$

$$\varphi^{(k)}(\varsigma^{*(g)}, r^{(g)}) = \left(r^{(g)}\right)^k \int_{t=0}^{\infty} (1-t)^k \left(-\frac{\mathrm{d}}{\mathrm{d}t}\right) \left[1 - P_{1;1}(r^{(g)} t \mid \varsigma^{*(g)}, r^{(g)})\right]^{\lambda} \mathrm{d}t. \quad (7.58)$$

Equation (7.58) suggests the introduction of

$$\boxed{\text{NORMALIZATION:} \qquad \varphi^{*(k)} := \varphi^{(k)} \left(\frac{N}{r^{(g)}} \right)^k.} \tag{7.59}$$

Hence, $\varphi^{*(k)}$ can be expressed as

$$\varphi^{*(k)}(\varsigma^{*(g)}) = N^k \lambda \int_{t=0}^{\infty} (1-t)^k \, p^*_{1;1}(t \mid \varsigma^{*(g)}) \left[1 - P^*_{1;1}(t \mid \varsigma^{*(g)}) \right]^{\lambda-1} \mathrm{d}t \tag{7.60}$$

with

$$P^*_{1;1}(t \mid \varsigma^{*(g)}) = \int_{\varsigma^*=0}^{\infty} \Phi\left(\frac{t - \sqrt{1+\varsigma^{*2}/N}}{\tilde{\varsigma}^*(\varsigma^*)/N} \right) p^*_\sigma(\varsigma^* \mid \varsigma^{*(g)}) \, \mathrm{d}\varsigma^*. \tag{7.61}$$

There is an alternative $\varphi^{*(k)}$ formula which can be obtained from (7.60) using integration by parts

$$\begin{aligned} \varphi^{*(k)}(\varsigma^{*(g)}) &= -N^k \int_{t=0}^{\infty} (1-t)^k \frac{\mathrm{d}}{\mathrm{d}t} \left[1 - P^*_{1;1}(t) \right]^\lambda \mathrm{d}t \\ \varphi^{*(k)}(\varsigma^{*(g)}) &= -N^k \left\{ \left[(1-t)^k \left[1 - P^*_{1;1}(t) \right]^\lambda \right]_0^\infty \right. \\ &\qquad \left. + k \int_{t=0}^{\infty} (1-t)^{k-1} \left[1 - P^*_{1;1}(t) \right]^\lambda \mathrm{d}t \right\}. \end{aligned} \tag{7.62}$$

Considering the effect of the substitution $r = r^{(g)}t$, it becomes clear that $P^*_{1;1}(t)\big|_{t=0} = 0$ and $P^*_{1;1}(t)\big|_{t=\infty} = 1$ hold. Additionally, $1 - P^*_{1;1}(t)$ vanishes exponentially for $t \to \infty$. Thus, only the lower limit in the first term of (7.62) contributes to the $\varphi^{*(k)}$ integral. One obtains

$$\varphi^{*(k)}(\varsigma^{*(g)}) = N^k \left\{ 1 - k \int_{t=0}^{\infty} (1-t)^{k-1} \left[1 - P^*_{1;1}(t \mid \varsigma^{*(g)}) \right]^\lambda \mathrm{d}t \right\}. \tag{7.63}$$

For the special case $k = 1$, $\varphi^* = \varphi^{*(1)}$, this reads

$$\varphi^*(\varsigma^{*(g)}) = N \left\{ 1 - \int_{t=0}^{\infty} \left[1 - P^*_{1;1}(t \mid \varsigma^{*(g)}) \right]^\lambda \mathrm{d}t \right\}. \tag{7.64}$$

The actual determination of the moments of $\varphi^{*(k)}$ follows in Sect. 7.3.1.

One has to evaluate the $\psi^{(k)}$ integrals to determine the self-adaptation response (SAR function). Considering the transition density (7.57), it follows from Definition (7.32) of the SAR function that

$$\begin{aligned} \psi^{(k)}(\varsigma^{(g)}, r^{(g)}) &= \lambda \int_{\varsigma=0}^{\infty} \left(\frac{\varsigma - \varsigma^{(g)}}{\varsigma^{(g)}} \right)^k p_\sigma(\varsigma \mid \varsigma^{(g)}) \\ &\quad \times \int_{r=0}^{\infty} p_r(r \mid \varsigma, r^{(g)}) \left[1 - P_{1;1}(r \mid \varsigma^{(g)}, r^{(g)}) \right]^{\lambda-1} \mathrm{d}r \, \mathrm{d}\varsigma. \end{aligned} \tag{7.65}$$

One obtains the alternative SAR formula after exchanging the integration order

$$\psi^{(k)}(\varsigma^{(g)}, r^{(g)}) = \lambda \int_{r=0}^{\infty} \left[1 - P_{1;1}(r \mid \varsigma^{(g)}, r^{(g)})\right]^{\lambda-1}$$
$$\times \int_{\varsigma=0}^{\infty} \left(\frac{\varsigma - \varsigma^{(g)}}{\varsigma^{(g)}}\right)^k p_r(r \mid \varsigma, r^{(g)})\, p_\sigma(\varsigma \mid \varsigma^{(g)})\, \mathrm{d}\varsigma\, \mathrm{d}r. \quad (7.66)$$

As in the case of $\varphi^{(k)}$ functions, normalized quantities are introduced. Using (7.27) one obtains

$$\psi^{(k)}(\varsigma^{*(g)}, r^{(g)}) = \lambda \int_{r=0}^{\infty} \left[1 - P_{1;1}(r \mid \varsigma^{*(g)}, r^{(g)})\right]^{\lambda-1}$$
$$\times \int_{\varsigma^*=0}^{\infty} \left(\frac{\varsigma^* - \varsigma^{*(g)}}{\varsigma^{*(g)}}\right)^k p_r(r \mid \varsigma^*, r^{(g)})\, p_\sigma^*(\varsigma^* \mid \varsigma^{*(g)})\, \mathrm{d}\varsigma^* \mathrm{d}r. \quad (7.67)$$

After the substitution $r = r^{(g)}t$, one finally gets

$$\psi^{(k)}(\varsigma^{*(g)}) = \lambda \int_{t=0}^{\infty} \left[1 - P_{1;1}^*(t \mid \varsigma^{*(g)})\right]^{\lambda-1}$$
$$\times \int_{\varsigma^*=0}^{\infty} \left(\frac{\varsigma^* - \varsigma^{*(g)}}{\varsigma^{*(g)}}\right)^k p_r^*(t \mid \varsigma^*)\, p_\sigma^*(\varsigma^* \mid \varsigma^{*(g)})\, \mathrm{d}\varsigma^*\, \mathrm{d}t. \quad (7.68)$$

The distribution $P_{1;1}^*$ is given by (7.61) and p_σ^* can be taken from the "collection" (7.48–7.52). After applying the substitution $r = r^{(g)}t$ to (7.41), one obtains for $p_r^*(t \mid \varsigma^*)$

$$p_r^*(t \mid \varsigma^*) = \frac{N}{\sqrt{2\pi}\tilde{\varsigma}^*(\varsigma^*)} \exp\left[-\frac{1}{2}\left(\frac{t - \sqrt{1 + \varsigma^{*2}/N}}{\tilde{\varsigma}^*(\varsigma^*)/N}\right)^2\right]. \quad (7.69)$$

Analytically closed expressions for the progress rate integrals (7.60) and the SAR functions (7.68) can only be found in exceptional cases. Such an exception concerns the $(1, 2)$-ES: for this algorithm, the integrations can be carried out for discrete mutation operators. For the case of the symmetric two-point distribution $p_\sigma^*(\varsigma^*) = p_{\text{II}}^*(\varsigma^*)$, the calculations follow in Sect. 7.3.1.2 and Sect. 7.3.2.2. In all other cases, one has to choose between two ways: numerical integration or derivation of approximation formulae. These two cases will make up the main part of the next section.

7.3 Determination of the Progress Rate and the SAR

7.3.1 Progress Integrals $\varphi^{*(k)}$

The results of numerical calculations of the φ^* function will be presented and discussed in the first part for different learning parameters. In the second part, $\varphi^*_{1,2}$ for the symmetric two-point operator is analytically calculated. The third part presents the derivation of D^*_φ for the case of small learning parameters.

7.3.1.1 Numerical Examples for the Progress Rate φ^*. In general, the progress rate integral (7.60) can only be calculated using numerical integration methods. This becomes clear after considering that the integrand in (7.60) contains the function $P^*_{1;1}(t)$, Eq. (7.61), as well as its derivative $p^*_{1;1}(t) = \frac{\mathrm{d}}{\mathrm{d}t} P^*_{1;1}(t)$, which are further defined by integrations over mutation densities of σ, i.e. $p^*_\sigma(\varsigma^*)$, in Eqs. (7.48–7.52).

As an example, the normalized progress rate ($k = 1$) $\varphi^* = \varphi^{*(1)}$ will be considered for a $(1, 10)$-σSA-ES with parameter space dimension $N = 30$. Figure 7.1 shows curves of φ^* that depend on the normalized mutation strength ς^*. The left figure is obtained when using the log-normal operator (7.8) with

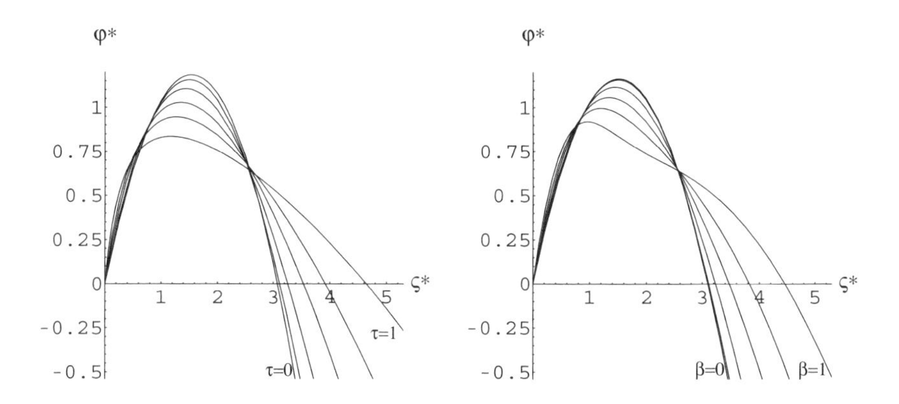

Fig. 7.1. The progress rate on the $(1, 10)$-σSA-ES, $N = 30$, for different learning parameters and different σ mutation operators. Left: log-normal operator with $\tau =$ 0, 0.1, 0.3, 0.5, 0.7, and 1.0; right: symmetric two-point operator with $\beta =$ 0, 0.1, 0.3, 0.5, 0.7, and 1.0

the density $p^*_\sigma(\varsigma^*) = p^*_{\mathrm{ln}}(\varsigma^*)$ (7.48). The right figure shows the relations when the symmetric two-point operator (7.14) is used with density $p^*_\sigma(\varsigma^*) = p^*_{\mathrm{II}}(\varsigma^*)$ (7.50).

Although the curves in Fig. 7.1 present the well-known quantity "progress rate φ^*" (cf., e.g., Fig. 3.4, p. 72), they require an additional explanation for the nonvanishing case of the learning parameter, i.e. for the cases $\tau \neq 0$ and $\beta = \alpha - 1 \neq 0$.

The progress rate $\varphi^* = \varphi^{*(1)}$ is one of the central quantities in ES theory. For a given parental state $(\varsigma^{(g)}, r^{(g)})^{\mathrm{T}}$, the quantity φ expresses the expected local change Δr of the residual distance from generation g to $g+1$. Positive φ values indicate that the algorithm exhibits local convergence behavior. It is interesting to realize that the normalized progress rate φ^* depends on the normalized mutation strength $\varsigma^{*(g)}$ only, not on the local curvature radius $r^{(g)}$.

The learning parameter τ and β influence the progress rate φ^*. As one sees, the φ^* curves deform as the learning parameter is increased. Since one is generally interested in large progress rates, too large values of the learning parameters should be avoided. Therefore, one might be tempted to choose $\tau,\ \beta = 0$; however, this case is principally excluded: as one can see in the formulae (7.8) and (7.14), for the $\tau,\ \beta = 0$ case, the mutation strength σ is not mutated at all. However, a $(1,\lambda)$-ES with a constant σ is principally *not convergent* (see Sect. 3.3.2.2, Eq. (3.182)). As a result, the fulfillment of the condition $\tau > 0$ or $\beta > 0$ is absolutely necessary. Otherwise the σSA algorithm cannot function as desired.

The actual choice of the learning parameter cannot be inferred from the investigation of φ^* only. The microscopic aspects of σSA are only one aspect of the analysis to be done. The self-adaptation is an *evolutionary learning process* that takes place in the time domain. That is why the entire σSA dynamics must be considered.

The derivation of analytical approximations for $\varphi^*(\varsigma^*, \tau, N)$ or $\varphi^*(\varsigma^*, \beta, N)$ is an open problem. Beyond the trivial case $\tau,\ \beta = 0$, which is already covered by formula (3.242)

$$\varphi^*(\varsigma^*)\big|_{\tau=0} = c_{1,\lambda}\,\varsigma^* \sqrt{\frac{1+\frac{\varsigma^{*2}}{2N}}{1+\frac{\varsigma^{*2}}{N}}} - N\left(\sqrt{1+\frac{\varsigma^{*2}}{N}} - 1\right) + \ldots, \tag{7.70}$$

no other sufficiently accurate *and simple* approximation formulae have been found up to now. Fortunately, the dependence of the progress rate is relatively weak for small values of the learning parameter, as one can see in Fig. 7.1. In the investigations of this chapter, it is assumed that the dependency of the progress rate on the learning parameter can be neglected if

$$0 < \tau \lessapprox 0.3 \qquad \text{and} \qquad 0 < \beta \lessapprox 0.3 \quad (\text{or } 1 < \alpha \lessapprox 1.3), \tag{7.71}$$

respectively, is fulfilled. Therefore, formula (7.70) and the asymptotic φ^* formula ($N \to \infty$), respectively, will be applied in the following. The derivation of more accurate formulae remains as a task for future research.

7.3.1.2 An Analytic φ^* Formula for $p_\sigma = p_{\mathrm{II}}$, $\lambda = 2$. As an exceptional case, the derivation of an analytical formula is possible for the discrete σ mutation operators in the case $\lambda = 2$. Although the immediate practical use of this formula is only minor for the general analysis of σSA, it can serve as a reference result. The accuracy of numerical methods and approximation approaches can be tested on it.

The derivation of $\varphi^*_{1,2}$ for the symmetric two-point operator (7.14) will be sketched here as an example. As the σ mutation density, Eq. (7.50) is used in integral (7.61). The integration yields

$$P^*_{1;1}(t) = \frac{1}{2}\left[\Phi\left(\frac{t-\sqrt{1+(s\alpha)^2/N}}{\tilde{\varsigma}^*(s\alpha)/N}\right) + \Phi\left(\frac{t-\sqrt{1+(s/\alpha)^2/N}}{\tilde{\varsigma}^*(s/\alpha)/N}\right)\right]. \tag{7.72}$$

For simplicity, $\varsigma^{*(g)}$ is denoted as s. Considering $p^*_{1;1}(t) = \frac{\mathrm{d}}{\mathrm{d}t}P^*_{1;1}(t)$, (7.60) becomes for $k=1$

$$\begin{aligned}\varphi^*_{1,2} = N\Bigg\{1 - \frac{1}{4}\frac{1}{\sqrt{2\pi}}\int_{t=0}^{\infty} t\Bigg[&\frac{N}{\tilde{\varsigma}^*(s\alpha)}\exp\left[-\frac{1}{2}\left(\frac{t-\sqrt{1+(s\alpha)^2/N}}{\tilde{\varsigma}^*(s\alpha)/N}\right)^2\right] \\ &+\frac{N}{\tilde{\varsigma}^*(s/\alpha)}\exp\left[-\frac{1}{2}\left(\frac{t-\sqrt{1+(s/\alpha)^2/N}}{\tilde{\varsigma}^*(s/\alpha)/N}\right)^2\right]\Bigg] \\ \times\left[2 - \Phi\left(\frac{t-\sqrt{1+(s\alpha)^2/N}}{\tilde{\varsigma}^*(s\alpha)/N}\right) - \Phi\left(\frac{t-\sqrt{1+(s/\alpha)^2/N}}{\tilde{\varsigma}^*(s/\alpha)/N}\right)\right]\mathrm{d}t\Bigg\}.\end{aligned}$$

After applying the substitutions

$$x = -\frac{t-\sqrt{1+(s\alpha)^2/N}}{\tilde{\varsigma}^*(s\alpha)/N} \quad\text{and}\quad y = -\frac{t-\sqrt{1+(s/\alpha)^2/N}}{\tilde{\varsigma}^*(s/\alpha)/N} \tag{7.73}$$

and expanding the upper integration limit to ∞ (which is admissible for sufficiently large N, see Sect. 3.4.5) one obtains

$$\begin{aligned}\varphi^*_{1,2} = N\Bigg\{1 &- \frac{1}{4}\frac{2}{\sqrt{2\pi}}\int_{-\infty}^{\infty}\left(\sqrt{1+(s\alpha)^2/N} - \frac{\tilde{\varsigma}^*(s\alpha)}{N}x\right)\mathrm{e}^{-\frac{1}{2}x^2}\Bigg[\Phi(x) \\ &+\Phi\left(\frac{\tilde{\varsigma}^*(s\alpha)}{\tilde{\varsigma}^*(s/\alpha)}x + N\frac{\sqrt{1+(s/\alpha)^2/N}-\sqrt{1+(s\alpha)^2/N}}{\tilde{\varsigma}^*(s/\alpha)}\right)\Bigg]\mathrm{d}x \\ &-\frac{1}{4}\frac{2}{\sqrt{2\pi}}\int_{-\infty}^{\infty}\left(\sqrt{1+(s/\alpha)^2/N} - \frac{\tilde{\varsigma}^*(s/\alpha)}{N}y\right)\mathrm{e}^{-\frac{1}{2}y^2}\Bigg[\Phi(y) \\ &+\Phi\left(\frac{\tilde{\varsigma}^*(s/\alpha)}{\tilde{\varsigma}^*(s\alpha)}y + N\frac{\sqrt{1+(s\alpha)^2/N}-\sqrt{1+(s/\alpha)^2/N}}{\tilde{\varsigma}^*(s\alpha)}\right)\Bigg]\mathrm{d}y\Bigg\}.\end{aligned} \tag{7.74}$$

The integrands in (7.74) are of the form $\mathrm{e}^{-\frac{1}{2}x^2}\Phi(ax+b)$ and $x\mathrm{e}^{-\frac{1}{2}x^2}\Phi(ax+b)$. Formulae (A.10) and (A.12) in Appendix A.1 can be used to carry out the integrations

$$\begin{aligned}
\varphi_{1,2}^* = N \Bigg\{ 1 & - \frac{1}{4}\sqrt{1+(s\alpha)^2/N} + \frac{1}{4}\frac{1}{\sqrt{\pi}}\frac{\tilde{\varsigma}^*(s\alpha)}{N} \\
& - \frac{1}{2}\sqrt{1+(s\alpha)^2/N}\;\Phi\left(N\frac{\sqrt{1+(s/\alpha)^2/N}-\sqrt{1+(s\alpha)^2/N}}{\sqrt{(\tilde{\varsigma}^*(s\alpha))^2+(\tilde{\varsigma}^*(s/\alpha))^2}}\right) \\
& + \frac{1}{2}\frac{1}{\sqrt{2\pi}}\frac{1}{N}\frac{(\tilde{\varsigma}^*(s\alpha))^2}{\sqrt{(\tilde{\varsigma}^*(s\alpha))^2+(\tilde{\varsigma}^*(s/\alpha))^2}} \\
& \qquad \times \exp\left[-\frac{1}{2}\left(N\frac{\sqrt{1+(s/\alpha)^2/N}-\sqrt{1+(s\alpha)^2/N}}{\sqrt{(\tilde{\varsigma}^*(s\alpha))^2+(\tilde{\varsigma}^*(s/\alpha))^2}}\right)^2\right] \\
& - \frac{1}{4}\sqrt{1+(s/\alpha)^2/N} + \frac{1}{4}\frac{1}{\sqrt{\pi}}\frac{\tilde{\varsigma}^*(s/\alpha)}{N} \\
& - \frac{1}{2}\sqrt{1+(s/\alpha)^2/N}\;\Phi\left(N\frac{\sqrt{1+(s\alpha)^2/N}-\sqrt{1+(s/\alpha)^2/N}}{\sqrt{(\tilde{\varsigma}^*(s/\alpha))^2+(\tilde{\varsigma}^*(s\alpha))^2}}\right) \\
& + \frac{1}{2}\frac{1}{\sqrt{2\pi}}\frac{1}{N}\frac{(\tilde{\varsigma}^*(s/\alpha))^2}{\sqrt{(\tilde{\varsigma}^*(s/\alpha))^2+(\tilde{\varsigma}^*(s\alpha))^2}} \\
& \qquad \times \exp\left[-\frac{1}{2}\left(\frac{\sqrt{1+(s\alpha)^2/N}-\sqrt{1+(s/\alpha)^2/N}}{\sqrt{(\tilde{\varsigma}^*(s/\alpha))^2+(\tilde{\varsigma}^*(s\alpha))^2}\;/N}\right)^2\right]\Bigg\}.
\end{aligned}$$

After collecting similar terms the result reads (note that s denotes $\varsigma^{*(g)}$)

$$\begin{aligned}
\varphi_{1,2}^*(s) = N\Bigg\{ 1 & - \frac{3}{4}\sqrt{1+(s\alpha)^2/N} - \frac{1}{4}\sqrt{1+(s/\alpha)^2/N} \\
& + \frac{1}{2}\left(\sqrt{1+(s\alpha)^2/N}-\sqrt{1+(s/\alpha)^2/N}\right) \\
& \qquad \times\Phi\left(N\frac{\sqrt{1+(s\alpha)^2/N}-\sqrt{1+(s/\alpha)^2/N}}{\sqrt{(\tilde{\varsigma}^*(s\alpha))^2+(\tilde{\varsigma}^*(s/\alpha))^2}}\right)\Bigg\} \\
& + \frac{1}{4}\frac{1}{\sqrt{\pi}}\left(\tilde{\varsigma}^*(s\alpha)+\tilde{\varsigma}^*(s/\alpha)\right) \\
& + \frac{1}{4}\frac{\sqrt{2}}{\sqrt{\pi}}\sqrt{(\tilde{\varsigma}^*(s\alpha))^2+(\tilde{\varsigma}^*(s/\alpha))^2} \\
& \qquad \times\exp\left[-\frac{1}{2}\left(\frac{\sqrt{1+(s\alpha)^2/N}-\sqrt{1+(s/\alpha)^2/N}}{\sqrt{(\tilde{\varsigma}^*(s\alpha))^2+(\tilde{\varsigma}^*(s/\alpha))^2}\;/N}\right)^2\right].
\end{aligned} \tag{7.75}$$

As one can simply verify, the special case $\alpha = 1$ ($\beta = 0$), Eq. (7.70), is included in (7.75), remembering that $c_{1,2} = 1/\sqrt{\pi}$ (see Eq. (3.109)).

7.3.1.3 The $\tau \to 0$ and $\beta \to 0$ Approximation for D^*_φ. The first-order progress rate, $\varphi^* = \varphi^{*(1)}$, describes the expected r change from generation g to $g+1$. Because of the stochastic nature of mutations, the actual change in r deviates from its expected value (see Eq. (7.25)). The variance of these deviations is denoted by D^2_φ in Definition (7.26). After considering the normalizations (7.27) and (7.59), the normalized standard deviation reads

$$D^*_\varphi = \sqrt{\varphi^{*(2)} - (\varphi^*)^2}. \tag{7.76}$$

Since $\varphi^{*(k)}$ depends on the learning parameters τ, β, etc., D^*_φ is also a function of learning parameters. All difficulties that emerged in the derivation of the analytical formulae for $\varphi^{*(k)}$ appear also in the determination of D^*_φ. Therefore, only the limit case τ, $\beta \to 0$ can be treated up to now. However, as can be seen in the results obtained using numerical integrations, the approximation τ, $\beta = 0$ yields satisfactory results, as long as one remains in the learning parameter intervals defined in (7.71).

For the case τ, $\beta = 0$, $\varphi^{*(2)}$ remains to be calculated, since φ^* already exists in (7.70). Using (7.60, 7.61), taking (7.50), $p^*_\sigma(\varsigma^* \mid \varsigma^{*(g)}) = \delta(\varsigma^* - \varsigma^{*(g)})$, and the notation $s := \varsigma^{*(g)}$ into account, one obtains

$$\varphi^{*(2)}(s) = N^2 \frac{\lambda}{\sqrt{2\pi}} \int_{t=0}^{\infty} \frac{(1-t)^2}{\tilde{\varsigma}^*(s)/N} \exp\left[-\frac{1}{2}\left(\frac{t-\sqrt{1+s^2/N}}{\tilde{\varsigma}^*(s)/N}\right)^2\right] \times \left[1-\Phi\left(\frac{t-\sqrt{1+s^2/N}}{\tilde{\varsigma}^*(s)/N}\right)\right]^{\lambda-1} \mathrm{d}t. \tag{7.77}$$

After the substitution

$$x = -\frac{t-\sqrt{1+s^2/N}}{\tilde{\varsigma}^*(s)/N} \quad \Rightarrow \quad -t = \frac{\tilde{\varsigma}^*(s)}{N}x - \sqrt{1+s^2/N} \tag{7.78}$$

and using the symmetry relation (C.8), it follows that

$$\varphi^{*(2)}(s) = N^2 \frac{\lambda}{\sqrt{2\pi}} \int_{x=-\infty}^{\infty} \left(1-\sqrt{1+s^2/N} + \frac{\tilde{\varsigma}^*(s)}{N}x\right)^2 \times \mathrm{e}^{-\frac{1}{2}x^2} \left[\Phi(x)\right]^{\lambda-1} \mathrm{d}x. \tag{7.79}$$

The upper limit was expanded to ∞ – as usual –, which is admissible for sufficiently large N (see Sect. 3.4.5). Considering the Definition (3.100) for the progress coefficient $c_{1,\lambda}$ and (4.41) for the second-order progress coefficient $d^{(2)}_{1,\lambda}$, one gets

$$\varphi^{*(2)} = N^2 \left[\left(1-\sqrt{1+s^2/N}\right)^2 + 2\left(1-\sqrt{1+s^2/N}\right)\frac{\tilde{\varsigma}^*(s)}{N} c_{1,\lambda} + \left(\frac{\tilde{\varsigma}^*(s)}{N}\right)^2 d^{(2)}_{1,\lambda}\right]. \tag{7.80}$$

Using Eq. (7.80), $(D_\varphi^*)^2$ can be completed. It results from (7.76), (7.70), and (7.80) that

$$(D_\varphi^*)^2 = \overline{\varphi^{*(2)}} - (\overline{\varphi^*})^2 = (\tilde{\varsigma}^*(s))^2 \left(d_{1,\lambda}^{(2)} - (c_{1,\lambda})^2\right). \tag{7.81}$$

Finally, one obtains therefore the standard deviation (note that $\varsigma^{*(g)} = s$)

$$D_\varphi^*(\varsigma^{*(g)}) = \tilde{\varsigma}^*(\varsigma^{*(g)})\sqrt{d_{1,\lambda}^{(2)} - c_{1,\lambda}^2}. \tag{7.82}$$

The square root in (7.82) is a slowly decreasing function of λ with values less than one (for $\lambda > 1$). It is tabulated in Appendix D.2. Considering (7.47), one realizes that $D_\varphi^*(\varsigma^{*(g)})$ is approximately proportional to the parental mutation strength $\varsigma^{*(g)}$. They are of the same order of magnitude. In other words, the random variable "change in the residual distance" fluctuates almost with the parental mutation strength.

7.3.2 The SAR Functions $\psi^{(k)}$

This subsection comprises five parts. In the first part, numerically obtained ψ functions are presented and compared with ES experiments. Using these curves, the principal functioning mechanism of the σSA is discussed. The second part is engaged with the derivation of an analytical $\psi_{1,2}$ formula. The results of numerical investigations from the previous part are verified by the $\psi_{1,2}$ formula. The ψ function is bounded: the third part provides bounds. The derivation of analytical approximations for ψ, $\psi^{(2)}$, and D_ψ follows in the fourth and fifth part for small values of the learning parameter and small ς^*. The results of the fifth part establish the "microscopic background" and the starting point for the investigations on the σSA dynamics in Sects. 7.4 and 7.5.

7.3.2.1 Numerical Examples for $\psi^{(1)}$, Discussion, and Comparison with Experiments. The central quantity of the σSA theory is the SAR function $\psi(\varsigma^*) = \psi^{(1)}(\varsigma^*)$. It quantifies the expected relative change in σ per generation, i.e. the self-adaptation response. This function is defined by the integral (7.68). It is difficult to determine. Even for the numerical integration, it presents a nontrivial problem for a computer algebra system, e.g., `Mathematica`. For a more efficient determination of (7.68), it is advisable to use special integration routines that take the special structure of the integrand into account.

In this section, the results of numerical integration of (7.68) are presented for $k = 1$ in the case of the (1, 10)-ES. The results give us knowledge of the essential behavior of the ψ function and a "feeling" as to the influence of the learning parameters and of the parameter space dimension N on the ψ curve.

Figure 7.2 (for the parameter space dimension $N = 30$) shows the ψ function for the log-normal operator (7.8) (left) and for the symmetric two-point operator (7.14) (right). It is striking and remarkable that both σ mutation operators yield similar curves for the corresponding choice of the learning

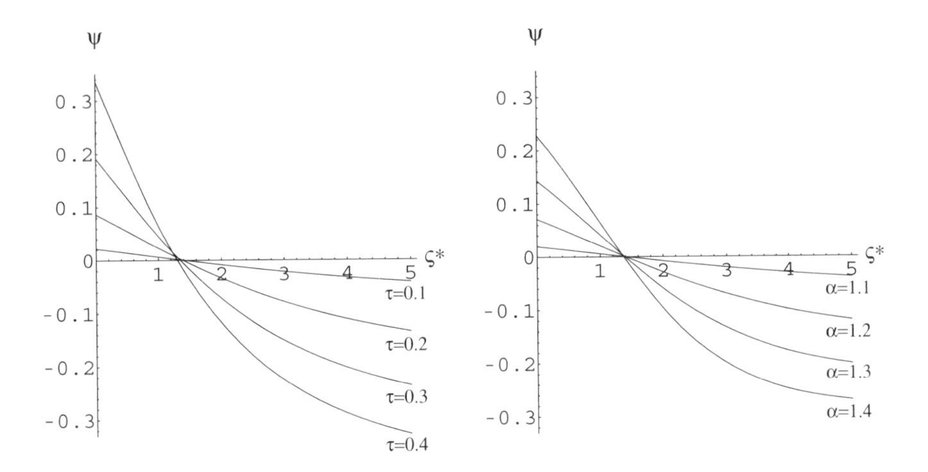

Fig. 7.2. The self-adaptation response $\psi(\varsigma^*) = \psi^{(1)}(\varsigma^*)$ for the $(1,10)$-σSA-ES with $N = 30$. Left: log-normal operator; right: symmetric two-point operator

parameter τ and $\beta = \alpha - 1$. As will be shown in Sect. 7.3.2.5, the effect is even the same for sufficiently small ς^* and small learning parameters.

Valuable information can be inferred from Fig. 7.2. Based on Definition (7.29), $\psi^{(1)}$ describes the self-adaptation response of the σSA algorithm. This can be positive or negative. At a specific $\varsigma^*_{\psi_0}$ value, ψ vanishes. If the parental ς^* is larger than this value, i.e. $\varsigma^{*(g)} > \varsigma^*_{\psi_0}$, then ψ is negative and the expected ς change is – because of (7.29) – toward the reduction of the parental ς value. In the opposite case $\varsigma^{*(g)} < \varsigma^*_{\psi_0}$, ψ is positive and the parental ς is increased in the statistical mean. In this manner, ς is altered so that it should be kept in the neighborhood of the zero point $\varsigma^*_{\psi_0}$ – at least for the *static case* considered here. Considering Fig. 7.1, one notes that $\varsigma^*_{\psi_0}$ surprisingly lies in the vicinity of the optimal ς^* value providing the maximal progress rate φ^*. Even if the ς^* values statistically fluctuate around $\varsigma^*_{\psi_0}$, Fig. 7.1 shows for the $(1,10)$-ES that the σSA algorithm should be able to control ς such that positive φ values are obtained. As a result, the convergence toward the optimum would be ensured.

However, this argumentation on the working mechanism of the σSA algorithm should be taken *cum grano salis*, since a *static picture* has been drawn. In the dynamic case, the relations are more complicated, since one has to differentiate between the dynamics of ς and ς^* because of the r evolution. We will come back to that in Sect. 7.4.

In addition to the first discussion of the σSA dynamics, the effect of the learning parameters on the self-adaptation response can be discussed using Fig. 7.2. One recognizes that the learning parameters τ and α influence the "learning strength" or the "learning rate" of the algorithm. For $\tau = 0$ (or $\alpha = 1$), there is no self-adaptation, i.e.

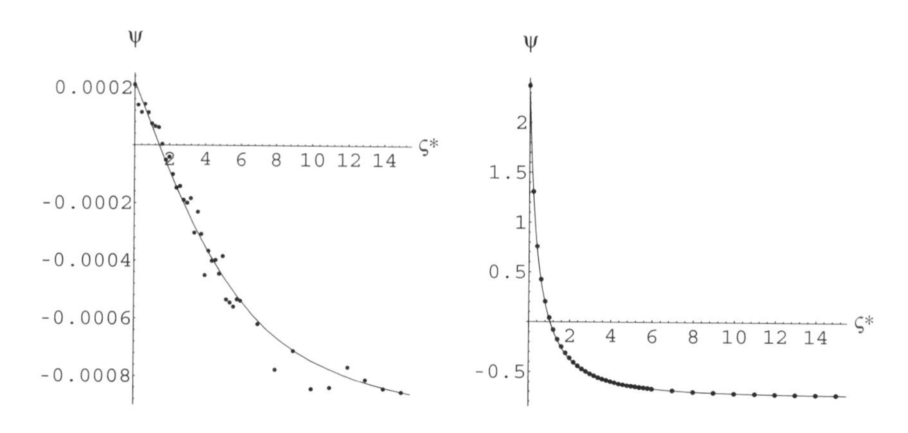

Fig. 7.3. The ψ function for the log-normal operator is compared with experiments. The simulation results are displayed as data points. Left: $\tau = 0.01$, $N = 30$; right: $\tau = 1$, $N = 30$

$$\psi(\varsigma^*, \tau)\big|_{\tau=0} \equiv 0, \qquad \psi(\varsigma^*, \alpha)\big|_{\alpha=1} \equiv 0. \tag{7.83}$$

This is immediately plausible since the parental ς remains unchanged by the σ mutation operators (7.8) and (7.14), respectively.

In general, larger learning parameters yield higher learning rates, since they give larger ψ values for a given ς^*. That is, using larger learning parameters allows for a faster adaptation in the case of a badly chosen initial mutation strength. This is a supporting argument for using large learning parameters. However, this is only one side of the coin. A look at Fig. 7.1 shows that an increase in the learning parameter results in a decrease of the maximal attainable progress rate. Therefore, one has to compromise between faster self-adaptation and a higher progress rate.

Since the concept of the SAR function is new, it is advisable to verify the predictions of the theory by ES experiments. Such simulations can be carried out after the scheme of *one-generation experiments* in Sect. 3.4.4. For this purpose, G experiments are carried out using the $(1, \lambda)$-ES for a given constant parent (r and ς are kept constant) for a single generation. The mean value of the relative ς change for the best descendant is calculated. This mean value is an estimation of the expected value (7.31) and thus for ψ.

Figures 7.3 and 7.4 show results of ES experiments for the $(1, 10)$-ES with log-normal operator. Every data point is obtained by averaging over $G = 40,000$ "one-generation experiments." Figure 7.3 (right) shows the case with a very large learning parameter $\tau = 1$. One observes a good agreement between prediction and experiment. The left-hand figure presents the case $\tau = 0.01$. The simulation results show considerable fluctuations around the predicted ψ values. The fluctuations can be explained theoretically. As will be shown in Sect. 7.3.2.5, the magnitude of the standard deviation D_ψ of the random variate $(\varsigma^{(g+1)} - \varsigma^{(g)})/\varsigma^{(g)}$ (see Eqs. (7.35, 7.36)) is almost equal

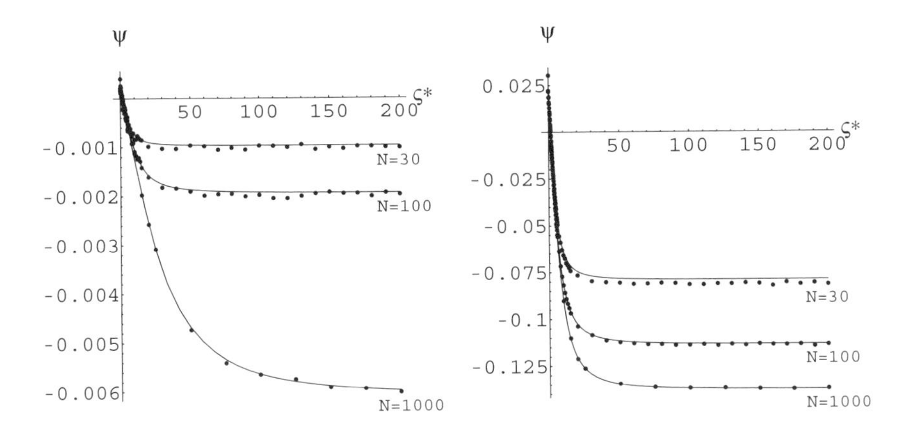

Fig. 7.4. Comparison with experiments and the influence of the dimension N on the saturation behavior of the ψ function in the case of the log-normal operator. Left: $\tau = 0.01$; right: $\tau = 0.1$

to the learning parameter τ. But its expected value, i.e. ψ, is of the order τ^2. That is, for $0 < \tau \ll 1$, the fluctuations dominate. By averaging over G generations, the standard deviation reduces to the order of magnitude of $\tau/\sqrt{G}$. In the concrete case considered $G = 40,000$, $\tau = 0.01$ was chosen; therefore, a standard deviation of 0.00005 should be expected. This is also roughly observed in Fig. 7.3.

Figure 7.3 (right) shows a further interesting property of the ψ function: one observes a saturation behavior for large values of ς^*, i.e. ψ becomes asymptotically constant

$$\psi_\infty := \lim_{\varsigma^* \to \infty} \psi(\varsigma^*). \tag{7.84}$$

Considering the definition of ψ (7.31), it becomes clear that ψ must have a lower bound. Independent of τ, λ, N, and ς^*,

$$\psi(\varsigma^*) \geq -1 \tag{7.85}$$

must hold. However, this bound is not very sharp, as one can see in Fig. 7.4. In this figure, the influence of τ and N on ψ_∞ becomes clear. There is a strong dependency of the saturation value ψ_∞ on N for small values of the learning parameter τ.

7.3.2.2 An Analytic ψ Formula for $p_\sigma = p_{\text{II}}$, $\lambda = 2$. Analogous to the case of the progress rate $\varphi^*_{1,2}$, it is also possible to derive an analytically closed expression for the self-adaptation response $\psi_{1,2}(\varsigma^*)$ of the two-point operator. The starting point is Eq. (7.68) for $k = 1$. The probability distribution $P^*_{1;1}(t)$ can be found in Eq. (7.72). Therefore, one can start with the treatment of the inner integral in (7.68). Using (7.69) and (7.50) one obtains (s is used instead of $\varsigma^{*(g)}$ to shorten the notation)

$$\int_{\varsigma^*=0}^{\infty} \frac{\varsigma^* - s}{s}\, p_r^*(t|\varsigma^*)\, p_{\mathrm{II}}^*(\varsigma^*|s)\, \mathrm{d}\varsigma^*$$
$$= \int_{\varsigma^*=0}^{\infty} \frac{\varsigma^* - s}{s} \frac{N}{\sqrt{2\pi}\tilde{\varsigma}^*(\varsigma^*)} \exp\left[-\frac{1}{2}\left(\frac{t - \sqrt{1+(\varsigma^*)^2/N}}{\tilde{\varsigma}^*(\varsigma^*)/N}\right)^2\right]$$
$$\times \frac{1}{2}\left[\delta\left(\varsigma^* - s\alpha\right) + \delta\left(\varsigma^* - s/\alpha\right)\right] \mathrm{d}\varsigma^*$$
$$= \frac{1}{2}\frac{1}{\sqrt{2\pi}}\left\{(\alpha - 1)\frac{N}{\tilde{\varsigma}^*(s\alpha)} \exp\left[-\frac{1}{2}\left(\frac{t - \sqrt{1+(s\alpha)^2/N}}{\tilde{\varsigma}^*(s\alpha)/N}\right)^2\right]\right.$$
$$\left. + \left(\frac{1}{\alpha} - 1\right)\frac{N}{\tilde{\varsigma}^*(s/\alpha)} \exp\left[-\frac{1}{2}\left(\frac{t - \sqrt{1+(s/\alpha)^2/N}}{\tilde{\varsigma}^*(s/\alpha)/N}\right)^2\right]\right\}. \quad (7.86)$$

Insertion of (7.86) and (7.72) in (7.68) yields

$$\psi_{1,2} = \frac{1}{2}\frac{1}{\sqrt{2\pi}} \int_{t=0}^{\infty} \left[2 - \Phi\left(\frac{t - \sqrt{1+(s\alpha)^2/N}}{\tilde{\varsigma}^*(s\alpha)/N}\right)\right.$$
$$\left. - \Phi\left(\frac{t - \sqrt{1+(s/\alpha)^2/N}}{\tilde{\varsigma}^*(s/\alpha)/N}\right)\right]$$
$$\times \left\{(\alpha - 1)\frac{N}{\tilde{\varsigma}^*(s\alpha)} \exp\left[-\frac{1}{2}\left(\frac{t - \sqrt{1+(s\alpha)^2/N}}{\tilde{\varsigma}^*(s\alpha)/N}\right)^2\right]\right.$$
$$\left. + \left(\frac{1}{\alpha} - 1\right)\frac{N}{\tilde{\varsigma}^*(s/\alpha)} \exp\left[-\frac{1}{2}\left(\frac{t - \sqrt{1+(s/\alpha)^2/N}}{\tilde{\varsigma}^*(s/\alpha)/N}\right)^2\right]\right\} \mathrm{d}t.$$

After carrying out the substitutions (7.73), analogous to (7.74) one gets

$$\psi_{1,2} = \frac{1}{2}(\alpha - 1)\frac{1}{\sqrt{2\pi}} \int_{x=-\infty}^{\infty} \mathrm{e}^{-\frac{1}{2}x^2} \left[\Phi(x)\right.$$
$$\left. + \Phi\left(\frac{\tilde{\varsigma}^*(s\alpha)}{\tilde{\varsigma}^*(s/\alpha)}\, x \,+\, N\frac{\sqrt{1+(s/\alpha)^2/N} - \sqrt{1+(s\alpha)^2/N}}{\tilde{\varsigma}^*(s/\alpha)}\right)\right] \mathrm{d}x$$
$$+ \frac{1}{2}\left(\frac{1}{\alpha} - 1\right)\frac{1}{\sqrt{2\pi}} \int_{y=-\infty}^{\infty} \mathrm{e}^{-\frac{1}{2}y^2} \left[\Phi(y)\right.$$
$$\left. + \Phi\left(\frac{\tilde{\varsigma}^*(s/\alpha)}{\tilde{\varsigma}^*(s\alpha)}\, y \,+\, N\frac{\sqrt{1+(s\alpha)^2/N} - \sqrt{1+(s/\alpha)^2/N}}{\tilde{\varsigma}^*(s\alpha)}\right)\right] \mathrm{d}y.$$

The application of the integral formula (A.10) yields

$$\psi_{1,2} = \frac{1}{2}(\alpha - 1)\left[\frac{1}{2} + \Phi\left(N \frac{\sqrt{1+(s/\alpha)^2/N} - \sqrt{1+(s\alpha)^2/N}}{\sqrt{(\tilde{\varsigma}^*(s\alpha))^2 + (\tilde{\varsigma}^*(s/\alpha))^2}}\right)\right]$$
$$+\frac{1}{2}\left(\frac{1}{\alpha} - 1\right)\left[\frac{1}{2} + \Phi\left(N \frac{\sqrt{1+(s\alpha)^2/N} - \sqrt{1+(s/\alpha)^2/N}}{\sqrt{(\tilde{\varsigma}^*(s\alpha))^2 + (\tilde{\varsigma}^*(s/\alpha))^2}}\right)\right].$$

Finally, after collecting the Φ terms ($\varsigma^* = s$) one gets

$$\psi_{1,2}(\varsigma^*) = \frac{1}{2}\frac{\alpha - 1}{\alpha}\left[\frac{3\alpha - 1}{2}\right.$$
$$\left. -(\alpha + 1)\,\Phi\left(N \frac{\sqrt{1+(\varsigma^*\alpha)^2/N} - \sqrt{1+(\varsigma^*/\alpha)^2/N}}{\sqrt{(\tilde{\varsigma}^*(\varsigma^*\alpha))^2 + (\tilde{\varsigma}^*(\varsigma^*/\alpha))^2}}\right)\right]. \quad (7.87)$$

This formula can be used to investigate the influence of the learning parameter α and the parameter dimension N on the characteristic properties of $\psi_{1,2}$.

The $\psi_{1,2}$ value for $\varsigma^* = 0$ is obtained in the simplest way by expanding the fraction in the Φ function of (7.87) in a Taylor series with respect to ς^*. Considering (7.47), the expansion yields under the condition $\varsigma^* \ll N$

$$\psi_{1,2}^{(1)}(\varsigma^*) \approx \frac{1}{2}\frac{\alpha - 1}{\alpha}\left[\frac{3\alpha - 1}{2} - (\alpha + 1)\,\Phi\left(\frac{\varsigma^*}{2}\frac{\alpha^2 - 1/\alpha^2}{\sqrt{\alpha^2 + 1/\alpha^2}}\right)\right]. \quad (7.88)$$

For $\varsigma^* \to 0$, one obtains directly ($\beta = \alpha - 1$)

$$\psi_{1,2}(\varsigma^*)\big|_{\varsigma^*=0} = \frac{1}{2}\frac{(\alpha - 1)^2}{\alpha} = \frac{1}{2}\frac{\beta^2}{1 + \beta}. \quad (7.89)$$

The asymptotic behavior $\varsigma^* \to \infty$ ($N < \infty$) of (7.87) can also be investigated easily. Because of (7.47), one has $\tilde{\varsigma}^*(\varsigma^*\alpha) \sim \varsigma^*\alpha/\sqrt{2}$ and $\tilde{\varsigma}^*(\varsigma^*/\alpha) \sim \varsigma^*/\alpha\sqrt{2}$. Hence, (7.87) behaves asymptotically ($\varsigma^* \to \infty$) like

$$\psi_{1,2\,\infty} = \lim_{\varsigma^* \to \infty} \psi_{1,2}(\varsigma^*)$$
$$= \frac{1}{2}\frac{\alpha - 1}{\alpha}\left[\frac{3\alpha - 1}{2} - (\alpha + 1)\,\Phi\left(\sqrt{2N}\,\frac{\alpha - 1/\alpha}{\sqrt{\alpha^2 + 1/\alpha^2}}\right)\right]. \quad (7.90)$$

As one can see, there is really a saturation value which depends on N and on the learning parameter α. The lower bound (7.85) can be made more accurate using (7.90). Since Φ cannot be larger than one,

$$\psi_{1,2} \geq \frac{1}{2}\frac{\alpha - 1}{\alpha}\left[\frac{3\alpha - 1}{2} - (\alpha + 1)\right] = \frac{1}{4}\frac{\alpha - 1}{\alpha}(\alpha - 3) \quad (7.91)$$

must hold.

The inequality (7.91) has an interesting application: it gives a necessary (but not sufficient) condition for the functioning of the σSA in the case of (1,2)-ES with a symmetric two-point operator. As we have seen in Figs. 7.2–7.4 for the sphere model, the ψ function should have a zero, i.e. it should enable an increase as well as a decrease of the parental ς value depending on the ς^* value. If such a zero does not exist, the σSA algorithm may not work. In particular, if the saturation value of ψ is larger than zero, the algorithm must necessarily fail. One can read from (7.91) that this is the case for $\alpha \geq 3$. Therefore we have obtained a necessary criterion for the functioning of the (1, 2)-ES with a two-point operator.[11]

However, one cannot conclude from this necessary criterion that the (1, 2)-σSA-ES works for $\alpha < 3$. A sufficient condition for self-adaptation requires the investigation of the whole σSA dynamics.

Beside the two characteristic ψ limit values for $\varsigma^* = 0$ and $\varsigma^* = \infty$, the zero $\varsigma^*_{\psi_0}$ is also interesting. For the case $\varsigma^* \ll N$, equating (7.88) to zero and resolving for ς^*, one obtains

$$\varsigma^*_{\psi_0} = 2\,\frac{\sqrt{\alpha^2 + 1/\alpha^2}}{\alpha^2 - 1/\alpha^2}\,\Phi^{-1}\left(\frac{3\alpha - 1}{2(\alpha + 1)}\right). \tag{7.92}$$

The zero $\varsigma^*_{\psi_0}$ is independent of N. One can further show that $\varsigma^*_{\psi_0}$ is a monotonously increasing function of α. The α interval is limited by the condition $\alpha \leq 3$.

For the log-normal operator on the (1, 10)-σSA-ES example, we have seen that the saturation value ψ_∞ is a decreasing function of N for a small value of the learning parameter τ (see Fig. 7.4). Similarly, this observation is verified on the (1, 2)-σSA-ES with a symmetric two-point operator using Eq. (7.90). As one can see, the ψ function shows on the (1, 2)-σSA-ES with a symmetric two-point operator a similar behavior as in the example using the (1, 10)-σSA-ES with a log-normal operator. This is obviously an indication – although not a proof – that the effects of actually very different σ mutation operators (two-point and log-normal operator) are similar.

7.3.2.3 Bounds for ψ.

As we have seen in examples, ψ is bounded. The lower bound (7.85) can be improved. Similarly, it is possible to find an upper bound for ψ, as well (see below). The investigations are carried out for two examples: the log-normal operator and the two-point operator. The starting point is Definition (7.31)

$$\psi(\varsigma^{*(g)}) = \mathrm{E}\left\{\frac{\varsigma^*_{1;\lambda} - \varsigma^{*(g)}}{\varsigma^{*(g)}}\right\}. \tag{7.93}$$

A lower bound is derived first. The symbol $\varsigma^*_{1;\lambda}$ in (7.93) stands for the ς^* value of the best descendant. However, this value cannot be smaller than the smallest ς^* of the λ descendants, which is expressed by the $\varsigma^*_{1:\lambda}$ order statistics notation. Hence, the inequality reads

[11] The necessary criterion $\alpha < 3$ does *not* depend on the approximation (7.88); it can also be derived directly from (7.87).

$$\psi = \mathrm{E}\left\{\frac{\varsigma^*_{1;\lambda} - \varsigma^{*(g)}}{\varsigma^{*(g)}}\right\} \geq \mathrm{E}\left\{\frac{\varsigma^*_{1:\lambda} - \varsigma^{*(g)}}{\varsigma^{*(g)}}\right\} = \frac{\overline{\varsigma^*_{1:\lambda}} - \varsigma^{*(g)}}{\varsigma^{*(g)}}. \tag{7.94}$$

The determination of $\overline{\varsigma^*_{1:\lambda}}$ is a standard task of order statistics. It has been carried out in the case of continuous distributions several times in this book. The integral expression reads

$$\overline{\varsigma^*_{1:\lambda}} = \lambda \int_{\varsigma^*=0}^{\infty} \varsigma^* p^*_\sigma(\varsigma^*) \left[\int_{\varsigma^{*\prime}=\varsigma^*}^{\infty} p^*_\sigma(\varsigma^{*\prime}) \mathrm{d}\varsigma^{*\prime}\right]^{\lambda-1} \mathrm{d}\varsigma^*. \tag{7.95}$$

In the case of the log-normal operator (7.8), the density $p^*_\sigma(\varsigma^*) = p^*_{\ln}(\varsigma^*)$ is specified by Eq. (7.48). Therefore, one obtains the integral

$$\overline{\varsigma^*_{1:\lambda}} = \frac{\lambda}{\sqrt{2\pi}\tau} \int_{\varsigma^*=0}^{\infty} \exp\left[-\frac{1}{2}\left(\frac{\ln(\varsigma^*/\varsigma^{*(g)})}{\tau}\right)^2\right] \times \left[1 - \Phi\left(\frac{\ln(\varsigma^*/\varsigma^{*(g)})}{\tau}\right)\right]^{\lambda-1} \mathrm{d}\varsigma^*. \tag{7.96}$$

After the substitution $x = -\ln(\varsigma^*/\varsigma^{*(g)})/\tau$, it follows for (7.94) that

$$\psi(\varsigma^*) \geq \frac{\lambda}{\sqrt{2\pi}} \int_{-\infty}^{\infty} \left(\mathrm{e}^{-\tau x} - 1\right) \mathrm{e}^{-\frac{1}{2}x^2} \left[\Phi(x)\right]^{\lambda-1} \mathrm{d}x. \tag{7.97}$$

Unfortunately, the integration in (7.97) cannot be carried out in an analytically closed form. However, it is possible to obtain a Taylor expansion for $(\mathrm{e}^{-\tau x} - 1)$ in τ. After this approximation, the result reads

$$\psi(\varsigma^*) \geq -c_{1,\lambda}\tau + \frac{d^{(2)}_{1,\lambda}}{2!}\tau^2 - \frac{d^{(3)}_{1,\lambda}}{3!}\tau^3 + \frac{d^{(4)}_{1,\lambda}}{4!}\tau^4 - \ldots. \tag{7.98}$$

Higher-order progress coefficients $d^{(k)}_{1,\lambda}$ appear in this result. They have been defined in (4.41). One notices after a numerical comparison of (7.98) with the saturation values ψ_∞ in Fig. 7.4 that (7.98) *cannot* be used in general as an estimation of the saturation value ψ_∞. As an interpretation, one can say that the σ-self-adaptation does not work perfectly. In other words, a deterministically "perfect" algorithm should *always* choose the individual with the smallest ς^* value (i.e. $\varsigma^*_{1:\lambda}$) if the parental ς^* values are very large. Obviously, this is not possible for the σ-SA algorithm because of the indirect selection.[12]

Eq. (7.95) is not appropriate in the case of the symmetric two-point operator for the determination of $\overline{\varsigma^*_{1:\lambda}}$. However, since there are only two possible states $\varsigma^{*(g)}/\alpha$ and $\varsigma^{*(g)}\alpha$, the expected value can be determined easily. The probability of obtaining an increase in ς^* for all λ trials, i.e. the probability of getting the state $\varsigma^{*(g)}\alpha$ λ times, is $(1/2)^\lambda$. The probability for (at least) a single decrease in ς^*, i.e. for the state $\varsigma^{*(g)}/\alpha$, is consequently $1 - (1/2)^\lambda$. Hence, the expected value of $\varsigma^*_{1:\lambda}$ is obtained

[12] The selection is based on the fitness values and not on the ς^* values which are unknown to the selection operator anyway.

$$\overline{\varsigma^*_{1:\lambda}} = \left(1 - \left(\frac{1}{2}\right)^{\lambda}\right) \frac{\varsigma^{*(g)}}{\alpha} + \left(\frac{1}{2}\right)^{\lambda} \varsigma^{*(g)}\alpha. \tag{7.99}$$

The bound for ψ results from (7.94)

$$\psi(\varsigma^*) \ \geq \ \left(\frac{1}{\alpha} - 1\right) + \left(\frac{1}{2}\right)^{\lambda} \left(\alpha - \frac{1}{\alpha}\right). \tag{7.100}$$

If λ is sufficiently large, this lower bound is dominated by the $\left(\frac{1}{\alpha} - 1\right)$ term.

The determination of the upper bound follows using the same scheme as in the lower bound. Since $\varsigma^*_{1;\lambda}$ cannot be larger than $\varsigma^*_{\lambda:\lambda}$, one obtains instead of (7.94)

$$\psi = \mathrm{E}\left\{\frac{\varsigma^*_{1;\lambda} - \varsigma^{*(g)}}{\varsigma^{*(g)}}\right\} \ \leq \ \mathrm{E}\left\{\frac{\varsigma^*_{\lambda:\lambda} - \varsigma^{*(g)}}{\varsigma^{*(g)}}\right\} = \frac{\overline{\varsigma^*_{\lambda:\lambda}} - \varsigma^{*(g)}}{\varsigma^{*(g)}}. \tag{7.101}$$

The value $\overline{\varsigma^*_{\lambda:\lambda}}$ is described by the integral

$$\overline{\varsigma^*_{\lambda:\lambda}} = \lambda \int_{\varsigma^*=0}^{\infty} \varsigma^* p^*_\sigma(\varsigma^*) \left[\int_{\varsigma^{*\prime}=0}^{\varsigma^{*\prime}=\varsigma^*} p^*_\sigma(\varsigma^{*\prime})\mathrm{d}\varsigma^{*\prime}\right]^{\lambda-1} \mathrm{d}\varsigma^*. \tag{7.102}$$

For the log-normal operator, one obtains with $p^*_\sigma(\varsigma^*) = p^*_{\mathtt{ln}}(\varsigma^*)$, Eq. (7.48), and applying the substitution $x = \ln(\varsigma^*/\varsigma^{*(g)})/\tau$, the following integral

$$\psi(\varsigma^*) \leq \frac{\lambda}{\sqrt{2\pi}} \int_{-\infty}^{\infty} (\mathrm{e}^{\tau x} - 1)\mathrm{e}^{-\frac{1}{2}x^2} \left[\Phi(x)\right]^{\lambda-1} \mathrm{d}x. \tag{7.103}$$

After Taylor expansion of $(\mathrm{e}^{\tau x} - 1)$ and integration one obtains the series

$$\psi(\varsigma^*) \ \leq \ c_{1,\lambda}\,\tau \ + \ \frac{d^{(2)}_{1,\lambda}}{2!}\,\tau^2 \ + \ \frac{d^{(3)}_{1,\lambda}}{3!}\,\tau^3 \ + \ \frac{d^{(4)}_{1,\lambda}}{4!}\,\tau^4 \ + \ \ldots. \tag{7.104}$$

Similar to the case of the lower bound (for $\varsigma^* \to \infty$), this upper bound also does not give a satisfactory estimate for the function value of ψ at $\varsigma^* = 0$. Hence, the statements made above about the "perfection" of the σSA algorithm also apply to this case: because of the indirect σ selection, the selection of the largest ς^* value out of λ descendants is not always guaranteed. As a result, $\psi(0)$ is smaller than $(\overline{\varsigma^*_{\lambda:\lambda}} - \varsigma^{*(g)})/\varsigma^{*(g)}$. (This is also the case for the symmetric two-point operator.)

In order to determine $\overline{\varsigma^*_{\lambda:\lambda}}$ for the symmetric two-point operator, the probability for the exclusive emergence of $\varsigma^{*(g)}/\alpha$ results must be determined. It is $(1/2)^\lambda$; consequently, the probability of at least one increase $\varsigma^{*(g)}\alpha$ is $1 - (1/2)^\lambda$. Therefore, the expected value reads

$$\overline{\varsigma^*_{\lambda:\lambda}} = \left(\frac{1}{2}\right)^{\lambda} \frac{\varsigma^{*(g)}}{\alpha} + \left(1 - \left(\frac{1}{2}\right)^{\lambda}\right) \varsigma^{*(g)}\alpha, \tag{7.105}$$

and the upper bound follows

$$\psi(\varsigma^*) \le (\alpha - 1) - \left(\frac{1}{2}\right)^\lambda \left(\alpha - \frac{1}{\alpha}\right). \tag{7.106}$$

Again, this bound is not very sharp in estimating $\psi(0)$, as can be seen after a comparison with Fig. 7.2 (right). More satisfactory analytical approximations of $\psi(\varsigma^*, \tau)$ will be derived for small ς^* and small τ in the next two points.

7.3.2.4 Analytic Approximation of $\psi^{(k)}$ for Small ς^* and τ – General Aspects. For the investigation of the evolutionary dynamics of the σSA, the interval $\varsigma^* \lessapprox 2c_{1,\lambda}$ is especially interesting, since one expects positive progress rates for the $(1,\lambda)$-ES for such ς^* values.[13] As can be seen at a glance in Figs. 7.2 and 7.3, the self-adaptation response ψ changes almost linearly with respect to ς^* for sufficiently small values of learning parameters. Therefore, it can be suggested that we describe the interval $\varsigma^* \lessapprox 2c_{1,\lambda}$ by a linear or quadratic approximation of ψ. Hence, this section and the next section aim at the derivation of such approximation formulae for the kth order SAR functions.

The derivation comprises three steps. In the first step, the $P^*_{1;1}(t \mid \varsigma^{*(g)})$ distribution (7.61) is simplified by an approximation. This distribution is needed in the $\psi^{(k)}$ integral (7.68). The calculations are carried out demonstratively for the log-normal operator. Interestingly, the case of the two-point operator yields the same results after a respective Taylor expansion. The derivation for the latter case is simpler as compared to the log-normal case; therefore, it will not be shown here.

In the second step, the inner ς^* integral of (7.68) is approximated, and the outer integration is carried out yielding expected value expressions that still depend on the type of the σ mutation distribution p^*_σ. They will be calculated in the third step in Sect. 7.3.2.5.

On approximating $P^*_{1;1}(t \mid \varsigma^{*(g)})$. The integrand of (7.61) and the density (7.69) will be simplified here under the condition $\varsigma^* \ll N$. For this purpose, $\sqrt{1+\varsigma^{*2}/N}$ and $\tilde{\varsigma}^*(\varsigma^*)$ (Eq. (7.47)) are expanded in a Taylor series. One obtains $\sqrt{1+\varsigma^{*2}/N} \approx 1 + \varsigma^{*2}/2N$ and $\tilde{\varsigma}^*(\varsigma^*) \approx \varsigma^*$, respectively. Consequently, $P^*_{1;1}(t \mid \varsigma^{*(g)})$ reads

$$P^*_{1;1}(t \mid \varsigma^{*(g)}) = \int_{\varsigma^*=0}^{\infty} \Phi\left(\frac{t - (1 + \varsigma^{*2}/2N)}{\varsigma^*/N}\right) p^*_\sigma(\varsigma^* \mid \varsigma^{*(g)})\, \mathrm{d}\varsigma^* + \ldots \tag{7.107}$$

and (7.69) becomes

$$p^*_r(t \mid \varsigma^*) = \frac{N}{\sqrt{2\pi}\varsigma^*} \exp\left[-\frac{1}{2}\left(\frac{t - (1 + \varsigma^{*2}/2N)}{\varsigma^*/N}\right)^2\right] + \ldots. \tag{7.108}$$

[13] For $\varsigma^* \ll N$, Eq. (7.70) can be approximated by $\varphi^*(\varsigma^*) = c_{1,\lambda}\varsigma^* - \varsigma^{*2}/2 + \ldots$. The interval $\varsigma^* \lessapprox 2c_{1,\lambda}$ follows immediately from the condition $\varphi^* \gtrapprox 0$.

The further simplification of the integral (7.107) concerns the dependency on the learning parameter. In the following, it will be shown that a Taylor expansion with respect to learning parameters τ exists for the case of the log-normal operator. The parameter τ is assumed to be small. Moreover, the expansion does not have a term linear in τ. One can conclude from this result (see below, Eq. (7.122)) that the τ dependency of $P^*_{1;1}$ can be neglected in the scope of further approximations.

For the derivation of this approximation, one should consider the expected value of a function $f(s)$ with respect to a log-normally distributed random variate s. The density of the log-normal distribution is given by Eq. (7.48). If one substitutes $s := \varsigma^*$ and $\sigma := \varsigma^{*(g)}$, the expected value integral reads

$$I(\sigma, \tau) := \mathrm{E}\{f(s)\} = \int_{s=0}^{\infty} f(s) \frac{1}{\sqrt{2\pi}\,\tau s} \exp\left[-\frac{1}{2}\left(\frac{\ln(s/\sigma)}{\tau}\right)^2\right] \mathrm{d}s. \quad (7.109)$$

Since τ is assumed to be small, $I(\sigma, \tau)$ can be expanded into a Taylor series at $\tau = 0$

$$I(\sigma, \tau) = I(\sigma, 0) + \left.\frac{\partial I}{\partial \tau}\right|_{\tau=0} \tau + \frac{1}{2} \left.\frac{\partial^2 I}{\partial \tau^2}\right|_{\tau=0} \tau^2 + \dots. \quad (7.110)$$

In order to determine the expansion coefficients in (7.110), $f(s)$ is expanded at $s = \sigma$. The idea behind this expansion is due to the fact that the "probability mass" is mainly concentrated at the state $s = \sigma$ for small values of τ.

$$f(s) = f(\sigma) + \left.\frac{\partial f}{\partial s}\right|_{s=\sigma} (s - \sigma) + \frac{1}{2} \left.\frac{\partial^2 f}{\partial s^2}\right|_{s=\sigma} (s - \sigma)^2 + \dots. \quad (7.111)$$

This series is inserted into the expansion coefficients of (7.110). The $I(\sigma, 0)$ term in (7.110) reads

$$I(\sigma, 0) = \lim_{\tau \to 0} \int_{s=0}^{\infty} \left[f(\sigma) + \left.\frac{\partial f}{\partial s}\right|_{s=\sigma} (s - \sigma) + \frac{1}{2} \left.\frac{\partial^2 f}{\partial s^2}\right|_{s=\sigma} (s - \sigma)^2 + \dots\right] \times \frac{1}{\sqrt{2\pi}\,\tau s} \exp\left[-\frac{1}{2}\left(\frac{\ln(s/\sigma)}{\tau}\right)^2\right] \mathrm{d}s. \quad (7.112)$$

The moments of the log-normal distribution appear in integral (7.112). They are given as (see, e.g., Johnson & Kotz, 1970)

$$\overline{s^k} = \int_{s=0}^{\infty} s^k \frac{1}{\sqrt{2\pi}\,\tau s} \exp\left[-\frac{1}{2}\left(\frac{\ln(s/\sigma)}{\tau}\right)^2\right] \mathrm{d}s = \sigma^k \mathrm{e}^{\frac{1}{2}k^2\tau^2}. \quad (7.113)$$

Hence, (7.112) becomes

$$I(\sigma, 0) = \lim_{\tau \to 0} \left[f(\sigma) + \frac{\partial f}{\partial s} \sigma \left(\mathrm{e}^{\frac{1}{2}\tau^2} - 1\right) + \frac{1}{2} \frac{\partial^2 f}{\partial s^2} \sigma^2 \left(\mathrm{e}^{2\tau^2} - 2\mathrm{e}^{\frac{1}{2}\tau^2} + 1\right) + \dots\right] = f(\sigma). \quad (7.114)$$

The second expansion coefficient in (7.110) is obtained after differentiating (7.109) and the substitution (7.111)

$$\left.\frac{\partial I}{\partial \tau}\right|_{\tau=0} = \lim_{\tau \to 0} \frac{1}{\tau} \int_{s=0}^{\infty} \left[f(\sigma) + \left.\frac{\partial f}{\partial s}\right|_{s=\sigma} (s-\sigma) + \frac{1}{2} \left.\frac{\partial^2 f}{\partial s^2}\right|_{s=\sigma} (s-\sigma)^2 + \ldots \right] \\ \times \left[\left(\frac{\ln(s/\sigma)}{\tau} \right)^2 - 1 \right] \frac{1}{\sqrt{2\pi}\,\tau s} \exp\left[-\frac{1}{2} \left(\frac{\ln(s/\sigma)}{\tau} \right)^2 \right] \mathrm{d}s. \quad (7.115)$$

This becomes after the substitution $x = \ln(s/\sigma)/\tau$, i.e. $s = \sigma \mathrm{e}^{\tau x}$,

$$\left.\frac{\partial I}{\partial \tau}\right|_{\tau=0} = \lim_{\tau \to 0} \frac{1}{\tau} \frac{1}{\sqrt{2\pi}} \int_{-\infty}^{\infty} \left[f(\sigma) + \frac{\partial f}{\partial s} \sigma \left(\mathrm{e}^{\tau x} - 1\right) + \frac{1}{2} \frac{\partial^2 f}{\partial s^2} \sigma^2 \left(\mathrm{e}^{\tau x} - 1\right)^2 \\ + \ldots \right] (x^2 - 1) \mathrm{e}^{-\frac{1}{2}x^2} \, \mathrm{d}x. \quad (7.116)$$

The expression $(\mathrm{e}^{\tau x} - 1)$ is written as a power series

$$\mathrm{e}^{\tau x} - 1 = \tau x + \frac{1}{2} \tau^2 x^2 + \frac{1}{3!} \tau^3 x^3 + \ldots \quad (7.117)$$

and inserted into (7.116)

$$\left.\frac{\partial I}{\partial \tau}\right|_{\tau=0} = \lim_{\tau \to 0} \frac{1}{\tau} \Bigg\{ f(\sigma) \frac{1}{\sqrt{2\pi}} \int_{-\infty}^{\infty} (x^2 - 1) \mathrm{e}^{-\frac{1}{2}x^2} \, \mathrm{d}x \\ + \frac{\partial f}{\partial s} \sigma \tau \frac{1}{\sqrt{2\pi}} \int_{-\infty}^{\infty} (x^3 - x) \mathrm{e}^{-\frac{1}{2}x^2} \, \mathrm{d}x \\ + \frac{\partial f}{\partial s} \sigma \tau^2 \frac{1}{\sqrt{2\pi}} \int_{-\infty}^{\infty} (x^2 - 1) \left(\frac{1}{2} x^2 + \frac{\tau}{3!} x^3 + \ldots \right) \mathrm{e}^{-\frac{1}{2}x^2} \, \mathrm{d}x \\ + \frac{1}{2} \frac{\partial^2 f}{\partial s^2} \sigma^2 \tau^2 \frac{1}{\sqrt{2\pi}} \int_{-\infty}^{\infty} (x^2 - 1) \left(x + \frac{\tau}{2} x^2 + \ldots \right)^2 \mathrm{e}^{-\frac{1}{2}x^2} \, \mathrm{d}x \Bigg\}. \quad (7.118)$$

The solutions to the integrals in (7.118) can be found in Table A.1 of Appendix A.1. After carrying out the integration, the limit $\tau \to 0$ is taken. It is obvious that the term (7.118) completely vanishes

$$\left.\frac{\partial I}{\partial \tau}\right|_{\tau=0} = 0. \quad (7.119)$$

The second derivative (i.e. the third expansion coefficient) in (7.110) is obtained in the simplest way by differentiating Eq. (7.118) with respect to τ (of course, before evaluating the limit)

$$\begin{aligned}
\left.\frac{\partial^2 I}{\partial \tau^2}\right|_{\tau=0} = \lim_{\tau\to 0}\Bigg\{ & -\frac{1}{\tau^2} f(\sigma)\frac{1}{\sqrt{2\pi}}\int_{-\infty}^{\infty}(x^2-1)\mathrm{e}^{-\frac{1}{2}x^2}\,\mathrm{d}x \\
& + \frac{\partial f}{\partial s}\,\sigma\frac{1}{\sqrt{2\pi}}\int_{-\infty}^{\infty}(x^2-1)\left(\frac{1}{2}x^2+\frac{\tau}{3!}x^3+\ldots\right)\mathrm{e}^{-\frac{1}{2}x^2}\,\mathrm{d}x \\
& + \frac{\partial f}{\partial s}\,\sigma\frac{\tau}{\sqrt{2\pi}}\int_{-\infty}^{\infty}(x^2-1)\frac{\partial}{\partial\tau}\left(\frac{1}{2}x^2+\frac{\tau}{3!}x^3+\ldots\right)\mathrm{e}^{-\frac{1}{2}x^2}\,\mathrm{d}x \\
& + \frac{1}{2}\frac{\partial^2 f}{\partial s^2}\,\sigma^2\frac{1}{\sqrt{2\pi}}\int_{-\infty}^{\infty}(x^2-1)\left(x+\frac{\tau}{2}x^2+\ldots\right)^2\mathrm{e}^{-\frac{1}{2}x^2}\,\mathrm{d}x \\
& + \frac{1}{2}\frac{\partial^2 f}{\partial s^2}\,\sigma^2\frac{\tau}{\sqrt{2\pi}}\int_{-\infty}^{\infty}(x^2-1)\frac{\partial}{\partial\tau}\left(x+\frac{\tau}{2}x^2+\ldots\right)^2\mathrm{e}^{-\frac{1}{2}x^2}\,\mathrm{d}x\Bigg\}.
\end{aligned}$$

After carrying out the integrations and calculating the limit $\tau\to 0$, this becomes

$$\left.\frac{\partial^2 I}{\partial \tau^2}\right|_{\tau=0} = \sigma\left.\frac{\partial f}{\partial s}\right|_{s=\sigma} + \sigma^2\left.\frac{\partial^2 f}{\partial s^2}\right|_{s=\sigma}. \tag{7.120}$$

After inserting the intermediate results (7.114, 7.119) and (7.120) into (7.110) one finally obtains

$$I(\sigma,\tau) = f(\sigma) + \frac{\tau^2}{2}\sigma\left(\left.\frac{\partial f}{\partial s}\right|_{s=\sigma} + \sigma\left.\frac{\partial^2 f}{\partial s^2}\right|_{s=\sigma}\right) + \mathcal{O}(\tau^4). \tag{7.121}$$

As one can see, there is no linear τ term in this expansion. Therefore, for sufficiently small values of τ, $f(\sigma)$ can be used for approximating $I(\sigma,\tau)$. As a result, the distribution (7.107) can be approximated as

$$P^*_{1;1}(t\mid\varsigma^{*(g)}) = \Phi\left(\frac{t-(1+(\varsigma^{*(g)})^2/2N)}{\varsigma^{*(g)}/N}\right) + \mathcal{O}(\tau^2). \tag{7.122}$$

As already noted, this approximation formula is also valid for the case of the symmetric two-point distribution $p^*_\sigma = p^*_{\mathrm{II}}$, Eq. (7.50).

Approximation of the $\psi^{(k)}$ integral. The approximation of (7.68) can be resumed now. After inserting (7.122) and (7.108), (7.68) reads (using the notation $s=\varsigma^{*(g)}$)

$$\begin{aligned}
\psi^{(k)} = \frac{\lambda}{\sqrt{2\pi}}\int_{t=0}^{\infty}&\left[1-\Phi\left(\frac{t-(1+s^2/2N)}{s/N}\right)\right]^{\lambda-1}\int_{\varsigma^*=0}^{\infty}\left(\frac{\varsigma^*-s}{s}\right)^k\frac{N}{\varsigma^*} \\
&\times\exp\left[-\frac{1}{2}\left(\frac{t-(1+(\varsigma^*)^2/2N)}{\varsigma^*/N}\right)^2\right]p^*_\sigma(\varsigma^*\mid s)\,\mathrm{d}\varsigma^*\,\mathrm{d}t.
\end{aligned} \tag{7.123}$$

Using the substitution $x=-(t-(1+s^2/2N))/(s/N)$, one obtains

$$\begin{aligned}
\psi^{(k)} = \frac{\lambda}{\sqrt{2\pi}}\int_{x=-\infty}^{\infty}&[\Phi(x)]^{\lambda-1} \\
&\times\int_{\varsigma^*=0}^{\infty}\frac{s}{\varsigma^*}\left(\frac{\varsigma^*-s}{s}\right)^k h(\varsigma^*,x,s)\,p^*_\sigma(\varsigma^*\mid s)\,\mathrm{d}\varsigma^*\,\mathrm{d}x.
\end{aligned} \tag{7.124}$$

The auxiliary function $h(\varsigma^*, x, s)$ reads

$$h(\varsigma^*, x, s) := \exp\left[-\frac{1}{2}\left(\frac{s}{\varsigma^*}x + \frac{(\varsigma^*)^2 - s^2}{2\varsigma^*}\right)^2\right]. \tag{7.125}$$

In order to approximate the inner integral of (7.124), h is expanded into a Taylor series at $\varsigma^* = s$. This step is justified by the smallness assumed for the learning parameters τ (or β in the case $p^*_\sigma = p_{\rm II}$). This smallness assumption guarantees that the "probability mass" of p^*_σ is mainly concentrated around the state s. The series reads

$$h(\varsigma^*, x, s) = h(\varsigma^*, x, s)\big|_{\varsigma^*=s} + \left.\frac{\partial h}{\partial \varsigma^*}\right|_{\varsigma^*=s}(\varsigma^* - s) + \left.\frac{1}{2}\frac{\partial^2 h}{\partial \varsigma^{*2}}\right|_{\varsigma^*=s}(\varsigma^* - s)^2 + \dots \tag{7.126}$$

with

$$h(\varsigma^*, x, s)\big|_{\varsigma^*=s} = \mathrm{e}^{-\frac{1}{2}x^2}. \tag{7.127}$$

The first and second derivative in (7.126) are obtained after lengthy calculations. The use of a computer algebra system is advisable. The results read

$$\left.\frac{\partial h}{\partial \varsigma^*}\right|_{\varsigma^*=s} = \mathrm{e}^{-\frac{1}{2}x^2}\frac{1}{s}\left(x^2 - sx\right) \tag{7.128}$$

and

$$\left.\frac{\partial^2 h}{\partial \varsigma^{*2}}\right|_{\varsigma^*=s} = \mathrm{e}^{-\frac{1}{2}x^2}\frac{1}{s^2}\left[x^4 - 2sx^3 + (s^2-3)x^2 + 3sx - s^2\right]. \tag{7.129}$$

After inserting these results into (7.126), and thereafter (7.126) into (7.124), one obtains the following integral for the $\psi^{(k)}$ function

$$\begin{aligned}\psi^{(k)}(s) = {} & \frac{\lambda}{\sqrt{2\pi}}\int_{x=-\infty}^{\infty}\mathrm{e}^{-\frac{1}{2}x^2}\left[\Phi(x)\right]^{\lambda-1} \\ & \times \int_{\varsigma^*=0}^{\infty} p^*_\sigma(\varsigma^* \,|\, s)\left\{\frac{s}{\varsigma^*}\left(\frac{\varsigma^*-s}{s}\right)^k + \frac{s}{\varsigma^*}\left(\frac{\varsigma^*-s}{s}\right)^{k+1}(x^2 - sx)\right. \\ & \left. + \frac{1}{2}\frac{s}{\varsigma^*}\left(\frac{\varsigma^*-s}{s}\right)^{k+2}\left[x^4 - 2sx^3 + (s^2-3)x^2 + 3sx - s^2\right]\right\}\mathrm{d}\varsigma^*\,\mathrm{d}x \\ & + \dots .\end{aligned} \tag{7.130}$$

The integration order is changed, and the integration over x is carried out. The definitions of the higher-order progress coefficients (4.41) and (4.42) are used ($d^{(1)}_{1,\lambda} = c_{1,\lambda}$ and $d^{(0)}_{1,\lambda} = 1$). One finally obtains the following expression for the kth *order self-adaptation response*

$$\boxed{\begin{aligned}\psi^{(k)}(s) = {} & \overline{\frac{s}{\varsigma^*}\left(\frac{\varsigma^* - s}{s}\right)^k} + \left(d_{1,\lambda}^{(2)} - s c_{1,\lambda}\right) \overline{\frac{s}{\varsigma^*}\left(\frac{\varsigma^* - s}{s}\right)^{k+1}} \\ & + \frac{1}{2}\left[d_{1,\lambda}^{(4)} - 2s d_{1,\lambda}^{(3)} + (s^2 - 3) d_{1,\lambda}^{(2)} + 3 s c_{1,\lambda} - s^2\right] \overline{\frac{s}{\varsigma^*}\left(\frac{\varsigma^* - s}{s}\right)^{k+2}} + \ldots\end{aligned}} \tag{7.131}$$

with

$$\boxed{\overline{\frac{s}{\varsigma^*}\left(\frac{\varsigma^* - s}{s}\right)^m} := \int_{\varsigma^*=0}^{\infty} p_\sigma^*(\varsigma^* \mid s)\, \frac{s}{\varsigma^*}\left(\frac{\varsigma^* - s}{s}\right)^m \mathrm{d}\varsigma^*.} \tag{7.132}$$

The expected values (7.132) are obtained depending on the σ mutation operator. This operator determines the density p_σ^*. In the next part, the SAR functions with $k = 1$ and $k = 2$ are investigated for the cases of the log-normal operator and symmetric two-point operator.

7.3.2.5 Approximations for ψ, $\psi^{(2)}$, and D_ψ. In order to determine ψ and $\psi^{(2)}$, the expected values (7.132) are to be calculated for $m = 1$ up to and including $m = 4$. This will be done for the two σ mutation densities p_{ln}^* and p_{II}^*. The same derivation can be carried out analogously for other densities, as e.g., $p_{\mathcal{N}}^*$ in Eq. (7.49), p_β^* in Eq. (7.51), or p_{III}^* in Eq. (7.52).

The case of the log-normal operator. The case $p_\sigma^* = p_{\mathrm{ln}}^*(\varsigma^*)$ will be considered first. The moments of the log-normal distribution are given by Eq. (7.113). Therefore, one obtains for (7.132)

$$\begin{aligned}\overline{\frac{s}{\varsigma^*}\left(\frac{\varsigma^* - s}{s}\right)^1} &= 1 - s\overline{\left(\frac{1}{\varsigma^*}\right)} = 1 - s\overline{(\varsigma^*)^{-1}} = 1 - s\left(s^{-1} \mathrm{e}^{\frac{1}{2}(-1)^2\tau^2}\right) \\ &= 1 - \mathrm{e}^{\frac{1}{2}\tau^2},\end{aligned} \tag{7.133}$$

$$\begin{aligned}\overline{\frac{s}{\varsigma^*}\left(\frac{\varsigma^* - s}{s}\right)^2} &= \frac{1}{s}\left[\overline{\varsigma^*} - 2s + s^2\overline{\left(\frac{1}{\varsigma^*}\right)}\right] = \mathrm{e}^{\frac{1}{2}\tau^2} - 2 + \mathrm{e}^{\frac{1}{2}\tau^2} \\ &= 2\left(\mathrm{e}^{\frac{1}{2}\tau^2} - 1\right),\end{aligned} \tag{7.134}$$

$$\begin{aligned}\overline{\frac{s}{\varsigma^*}\left(\frac{\varsigma^* - s}{s}\right)^3} &= \frac{1}{s^2}\left[\overline{(\varsigma^*)^2} - 3s\overline{\varsigma^*} + 3s^2 - s^3\overline{\left(\frac{1}{\varsigma^*}\right)}\right] \\ &= \mathrm{e}^{2\tau^2} - 4\mathrm{e}^{\frac{1}{2}\tau^2} + 3,\end{aligned} \tag{7.135}$$

$$\begin{aligned}\overline{\frac{s}{\varsigma^*}\left(\frac{\varsigma^* - s}{s}\right)^4} &= \frac{1}{s^3}\left[\overline{(\varsigma^*)^3} - 4s\overline{(\varsigma^*)^2} + 6s^2\overline{\varsigma^*} - 4s^3 + s^4\overline{\left(\frac{1}{\varsigma^*}\right)}\right] \\ &= \mathrm{e}^{\frac{9}{2}\tau^2} - 4\mathrm{e}^{2\tau^2} + 7\mathrm{e}^{\frac{1}{2}\tau^2} - 4.\end{aligned} \tag{7.136}$$

The derivations in Sect. 7.3.2.4 have been done assuming small values for τ. Therefore, it is sufficient to approximate (7.133–7.136) by respective power series. Including third-order τ expressions one obtains

$$\overline{\frac{s}{\varsigma^*}\left(\frac{\varsigma^*-s}{s}\right)^1} = 1 - \left(1+\frac{1}{2}\tau^2+\frac{1}{4}\tau^4+\dots\right) = -\frac{1}{2}\tau^2 + \mathcal{O}(\tau^4), \quad (7.137)$$

$$\overline{\frac{s}{\varsigma^*}\left(\frac{\varsigma^*-s}{s}\right)^2} = 2\left(\left(1+\frac{1}{2}\tau^2+\frac{1}{4}\tau^4+\dots\right)-1\right) = \tau^2 + \mathcal{O}(\tau^4), \quad (7.138)$$

$$\overline{\frac{s}{\varsigma^*}\left(\frac{\varsigma^*-s}{s}\right)^3} = 1+2\tau^2-4-4\frac{1}{2}\tau^2+3+\mathcal{O}(\tau^4) = 0+\mathcal{O}(\tau^4), \quad (7.139)$$

$$\begin{aligned}\overline{\frac{s}{\varsigma^*}\left(\frac{\varsigma^*-s}{s}\right)^4} &= 1+\frac{9}{2}\tau^2-4-8\tau^2+7+\frac{7}{2}\tau^2-4+\mathcal{O}(\tau^4)\\ &= 0+\mathcal{O}(\tau^4). \end{aligned} \quad (7.140)$$

Using these results, the expressions for the SAR function can be simplified considerably for ψ and $\psi^{(2)}$. By inserting (7.137–7.140) into (7.131), the *first-order self-adaptation response* reads for the log-normal operator (7.8) (back-substitution: $\varsigma^{*(g)} = s$)

$$\text{LOG-NORMAL OPERATOR:} \quad \psi_{1,\lambda}(\varsigma^{*(g)}) = \tau^2\left[\left(d_{1,\lambda}^{(2)}-\frac{1}{2}\right) - c_{1,\lambda}\varsigma^{*(g)}\right] + \dots \quad (7.141)$$

and for the *second-order SAR function*

$$\psi_{1,\lambda}^{(2)}(\varsigma^{*(g)}) = \tau^2 \; + \; \dots . \quad (7.142)$$

For small values of ς^*, the error terms of (7.141) and (7.142) are of order $\mathcal{O}(\tau^4)$.

Using these results, one can calculate the deviation strength of the self-adaptation response. According to (7.36), the standard deviation of the σ fluctuations are proportional to D_ψ. The quantity D_ψ can be obtained immediately from (7.141) and (7.142)

$$D_\psi = \sqrt{\psi^{(2)} - \psi^2} = \sqrt{\tau^2 + \mathcal{O}(\tau^4)} = \tau \; + \; \mathcal{O}(\tau^2). \quad (7.143)$$

It is interesting to note that the deviation strength of the self-adaptation response ψ (i.e. its standard deviation) has the same order of magnitude as the learning parameter τ. This result is in agreement with the ES simulations (see also the remarks for Fig. 7.3 on p. 282).

As has been pointed out several times, these approximation results for the first- and second-order SAR functions are only valid for sufficiently small values of τ ($\tau \lessapprox 0.3$) and for small (normalized) mutation strengths $\varsigma^{*(g)} \lessapprox 2c_{1,\lambda}$.

The latter condition is mainly based on the series cut off in (7.126) and (7.137–7.140). These approximations can be improved without any technical difficulties. However, they are not carried out here and remain as a task for the future. Figure 7.5 shows the approximation quality of the ψ formula (7.141). Since (7.141) is a linear function of $\varsigma^{*(g)}$, its validity range is limited. However, its prediction for the zero $\varsigma^*_{\psi_0}$ of ψ is satisfactory. Moreover, the approximation error is tolerable up to the mutation strength $\varsigma^* \approx 2\varsigma^*_{\psi_0}$.

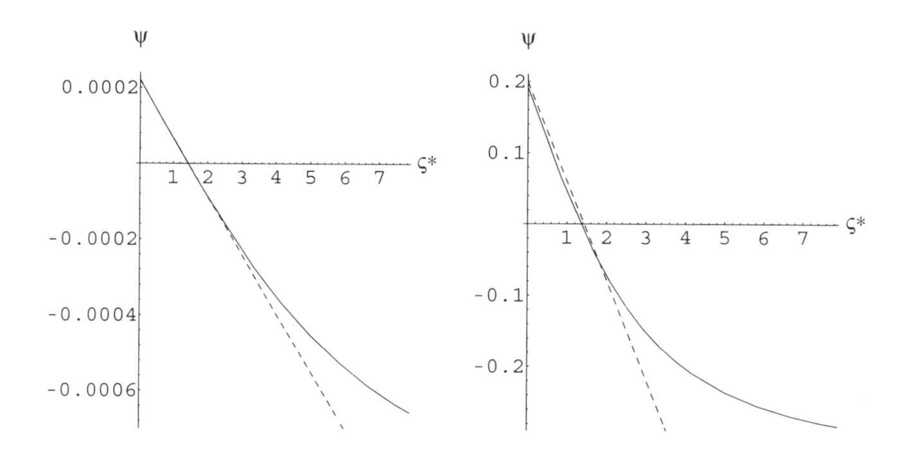

Fig. 7.5. The verification of the approximation quality of the linear ψ function (7.141), dashed line, with the results obtained from numerical integrations for the $(1,10)$-ES, $N = 30$. Left: $\tau = 0.01$; right: $\tau = 0.3$

The case of the two-point operator. The moments of the two-point density $p^*_\sigma = p^*_{\text{II}}$ are needed to determine (7.132). Considering (7.50) and substituting $s = \varsigma^{*(g)}$, one obtains

$$\overline{(\varsigma^*)^m} = \int_{\varsigma^*=0}^{\infty} (\varsigma^*)^m \frac{1}{2}\left[\delta(\varsigma^* - s\alpha) + \delta(\varsigma^* - s/\alpha)\right] \mathrm{d}\varsigma^*$$

$$\overline{(\varsigma^*)^m} = \frac{s^m}{2}\left(\alpha^m + \frac{1}{\alpha^m}\right). \tag{7.144}$$

Instead of (7.133–7.136), one gets

$$\overline{\frac{s}{\varsigma^*}\left(\frac{\varsigma^* - s}{s}\right)} = 1 - \frac{1}{2}\left(\frac{1}{\alpha} + \alpha\right), \tag{7.145}$$

$$\overline{\frac{s}{\varsigma^*}\left(\frac{\varsigma^* - s}{s}\right)^2} = \frac{1}{2}\left(\alpha + \frac{1}{\alpha}\right) - 2 + \frac{1}{2}\left(\frac{1}{\alpha} + \alpha\right) = \alpha + \frac{1}{\alpha} - 2, \tag{7.146}$$

$$\overline{\frac{s}{\varsigma^*}\left(\frac{\varsigma^*-s}{s}\right)^3} = \frac{1}{2}\left(\alpha^2+\frac{1}{\alpha^2}\right) - \frac{3}{2}\left(\alpha+\frac{1}{\alpha}\right) + 3 - \frac{1}{2}\left(\frac{1}{\alpha}+\alpha\right)$$
$$= \frac{1}{2}\left(\alpha^2+\frac{1}{\alpha^2}\right) - 2\left(\alpha+\frac{1}{\alpha}\right) + 3, \qquad (7.147)$$

$$\overline{\frac{s}{\varsigma^*}\left(\frac{\varsigma^*-s}{s}\right)^4} = \frac{1}{2}\left(\alpha^3+\frac{1}{\alpha^3}\right) - 2\left(\alpha^2+\frac{1}{\alpha^2}\right) + 3\left(\alpha+\frac{1}{\alpha}\right) - 4$$
$$+\frac{1}{2}\left(\frac{1}{\alpha}+\alpha\right)$$
$$= \frac{1}{2}\left(\alpha^3+\frac{1}{\alpha^3}\right) - 2\left(\alpha^2+\frac{1}{\alpha^2}\right) + \frac{7}{2}\left(\alpha+\frac{1}{\alpha}\right) - 4. \qquad (7.148)$$

As in the case of the p^*_{1n} variant, the Eqs. (7.145–7.148) are approximated for small values of the learning parameter $\alpha \approx 1$. Since $\alpha = 1 + \beta$ holds because of (7.13), the formulae for small β values ($\beta > 0$) are expanded up to and including β^3. One obtains as intermediate results

$$\frac{1}{\alpha} = 1 - \beta + \beta^2 - \beta^3 + \mathcal{O}(\beta^4),$$

$$\frac{1}{\alpha^2} = 1 - 2\beta + 3\beta^2 - 4\beta^3 + \mathcal{O}(\beta^4),$$

$$\frac{1}{\alpha^3} = 1 - 3\beta + 6\beta^2 - 10\beta^3 + \mathcal{O}(\beta^4),$$

$$\alpha^2 = 1 + 2\beta + \beta^2,$$

$$\alpha^3 = 1 + 3\beta + 3\beta^2 + \beta^3.$$

These are inserted into (7.145–7.148)

$$\overline{\frac{s}{\varsigma^*}\left(\frac{\varsigma^*-s}{s}\right)} = 1 - \frac{1}{2} + \frac{1}{2}\beta - \frac{1}{2}\beta^2 + \frac{1}{2}\beta^3 - \frac{1}{2} - \frac{1}{2}\beta + \mathcal{O}(\beta^4)$$
$$= -\frac{1}{2}\beta^2 + \frac{1}{2}\beta^3 + \mathcal{O}(\beta^4), \qquad (7.149)$$

$$\overline{\frac{s}{\varsigma^*}\left(\frac{\varsigma^*-s}{s}\right)^2} = \beta^2 - \beta^3 + \mathcal{O}(\beta^4), \qquad (7.150)$$

$$\overline{\frac{s}{\varsigma^*}\left(\frac{\varsigma^*-s}{s}\right)^3} = \frac{1}{2} + \beta + \frac{1}{2}\beta^2 + \frac{1}{2} - \beta + \frac{3}{2}\beta^2 - 2\beta^3$$
$$-2 - 2\beta - 2 + 2\beta - 2\beta^2 + 2\beta^3 + 3 + \mathcal{O}(\beta^4)$$
$$= 0 + 0\beta + 0\beta^2 + 0\beta^3 + \mathcal{O}(\beta^4), \qquad (7.151)$$

$$
\begin{aligned}
\overline{\frac{s}{\varsigma^*}\left(\frac{\varsigma^*-s}{s}\right)^4} &= \frac{1}{2}+\frac{3}{2}\beta+\frac{3}{2}\beta^2+\frac{1}{2}\beta^3+\frac{1}{2}-\frac{3}{2}\beta+3\beta^2-5\beta^3-2 \\
&\quad -4\beta-2\beta^2-2+4\beta-6\beta^2+8\beta^3+\frac{7}{2}+\frac{7}{2}\beta+\frac{7}{2} \\
&\quad -\frac{7}{2}\beta+\frac{7}{2}\beta^2-\frac{7}{2}\beta^3-4+\mathcal{O}(\beta^4) \\
&= 0+0\beta+0\beta^2+0\beta^3+\mathcal{O}(\beta^4).
\end{aligned}
\tag{7.152}
$$

After inserting (7.149–7.152) into (7.131), the *first-order self-adaptation response* reads for the case of the symmetric two-point operator (7.14) (substituting back $\varsigma^{*(g)}=s$)

TWO-POINT OPERATOR:
$\psi_{1,\lambda}(\varsigma^{*(g)}) = \beta^2(1-\beta)\left[\left(d_{1,\lambda}^{(2)}-\frac{1}{2}\right)-c_{1,\lambda}\varsigma^{*(g)}\right] + \ldots$

(7.153)

and the *second-order SAR function*

$$
\psi_{1,\lambda}^{(2)}(\varsigma^{*(g)}) = \beta^2(1-\beta) + \ldots . \tag{7.154}
$$

For small ς^*, the error terms of (7.153) and (7.154) are of order $\mathcal{O}(\beta^4)$.

Comparison of the effects of the σ mutation operators. The microscopic effects of the σ mutation operators can be compared for the cases of the log-normal operator and the two-point operator. Their first- and second-order SAR functions are used in this comparison. One finds a remarkable correspondence: for any given learning parameter β there is a learning parameter τ leading to the same first- and second-order SAR function. That is, these two σ mutation rules correspond to each other with their first two SAR functions. In other words, if

$p_{\text{ln}} \leftrightarrow p_{\text{II}}$ - CORRESPONDENCE:	$\tau^2 = \beta^2(1-\beta)$

(7.155)

is fulfilled, both σ mutation rules exhibit the same microscopic behavior. Since the macroscopic evolution is determined by the microscopic forces, this result means that both self-adaptation strategies should show the same dynamic behavior for the respective choice of the learning parameters. This is a very remarkable – and unexpected – prediction, if one takes into account that the log-normal operator (7.8) and the two-point operator (7.14) are very different σ mutation rules.

Naturally, one should recall the condition that this prediction was obtained in the framework of the approximations used. Thorough experimental comparisons of both σ mutation rules on real ES algorithms remain still to be done. For our further investigations on the ES dynamics, however, this question will not play a role because the analysis in this book will rely on the respective approximations made. Experiments on the $(1, \lambda)$-σSA-ES with log-normal operators have confirmed the admissibility of these approximations. The results will be reported below.

7.4 The $(1, \lambda)$-σSA Evolution (I) – Dynamics in the Deterministic Approximation

This section and the next one aim at explaining how and why the σSA algorithm works in the time domain. Because of the $p_{\ln} \leftrightarrow p_{\text{II}}$ correspondence principle (7.155), one can restrict the investigations to one of these σ mutation rules. In the following, the SAR functions of the log-normal operator will be used in the investigations. The results can be immediately transferred to the two-point operator by substituting the learning parameter τ according to $\tau \to \beta\sqrt{1-\beta}$.

7.4.1 The Evolution Equations of the $(1, \lambda)$-σSA-ES

The macroscopic evolution, i.e. the dynamics of the $(1, \lambda)$-σSA-ES, is determined by the two evolution equations (7.28, 7.39):

$$r^{(g+1)} = r^{(g)} \left(1 - \frac{1}{N} \varphi^*(\varsigma^{*(g)}, N)\right) + \frac{r^{(g)}}{N} D_\varphi^*(\varsigma^{*(g)}, N) \mathcal{N}(0,1) \quad (7.156)$$

and

$$\varsigma^{*(g+1)} = \varsigma^{*(g)} \frac{1 + \psi^{(1)}(\varsigma^{*(g)}, N) + D_\psi(\varsigma^{*(g)}, N) \mathcal{N}_\psi(0,1)}{1 - \frac{1}{N}\varphi^*(\varsigma^{*(g)}, N) + \frac{1}{N} D_\varphi^*(\varsigma^{*(g)}, N) \mathcal{N}_\varphi(0,1)}. \quad (7.157)$$

From the mathematical perspective, these are *stochastic iterative mappings* (noise-perturbed difference equations). The analytical treatment of such systems is difficult. Therefore, it will be carried out using approximations. The analysis is simplified by the fact that the evolution equation (7.157) is decoupled from (7.156) in one way as a result of the transformation from ς evolution to ς^* evolution. In other words, the evolution of r is governed by the variable ς^*; however, there is *no* feedback from r to ς^*. Thus, Eq. (7.157) can be solved separately. In spite of this simplification, the analytical treatment of (7.157) is still a difficult problem. Further simplifications are necessary.

The roughest but logical simplification is the assumption that the stochastic terms in (7.156) and (7.157) do not influence the evolution of the mean values of r and ς^*. The idea is to simply "discard" the fluctuations

$$D_\psi \stackrel{!}{=} 0, \qquad D_\varphi^* \stackrel{!}{=} 0. \quad (7.158)$$

In this way, the system (7.156, 7.157) reduces to *deterministic* equations

$$r^{(g+1)} = r^{(g)} \left(1 - \frac{1}{N} \varphi^*(\varsigma^{*(g)}, N)\right), \quad (7.159)$$

$$\varsigma^{*(g+1)} = \varsigma^{*(g)} \frac{1 + \psi^{(1)}(\varsigma^{*(g)}, N)}{1 - \frac{1}{N}\varphi^*(\varsigma^{*(g)}, N)}. \qquad (7.160)$$

The investigation of their dynamics is carried out in the following parts of this section. The stochastic terms will be included in the analysis in Sect. 7.5.

7.4.2 The ES in the Stationary State

7.4.2.1 Determining the Stationary State. The system state of the $(1, \lambda)$-σSA-ES on the sphere model is completely determined by specifying $(r^{(g)}, \varsigma^{*(g)})^{\mathrm{T}}$ (for constant exogenous strategy parameters). The *stationary state* of the ES system, also referred to as *steady-state*, is the state obtained for large time scales, i.e. for a sufficiently large number of the generation counter g. It is clear that this state is necessarily nonstatic in time: the prerequisite for a correct working of the ES as an optimization algorithm is the permanent decrease of the residual distance $r^{(g)}$. Hence, "stationariness" means in this context the evolution of a normalized mutation strength which is constant in time.[14] That is, one asymptotically expects from a correctly working σSA algorithm a stationary mutation strength ς^*_{ss}

$$\varsigma^*_{ss} := \lim_{g \to \infty} \varsigma^{*(g)}. \qquad (7.161)$$

In other words, for sufficiently large g, $\varsigma^{*(g)} = \varsigma^{*(g+1)} = \varsigma^*_{ss}$ should hold. If one inserts this condition into (7.160), one obtains the *criterion* for the stationary state

$$\text{Steady-State Condition:} \quad \frac{1}{N}\varphi^*(\varsigma^*_{ss}, N) = -\psi(\varsigma^*_{ss}, \tau, N). \qquad (7.162)$$

The σ-state ς^*_{ss} defined by this condition determines (and controls) the r evolution after a certain adaptation time g_A (see Sect. 7.4.2.2).

To have evolutionary progress, i.e. $r^{(g+1)} < r^{(g)}$, one has to require $\varphi^*(\varsigma^*) > 0$ because of (7.159). Therefore, the stationary mutation strength ς^*_{ss} must obey the

$$\text{Necessary Evolution Condition:} \quad 0 < \varsigma^*_{ss} < \varsigma^*_{\varphi_0}. \qquad (7.163)$$

Here, the symbol $\varsigma^*_{\varphi_0}$ denotes the second zero[15] of $\varphi^*(\varsigma^*)$

$$\varsigma^*_{\varphi_0}: \quad \varphi^*(\varsigma^*) = 0 \quad \Leftrightarrow \quad \varsigma^* = \varsigma^*_{\varphi_0} \quad \wedge \quad \varsigma^* > 0. \qquad (7.164)$$

For large N ($N \to \infty$), the stationary state ς^*_{ss} is determined by the zero $\varsigma^*_{\psi_0}$ of $\psi(\varsigma^*)$. The reason for this is because $-\psi(\varsigma^*)$ is a monotonically increasing function of ς^*. The zero of this function is at $\varsigma^*_{\psi_0}$ (see the explanations and figures in Sect. 7.3.2.1). The left side of (7.162) is a function with a non-negative maximum. Thus, the stationary state ς^*_{ss} is determined as the

[14] In the stochastic case, this "constancy of ς^* condition" must be replaced by a "constancy of $p(\varsigma^*)$ distribution condition."

[15] The first zero of $\varphi^*(\varsigma^*)$ is the trivial case $\varsigma^* = 0$.

intersection of the functions φ^*/N and $-\psi$ according to (7.162). Therefore, it must be located to the right of $\varsigma^*_{\psi_0}$. In other words, $\varsigma^*_{ss} \ge \varsigma^*_{\psi_0}$ holds. In the limit case $N \to \infty$, the left side of (7.162) becomes zero. Consequently, the stationary state ς^*_{ss} approaches $\varsigma^*_{\psi_0}$. In summary, one obtains as the "operation domain" for convergence of the $(1,\lambda)$-σSA-ES in the framework of the deterministic approximation (7.158–7.160)

$$\varsigma^*_{\psi_0} < \varsigma^*_{ss} < \varsigma^*_{\varphi_0}. \tag{7.165}$$

The inequality (7.165), in particular $\varsigma^*_{\psi_0} < \varsigma^*_{\varphi_0}$, can be interpreted as a *necessary condition* (design condition). The probability density $p^*_\sigma(\varsigma^* \,|\, \varsigma^{*(g)})$ of the σ mutation operator must fulfill this condition in order to be a suitable "candidate" for the σSA algorithm.

One can easily verify that the condition $\varsigma^*_{\psi_0} < \varsigma^*_{\varphi_0}$ is fulfilled for the log-normal operator (7.8) with the density $p^*_\sigma = p^*_{\text{ln}}(\varsigma^* \,|\, \varsigma^{*(g)})$ – at least for sufficiently small learning parameters τ (but $\tau > 0$). In this case, $\varphi^*(\varsigma^*)$ can be approximated by (7.70). Moreover, for $N \to \infty$, the quadratic progress law

$$\varphi^*(\varsigma^*) = c_{1,\lambda}\varsigma^* - \frac{1}{2}\varsigma^{*2} \tag{7.166}$$

holds. Consequently, one obtains for the second zero of φ^*

$$\varsigma^*_{\varphi_0} = 2c_{1,\lambda}. \tag{7.167}$$

The zero of the SAR function is obtained by equating (7.141) to zero

$$\varsigma^*_{\psi_0} = \left(d^{(2)}_{1,\lambda} - \frac{1}{2}\right) \Big/ \, c_{1,\lambda}. \tag{7.168}$$

The zeros $\varsigma^*_{\psi_0}$ and the progress coefficients $c_{1,\lambda}$ are listed in Table D.1 of Appendix D.2. A numerical comparison confirms the validity of $\varsigma^*_{\psi_0} < \varsigma^*_{\varphi_0}$. For the asymptotic limit case $\lambda \to \infty$, the validity of the inequality can even be shown mathematically. To this end, one should recall that asymptotically

$$\lim_{\lambda\to\infty} d^{(2)}_{1,\lambda} \Big/ \, c^2_{1,\lambda} = 1 \tag{7.169}$$

holds (see Eq. (D.10) on p. 360). Therefore, for $\varsigma^*_{\psi_0}$ one obtains using (7.168) and (7.167)

$$\lambda \to \infty: \qquad \varsigma^*_{\psi_0} \simeq c_{1,\lambda} \qquad \Longrightarrow \qquad 0 < \varsigma^*_{\psi_0} < \varsigma^*_{\varphi_0}. \tag{7.170}$$

The steady-state mutation strength ς^*_{ss} is a function of the learning parameter τ. This can also be inferred from the stationarity condition (7.162): ψ depends on τ. If one resolves (7.162) for ς^*_{ss} by using (7.166) and (7.141), one gets

$$\varsigma^*_{ss} = c_{1,\lambda}(1 - N\tau^2) \; + \; \sqrt{c^2_{1,\lambda}(1 - N\tau^2)^2 + N\tau^2\left(2d^{(2)}_{1,\lambda} - 1\right)}. \tag{7.171}$$

Consequently, the ς^*_{ss} value depends on the parameter space dimension N, on the number of descendants λ, and the learning parameter τ. If N and λ are given as fixed (exogenous) parameters, the stationary mutation strength depends on τ only. Actually, the whole admissible ς^*_{ss} interval (7.165) can be controlled by a respective tuning of τ. One obtains from Eq. (7.171) the two extremes

$$\varsigma^*_{ss} = \frac{1}{2}\frac{2d^{(2)}_{1,\lambda} - 1}{c_{1,\lambda}} = \varsigma^*_{\psi_0}, \qquad \text{if} \quad N\tau^2 \gg 1, \tag{7.172}$$

$$\varsigma^*_{ss} = 2c_{1,\lambda} = \varsigma^*_{\varphi_0}, \qquad \text{if} \quad N\tau^2 \to 0. \tag{7.173}$$

The question now is how to choose the τ value such that a reasonable performance of the σSA algorithm can be obtained.

7.4.2.2 Optimal ES Performance and the $1/\sqrt{N}$ Rule. The choice of the learning parameter can directly influence the performance of the ES algorithm. This was recognized already in the early period of ES theory. Scheel (1985, pp. 57f) suggested for the choice of the α value in the two-point operator (7.14) the proportionality

$$\alpha \sim (1 - c^2_{1,\lambda}/2N)^{-1} \quad \Longrightarrow \quad \beta = \alpha - 1 \sim c^2_{1,\lambda}/2N \propto 1/N.$$

A seemingly correct justification for this suggestion follows directly from Eq. (7.5). However, as Scheel noticed, this choice of α proved itself to be too small in numerical simulations.

Indeed, one can immediately show that the $1/N$ scaling rule must fail. The log-normal operator is considered here; however, the result is also valid for the two-point operator because of the correspondence rule (7.155). The value $\tau \propto 1/N$ for the learning parameter yields $N\tau^2 \propto N(1/N)^2 = 1/N \to 0$ for large N. Therefore, the stationary mutation strength ς^*_{ss} is obtained using Eq. (7.173), i.e. ς^*_{ss} is close to $\varsigma^*_{\varphi_0}$. In other words, the progress rate becomes zero, $\varphi^* \to 0$.

Naturally, one aims at a maximal progress rate. This is obtained for the approximation (7.166) at a mutation strength $\varsigma^* = c_{1,\lambda}$. Hence, in the scope of the deterministic approximation considered here, one requires for

$$\text{MAXIMAL PERFORMANCE:} \qquad \varsigma^*_{ss} = c_{1,\lambda}. \tag{7.174}$$

As a result, the steady-state condition (7.162) yields

$$\frac{1}{N}\frac{c^2_{1,\lambda}}{2} = -\tau^2\left[\left(d^{(2)}_{1,\lambda} - \frac{1}{2}\right) - c^2_{1,\lambda}\right]. \tag{7.175}$$

Resolved for τ, one immediately obtains

$$\tau = \frac{1}{\sqrt{N}}\frac{c_{1,\lambda}}{\sqrt{2c^2_{1,\lambda} + 1 - 2d^{(2)}_{1,\lambda}}} \qquad \Longrightarrow \qquad \tau \propto \frac{1}{\sqrt{N}}. \tag{7.176}$$

We have found therefore the scaling rule $\tau \propto 1/\sqrt{N}$ proposed by Schwefel (1974). The value $c_{1,\lambda}$ can be used as the proportionality factor as long as λ is sufficiently large. As one can easily show using the asymptotic property (7.169), the square root $\sqrt{2c_{1,\lambda}^2 + 1 - 2d_{1,\lambda}^{(2)}}$ in (7.176) becomes one under this condition. Thus one obtains

$$\tau\text{-SCALING RULE:} \qquad \tau \approx \frac{c_{1,\lambda}}{\sqrt{N}}. \tag{7.177}$$

It must be emphasized that formula (7.177) is just a *rule*. It has been derived based on the (approximate) validity of the deterministic equations of dynamics (7.159, 7.160) leading to (7.176); additionally, the square root $\sqrt{2c_{1,\lambda}^2 + 1 - 2d_{1,\lambda}^{(2)}}$ was simplified. As a remark, the left formula in (7.176) is *not* defined for $\lambda < 4$: the square root becomes imaginary. However, the $\tau \propto 1/\sqrt{N}$ rule holds also for the stochastic case of the equation of dynamics, and it is independent of the choice of λ. This will be shown below in the investigations considering the fluctuations (see Sect. 7.5 and especially Sect. 7.5.6.3).

As already mentioned, the left formula in (7.176) fails for $\lambda < 4$. It is worth noting that the σSA algorithm is not able to tune the optimal mutation strength $\varsigma = c_{1,\lambda}$, at least in the scope of the deterministic approximation. This follows already from the violation of the inequality (7.165) for $\varsigma_{ss}^* = c_{1,\lambda}$: one finds in Table D.1 (Appendix D.2) $\varsigma_{\psi_0}^* \not< \varsigma_{ss}^* = c_{1,\lambda}$ for $\lambda < 4$. However, one *cannot* conclude by this that the σSA algorithm principally fails for $\lambda < 4$ just because it cannot tune the optimal state for the mutation strength. The self-adaptation also functions for the case $\lambda < 4$. The stationary state reached can be determined using Eq. (7.171). Considering the scaling rule (7.177), one gets for $\lambda = 2$ a stationary value $\varsigma_{ss}^* \approx 1.892c_{1,2} \approx 1.067$: it corresponds to a progress rate $\varphi_{1,2}^* \approx 0.0325$. For $\lambda = 3$, one gets $\varsigma_{ss}^* \approx 1.561c_{1,3} \approx 1.321$ and $\varphi_{1,3}^* \approx 0.245$. In both cases, one obtains $\varsigma_{ss}^* < 2c_{1,\lambda}$; therefore, the algorithm can proceed with self-adaptation. However, the progress rates obtained are very poor as compared to the corresponding maximal theoretical values $\max[\varphi_{1,2}^*] \approx 0.159$ and $\max[\varphi_{1,3}^*] \approx 0.358$. Therefore, it is not recommended to use a $(1, 2)$-σSA-ES or a $(1, 3)$-σSA-ES.

7.4.2.3 The Differential Equation of the σ Evolution, the Transient Behavior for Small $\varsigma^{*(0)} < \varsigma_{ss}^*$, and the Stationary r Dynamics. The choice of the learning parameter τ with respect to the optimal stationary ς^*-state is one aspect of σSA. Another aspect is the expected *adaptation time* g_A. This time period denotes the number of generations needed by an algorithm with a wrongly tuned initial σ value to approach the vicinity of the ς_{ss}^* state.

Starting from an initial state $(\varsigma^{(0)}, r^{(0)})^{\mathrm{T}}$ and therefore $(\varsigma^{*(0)}, r^{(0)})^{\mathrm{T}}$, the dynamics of the ES system is determined by the iterative mapping (7.159, 7.160). Two cases will be considered in order to gain an insight into the evolution dynamics of self-adaptation. The case with too small an initial mutation strength, i.e. $\varsigma^{*(0)} < \varsigma^*_{ss}$, is considered in this section. In the next section, the extreme case with a too large initial mutation strength $\varsigma^{*(0)} \gg \varsigma^*_{ss}$ will be discussed.

In order to simplify the analysis, the denominator of the difference equation (7.160) is expanded into a Taylor series, and the series is cut after the linear term. This is admissible as long as the condition

$$\left| \frac{\varphi^*(\varsigma^*)}{N} \right| \ll 1 \tag{7.178}$$

holds. Since the case $\varsigma^{*(0)} < \varsigma^*_{ss}$ is investigated here, we will assume that $\varsigma^* < 2c_{1,\lambda}$. Therefore, one obtains $c^2_{1,\lambda}/2 \ge \varphi^* > 0$ such that the condition (7.178) is not very restrictive. The Taylor expansion yields (see also Eq. (7.40))

$$\varsigma^{*(g+1)} = \varsigma^{*(g)} \left(1 + \psi^{(1)}(\varsigma^{*(g)})\right) \left(1 + \frac{1}{N}\varphi^*(\varsigma^{*(g)})\right) + \mathcal{O}\left(\left(\frac{\varphi^*}{N}\right)^2\right). \tag{7.179}$$

After inserting (7.166) and (7.141) into (7.179), one obtains

$$\varsigma^{*(g+1)} = \varsigma^{*(g)} + \tau^2 \varsigma^{*(g)} \left[\left(\left(d^{(2)}_{1,\lambda} - \frac{1}{2} \right) - c_{1,\lambda}\varsigma^{*(g)} \right) \left(1 + \frac{\varphi^*(\varsigma^{*(g)})}{N} \right) + \frac{1}{N\tau^2} \left(c_{1,\lambda}\varsigma^{*(g)} - \frac{1}{2}(\varsigma^{*(g)})^2 \right) \right]. \tag{7.180}$$

The term φ^*/N can be neglected as compared to 1 in this expression. Furthermore, one can assume that τ^2 is sufficiently small. As a result, the difference $\varsigma^{*(g+1)} - \varsigma^{*(g)}$ can be interpreted as a differential quotient (see also the derivation of the differential equation of R in Sect. 2.4.1). Thus, one obtains the *differential equation of the σ^* evolution.* This equation serves as an approximation to (7.180). The notation $\varsigma^{*(g)} = \sigma^*(g)$ will be used in the following in order to distinguish it from the discrete time dynamics

$$\frac{\mathrm{d}}{\mathrm{d}g}\sigma^*(g) = \tau^2\sigma^*(g) \left[\left(d^{(2)}_{1,\lambda} - \frac{1}{2} \right) - c_{1,\lambda}\left(1 - \frac{1}{N\tau^2}\right)\sigma^*(g) - \frac{1}{2}\frac{1}{N\tau^2}(\sigma^*(g))^2 \right]. \tag{7.181}$$

This differential equation can be solved for g by separation. After using the abbreviations

$$a := -\frac{1}{2}\frac{1}{N\tau^2}, \qquad b := -c_{1,\lambda}\left(1 - \frac{1}{N\tau^2}\right), \qquad c := d^{(2)}_{1,\lambda} - \frac{1}{2}, \tag{7.182}$$

one obtains using an integration table (Bronstein & Semendjajew, 1981, p. 89) under the condition $\sigma^*(g) < \varsigma^*_{ss}$

$$g = \frac{1}{\tau^2} \frac{1}{2d_{1,\lambda}^{(2)} - 1} \left\{ \ln\left(\frac{\sigma^{*2}(g)}{\sigma^{*2}(0)} \frac{c + b\sigma^*(0) + a\sigma^{*2}(0)}{c + b\sigma^*(g) + a\sigma^{*2}(g)} \right) \right.$$
$$\left. + \frac{b}{\sqrt{b^2 - 4ac}} \ln\left(\frac{2a\sigma^*(g) + b + \sqrt{b^2 - 4ac}}{2a\sigma^*(g) + b - \sqrt{b^2 - 4ac}} \frac{2a\sigma^*(0) + b - \sqrt{b^2 - 4ac}}{2a\sigma^*(0) + b + \sqrt{b^2 - 4ac}} \right) \right\}. \tag{7.183}$$

There is no analytically closed formula for the $\sigma^*(g)$ dynamics itself, since (7.183) cannot be solved for $\sigma^*(g)$. Equation (7.183), however, can be used for estimating the number of generations for the adaptation period. This quantity estimates the number of generations needed to reach a "certain vicinity" of the stationary state ς^*_{ss}.

It is interesting to investigate the scaling behavior of the *adaptation time* g_A (transient period). The special case $\tau \propto 1/\sqrt{N}$ is considered here, i.e. the learning parameter τ is chosen according to this scaling rule. Under this condition, the contents of the brace in (7.183) is a constant given fixed values of $\sigma^*(0)$ and $\sigma^*(g)$. The number of generations necessary to attain a specific $\sigma^*(g)$, $0 < \sigma^*(0) < \sigma^*(g) < \varsigma^*_{ss}$, is consequently inversely proportional to τ^2. Hence, g is proportional to N because of the scaling rule $\tau \propto 1/\sqrt{N}$. One finds for the scaling behavior of the adaptation time g_A on the $(1, \lambda)$-σSA-ES working in the stationary maximal performance regime

$$\text{ADAPTATION TIME:} \qquad g_A \propto N, \qquad \text{if} \qquad \tau \propto \frac{1}{\sqrt{N}}. \tag{7.184}$$

Apparently, this is the other side of the coin: an σSA algorithm that operates stationarily optimal has an adaptation time g_A proportional to the problem size N. For most applications, this should not be critical since the parameter space dimension of nonlinear optimization problems are seldom greater than, say, $N = 200$. However, the property (7.184) will cause problems for large-scale optimization tasks.

As an example, the adaptation behavior of a $(1, 10)$-ES is tested with $N = 10,000$. The optimal τ value for maximal stationary performance is $\tau \approx 0.015$ according to (7.177). The ES is initialized at a residual distance $r^{(0)} = 10^4$, however, with a totally badly chosen mutation strength $\sigma = 10^{-3}$ (i.e. $\sigma^{*(0)} = 10^{-3}$). The normalized σ values of a real ES simulation run were recorded for $100,000$ generations. The evolution of the $\sigma^{*(g)}$ values is plotted in the left-hand side of Fig. 7.6 over the first $30,000$ generations. The solid smooth curve was obtained by numerical calculation of the inverse function to (7.183). As one can see, the σSA algorithm requires approximately $20,000$ generations to attain the stationary state. In the right-hand side, the value $\tau = 0.15$ is chosen in contrast. The theory (see Eq. (7.183)) predicts a reduction of the adaptation time by a factor of $1/100$ for a 10-fold increase

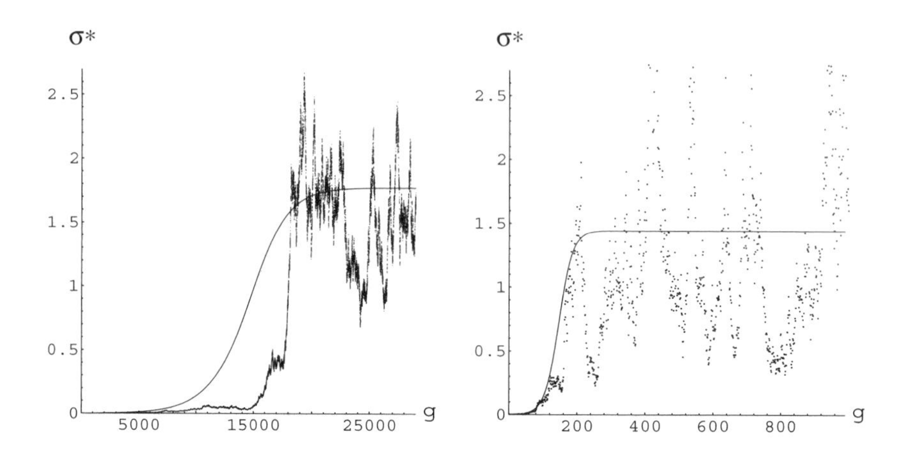

Fig. 7.6. On the dependence of the adaptation time on the learning parameter τ. The $\sigma^*(g)$ dynamics of the deterministic approximation (smooth curve) is compared with the results of a real ES simulation run (data points). Left: optimal $\tau = 0.015$ ($N = 10,000$, $(1,10)$-ES); right: the same ES, but with $\tau = 0.15$

in τ. As can be seen in Fig. 7.6, this is at least the case in the order of magnitude.

On the stationary r dynamics. The dynamics of r can be described for the deterministic approximation (7.159) under Condition (7.178) by using the differential equation (2.101). Its formal integration was given by Eq. (2.102):

$$r(g) = r(0) \exp\left[-\frac{1}{N}\int_{g'=0}^{g'=g} \varphi^*(\sigma^*(g'))\, \mathrm{d}g'\right]. \tag{7.185}$$

For $\sigma^*(g) = \text{const}$, the integration in (7.185) can be carried out. Therefore, it is possible to calculate the dynamics for the stationary case $\sigma^*(g) = \varsigma^*_{ss}$ (neglecting the fluctuations). Considering (7.166), one obtains[16]

$$r(g) = r(g_1) \exp\left[-\frac{g-g_1}{N}\left(c_{1,\lambda} - \frac{1}{2}\varsigma^*_{ss}\right)\varsigma^*_{ss}\right]. \tag{7.186}$$

On the decadic (base 10) logarithmic scale this equation becomes

$$\lg(r(g)) = \lg(r(g_1)) \;-\; \lg(\mathrm{e})\,\frac{\varsigma^*_{ss}}{N}\left(c_{1,\lambda} - \frac{1}{2}\varsigma^*_{ss}\right)(g-g_1). \tag{7.187}$$

As one can see, the stationary state of the σSA-ES is characterized by a constant negative slope $-\lg(\mathrm{e})\,\varphi^*(\varsigma^*_{ss})/N$ of the logarithmic $r(g)$ function. Consequently, the σSA-ES exhibits a *linear convergence behavior* (see Sect. 2.4.2).

[16] This formula presumes that the σSA-ES has reached the vicinity of the stationary state $\sigma^*(g_1) \approx \varsigma^*_{ss}$ after $g = g_1$ generations.

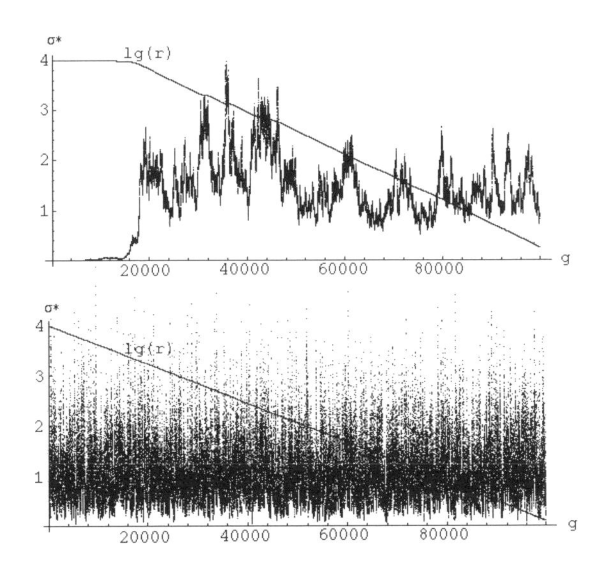

Fig. 7.7. The ES experiment on the dynamics of the $(1, 10)$-ES with $N = 10,000$, $\sigma(0) = 10^{-3}$, $r(0) = 10^4$. Top: $\tau = 0.015$; bottom: $\tau = 0.15$

In Fig. 7.7, the complete record of the ES simulation run described above is depicted over the period of $100,000$ generations. In addition to Fig. 7.6, the $\lg(r) = f(g)$ curve is also plotted.[17] In the upper part of Fig. 7.7, a stationary progress rate $\varphi^*_\infty \approx 1.03$ is measured (from $g = 20,000$ to $g = 100,000$). In contrast, only a $\varphi^*_\infty \approx 0.90$ is attained in the lower part. Because of the long adaptation time in the upper variant ($\tau = 0.015$), however, more than $100,000$ generations are necessary to surpass the variant with the nonoptimal learning parameter $\tau = 0.15$. (Note, the actual values obtained here are valid for the wrongly chosen start condition $\sigma(0) = 10^{-3}$.)

The results obtained above can be summarized as follows: fast self-adaptation and optimal performance are mutually exclusive goals. In other words, there is no *single* learning parameter that fulfills both objectives. However, this observation already yields a solution: one has to alter the learning parameter over the generations: in the initial phase, e.g., for the first, say $1,000$, generations, an N-independent learning parameter should be used, such as $\tau = 0.3$. The σSA process should be almost finished in this time interval, and the system should be in the neighborhood of the stationary state. After this interval, the learning parameter value is changed according to the N-dependent rule $\tau = c_{1,\lambda}/\sqrt{N}$. In this way, it should be roughly possible to combine fast self-adaptation with maximal ES performance.

[17] It is remarkable that $\lg(r)$ is almost linear and smooth in spite of enormous fluctuations in ς^* (even in the stationary region). This can be explained by the smoothing effect of the integration in (7.185).

7.4.2.4 Approaching the Steady-State from $\varsigma^* \gg \varsigma^*_{ss}$. Differing from the case $\varsigma^* \lessapprox \varsigma^*_{ss}$, the evolution equation (7.179) cannot be used directly in the derivation: because of $\varsigma^* \gg \varsigma^*_{ss}$, condition (7.178) is violated. If ς^* is very large, i.e.

$$\varsigma^{*(0)} \gg N, \tag{7.188}$$

the asymptotic expressions for ψ and φ^* can be used

$$\psi(\varsigma^*) = \psi_\infty = \text{const} \tag{7.189}$$

$$\varphi^*(\varsigma^*) \sim -\sqrt{N}\,\varsigma^*. \tag{7.190}$$

Equation (7.189) follows immediately from the definition of the saturation value of ψ (7.84), and (7.190) can be derived from (7.70) or directly taken from (3.243). After inserting (7.189, 7.190) into (7.160) and considering (7.188), the expected ς^* value of the first generation reads

$$\varsigma^{*(1)} \sim \varsigma^{*(0)} \frac{1+\psi_\infty}{\varsigma^{*(0)}/\sqrt{N}} = (1+\psi_\infty)\sqrt{N}. \tag{7.191}$$

The result is noteworthy: already in the first generation, the initial mutation strength $\varsigma^* = \varsigma^{*(0)} \gg N$ is reduced to a value *smaller* than $\sqrt{N}$. However, this "adaptation" is bought at a high price as to the r evolution. Using (7.190), one obtains from (7.159)

$$r^{(1)} \sim r^{(0)}\,\varsigma^{*(0)}/\sqrt{N}. \tag{7.192}$$

This means that the individuals are driven off the optimum just in the first generation by a factor of about $\varsigma^{*(0)}/\sqrt{N}$ in the statistical mean. Of course, this can be interpreted as a desired step into a global search. However, such a behavior is usually unwanted. The initial state $\mathbf{y}^{(0)}$ is usually chosen with prior knowledge about the location of the optimum, which would get lost already in the first generation. In such cases, it is better to start with a sufficiently small $\varsigma^{*(0)}$; however, this again requires a corresponding prior knowledge.

We may come back again to the evolution dynamics. There is only one "large" ς^* step from generation $g=0$ to generation $g=1$. All other changes in ς^* are again moderate. For example, the relative change in ς^* from $g=1$ to $g=2$ has the order of magnitude $1/\sqrt{2}$, and $|\varphi(\varsigma^*)/N| < 1$ is fulfilled. Therefore the difference equation (7.179) can be used for (roughly) approximating the ES dynamics. For sufficiently small τ, the differential equation (7.181) can serve as a continuous approximation. Given the initial condition $\sigma^*(1) \approx \sqrt{N}$ (see Eq. (7.191)), this differential equation describes the self-adaptation process starting from generation $g=1$. As one can see, the adaptation time g_A exhibits a scaling behavior according to (7.184) also for this case, provided that $\tau \propto 1/\sqrt{N}$ is used. Furthermore, the influence of the learning parameter τ is almost *independent* of the choice of the initial conditions.

7.5 The $(1, \lambda)$-σSA Evolution (II) – Dynamics with Fluctuations

7.5.1 Motivation

The essential aspects of the σSA process can be investigated using the deterministic approach. Nevertheless, there is a certain necessity to include fluctuations in the considerations of the evolution dynamics. In particular, this becomes clear when looking at Fig. 7.7. The fluctuations of the mutation strength are very large. Hence, these fluctuations are expected to influence the mean value dynamics.

The numerical investigation of the ES runs corresponding to Fig. 7.7 confirm this supposition. The difference between the stationary mutation strength ς^*_{ss} obtained from Eq. (7.171) and that (σ^*) obtained from the real ES runs should not be neglected. For $\tau = 0.015$, Eq. (7.171) yields $\varsigma^*_{ss} \approx 1.77$. In contrast to that, one obtains the stationary mean value of σ^* as $\sigma^*_\infty \approx 1.50$ after averaging from $g = 20,000$ to $g = 100,000$. Similarly, for $\tau = 0.15$ one gets the prediction $\varsigma^*_{ss} \approx 1.44$ from Eq. (7.171) and $\sigma^*_\infty \approx 1.08$ from the simulation.

The fluctuations around the stationary state σ^*_∞ are large compared to the fluctuations of their "driving forces" (microscopic forces). The standard deviation D_ψ of these driving forces is of order τ (see (7.143) and the following discussion). Moreover, it appears remarkable that the value of the learning parameter has a relatively small influence on the standard deviation of the stationary ς^* density. The following conclusion can be drawn from these observations: the microscopic fluctuations drive a *nonlinear* stochastic process that causes a shift of the stationary mean value ς^* compared to the deterministic prediction.

This section aims at extracting the essential nonlinearities of the stochastic process that influence the ς^* density. Firstly, the Chapman-Kolmogorov equations will be derived for the stochastic mapping of the $(\varsigma^*, r)^{\mathrm{T}}$ evolution in Sect. 7.5.2. To this end, the transition densities must be determined. The approximate treatment of these equations with the momentum method follows in Sect. 7.5.3 up to Sect. 7.5.5. The results will be discussed in Sect. 7.5.6.

7.5.2 Chapman-Kolmogorov Equation and Transition Densities

The description of the σSA evolution as a stochastic process necessitate the extension of the state conception. In the framework of the deterministic approximation, it was sufficient to specify $(\varsigma^{(g)}, r^{(g)})^{\mathrm{T}}$ or $(\varsigma^{*(g)}, r^{(g)})^{\mathrm{T}}$, respectively, to describe the evolution of the next generation on the sphere model. In contrast to that, only probabilistic statements on the state at generation g can be made in the general case. In other words, the state is described just by a probability density of the general form $p(\varsigma^*, r; g)$. However, since ς and r

are mutated independently in Algorithm 4, this probability density simplifies to

$$p(\varsigma^*, r; g) = p(\varsigma^*; g)\, p(r; g) = p(\varsigma^{*(g)})\, p(r^{(g)}). \tag{7.193}$$

The transition from generation g to $g+1$ is described in general by the transition density $p(\varsigma^{*\prime}, r' \,|\, \varsigma^*, r; g)$. Since $\varsigma^{*\prime}$ and r' are statistically independent, this density decomposes to a transition from ς^* to $\varsigma^{*\prime}$ and to a further one for the transition of r.

The transition densities for the state variables $(\varsigma^{(g)}, r^{(g)})^{\mathrm{T}}$ have already been determined. They can be found in (7.53) and (7.57), respectively. However, they cannot be directly used in the investigation of stochastic dynamics since they have a complicated structure. Therefore, one has to find transition densities containing the normalized mutation strength and which are of the form $p_{1;\lambda}(r^{(g+1)} \,|\, \varsigma^{*(g)}, r^{(g)})$ and $p_{1;\lambda}(\varsigma^{*(g+1)} \,|\, \varsigma^{*(g)})$. The resulting Chapman-Kolmogorov equation for the stochastic evolution of r reads

$$p(r^{(g+1)}) = \int_{\varsigma^{*(g)}=0}^{\infty} \int_{r^{(g)}=0}^{\infty} p_{1;\lambda}(r^{(g+1)} \,|\, \varsigma^{*(g)}, r^{(g)})\, p(r^{(g)}) \times p(\varsigma^{*(g)})\, \mathrm{d}r^{(g)}\, \mathrm{d}\varsigma^{*(g)} \tag{7.194}$$

and of ς^*

$$p(\varsigma^{*(g+1)}) = \int_{\varsigma^{*(g)}=0}^{\infty} p_{1;\lambda}(\varsigma^{*(g+1)} \,|\, \varsigma^{*(g)})\, p(\varsigma^{*(g)})\, \mathrm{d}\varsigma^{*(g)}. \tag{7.195}$$

These equations describe the evolution of the density functions of ς^* and r. Their form is already adapted to the concrete model of the $(1,\lambda)$-σSA-ES: the evolution of ς^* is autonomous. However, the evolution of r is controlled by ς^*. It is a more or less formal act to write down these integral equations, besides the fact that the structure of the transition densities has already been taken into account. The difficult question is how these transition densities should be determined.

Based on previous results, these transition densities can be determined approximately in a relatively easy manner. We first consider the case of the transition density of r, $p_{1;\lambda}(r^{(g+1)} \,|\, \varsigma^{*(g)}, r^{(g)})$. This conditional density can be obtained from the conditional cumulative distribution function

$$P_{1;\lambda}(r \,|\, \varsigma^{*(g)}, r^{(g)}) = \Pr(r^{(g+1)} < r \,|\, \varsigma^{*(g)}, r^{(g)}) \tag{7.196}$$

by differentiating with respect to r

$$p_{1;\lambda}(r^{(g+1)} \,|\, \varsigma^{*(g)}, r^{(g)}) = \frac{\mathrm{d}}{\mathrm{d}r} P_{1;\lambda}(r \,|\, \varsigma^{*(g)}, r^{(g)}) \Big|_{r=r^{(g+1)}}. \tag{7.197}$$

For the calculation of $\Pr(r^{(g+1)} < r \mid \varsigma^{*(g)}, r^{(g)})$, one should recall that the evolution of r can be described by the difference equation (7.28) with an additional stochastic term. The approximate character in (7.28) rests only on the assumption that the fluctuations can be approximated by Gaussian noise.[18] Consequently, the inequality in (7.196) reads

$$r^{(g+1)} = r^{(g)} \left(1 - \frac{1}{N} \varphi^*(\varsigma^{*(g)})\right) + \frac{r^{(g)}}{N} D_\varphi^*(\varsigma^{*(g)}) \mathcal{N}(0,1) \;<\; r. \qquad (7.198)$$

By solving for the random variate $\mathcal{N}(0,1)$, one obtains

$$\mathcal{N}(0,1) \;<\; \frac{N}{D_\varphi^*(\varsigma^{*(g)}) r^{(g)}} \left[r - r^{(g)} \left(1 - \frac{1}{N} \varphi^*(\varsigma^{*(g)})\right)\right] \qquad (7.199)$$

and consequently for the distribution function

$$\begin{aligned} P_{1;\lambda}(r \,|\, \varsigma^{*(g)}, r^{(g)}) &= \Pr(r^{(g+1)} < r \,|\, \varsigma^{*(g)}, r^{(g)}) \\ &= \Phi\left[\frac{N}{D_\varphi^*(\varsigma^{*(g)}) r^{(g)}} \left[r - r^{(g)} \left(1 - \frac{1}{N} \varphi^*(\varsigma^{*(g)})\right)\right]\right]. \end{aligned}$$

Finally, the differentiation in (7.197) yields

$$\begin{aligned} & p_{1;\lambda}(r^{(g+1)} \,|\, \varsigma^{*(g)}, r^{(g)}) \\ &= \frac{1}{\sqrt{2\pi}} \frac{N}{D_\varphi^*(\varsigma^{*(g)}) r^{(g)}} \exp\left\{ -\frac{1}{2} \left[\frac{N}{D_\varphi^*(\varsigma^{*(g)}) r^{(g)}} \left[r^{(g+1)} \right.\right.\right. \\ &\qquad \left.\left.\left. - r^{(g)} \left(1 - \frac{1}{N} \varphi^*(\varsigma^{*(g)})\right)\right]\right]^2 \right\}. \qquad (7.200) \end{aligned}$$

It is worth noting that the derivation of $p_{1;\lambda}(r^{(g+1)} \,|\, \varsigma^{*(g)}, r^{(g)})$ presented could be avoided since the result can directly be read from Eq. (7.198): the random variable $r^{(g+1)}$ is simply normally distributed. However, it was important to demonstrate the derivation on this simple example since the same method will also be used in the determination of the transition density $p_{1;\lambda}(\varsigma^{(g+1)} \,|\, \varsigma^{*(g)})$. Its starting point is (7.40), which is an approximation to the more complicated Eq. (7.39). The transition density is obtained analogous to (7.197) after differentiating the respective conditional distribution function

$$p_{1;\lambda}(\varsigma^{*(g+1)} \,|\, \varsigma^{*(g)}) = \frac{\mathrm{d}}{\mathrm{d}\varsigma^*} P_{1;\lambda}(\varsigma^* \,|\, \varsigma^{*(g)}) \bigg|_{\varsigma^* = \varsigma^{*(g+1)}}, \qquad (7.201)$$

where

$$P_{1;\lambda}(\varsigma^* \,|\, \varsigma^{*(g)}) = \Pr\left(\varsigma^{*(g+1)} < \varsigma^* \,|\, \varsigma^{*(g)}\right). \qquad (7.202)$$

After carrying out some multiplications in (7.40) and rearranging terms, one gets

[18] It is in principle possible to approximate this distribution to a desired accuracy using a series expansion of Hermite polynomials. Such a series considers the deviations in the higher-order moments from the normal distribution (see Appendix B.3.2).

$$\begin{aligned}\varsigma^{*(g+1)} = \varsigma^{*(g)} \Bigg[&\left(1 + \psi(\varsigma^{*(g)})\right) \left(1 + \frac{\varphi^*(\varsigma^{*(g)})}{N}\right) \\ &- \left(1 + \psi(\varsigma^{*(g)})\right) \frac{D_\varphi^*(\varsigma^{*(g)})}{N} \mathcal{N}_\varphi(0,1) \\ &+ \left(1 + \frac{\varphi^*(\varsigma^{*(g)})}{N} - \frac{D_\varphi^*(\varsigma^{*(g)})}{N} \mathcal{N}_\varphi(0,1)\right) D_\psi \mathcal{N}_\psi(0,1) \Bigg] \\ + \ldots. \end{aligned} \tag{7.203}$$

For further simplifications, $|\psi(\varsigma^*)| \ll 1$ is assumed. This condition is fulfilled for sufficiently small learning parameters τ because of (7.141). One obtains

$$\begin{aligned}\varsigma^{*(g+1)} = \varsigma^{*(g)} \Bigg[&1 + \psi(\varsigma^{*(g)}) + \frac{\varphi^*(\varsigma^{*(g)})}{N} \\ &- \frac{D_\varphi^*(\varsigma^{*(g)})}{N} \mathcal{N}_\varphi(0,1) + D_\psi \mathcal{N}_\psi(0,1) \Bigg] + \ldots. \end{aligned} \tag{7.204}$$

The two independent normally distributed random variates $\mathcal{N}_\varphi$ and $\mathcal{N}_\psi$ can be combined in a new one, denoted by $\mathcal{N}$. The variance D^2 of this new random variate is obtained as the sum of the previous two variances $D^2 = D_\psi^2 + (D_\varphi^*(\varsigma^{*(g)}))^2/N^2$. Taking (7.81) and (7.143) into account for these two variances, D^2 reads

$$\begin{aligned} D^2 &= \tau^2 + \left(\tilde{\varsigma}^*(\varsigma^{*(g)})\right)^2 \left(d_{1,\lambda}^{(2)} - c_{1,\lambda}^2\right) \Big/ N^2 \\ D^2 &= \tau^2 \left(1 + \frac{\tilde{\varsigma}^{*2}\left(d_{1,\lambda}^{(2)} - c_{1,\lambda}^2\right)}{N^2\tau^2}\right). \end{aligned} \tag{7.205}$$

The second term in the parentheses of (7.205) can be neglected if the learning parameter τ is chosen according to the scaling rule (7.177), or if one ensures that $\tau \gtrapprox 1/\sqrt{N}$ is fulfilled. Additionally, the ς^* values must be sufficiently small, $\varsigma^{*2} \ll N$.[19] Thus, one obtains $D \approx \tau$ and Eq. (7.204) simplifies to

$$\varsigma^{*(g+1)} = \varsigma^{*(g)} \left[1 + \psi(\varsigma^{*(g)}) + \frac{\varphi^*(\varsigma^{*(g)})}{N} + \tau\mathcal{N}(0,1)\right] + \ldots. \tag{7.206}$$

In order to derive the transition density $p_{1;\lambda}(\varsigma^{*(g+1)} \,|\, \varsigma^{*(g)})$, the probability for $\varsigma^{*(g+1)} < \varsigma^*$ is determined next. Considering (7.206), the inequality can be reordered for $\mathcal{N}(0,1)$

$$\mathcal{N}(0,1) \;<\; \frac{1}{\tau\varsigma^{*(g)}} \left[\varsigma^* - \varsigma^{*(g)} \left(1 + \psi(\varsigma^{*(g)}) + \frac{\varphi^*(\varsigma^{*(g)})}{N}\right)\right]. \tag{7.207}$$

[19] One can easily verify that $0 < d_{1,\lambda}^{(2)} - c_{1,\lambda}^2 \le 1$ holds by using Table D.1.

The error terms denoted by "$+\ldots$" have been neglected in this step. Comparing (7.207) with (7.199) and repeating the analogous steps from (7.199) to (7.200), one finally obtains the transition density

$$p_{1;\lambda}(\varsigma^{*(g+1)} \,|\, \varsigma^{*(g)}) = \frac{1}{\sqrt{2\pi}} \frac{1}{\tau\varsigma^{*(g)}} \exp\left\{ -\frac{1}{2} \frac{1}{\left(\tau\varsigma^{*(g)}\right)^2} \left[\varsigma^{*(g+1)} - \varsigma^{*(g)} \left(1 + \psi(\varsigma^{*(g)}) + \frac{\varphi^*(\varsigma^{*(g)})}{N}\right)\right]^2 \right\}. \quad (7.208)$$

7.5.3 Mean Value Dynamics of the r Evolution

The system of integral equations (7.194, 7.195) with transition densities (7.200) and (7.208) contains the complete information on the σSA process. For a given initial $(\varsigma^{*(0)}, r^{(0)})^{\mathrm{T}}$ distribution, e.g., in terms of the initial values $p(\varsigma^{*(0)}) = \delta(\varsigma^{*(0)} - \sigma_0^*)$, $p(r^{(0)}) = \delta(r^{(0)} - R_0)$, one can describe the evolution in the "space of probabilities" using *path integrals*. These are obtained by treating Eqs. (7.194) and (7.195) as iteration instructions. Thus, one obtains $2g$-fold integral equations. However, the description of the evolution by path integrals is not appropriate for practical purposes. Other approximate approaches, such as the *Fokker-Planck equation*, allow for the approximate determination of the probability densities over time (see, e.g., Ahlbehrendt & Kempe, 1984).

The explicit knowledge of the $p(\varsigma^{*(g)})$ and $p(r^{(g)})$ densities is mostly not required. The evolution of certain statistical parameters of the distributions suffices for most questions. For example, the expected value $R(g) := \mathrm{E}\{r^{(g)}\}$ is the principal quantity for the evolution of r

$$\boxed{R(g) := \overline{r^{(g)}} = \int_{r^{(g)}=0}^{\infty} r^{(g)} p(r^{(g)}) \, \mathrm{d}r^{(g)}.} \quad (7.209)$$

Taking the Eqs. (7.209, 7.194, 7.200) into account, one obtains for the expected value of $R(g+1)$

$$R(g+1) = \int_{\varsigma^{*(g)}=0}^{\infty} \int_{r^{(g)}=0}^{\infty} r^{(g)} \left(1 - \frac{1}{N}\varphi^*(\varsigma^{*(g)})\right) p(r^{(g)})\, p(\varsigma^{*(g)}) \, \mathrm{d}r^{(g)} \, \mathrm{d}\varsigma^{*(g)},$$

$$R(g+1) = R(g) \left(1 - \frac{1}{N} \overline{\varphi^*(\varsigma^{*(g)})}\right) \quad (7.210)$$

where

$$\overline{\varphi^*(\varsigma^{*(g)})} = \int_{\varsigma^{*(g)}=0}^{\infty} \varphi^*(\varsigma^{*(g)})\, p(\varsigma^{*(g)}) \, \mathrm{d}\varsigma^{*(g)}. \quad (7.211)$$

Equation (7.210) has the same structure as the dynamic Eq. (7.159) in the deterministic approximation. Therefore, the mean value dynamics of the evolution of r can be approximated by a *linear* differential equation similar to (2.101) as in the case of the deterministic approximation

$$\frac{\mathrm{d}R}{\mathrm{d}g} = -R\,\frac{\overline{\varphi^*(\varsigma^{*(g)})}}{N}. \tag{7.212}$$

Consequently, the formal solutions (2.102) and (7.185), respectively, hold also for the case of stochastic dynamics.

Provided that the evolution of ς^* possesses a stationary $p_\infty(\varsigma^*)$ density for sufficiently large $g \geq g_1$ or that at least the (first) moments of ς^* become stationary, such that the expected value $\overline{\varphi^*}$ becomes stationary, $\overline{\varphi^*} \to \varphi^*_\infty$, then an elementary integration of (7.212) is possible. Thus, one obtains

$$R(g) = R(g_1)\,\exp\!\left(-\frac{g-g_1}{N}\,\varphi^*_\infty\right). \tag{7.213}$$

Similar to the deterministic approximation, one observes a *linear convergence order* also in the stochastic model. The slope on the decadic (base 10) logarithmic scale $\lg(R(g)) = f(g)$ is therefore a constant with the value $-\lg(\mathrm{e})\varphi^*_\infty/N$. The only difference to the result of the deterministic model is in the expected value φ^*_∞ (see the discussion after Eq. (7.187)).

The integral (7.211) has to be evaluated to determine the expected value of the progress rate $\overline{\varphi^*(\varsigma^{*(g)})}$, which is needed in the differential equation (7.212). It is also needed for the stationary progress rate φ^*_∞. Its determination requires the knowledge of the generation-dependent density $p(\varsigma^{*(g)})$. However, one can also derive principal statements on $\overline{\varphi^*(\varsigma^{*(g)})}$ without knowing $p(\varsigma^{*(g)})$. The formula of the (microscopic) progress rate (7.166) inserted into (7.211) is used for that purpose. One gets

$$\overline{\varphi^*(\varsigma^{*(g)})} = c_{1,\lambda}\overline{\varsigma^{*(g)}} - \frac{1}{2}\overline{(\varsigma^{*(g)})^2}, \tag{7.214}$$

with the kth order moments of ς^*

$$\overline{(\varsigma^{*(g)})^k} := \int_{\varsigma^{*(g)}=0}^{\infty} (\varsigma^{*(g)})^k\, p(\varsigma^{*(g)})\,\mathrm{d}\varsigma^{*(g)}. \tag{7.215}$$

We denote the first moment as

$$\boxed{\sigma^*(g) := \overline{\varsigma^{*(g)}}.} \tag{7.216}$$

The second moment in (7.214) can be expressed by the variance of the random variable $\varsigma^{*(g)}$

$$\mathrm{D}^2\{\varsigma^{*(g)}\} = \overline{(\varsigma^{*(g)})^2} - \left(\sigma^*(g)\right)^2. \tag{7.217}$$

After reordering terms and inserting this into the asymptotically exact ($N \to \infty$, $\tau \to 0$) progress rate formula (7.214), one obtains for the *average progress rate* of the $(1, \lambda)$-σSA-ES

$$\boxed{\overline{\varphi^*(\varsigma^{*(g)})} = c_{1,\lambda}\sigma^*(g) - \frac{1}{2}\left(\sigma^*(g)\right)^2 - \frac{1}{2}\,\mathrm{D}^2\{\varsigma^{*(g)}\}.} \tag{7.218}$$

This result is important. It explains why the observed maximal average progress rate of real $(1, \lambda)$-σSA-ES runs is always smaller than the theoretical maximum $\hat{\varphi}^* = \frac{1}{2}\, c_{1,\lambda}^2$: since the variance of the σSA-ES must necessarily be larger than zero, i.e. $\mathrm{D}^2\{\varsigma^*\} > 0$, it follows from (7.218) that

$$\begin{aligned} \overline{\varphi^*} = c_{1,\lambda}\sigma^* - \frac{\sigma^{*2}}{2} - \frac{\mathrm{D}^2\{\varsigma^*\}}{2} &< c_{1,\lambda}\sigma^* - \frac{\sigma^{*2}}{2} \\ &\le \max_{\sigma^*}\left[c_{1,\lambda}\sigma^* - \frac{\sigma^{*2}}{2}\right] = \frac{c_{1,\lambda}^2}{2}. \end{aligned} \tag{7.219}$$

As one can see, the variance of ς^* reduces the progress rate. Indeed, the measured average progress rate is predicted very well by Eq. (7.218). This can be verified using the ES experiments in Sect. 7.4.2.3, Figs. 7.6 and 7.7. One obtains for the learning parameter $\tau = 0.015$ a stationary mean value $\sigma^*_\infty \approx 1.503$, and the standard deviation $D_\infty := \mathrm{D}\{\varsigma^*_\infty\} \approx 0.548$ ($g_1 = 20,000$). After inserting these into (7.218), one obtains a stationary average progress rate $\varphi^*_\infty \approx 1.03$. This prediction accords very well with the experimental value. The accordance is also very good for the case $\tau = 0.15$. One measures $\sigma^*_\infty \approx 1.081$ and $D_\infty = \mathrm{D}\{\varsigma^*_\infty\} \approx 0.594$ in the experiments, yielding a stationary progress rate $\varphi^*_\infty \approx 0.90$.

7.5.4 Mean Value Dynamics of the ς^* Evolution

Equation (7.218) allows for an accurate determination of the average progress rate provided that the mean value of $\varsigma^{*(g)}$ and the variance $\mathrm{D}^2\{\varsigma^{*(g)}\}$ are known. Therefore, the next question is how $\sigma^*(g)$ and $\mathrm{D}\{\varsigma^{*(g)}\}$ can be calculated. This task is more difficult than the treatment of the R dynamics. For the dynamics of R, a simple linear difference equation (7.211) has been obtained using the method of moments. In contrast, there exists an infinite system of momentum equations for the case of $\sigma^*(g)$ as we will see below.

The first moment (7.216) of $\varsigma^{*(g)}$ is obtained by applying the Definition (7.215) and using (7.195) with the transition density (7.208)

$$\sigma^*(g+1) = \int_{\varsigma^{*(g)}=0}^{\infty} \varsigma^{*(g)}\left(1 + \psi(\varsigma^{*(g)}) + \frac{\varphi^*(\varsigma^{*(g)})}{N}\right) p(\varsigma^{*(g)})\,\mathrm{d}\varsigma^{*(g)},$$

$$\sigma^*(g+1) = \sigma^*(g) + \overline{\varsigma^{*(g)}\psi(\varsigma^{*(g)})} + \frac{1}{N}\,\overline{\varsigma^{*(g)}\varphi^*(\varsigma^{*(g)})}. \tag{7.220}$$

The substitution of ψ using (7.141) and φ^* using (7.166) yields

$$\sigma^*(g+1) = \sigma^*(g)\left[1 + \tau^2\left(d_{1,\lambda}^{(2)} - \frac{1}{2}\right)\right] - \overline{(\varsigma^{*(g)})^2}\, c_{1,\lambda}\tau^2\left(1 - \frac{1}{\tau^2 N}\right) - \overline{(\varsigma^{*(g)})^3}\,\frac{1}{2N}. \qquad (7.221)$$

As one can see, the first moment $\sigma^*(g+1)$ at generation $g+1$ depends on the higher-order moments of $\varsigma^{*(g)}$ at generation g. One can neglect the last term in (7.221) for sufficiently large N. However, the second-order moment *must* be considered: No stationary $\sigma_\infty = \sigma(g+1) = \sigma(g)$ can exist if the second moment would be neglected.

An evolution equation must be derived for the second moment of $\varsigma^{*(g)}$ since it occurs in (7.221). Using Definition (7.215) and considering (7.195) one obtains

$$\overline{(\varsigma^{*(g+1)})^2} = \int_{\varsigma^{*(g)}=0}^{\infty}\int_{\varsigma^{*(g+1)}=0}^{\infty} (\varsigma^{*(g+1)})^2 p_{1;\lambda}(\varsigma^{*(g+1)} \mid \varsigma^{*(g)}) \times p(\varsigma^{*(g)})\,\mathrm{d}\varsigma^{*(g+1)}\,\mathrm{d}\varsigma^{*(g)}. \qquad (7.222)$$

With (7.208) and using the substitution

$$x = \left[\varsigma^{*(g+1)} - \varsigma^{*(g)}\left(1 + \psi(\varsigma^{*(g)}) + \frac{\varphi^*(\varsigma^{*(g)})}{N}\right)\right] \Big/ \left(\tau\varsigma^{*(g)}\right) \qquad (7.223)$$

the second moment becomes[20]

$$\overline{(\varsigma^{*(g+1)})^2} = \int_{\varsigma^{*(g)}=0}^{\infty}\frac{(\varsigma^{*(g)})^2}{\sqrt{2\pi}}\int_{x=-\infty}^{\infty}\left[\tau x + \left(1 + \psi(\varsigma^{*(g)}) + \frac{\varphi^*(\varsigma^{*(g)})}{N}\right)\right]^2 \times \mathrm{e}^{-\frac{1}{2}x^2}\,\mathrm{d}x\, p(\varsigma^{*(g)})\mathrm{d}\varsigma^{*(g)}. \qquad (7.224)$$

The integration over x can be carried out next, and, finally, the outer integration yields

$$\begin{aligned}\overline{(\varsigma^{*(g+1)})^2} &= \tau^2\,\overline{(\varsigma^{*(g)})^2} + \overline{\left[\varsigma^{*(g)}\left(1 + \psi(\varsigma^{*(g)}) + \frac{\varphi^*(\varsigma^{*(g)})}{N}\right)\right]^2} \\ &= \overline{(\varsigma^{*(g)})^2}\,(1+\tau^2) + 2\,\overline{(\varsigma^{*(g)})^2\left(\psi(\varsigma^{*(g)}) + \frac{\varphi^*(\varsigma^{*(g)})}{N}\right)} \\ &\quad + \overline{(\varsigma^{*(g)})^2\left(\psi(\varsigma^{*(g)}) + \frac{\varphi^*(\varsigma^{*(g)})}{N}\right)^2}. \end{aligned} \qquad (7.225)$$

The quantities ψ and φ^* are expressed by (7.141) and (7.166) in (7.225). After considering the τ-scaling rule (7.177), one finds that the third term in (7.225) is of order $\mathcal{O}(\tau^4)$, i.e. of order $\mathcal{O}(1/N^2)$. Therefore, it can be neglected

[20] The parameter τ has been assumed to be small and the probability density $\exp(-x^2/2)/\sqrt{2\pi}$ is concentrated around the origin. Therefore, the lower integration limit of x can be shifted to $-\infty$.

$$\overline{(\varsigma^{*(g+1)})^2} = \overline{(\varsigma^{*(g)})^2}\left(1 + 2\tau^2 d_{1,\lambda}^{(2)}\right) - \overline{(\varsigma^{*(g)})^3}\,\tau^2 c_{1,\lambda}\, 2\left(1 - \frac{1}{N\tau^2}\right) - \overline{(\varsigma^{*(g)})^4}\,\frac{1}{N} + \mathcal{O}(\tau^4). \qquad (7.226)$$

As in the case of Eq. (7.220) for the first moment, this equation depends on higher-order moments. This makes the determination of the dynamics of $\sigma^*(g)$ a difficult task, even in the stationary case $\sigma^*_\infty = \sigma^*(g \to \infty)$. A rough approximation will be developed in the next subsection.

A remark on the adaptation time g_A should be added at this point. The transient behavior of the $(1,\lambda)$-σSA-ES considering the fluctuations is similar to that of the deterministic approximation (7.180, 7.181). This can immediately be seen if (7.221) and (7.226) are written as difference equations

$$\sigma^*(g+1) - \sigma^*(g) = \tau^2\left[\left(d_{1,\lambda}^{(2)} - \frac{1}{2}\right)\sigma^*(g) - c_{1,\lambda}(1-T)\,\overline{(\varsigma^{*(g)})^2} - \frac{T}{2}\,\overline{(\varsigma^{*(g)})^3}\right], \qquad (7.227)$$

$$\overline{(\varsigma^{*(g+1)})^2} - \overline{(\varsigma^{*(g)})^2} = 2\tau^2\left[d_{1,\lambda}^{(2)}\,\overline{(\varsigma^{*(g)})^2} - c_{1,\lambda}(1-T)\,\overline{(\varsigma^{*(g)})^3} - \frac{T}{2}\,\overline{(\varsigma^{*(g)})^4}\right], \qquad (7.228)$$

using the abbreviation

$$T := \frac{1}{N\tau^2}. \qquad (7.229)$$

A comparison with (7.180, 7.181) shows that the time constants of the difference equations are mainly determined by $1/\tau^2$, i.e. the adaptation time g_A is proportional to $1/\tau^2$. Furthermore, one notes that Eq. (7.228) yields a time constant which is by a factor $\frac{1}{2}$ smaller as compared to (7.227). Consequently, the second-order moment attains the stationary state faster than the first moment σ^*. This can be regarded as an indicator for the stability of the stationary state.

7.5.5 Approximate Equations for the Stationary σ^*_∞ State

7.5.5.1 The Integral Equation of the Stationary $p(\varsigma^*)$ Density. The stationary state σ^*_∞ cannot be determined directly from moment equations (7.227, 7.228) since these equations contain higher-order (stationary) moments. However, if the stationary $p(\varsigma^*)$ density were known, the values σ^*_∞ and D_∞ could be calculated directly. From the mathematical point of view, the determination of the density $p_\infty(\varsigma^*) := p(\varsigma^*_\infty)$ can be regarded as an eigenvalue problem. This eigenvalue problem is obtained from the Chapman-Kolmogorov equation (7.195) for the case $g \to \infty$. The stationary density must fulfill an integral equation of Fredholm type

$$c\, p_\infty(\varsigma^*) = \int_{\varsigma^{*\prime}=0}^{\infty} p_{1;\lambda}(\varsigma^* \,|\, \varsigma^{*\prime})\, p_\infty(\varsigma^{*\prime})\, d\varsigma^{*\prime} \qquad (7.230)$$

with the eigenvalue $c = 1$. This equation immediately follows from the Chapman-Kolmogorov equation (7.195): if one postulates the existence of a stationary distribution $p(\varsigma^{*(g)}) = p(\varsigma^{*(g+1)}) = p(\varsigma^*_\infty)$ for $g \to \infty$ and inserts this in (7.195).

However, since the integral kernel (7.208) is complicated, the eigenvalue problem (7.230) cannot be solved analytically. One possibility for finding the stationary distribution is by successive numerical integration. The idea is to integrate (7.195) numerically and to reinsert the result as an iteration in the Chapman-Kolmogorov equation. One can simply start with a Dirac distribution, $p(\varsigma^{*(0)}) = \delta(\varsigma^{*(0)} - \sigma^*(0))$, with, e.g., $\sigma^*(0) = c_{1,\lambda}$. This corresponds to the choice of a fixed starting value for σ^*.

Figure 7.8 shows the result of such a numerical $p(\varsigma^*_\infty)$ calculation. The ES experiment in Sect. 7.4.2.3, p. 306, with the learning parameter $\tau = 0.15$ was used for this purpose. The result was obtained using 218 iterations of the numerical integration and is displayed as the continuous curve. The data points

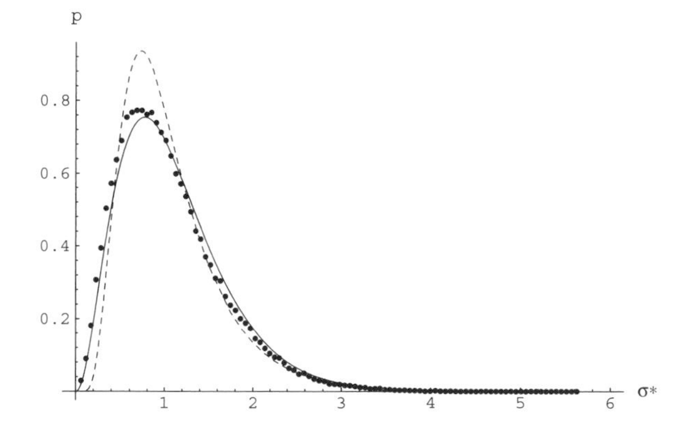

Fig. 7.8. The stationary $p(\varsigma^*_\infty)$ density for the $(1, 10)$-σSA-ES with $N = 10{,}000$ and $\tau = 0.15$. The dashed curve is a log-normal approximation of the $p(\varsigma^*_\infty)$ density.

are obtained from ES experiments by determining the actual bin frequencies of the $\varsigma^{*(g)}$ values (number of bins: 100). A comparison of the theoretical curve with the experimental results shows a satisfactory accordance. As a consequence, this accordance justifies the application of (7.208) as a normal approximation for the transition density $p_{1;\lambda}(\varsigma^{*(g+1)} \,|\, \varsigma^{*(g)})$ (at least for this special case).

The determination of the stationary density $p(\varsigma^*_\infty)$ by iterating numerical integrations is very time consuming. Therefore, it is practically applicable for larger τ values only. This method simulates to some extent the $(1, \lambda)$-σSA process numerically. Therefore, the respective approximations for the adaptation time also hold here. In other words, the necessary number of iterations increases by a factor of 100 (!) for the determination of the $p(\varsigma^*_\infty)$ density in the $\tau = 0.015$ case. Such lengthy simulations are not acceptable.

Analytic and semi-analytic approximation methods alternatively exist for solving the integral equation (7.230). A direct method would be the expansion of $p(\varsigma^*_\infty)$ in a complete function system. However, the difficulty lies in the necessary integrations thereafter, which cannot be carried out analytically because of the complicated structure of (7.208).

7.5.5.2 An Approach for Solving the Momentum Equations. Because of the obvious problems caused with the solution of (7.230), we come back to the momentum method. There is the possibility of reducing the infinite number of momentum equations: one has to use an *Ansatz* for the $p(\varsigma^*_\infty)$ density. The $p(\varsigma^*_\infty)$ *Ansatz* contains a number of free parameters to be determined by a finite set of momentum equations. The case with two parameters will be considered here. One needs two momentum equations to determine them. Since the conditions for the first and second moment of ς^* are

$$g \to \infty: \quad \sigma^*(g+1) = \sigma^*(g) = \sigma^*_\infty, \quad \overline{(\varsigma^{*(g+1)})^2} = \overline{(\varsigma^{*(g)})^2} = \overline{\varsigma^{*2}_\infty} \quad (7.231)$$

for the stationary case, the following system of equations results from the momentum equations (7.227, 7.228)

$$\left(d^{(2)}_{1,\lambda} - \frac{1}{2}\right)\sigma^*_\infty = c_{1,\lambda}\,(1-T)\,\overline{\varsigma^{*2}_\infty} + \frac{T}{2}\,\overline{\varsigma^{*3}_\infty}, \quad (7.232a)$$

$$d^{(2)}_{1,\lambda}\,\overline{\varsigma^{*2}_\infty} = c_{1,\lambda}\,(1-T)\,\overline{\varsigma^{*3}_\infty} + \frac{T}{2}\,\overline{\varsigma^{*4}_\infty}. \quad (7.232b)$$

This system implicitly contains the *Ansatz* parameters. If the functional dependency to the parameters of the *Ansatz* is known, then the system (7.232a, b) presents a (nonlinear) system of equations for the determination of the parameters.

In addition to the stationary solution, the method of *Ansatz* functions can also be used to investigate the dynamic momentum Eqs. (7.227, 7.228). One obtains in this manner a natural iteration procedure to solve the stationary system (7.232a, b).

It is clear that the quality of the $p(\varsigma^*_\infty)$-approximation essentially depends on the *Ansatz* used. The prior knowledge on the special shape of the distribution is therefore very helpful in the selection of an appropriate *Ansatz*. A glance at Fig. 7.8 shows that the *log-normal distribution* is an appropriate candidate

$$p(\varsigma^*_\infty) = \frac{1}{\sqrt{2\pi}\,s}\,\frac{1}{\varsigma^*_\infty}\exp\left[-\frac{1}{2}\left(\frac{\ln(\varsigma^*_\infty/\sigma^*_0)}{s}\right)^2\right]. \quad (7.233)$$

This *Ansatz* contains two parameters σ^*_0 and s. The approximation quality is moderate. Therefore, finding better *Ansatz* functions remains for future research.

The first four moments of the log-normal distribution (7.233) are needed to carry out the moment method in Eqs. (7.232a, b) with the *Ansatz* (7.233). They can be taken from (7.113), after adapting the symbols

$$\sigma_\infty^* = \sigma_0^* \, \mathrm{e}^{\frac{1}{2}s^2}, \quad \overline{\varsigma_\infty^{*2}} = \sigma_0^{*2} \, \mathrm{e}^{2s^2}, \quad \overline{\varsigma_\infty^{*3}} = \sigma_0^{*3} \, \mathrm{e}^{\frac{9}{2}s^2}, \quad \overline{\varsigma_\infty^{*4}} = \sigma_0^{*4} \, \mathrm{e}^{8s^2}. \qquad (7.234)$$

The symbol σ_0^* can be eliminated in (7.234), with the result

$$\overline{\varsigma_\infty^{*2}} = \sigma_\infty^{*2} \, \mathrm{e}^{s^2}, \quad \overline{\varsigma_\infty^{*3}} = \sigma_\infty^{*3} \, \mathrm{e}^{3s^2}, \quad \overline{\varsigma_\infty^{*4}} = \sigma_\infty^{*4} \, \mathrm{e}^{6s^2}. \qquad (7.235)$$

After inserting these moments into (7.232a, b) one obtains

$$\left(d_{1,\lambda}^{(2)} - \frac{1}{2}\right) = c_{1,\lambda} \, (1-T) \, \sigma_\infty^* \, \mathrm{e}^{s^2} + \frac{1}{2} T \, \sigma_\infty^{*2} \, \mathrm{e}^{3s^2}, \qquad (7.236\mathrm{a})$$

$$d_{1,\lambda}^{(2)} = c_{1,\lambda} \, (1-T) \, \sigma_\infty^* \, \mathrm{e}^{2s^2} + \frac{1}{2} T \, \sigma_\infty^{*2} \, \mathrm{e}^{5s^2}. \qquad (7.236\mathrm{b})$$

This is a nonlinear equation system that can be used to determine the stationary mean value of ς^*, i.e. σ_∞^*, as well as s. Using the solution, one can determine the standard deviation by $D_\infty = \mathrm{D}\{\varsigma_\infty^*\}$

$$D_\infty = \sigma_\infty^* \sqrt{\mathrm{e}^{s^2} - 1}. \qquad (7.237)$$

7.5.6 Discussion of the Stationary State and the $1/\sqrt{N}$ Rule

7.5.6.1 Comparison with ES Experiments. The numerical solutions to the system (7.236a, b) can be obtained by iterating the nonstationary system (7.228, 7.229). This enables us to perform comparisons with the stationary performance of the real $(1, \lambda)$-σSA-ES. The results from the $(1, 10)$-ES experiments are shown in Fig. 7.9. The experiments follow the same pattern of Sect. 7.4.2.3, p. 306. Every data point is the result of a complete ES run with over 90,000 generations. The stationary averaged mutation strength σ_∞^* as well as the average progress rate $\overline{\varphi^*}$ are displayed in Fig. 7.9. Both quantities are plotted as functions of the learning parameter τ and parameter space dimension N. The average progress rate $\varphi_\infty^* = \overline{\varphi^*(\varsigma_\infty^*)}$ is approximately determined by formula (7.218) using the stationary solution of (7.236a, b) and (7.237).

The results of the numerical approximation are in relative agreement with the experimental results as long as the parameter space dimension is sufficiently large ($N \gtrapprox 100$). Larger deviations are observed for the case $N = 30$ at smaller values of τ. Such parameter constellations yield a larger T value (see Eq. (7.229)). The effect of a large T value becomes clear when considering the nonstationary momentum system (7.227, 7.228): the higher-order moments have a larger influence on the evolution process. The log-normal density (7.233) used does not seem to be a suitable *Ansatz* for large T. The normal distribution will probably yield better results for this case.

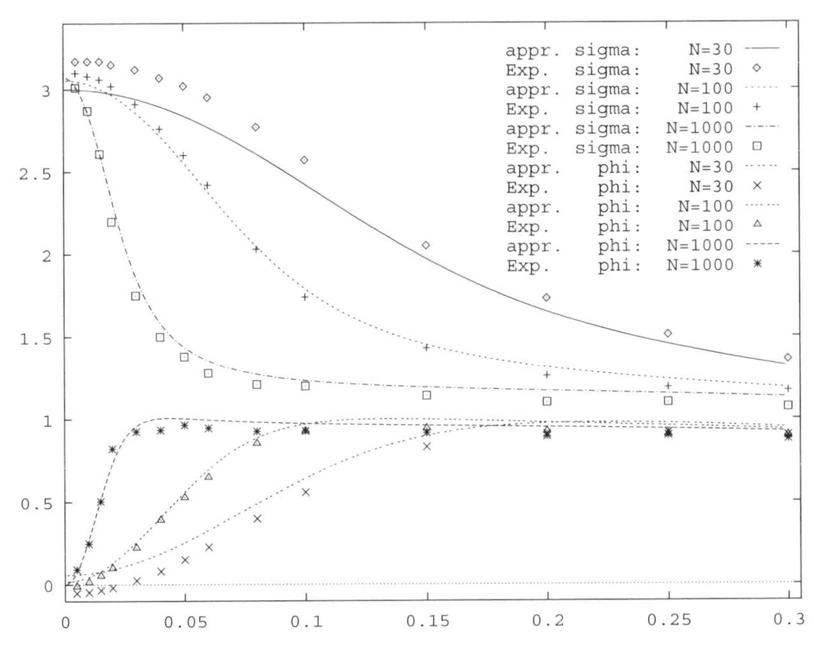

Fig. 7.9. The $(1, 10)$-ES experiments on the stationary performance (labeled with "Exp.") are compared with numerical solutions of the system (7.236a, b) (displayed as curves and labeled with "appr."). The learning parameter τ is indicated on the horizontal axis. The vertical axis shows the values of σ^*_∞ and $\overline{\varphi^*}$ for the parameter space dimensions $N = 30$, $N = 100$, and $N = 1,000$.

7.5.6.2 Special Analytical Cases. A general analytic solution to the system (7.236a, b) is practically excluded. However, one can derive certain properties of the stationary state σ^*_∞ using analytical methods.

Equation (7.236a) can be solved for σ^*_∞ if s is known. Considering (7.229) one obtains

$$\sigma^*_\infty = \left[c_{1,\lambda}(1 - N\tau^2) \; + \; \sqrt{c^2_{1,\lambda}(1 - N\tau^2)^2 \; + \; N\tau^2(2d^{(2)}_{1,\lambda} - 1)\mathrm{e}^{s^2}} \right] \mathrm{e}^{-2s^2}. \tag{7.238}$$

Equation (7.238) can be used to demonstrate the effect of the "$\varsigma^*_{ss} \to \sigma^*_\infty$ shift." This shift is responsible for the discrepancy between the real ES and its deterministic modeling (7.160). The parameter s in the *Ansatz* (7.233) is a measure for the strength of ς^* fluctuations. For $s \to 0$, the density (7.233) degenerates to a Dirac delta function and $\sigma^*_\infty \to \sigma^*_0$. Under this condition, the stationary mean value σ^*_∞ of Eq. (7.238) becomes equivalent to the stationary value ς^*_{ss} of the deterministic approximation (7.171). Consequently, (7.171) is contained as a special case in (7.238) for $s = 0$. If the effect of the factor e^{s^2} in the square root of (7.238) is neglected in a first approximation, the expression in the brackets becomes $\approx \varsigma^*_{ss}$. Therefore, one can say that the influences of the fluctuations are mainly determined by the factor e^{-2s^2}. An increase in s results in a multiplicative decrease of the deterministic ς^*_{ss} value. Hence, one observes a smaller stationary mean value of ς^* than that predicted by the deterministic equations of Sects. 7.4.1 and 7.4.2.

The system (7.236a, b) can be solved for $T \to 0$. The division of (7.236a) by (7.236b) yields

$$e^{-s^2} = \frac{d^{(2)}_{1,\lambda} - \frac{1}{2}}{d^{(2)}_{1,\lambda}}. \qquad (7.239)$$

Equation (7.236a) can be solved for σ^*_∞

$$\sigma^*_\infty = e^{-s^2} \frac{d^{(2)}_{1,\lambda} - \frac{1}{2}}{c_{1,\lambda}}. \qquad (7.240)$$

Finally, the insertion of (7.239) into (7.240) yields

$$T \to 0: \qquad \sigma^*_\infty = \frac{\left(d^{(2)}_{1,\lambda} - \frac{1}{2}\right)^2}{d^{(2)}_{1,\lambda} c_{1,\lambda}}. \qquad (7.241)$$

The condition $T \to 0$ is equivalent to $N\tau^2 \to \infty$ because of Definition (7.229). In other words, the stationary solution (7.241) should be applicable for sufficiently large values of the product $N\tau^2$, i.e. $N\tau^2 \gtrapprox 10$. This is possible when the learning parameter is sufficiently small ($\tau \lessapprox 0.3$) so that the approximation (7.141) for $\psi(\varsigma^*)$ is still valid. For example, this is the case for $N = 10,000$, $\tau = 0.15$ in Sect. 7.4.2.3, p. 306. The measured stationary mean value of ς^* is $\sigma^*_\infty \approx 1.08$. Equation (7.241) yields $\sigma^*_\infty \approx 1.17$, the relative error is 8%.

The mean value of the stationary progress rate φ^*_∞ is obtained by inserting (7.216, 7.231, 7.235) into (7.214)

$$\varphi^*_\infty = c_{1,\lambda}\sigma^*_\infty - \frac{1}{2}\sigma^{*2}_\infty e^{s^2}, \qquad (7.242)$$

one obtains with (7.241) for the case

$$T \to 0: \qquad \varphi^*_\infty = \frac{\left(d^{(2)}_{1,\lambda} - \frac{1}{2}\right)^2}{d^{(2)}_{1,\lambda}} \left[1 - \frac{d^{(2)}_{1,\lambda} - \frac{1}{2}}{2c^2_{1,\lambda}}\right]. \qquad (7.243)$$

For the ES example above ($N = 10,000$, $\tau = 0.15$), one calculates with this formula $\varphi^*_\infty \approx 0.96$. The corresponding ES experiment gives $\varphi^*_\infty \approx 0.90$. As a consequence, one can say that the stochastic model based on the density approach (7.233) gives more accurate results than the deterministic model which predicts the optimal performance as $\hat{\varphi}^*_{1,10} \approx 1.183$.

7.5.6.3 The τ-Scaling Rule. As one can see in Fig. 7.9 and verify using Eq. (7.243), the choice of a small T is more reliable. The obtained performance is satisfactory, and the σSA is certain to function. Therefore, a *constant* τ such as $\tau = 0.3$ is sufficient for most of the real applications.

However, the performance can be improved slightly if the learning parameter is chosen according to the scaling law $\tau \propto 1/\sqrt{N}$. This can be verified in Fig. 7.9 by using the scaling rule (7.177), $\tau = c_{1,\lambda}/\sqrt{N}$. There exists an optimal τ for a constant N giving a maximal average performance.

The next question is: can the validity of the $\tau \propto 1/\sqrt{N}$ rule also be proven in the framework of the stochastic model based on the log-normal *Ansatz* (7.233)? This is indeed possible.

Assuming a fixed λ value, the stationary solution (σ^*_∞, s) of the system (7.236a, 7.236b) is a function depending on T *only*

$$\sigma^*_\infty = \sigma^*_\infty(T), \qquad s = s(T). \tag{7.244}$$

Inserting this solution into the average progress rate (7.242), φ^*_∞ becomes a function of T only

$$\varphi^*_\infty = \varphi^*_\infty(T). \tag{7.245}$$

One obtains a maximal performance if $\varphi^*_\infty(T)$ is maximal. Let $\hat{T}$ denote the T value at which φ^*_∞ attains its maximum

$$\hat{\varphi}^*_\infty = \max_T \left[\varphi^*_\infty(T)\right] \qquad \Longleftrightarrow \qquad T = \hat{T} < \infty. \tag{7.246}$$

In other words, if $T = \hat{T}$ is chosen, the ES algorithm exhibits maximal performance (provided that a positive maximum exists). According to (7.229), this is equivalent to

$$\begin{aligned} \varphi^*_\infty = \hat{\varphi}^*_\infty \quad &\Leftrightarrow \quad T = \hat{T} \quad \Leftrightarrow \quad \hat{T} = \frac{1}{N\tau^2} \\ &\Rightarrow \quad \tau = \frac{1}{\sqrt{\hat{T}}}\frac{1}{\sqrt{N}} \quad \Rightarrow \quad \tau \propto \frac{1}{\sqrt{N}}, \end{aligned} \tag{7.247}$$

which proves the validity of the τ scaling rule.

7.6 Final Remarks on the σ Self-Adaptation

The most important results obtained from the analysis of the $(1, \lambda)$-σSA-ES will be summarized here. Some remarks on general self-adaptation algorithms are made thereafter.

The σ mutation operators. One of the most remarkable results obtained from the analysis is that the $(1, \lambda)$-σSA-ES works equally well with different σ mutation operators, at least in the framework of the approximations used. In addition to the log-normal operator which is probably the most widespread, the two-point operator can be applied alternatively. The characteristically different learning parameters τ and $\beta = \alpha - 1$ can be converted to each other using the equation $\tau = \beta\sqrt{1-\beta}$. Furthermore, the Meta-EP rule should exhibit the same behavior, since the log-normal transformation (7.8) can be approximated by a normal distribution for at least small values of τ. A more detailed investigation into the dynamics produced by the different σ mutation operators, in particular the validity of the correspondence principle (7.155) of the learning parameters, remains to be done.

The σ mutation operators investigated here have a common characteristic property: their influence is symmetrical in a certain sense. The probabilities of an increase or decrease are the same (i.e. $\frac{1}{2}$) and their median is σ. The *nonsymmetric* two-point operator (7.15) suggested by the author forms an exception. The operator design rule "median= σ" is usually justified by the argument that the mutation process should not favor any direction (postulate of isotropy). However, a complete isotropy is neither always realizable, nor desirable in some cases. For example, the expected value of the log-normal operator (7.8) is not σ (the parental mutation strength), but $\sigma e^{\tau^2/2}$ (see Eq. (7.113)).

A "Median= σ" postulate is not absolutely necessary from the viewpoint of the theory developed in this chapter. Instead, one must ensure that a zero $\varsigma^*_{\psi_0}$ of the SAR function exists. Furthermore, this zero must be smaller than the second zero $\varsigma^*_{\varphi_0}$ of the progress rate, so that a positive progress rate is guaranteed. Astonishingly, this condition is not only fulfilled by the σ mutation operators, but this zero is in the neighborhood of that ς^* which maximizes φ^* on the sphere model. The analysis was carried out here for the asymptotic limit case. It provided both the result and the proof; however, it was not able to serve as a descriptive explanation. Hence, some explanatory research is required at this point.

On the choice of the learning parameters. If τ is chosen as $\tau = c_{1,\lambda}/\sqrt{N}$, one obtains an $(1, \lambda)$-σSA-ES which works approximately optimally. According to the results of the microscopic analysis, the same performance can be obtained by the two-point operator if one ensures $\tau^2 \approx \beta^2(1-\beta)$. For sufficiently small $\beta = \alpha - 1$, this means that the learning parameter α should be chosen according to $\alpha \approx 1 + c_{1,\lambda}/\sqrt{N}$. This formula is therefore valid under the (simplifying) assumption $\tau \approx \beta$.

In general, one only observes a slight dependency of the stationary performance φ^*_∞ on the learning parameters, as long as $\tau, \beta \gtrapprox 1/\sqrt{N}$ is fulfilled. Consequently, a constant N-independent learning parameter should suffice for most cases.

The learning parameter influences the adaptation time g_A (learning rate, learning time). The symbol g_A denotes the time required to adapt the mutation strength to the local topology if σ had a wrong initial value. The adaptation time is generally inversely proportional to the square of the learning parameter τ, $g_A \propto \frac{1}{\tau^2}$. The adaptation time becomes proportional to the parameter space dimension N if τ is chosen according to the τ-scaling rule, $\tau \propto 1/\sqrt{N}$. Therefore, one should work with a *generation-dependent* learning parameter on optimization and adaptation problems with very large N ($N > 1,000$). One should start with an N-independent τ, e.g., $\tau = 0.3$, and use it for a certain number of generations. This number of generations must be chosen so that the σSA algorithm reaches the vicinity of the stationary state. Thereafter one can change to the $\tau = c_{1,\lambda}/\sqrt{N}$ rule.

On the influence of the σ fluctuations. The σ fluctuations and therefore those of σ^* degrade the performance of the $(1, \lambda)$-ES. The practically attainable progress rate φ^*_∞ is always smaller than the theoretical maximum $c^2_{1,\lambda}/2$. The performance degradation depends on the variance $\mathrm{D}^2\{\varsigma^*\}$ of the ς^* state (see Eq. (7.218)). Any method for the reduction of the σ fluctuations contributes to the improvement of the performance.

One *could* suggest as a first idea using smaller learning parameters to reduce $\mathrm{D}^2\{\varsigma^*\}$. However, as has been shown in the analysis, τ influences the *stationary* variance ς^* rather slightly (see the remarks below Eq. (7.219)). The fluctuations in ς^* are *not* the result of a *single* generation, but an effect of the stochastic dynamics.

For the $(1, \lambda)$-σSA-ES, there is a technique that can be used to reduce $\mathrm{D}^2\{\varsigma^*\}$. The *moving σ-average* technique can successfully be applied to the parental σ values of the last κ generations.

A more sophisticated method was suggested by Ostermeier et al. (1994). In this method, the moving average values are weighted such that they represent a *fading memory.* This method was developed for the self-adaptation of N coordinate-specific mutation strengths σ_i. It should also be an effective technique to reduce $\mathrm{D}^2\{\varsigma^*\}$.

It is difficult to estimate whether or not new methods can be obtained to reduce the variance of ς^* in ES algorithms with more than one parent ($\mu > 1$). The effect of (multi-)recombination in the strategy parameter space is not investigated yet, neither theoretically nor empirically in a systematic way. In particular, it is not clear whether the genetic repair mechanism extracted in Chap. 6 possesses a significant effect in reducing fluctuations in the strategy parameter space.

The σSA strategies with recombination as well as the methods for the reduction of $\mathrm{D}^2\{\varsigma^*\}$ are not theoretically investigated yet. However, they are expected to be an important part of the future research.

On the self-adaptation of strategy parameter sets. The seventh chapter contained the analysis of the self-adaptation of a *single* strategy parameter – of the mutation strength σ of isotropic Gaussian mutations. However, the principle of self-adaptation functions also for more general cases, e.g., for the adaptation of coordinate-dependent mutation strengths σ_i or for the learning of correlated mutation distributions (Schwefel, 1987). These techniques work with two learning parameters and the effects of the σ mutation operators interact nonlinearly. Therefore, the results obtained here cannot be transferred directly to the self-adaptation case with correlated mutations. A considerable amount of research is needed at this point in the theoretical analysis.

Although the theory developed cannot be transferred directly to the cases with more complex learning parameter sets, certain principal statements can be made on the behavior of such strategies. In particular, the variance of ς^* should influence the performance negatively, i.e. $\mathrm{D}^2\{\varsigma^*\}$ should have a decreasing effect on the performance of any $(\mu/\rho, \lambda)$-SA-ES algorithm. This is obvious since the local performance is defined as a function of the mutation strength(s), no matter whether it is measured as a progress rate φ or as a quality gain $\overline{Q}$.

For example, if the progress rate of an ES with N individual mutation strengths σ_i is considered, φ is a function of these strategy parameters

$$\varphi^{(g)} = \varphi(\varsigma_1^{(g)}, \dots, \varsigma_N^{(g)}). \tag{7.248}$$

The SA algorithm aims at evolving the strategy so that the progress rate (or the quality gain) is maximal at any generation g, $\varphi \to \hat{\varphi}$. Assuming that (at least) one optimal choice of strategy parameters exists for this

$$\hat{\varphi}^{(g)} = \varphi(\hat{\varsigma}_1^{(g)}, \dots, \hat{\varsigma}_N^{(g)}), \tag{7.249}$$

the algorithm must tune the strategy parameter values $\hat{\varsigma}_1^{(g)}, \dots, \hat{\varsigma}_N^{(g)}$ as accurately as possible. Consequently, any fluctuation from this optimal state decreases the *average* progress rate $\overline{\varphi^{(g)}}$. As a result, $\overline{\varphi^{(g)}} < \hat{\varphi}^{(g)}$ always holds. Methods reducing the fluctuation variances $\mathrm{D}^2\{\varsigma_i^*\}$ influence the performance of the SA algorithm positively. A more thorough investigation of the influence of these fluctuations on the performance of $(1, \lambda)$-strategies can be made using the quality gain. As long as the learning parameter is sufficiently small, the results in Chap. 4 remain valid. Therefore, one can approximately calculate the influence of the ς_i fluctuations at the maximum of $\overline{Q}$ using Taylor expansions. Such investigations should be part of the future research.

Appendices

A. Integrals

A.1 Definite Integrals of the Normal Distribution

The moments $\overline{x^k}$ of the standard normal distribution $x = \mathcal{N}(0,1)$ present the simplest integrals over the Gaussian density. Because of the symmetry of the Gaussian distribution, all odd moments vanish. For the even ones, one finds by successive differentiation of the identity

$$\frac{1}{\beta} = \frac{1}{\sqrt{2\pi}} \int_{-\infty}^{\infty} \mathrm{e}^{-\frac{1}{2}(\beta x)^2} \mathrm{d}x \tag{A.1}$$

with respect to β, at $\beta = 1$, the desired moments

$$\overline{x^k} = \frac{1}{\sqrt{2\pi}} \int_{-\infty}^{\infty} x^k \mathrm{e}^{-\frac{1}{2}x^2} \mathrm{d}x = \begin{cases} 1 \cdot 3 \cdot \ldots (k-1), & \text{if } k \text{ even} \\ 0, & \text{if } k \text{ odd.} \end{cases} \tag{A.2}$$

For the even moments of the $\mathcal{N}(0, \sigma^2)$ normal distribution $p_x(x)$ (3.12), one obtains consequently

Table A.1. The even moments of the $\mathcal{N}(0, \sigma^2)$ normal distribution

$\overline{x^2}$	$\overline{x^4}$	$\overline{x^6}$	$\overline{x^8}$	$\overline{x^{10}}$	$\overline{x^{12}}$	$\overline{x^{14}}$
σ^2	$3\,\sigma^4$	$15\,\sigma^6$	$105\,\sigma^8$	$945\,\sigma^{10}$	$10\,395\,\sigma^{12}$	$135\,135\,\sigma^{14}$

Combinations of the Gaussian densities in the integrand can be treated by quadratic completion. In the following integral

$$I_k = \frac{1}{\sqrt{2\pi}} \int_{-\infty}^{\infty} t^k \mathrm{e}^{-\frac{1}{2}t^2} \mathrm{e}^{-\frac{1}{2}(at+b)^2} \mathrm{d}t, \tag{A.3}$$

the e exponents are combined first by quadratic completion

$$t^2 + (at+b)^2 = (1+a^2)\left(t + \frac{ab}{1+a^2}\right)^2 + \frac{b^2}{1+a^2}. \tag{A.4}$$

Hence, for the $k = 0$ case, I_0 reads

$$I_0 = \frac{1}{\sqrt{2\pi}} \int_{-\infty}^{\infty} \exp\left[-\frac{1}{2}(1+a^2)\left(t + \frac{ab}{1+a^2}\right)^2\right] \mathrm{d}t \, \exp\left(-\frac{1}{2}\frac{b^2}{1+a^2}\right) \tag{A.5}$$

and one obtains after the integration

$$I_0 = \frac{1}{\sqrt{2\pi}} \int_{-\infty}^{\infty} \mathrm{e}^{-\frac{1}{2}t^2} \mathrm{e}^{-\frac{1}{2}(at+b)^2} \mathrm{d}t = \frac{1}{\sqrt{1+a^2}} \exp\left(-\frac{1}{2}\frac{b^2}{1+a^2}\right). \tag{A.6}$$

In order to calculate I_1, (A.6) is differentiated with respect to b, and one obtains

$$\frac{-a}{\sqrt{2\pi}} \int_{-\infty}^{\infty} t\,\mathrm{e}^{-\frac{1}{2}t^2} \mathrm{e}^{-\frac{1}{2}(at+b)^2} \mathrm{d}t - \frac{b}{\sqrt{2\pi}} \int_{-\infty}^{\infty} \mathrm{e}^{-\frac{1}{2}t^2} \mathrm{e}^{-\frac{1}{2}(at+b)^2} \mathrm{d}t$$
$$= \frac{-b \exp\left(-\frac{1}{2}\frac{b^2}{1+a^2}\right)}{\sqrt{1+a^2}\,(1+a^2)}. \tag{A.7}$$

The second integral in (A.7) is substituted by (A.6), and the terms are reordered so that the first integral remains on the left side. One gets

$$I_1 = \frac{1}{\sqrt{2\pi}} \int_{-\infty}^{\infty} t\,\mathrm{e}^{-\frac{1}{2}t^2} \mathrm{e}^{-\frac{1}{2}(at+b)^2} \mathrm{d}t = \frac{-ab \exp\left(-\frac{1}{2}\frac{b^2}{1+a^2}\right)}{\sqrt{1+a^2}\,(1+a^2)}. \tag{A.8}$$

One can calculate I_2 using the same method; I_1 is differentiated by b for this purpose. The result reads

$$I_2 = \frac{1}{\sqrt{2\pi}} \int_{-\infty}^{\infty} t^2 \mathrm{e}^{-\frac{1}{2}t^2} \mathrm{e}^{-\frac{1}{2}(at+b)^2} \mathrm{d}t$$
$$= \frac{1+a^2+a^2b^2}{\sqrt{1+a^2}\,(1+a^2)^2} \exp\left(-\frac{1}{2}\frac{b^2}{1+a^2}\right). \tag{A.9}$$

The I_k terms of higher order can be calculated successively. However, they are not necessary for further considerations.

The indefinite integration of $I_0 = f(b)$ with respect to b yields an important integral for the $(\tilde{1} \,\dot{+}\, \tilde{\lambda})$-ES

$$\int I_0(b)\,\mathrm{d}b = \int_{-\infty}^{\infty} \mathrm{e}^{-\frac{1}{2}t^2} \Phi(at+b)\,\mathrm{d}t = \sqrt{2\pi}\,\Phi\left(\frac{b}{\sqrt{1+a^2}}\right). \tag{A.10}$$

This relation can be proven easily by the differentiation of (A.10) with respect to b. Using the same procedure, also the indefinite integral of $I_1(b)$ can be obtained. First we note for the right-hand side of (A.8)

$$I_1 = \frac{-ab \exp\left(-\frac{1}{2}\frac{b^2}{1+a^2}\right)}{\sqrt{1+a^2}\,(1+a^2)} = \frac{\mathrm{d}}{\mathrm{d}b}\left\{\frac{a}{\sqrt{1+a^2}} \exp\left(-\frac{1}{2}\frac{b^2}{1+a^2}\right)\right\}. \tag{A.11}$$

Therefore, the integration of (A.8) with respect to b yields

$$\int_{-\infty}^{\infty} t\,\mathrm{e}^{-\frac{1}{2}t^2} \Phi(at+b)\,\mathrm{d}t = \frac{a}{\sqrt{1+a^2}} \exp\left(-\frac{1}{2}\frac{b^2}{1+a^2}\right). \tag{A.12}$$

A.2 Indefinite Integrals of the Normal Distribution

A.2.1 Integrals of the Form $I^{\alpha,\beta}(x) = \int_{-\infty}^{t=x} t^\beta \, e^{-\frac{1+\alpha}{2}t^2} dt$

The case $I^{0,\beta}(x)$ is considered first. Using the partial integration for $\beta = m$, one obtains

$$I^{0,m}(x) = \int_{-\infty}^{x} t^m e^{-\frac{1}{2}t^2} dt = \frac{x^{m+1}}{m+1} e^{-\frac{1}{2}x^2} + \int_{-\infty}^{x} \frac{t^{m+1}}{m+1} t \, e^{-\frac{1}{2}t^2} dt. \quad \text{(A.13)}$$

Substituting $m = \beta - 2$, the recursion formula reads

$$I^{0,\beta}(x) = \int_{-\infty}^{x} t^\beta e^{-\frac{1}{2}t^2} dt = (\beta - 1) \int_{-\infty}^{x} t^{\beta-2} e^{-\frac{1}{2}t^2} dt - x^{\beta-1} e^{-\frac{1}{2}x^2}$$

$$I^{0,\beta}(x) = (\beta - 1) I_{\beta-2}(x) - x^{\beta-1} e^{-\frac{1}{2}x^2}. \quad \text{(A.14)}$$

Since

$$I^{0,0}(x) = \sqrt{2\pi} \frac{1}{\sqrt{2\pi}} \int_{-\infty}^{x} e^{-\frac{1}{2}t^2} dt = \sqrt{2\pi} \Phi(x) \quad \text{(A.15)}$$

and

$$I^{0,1}(x) = \int_{-\infty}^{x} t \, e^{-\frac{1}{2}t^2} dt = \int_{-\infty}^{x} \frac{d}{dt} \left(e^{-\frac{1}{2}t^2} \right) dt = -e^{-\frac{1}{2}x^2}, \quad \text{(A.16)}$$

the application of (A.14) yields, stepwise

$$\begin{aligned}
I^{0,2}(x) &= \sqrt{2\pi}\Phi(x) - x e^{-\frac{1}{2}x^2}, \\
I^{0,3}(x) &= -e^{-\frac{1}{2}x^2} \left(2 + x^2\right), \\
I^{0,4}(x) &= 3\sqrt{2\pi}\Phi(x) - e^{-\frac{1}{2}x^2} \left(3x + x^3\right), \\
I^{0,5}(x) &= -e^{-\frac{1}{2}x^2} \left(8 + 4x^2 + x^4\right), \\
I^{0,6}(x) &= 15\sqrt{2\pi}\Phi(x) - e^{-\frac{1}{2}x^2} \left(15x + 5x^3 + x^5\right).
\end{aligned} \quad \text{(A.17)}$$

The general case $I^{\alpha,\beta}(x)$ can be obtained by the substitution of $t = y/\sqrt{1+\alpha}$ for the case $I^{0,\beta}(x)$

$$I^{\alpha,\beta}(x) = \int_{-\infty}^{t=x} t^\beta \, e^{-\frac{1+\alpha}{2}t^2} dt = \int_{-\infty}^{y=\sqrt{1+\alpha}\,x} \left(\frac{y}{\sqrt{1+\alpha}} \right)^\beta e^{-\frac{1}{2}y^2} \frac{dy}{\sqrt{1+\alpha}}$$

$$I^{\alpha,\beta}(x) = \left(\sqrt{1+\alpha}\right)^{-(\beta+1)} I^{0,\beta}\left(\sqrt{1+\alpha}\,x\right). \quad \text{(A.18)}$$

Consequently, for $\alpha = 1$ one gets

$$
\begin{aligned}
I^{1,0}(x) &= \sqrt{\pi}\Phi(\sqrt{2}x), \\
I^{1,1}(x) &= -\frac{1}{2}\mathrm{e}^{-x^2}, \\
I^{1,2}(x) &= \frac{\sqrt{\pi}}{2}\Phi(\sqrt{2}x) - \frac{1}{2}x\mathrm{e}^{-x^2}, \\
I^{1,3}(x) &= -\frac{1}{2}\mathrm{e}^{-x^2}\left(1+x^2\right), \\
I^{1,4}(x) &= \sqrt{\pi}\frac{3}{4}\,\Phi(\sqrt{2}x) - \frac{1}{4}\mathrm{e}^{-x^2}\left(3x+2x^3\right), \\
I^{1,5}(x) &= -\frac{1}{2}\mathrm{e}^{-x^2}\left(2+2x^2+x^4\right),
\end{aligned}
\tag{A.19}
$$

and for $\alpha = 2$

$$
\begin{aligned}
I^{2,0}(x) &= \frac{\sqrt{2\pi}}{\sqrt{3}}\Phi(\sqrt{3}x), \\
I^{2,1}(x) &= -\frac{1}{3}\mathrm{e}^{-\frac{3}{2}x^2}, \\
I^{2,2}(x) &= \frac{1}{3}\,\frac{\sqrt{2\pi}}{\sqrt{3}}\Phi(\sqrt{3}x) - \frac{1}{3}x\mathrm{e}^{-\frac{3}{2}x^2}, \\
I^{2,3}(x) &= -\frac{1}{9}\mathrm{e}^{-\frac{3}{2}x^2}\left(2+3x^2\right), \\
I^{2,4}(x) &= \frac{1}{3}\,\frac{\sqrt{2\pi}}{\sqrt{3}}\Phi(\sqrt{3}x) - \frac{1}{3}\mathrm{e}^{-\frac{3}{2}x^2}\left(x+x^3\right).
\end{aligned}
\tag{A.20}
$$

A.2.2 Integrals of the Form $I_\phi^\beta(x) = \int_{-\infty}^{t=x} t^\beta \mathrm{e}^{-\frac{1}{2}t^2}\Phi(t)\,\mathrm{d}t$

The integral $I_\phi^\beta(x)$ can be handled by partial integration. One obtains

$$
\begin{aligned}
I_\phi^\beta(x) &= \int_{-\infty}^{x} t^\beta \mathrm{e}^{-\frac{1}{2}t^2}\Phi(t)\,\mathrm{d}t = \int_{-\infty}^{x} t\, t^{\beta-1}\mathrm{e}^{-\frac{1}{2}t^2}\Phi(t)\,\mathrm{d}t \\
&= \left[t\int_{z=-\infty}^{z=t} z^{\beta-1}\mathrm{e}^{-\frac{1}{2}z^2}\Phi(z)\,\mathrm{d}z\right]_{t=-\infty}^{t=x} \\
&\quad - \int_{t=-\infty}^{t=x}\int_{z=-\infty}^{z=t} z^{\beta-1}\mathrm{e}^{-\frac{1}{2}z^2}\Phi(z)\,\mathrm{d}z\,\mathrm{d}t
\end{aligned}
\tag{A.21}
$$

and, therefore, the recursive equation

$$
I_\phi^\beta(x) = x\, I_\phi^{\beta-1}(x) - \int_{z=-\infty}^{z=x} I_\phi^{\beta-1}(z)\,\mathrm{d}z. \tag{A.22}
$$

For $\beta = 0$, one finds

$$
\begin{aligned}
I_\phi^0(x) &= \int_{-\infty}^{x} \mathrm{e}^{-\frac{1}{2}t^2}\Phi(t)\,\mathrm{d}t = \frac{\sqrt{2\pi}}{2}\int_{-\infty}^{x}\frac{\mathrm{d}}{\mathrm{d}t}\left(\Phi(t)\right)^2\,\mathrm{d}t \\
&= \sqrt{\frac{\pi}{2}}\left(\Phi(x)\right)^2.
\end{aligned}
\tag{A.23}
$$

The case $\beta = 1$ is treated as follows

$$\begin{aligned} I_\phi^1(x) &= \int_{-\infty}^{x} t\,\mathrm{e}^{-\frac{1}{2}t^2}\,\Phi(t)\,\mathrm{d}t = -\int_{-\infty}^{x} \frac{\mathrm{d}}{\mathrm{d}t}\left(\mathrm{e}^{-\frac{1}{2}t^2}\right)\Phi(t)\,\mathrm{d}t \\ &= -\mathrm{e}^{-\frac{1}{2}x^2}\,\Phi(x) + \frac{1}{\sqrt{2\pi}}\int_{-\infty}^{x} \mathrm{e}^{-t^2}\,\mathrm{d}t \\ &= \frac{1}{\sqrt{2}}\,\Phi(\sqrt{2}x) \;-\; \mathrm{e}^{-\frac{1}{2}x^2}\,\Phi(x). \end{aligned} \tag{A.24}$$

The cases for $\beta > 1$ can be solved by using the recursive equation (A.22)

$$I_\phi^2(x) = \frac{1}{2}\sqrt{2\pi}\,(\Phi(x))^2 - x\,\mathrm{e}^{-\frac{1}{2}x^2}\,\Phi(x) - \frac{1}{2}\,\frac{1}{\sqrt{2\pi}}\,\mathrm{e}^{-x^2} \tag{A.25}$$

$$I_\phi^3(x) = \frac{5}{2}\,\frac{1}{\sqrt{2}}\,\Phi(\sqrt{2}x) - \left(2 + x^2\right)\mathrm{e}^{-\frac{1}{2}x^2}\,\Phi(x) - \frac{1}{2}\,\frac{1}{\sqrt{2\pi}}\,x\,\mathrm{e}^{-x^2} \tag{A.26}$$

$$I_\phi^4(x) = \frac{3}{2}\sqrt{2\pi}\,(\Phi(x))^2 - \left(3x + x^3\right)\mathrm{e}^{-\frac{1}{2}x^2}\,\Phi(x) - \frac{1}{2}\,\frac{1}{\sqrt{2\pi}}\left(4 + x^2\right)\mathrm{e}^{-x^2}. \tag{A.27}$$

A.3 Some Integral Identities

Suppose that an integral of the form

$$I = \frac{1}{\left(\sqrt{2\pi}\right)^{\alpha+1}} \int_{-\infty}^{\infty} \mathrm{e}^{-\frac{\alpha+1}{2}x^2} \left(1 - \Phi(x)\right)^a \left(\Phi(x)\right)^b f(x)\, \mathrm{d}x \tag{A.28}$$

is given, where $f(x)$ stands for an ordinary polynomial of x. Instead of (A.28), one can also write

$$I = \frac{1}{\left(\sqrt{2\pi}\right)^{\alpha}} \int_{-\infty}^{\infty} \left(\frac{\mathrm{e}^{-\frac{1}{2}x^2}}{\sqrt{2\pi}}\right) \mathrm{e}^{-\frac{\alpha}{2}x^2} \left(1 - \Phi(x)\right)^a \left(\Phi(x)\right)^b f(x)\, \mathrm{d}x. \tag{A.29}$$

The application of the partial integration yields

$$I = -\frac{1}{\left(\sqrt{2\pi}\right)^{\alpha}} \int_{-\infty}^{\infty} \Phi(x) \frac{\mathrm{d}}{\mathrm{d}x} \left[\mathrm{e}^{-\frac{\alpha}{2}x^2} \left(1 - \Phi(x)\right)^a \left(\Phi(x)\right)^b f(x)\right] \mathrm{d}x$$

$$\begin{aligned} I = &-\frac{1}{\left(\sqrt{2\pi}\right)^{\alpha}} \int_{-\infty}^{\infty} \mathrm{e}^{-\frac{\alpha}{2}x^2} \left(1 - \Phi(x)\right)^a \left(\Phi(x)\right)^{b+1} \left[-\alpha x f(x) + \frac{\mathrm{d}f(x)}{\mathrm{d}x}\right] \mathrm{d}x \\ &+ \frac{a}{\left(\sqrt{2\pi}\right)^{\alpha+1}} \int_{-\infty}^{\infty} \mathrm{e}^{-\frac{\alpha+1}{2}x^2} \left(1 - \Phi(x)\right)^{a-1} \left(\Phi(x)\right)^{b+1} f(x)\, \mathrm{d}x \\ &- \frac{b}{\left(\sqrt{2\pi}\right)^{\alpha+1}} \int_{-\infty}^{\infty} \mathrm{e}^{-\frac{\alpha+1}{2}x^2} \left(1 - \Phi(x)\right)^a \left(\Phi(x)\right)^b f(x)\, \mathrm{d}x. \end{aligned} \tag{A.30}$$

Substituting the integral (A.28) from (A.30), one obtains

$$\begin{aligned} 0 = &\frac{1}{\left(\sqrt{2\pi}\right)^{\alpha}} \int_{-\infty}^{\infty} \mathrm{e}^{-\frac{\alpha}{2}x^2} \left(1 - \Phi(x)\right)^a \left(\Phi(x)\right)^{b+1} \left[\alpha x f(x) - \frac{\mathrm{d}f(x)}{\mathrm{d}x}\right] \mathrm{d}x \\ &+ \frac{a}{\left(\sqrt{2\pi}\right)^{\alpha+1}} \int_{-\infty}^{\infty} \mathrm{e}^{-\frac{\alpha+1}{2}x^2} \left(1 - \Phi(x)\right)^{a-1} \left(\Phi(x)\right)^{b+1} f(x)\, \mathrm{d}x \\ &- \frac{b+1}{\left(\sqrt{2\pi}\right)^{\alpha+1}} \int_{-\infty}^{\infty} \mathrm{e}^{-\frac{\alpha+1}{2}x^2} \left(1 - \Phi(x)\right)^a \left(\Phi(x)\right)^b f(x)\, \mathrm{d}x. \end{aligned} \tag{A.31}$$

Therefore, one finds for the case $a = \lambda - \mu - 1$, $b = \mu - \alpha$

$$\begin{aligned} 0 = &\frac{1}{\left(\sqrt{2\pi}\right)^{\alpha}} \int_{-\infty}^{\infty} \mathrm{e}^{-\frac{\alpha}{2}x^2} \left(1 - \Phi(x)\right)^{\lambda-\mu-1} \left(\Phi(x)\right)^{\mu+1-\alpha} \left[\alpha x f(x) - \frac{\mathrm{d}f(x)}{\mathrm{d}x}\right] \mathrm{d}x \\ &+ \frac{\lambda - \mu - 1}{\left(\sqrt{2\pi}\right)^{\alpha+1}} \int_{-\infty}^{\infty} \mathrm{e}^{-\frac{\alpha+1}{2}x^2} \left(1 - \Phi(x)\right)^{\lambda-\mu-2} \left(\Phi(x)\right)^{\mu+1-\alpha} f(x)\, \mathrm{d}x \\ &- \frac{\mu + 1 - \alpha}{\left(\sqrt{2\pi}\right)^{\alpha+1}} \int_{-\infty}^{\infty} \mathrm{e}^{-\frac{\alpha+1}{2}x^2} \left(1 - \Phi(x)\right)^{\lambda-\mu-1} \left(\Phi(x)\right)^{\mu-\alpha} f(x)\, \mathrm{d}x. \end{aligned} \tag{A.32}$$

Some $f(x)$ polynomials of interest for the (μ, λ)-ES are given as follows

$$f(x) = x^2 - 1, \qquad \alpha = 1 \Rightarrow \left[\alpha\, x f(x) - \frac{\mathrm{d}\, f}{\mathrm{d}x}\right] = x^3 - 3x, \tag{A.33}$$

$$f(x) = x^2 - 1, \qquad \alpha = 2 \Rightarrow \left[\alpha\, x f(x) - \frac{\mathrm{d}\, f}{\mathrm{d}x}\right] = 2x^3 - 4x, \tag{A.34}$$

$$f(x) = x^2 - 1, \qquad \alpha = 3 \Rightarrow \left[\alpha\, x f(x) - \frac{\mathrm{d}\, f}{\mathrm{d}x}\right] = 3x^3 - 5x, \tag{A.35}$$

$$f(x) = x^2 - 1, \qquad \alpha = 4 \Rightarrow \left[\alpha\, x f(x) - \frac{\mathrm{d}\, f}{\mathrm{d}x}\right] = 4x^3 - 6x, \tag{A.36}$$

$$f(x) = x^3 - x, \qquad \alpha = 2 \Rightarrow \left[\alpha\, x f(x) - \frac{\mathrm{d}\, f}{\mathrm{d}x}\right] = 2x^4 - 5x^2 + 1, \tag{A.37}$$

$$f(x) = x^3 - x, \qquad \alpha = 3 \Rightarrow \left[\alpha\, x f(x) - \frac{\mathrm{d}\, f}{\mathrm{d}x}\right] = 3x^4 - 6x^2 + 1, \tag{A.38}$$

$$f(x) = 2 - x^2 - x^4, \quad \alpha = 2 \Rightarrow \left[\alpha\, x f(x) - \frac{\mathrm{d}\, f}{\mathrm{d}x}\right] = -2x^5 + 2x^3 + 6x. \tag{A.39}$$

B. Approximations

B.1 Frequently Used Taylor Expansions

The expansion of functions with respect to a parameter in a power series plays an important role in the derivations of the progress rate formulae. Some of the expansions used will be presented here.

The derivation of functions with a square root and small x parameter is frequently used in this book; the expansion at $x = 0$ yields (Bronstein & Semendjajew, 1981, p. 83)

$$\sqrt{1+x} = 1 + \frac{1}{2}x - \frac{1}{8}x^2 + \frac{1}{16}x^3 - \frac{5}{128}x^4 + \mathcal{O}\left(x^5\right), \qquad |x| < 1. \tag{B.1}$$

Substituting x by $-2x$ in this expansion, it follows for $1 - \sqrt{1-2x}$ that

$$1 - \sqrt{1-2x} = x + \frac{1}{2}x^2 + \frac{1}{2}x^3 + \frac{5}{8}x^4 + \mathcal{O}\left(x^5\right), \qquad |x| < \frac{1}{2}. \tag{B.2}$$

The expansion of the natural logarithm for small x yields

$$\ln(1+x) = x - \frac{1}{2}x^2 + \frac{1}{3}x^3 - \frac{1}{4}x^4 + \mathcal{O}\left(x^5\right), \qquad -1 < x \leq 1. \tag{B.3}$$

The expansion of the density function $\Phi(A+B)$ of the Gaussian normal distribution at A will be demonstrated now in detail. First of all, the application of the Taylor expansion formula yields

$$\begin{aligned}
\Phi(A+B) &= \Phi(A) + \frac{\mathrm{d}\Phi}{\mathrm{d}A}B + \sum_{k=2}^{\infty} \frac{1}{k!}\frac{\mathrm{d}^k\Phi}{\mathrm{d}A^k}B^k \\
&= \Phi(A) + \frac{\mathrm{d}\Phi}{\mathrm{d}A}B + B\sum_{k=2}^{\infty} \frac{B^{k-1}}{k\,(k-1)!}\frac{\mathrm{d}^{k-1}}{\mathrm{d}A^{k-1}}\frac{\mathrm{d}\Phi}{\mathrm{d}A} \\
&= \Phi(A) + \frac{B}{\sqrt{2\pi}}\mathrm{e}^{-\frac{1}{2}A^2} + \frac{B}{\sqrt{2\pi}}\sum_{k=2}^{\infty} \frac{B^{k-1}}{k\,(k-1)!}\frac{\mathrm{d}^{k-1}}{\mathrm{d}A^{k-1}}\mathrm{e}^{-\frac{1}{2}A^2}.
\end{aligned} \tag{B.4}$$

At this point, the relation between the density and distribution functions of the $\mathcal{N}(0,1)$ standard normal distribution (C.2) is considered. Using the definition of the Hermite polynomials (B.7), it follows that

$$\Phi(A+B) = \Phi(A) + \frac{B}{\sqrt{2\pi}}\,\mathrm{e}^{-\frac{1}{2}A^2} + \frac{B}{\sqrt{2\pi}}\,\mathrm{e}^{-\frac{1}{2}A^2} \sum_{k=2}^{\infty} \frac{(-1)^{k-1}B^{k-1}}{k\,(k-1)!}\,\mathrm{He}_{k-1}(A). \tag{B.5}$$

After renumbering $k-1=i$, using $\mathrm{He}_0(x)=1$, (B.8), and renaming $i \to k$ thereafter, one obtains

$$\Phi(A+B) = \Phi(A) + \frac{B}{\sqrt{2\pi}}\,\mathrm{e}^{-\frac{1}{2}A^2} \sum_{k=0}^{\infty} \frac{(-1)^k B^k}{(k+1)!}\,\mathrm{He}_k(A). \tag{B.6}$$

B.2 The Hermite Polynomials $\mathrm{He}_k(x)$

The Hermite polynomials are defined as follows

$$\mathrm{He}_k(x) := (-1)^k \mathrm{e}^{\frac{1}{2}x^2} \frac{\mathrm{d}^k}{\mathrm{d}x^k} \left(\mathrm{e}^{-\frac{1}{2}x^2} \right) . \tag{B.7}$$

One finds by simple direct computation the first eight members

$$\begin{aligned} \mathrm{He}_0(x) &= 1, & \mathrm{He}_4(x) &= x^4 - 6x^2 + 3 \\ \mathrm{He}_1(x) &= x, & \mathrm{He}_5(x) &= x^5 - 10x^3 + 15x, \\ \mathrm{He}_2(x) &= x^2 - 1, & \mathrm{He}_6(x) &= x^6 - 15x^4 + 45x^2 - 15, \\ \mathrm{He}_3(x) &= x^3 - 3x, & \mathrm{He}_7(x) &= x^7 - 21x^5 + 105x^3 - 105x . \end{aligned} \tag{B.8}$$

The general case can be expressed as

$$\mathrm{He}_k(x) = x^k - 1 \cdot \binom{k}{2} x^{k-2} + 1 \cdot 3 \cdot \binom{k}{4} x^{k-4} - 1 \cdot 3 \cdot 5 \cdot \binom{k}{6} x^{k-6} + \dots . \tag{B.9}$$

The $\mathrm{He}_k(x)$ polynomial is therefore of kth order in x. As one can easily show by differentiating (B.7), the following relations are valid:

$$\mathrm{He}_{k+1}(x) = x\mathrm{He}_k(x) - \frac{\mathrm{d}}{\mathrm{d}x}\mathrm{He}_k(x) \quad \text{and} \quad \frac{\mathrm{d}}{\mathrm{d}x}\mathrm{He}_k(x) = k\mathrm{He}_{k-1} . \tag{B.10}$$

The exponents of x can be expressed by successive reorganization of (B.8) as a linear combination of the $\mathrm{He}_k(x)$

$$\begin{aligned} x^1 &= \mathrm{He}_1(x), \\ x^2 &= \mathrm{He}_2(x) + \mathrm{He}_0(x), \\ x^3 &= \mathrm{He}_3(x) + 3\mathrm{He}_1(x), \\ x^4 &= \mathrm{He}_4(x) + 6\mathrm{He}_2(x) + 3\mathrm{He}_0(x), \\ x^5 &= \mathrm{He}_5(x) + 10\mathrm{He}_3(x) + 15\mathrm{He}_1(x), \\ x^6 &= \mathrm{He}_6(x) + 15\mathrm{He}_4(x) + 45\mathrm{He}_2(x) + 15\mathrm{He}_0(x). \end{aligned} \tag{B.11}$$

The Hermite polynomials are a complete function system. Hence, they can be used for the regular approximation of functions. As can be shown easily, the following orthogonality relation is valid for the scalar product of the polynomials $\mathrm{He}_m(x)$ and $\mathrm{He}_k(x)$, which are weighted by $\mathrm{e}^{-x^2/2}$ (see, e.g., Abramowitz & Stegun, 1984, p. 334)

$$\frac{1}{\sqrt{2\pi}} \int_{-\infty}^{\infty} \mathrm{e}^{-\frac{1}{2}x^2} \mathrm{He}_m(x)\mathrm{He}_k(x)\,\mathrm{d}x = m!\,\delta_{mk} . \tag{B.12}$$

Any function $f(x)$ can be expanded with respect to this function system. Aiming at the approximation of density functions, we consider the special case

$$f(x) = \frac{\mathrm{e}^{-\frac{1}{2}x^2}}{\sqrt{2\pi}} \sum_{m=0}^{\infty} \frac{b_m}{m!} \mathrm{He}_m(x). \tag{B.13}$$

The Fourier coefficients b_k follow from the scalar product operation of $f(x)$ with $\mathrm{He}_k(x)$

$$\begin{aligned} &\int_{-\infty}^{\infty} f(x)\mathrm{He}_k(x)\,\mathrm{d}x \\ &= \sum_{m=0}^{\infty} \frac{b_m}{m!} \frac{1}{\sqrt{2\pi}} \int_{-\infty}^{\infty} \mathrm{e}^{-\frac{1}{2}x^2} \mathrm{He}_m(x)\mathrm{He}_k(x)\,\mathrm{d}x = \sum_{m=0}^{\infty} b_m \delta_{mk} = b_k. \end{aligned} \tag{B.14}$$

B.3 Cumulants, Moments, and Approximations

B.3.1 Fundamental Relations

In this section, some essential properties and relations are presented. These are helpful for understanding Chaps. 4 and 5; however, they are not strictly necessary.

Definitions. We start with the definition of the *characteristic function* $\varphi(t)$ (not to be mixed up with the progress rate) and the *cumulant-generating function* $\psi(t)$ (not to be mixed up with the SAR function) of the random variable x

$$\varphi(t) := \mathrm{E}\left\{e^{\imath t x}\right\} = \int_{-\infty}^{\infty} e^{\imath t x} p(x)\,\mathrm{d}x, \tag{B.15a}$$

$$\psi(t) := \ln \varphi(t) = \ln\left(\mathrm{E}\left\{e^{\imath t x}\right\}\right). \tag{B.15b}$$

By differentiating $\varphi(t)$ at t, one finds the relation to the moments[1] (Fisz, 1971)

$$\overline{x^k} = \mathrm{E}\left\{x^k\right\} = \frac{1}{\imath^k} \frac{\mathrm{d}^k}{\mathrm{d}t^k} \varphi(t)\bigg|_{t=0}. \tag{B.16}$$

Therefore, the characteristic function has the Taylor expansion

$$\varphi(t) = \sum_{k=0}^{\infty} \frac{\mathrm{d}^k \varphi(t)}{\mathrm{d}t^k}\bigg|_{t=0} \frac{t^k}{k!} = \sum_{k=0}^{\infty} \overline{x^k} \frac{(\imath t)^k}{k!}. \tag{B.17}$$

Analogous to (B.17), it is postulated that there exists also a Taylor expansion for the cumulant-generating function $\psi(t)$, corresponding to the pattern (B.17)

$$\psi(t) = \sum_{k=0}^{\infty} \frac{\mathrm{d}^k \psi(t)}{\mathrm{d}t^k}\bigg|_{t=0} \frac{t^k}{k!} \stackrel{!}{=} \sum_{k=0}^{\infty} \kappa_k \frac{(\imath t)^k}{k!}. \tag{B.18}$$

The weight coefficients κ_k are also called *cumulants* as well as *semi-invariants* (see below). They are defined as

$$\kappa_k := \frac{1}{\imath^k} \frac{\mathrm{d}^k}{\mathrm{d}t^k} \psi(t)\bigg|_{t=0}. \tag{B.19}$$

By directly applying (B.19), considering (B.15b), one obtains

$$\psi(0) = 0 \qquad \text{and} \qquad \kappa_0 = 1 \tag{B.20}$$

as well as, using (B.16),

$$\kappa_1 = \frac{1}{\imath} \frac{\mathrm{d}}{\mathrm{d}t} \ln(\varphi(t))\big|_{t=0} = \frac{1}{\imath} \frac{1}{\varphi(0)} \frac{\mathrm{d}\varphi}{\mathrm{d}t}\bigg|_{t=0} = \overline{x}, \tag{B.21}$$

[1] The quantity $\imath$ is the *imaginary unit* $\imath := \sqrt{-1}$.

$$\kappa_2 = -\left.\frac{\mathrm{d}^2\psi}{\mathrm{d}t^2}\right|_{t=0} = -\left.\left[-\left(\frac{1}{\varphi}\frac{\mathrm{d}\varphi}{\mathrm{d}t}\right)^2 + \frac{1}{\varphi}\frac{\mathrm{d}^2\varphi}{\mathrm{d}t^2}\right]\right|_{t=0} = -\left[-(\imath\overline{x})^2 + \imath^2\overline{x^2}\right]$$
$$= \mathrm{D}^2\{x\}. \tag{B.22}$$

In the last formula, the definition of the variance $\mathrm{D}^2\{x\} = \overline{x^2} - \overline{x}^2$ is used.

Transformation properties. The linear transformation $y = ax+c$ is considered here. The cumulant-generating function for that is

$$\psi_y(t) = \ln\Big(\mathrm{E}\left\{\mathrm{e}^{\imath(ax+c)t}\right\}\Big) = \ln\left(\mathrm{e}^{\imath ct}\,\mathrm{E}\left\{\mathrm{e}^{\imath axt}\right\}\right)$$
$$= \imath ct + \ln\left(\mathrm{E}\left\{\mathrm{e}^{\imath x\,at}\right\}\right). \tag{B.23}$$

Using the definition (B.19), the cumulants of the random variable y are obtained as

$$\kappa_k\{y\} = \left.\frac{1}{\imath^k}\frac{\mathrm{d}^k}{\mathrm{d}t^k}\left[\imath ct + \ln\mathrm{E}\left\{\mathrm{e}^{\imath x\,at}\right\}\right]\right|_{t=0}$$
$$= \left.\frac{1}{\imath^k}\frac{\mathrm{d}^k}{\mathrm{d}t^k}\imath ct\right|_{t=0} + \left.\frac{1}{\imath^k}a^k\frac{\mathrm{d}^k}{\mathrm{d}(at)^k}\ln\mathrm{E}\left\{\mathrm{e}^{\imath x\,at}\right\}\right|_{t=0}, \tag{B.24}$$

where $\kappa_k\{y\} = \left.\frac{1}{\imath^k}\frac{\mathrm{d}^k}{\mathrm{d}t^k}\imath ct\right|_{t=0} + a^k\kappa_k\{x\}$ is valid. Particularly, one finds for the random variable $y = ax + c$

$$\kappa_1\{y\} = c + a\kappa_1\{x\} \qquad \text{and for } k > 1: \qquad \kappa_k\{y\} = a^k\kappa_k\{x\}. \tag{B.25}$$

As one can see, a *translation* of the probability distribution ($c \neq 0$, $a = 1$) influences the cumulant κ_1. All other cumulants are not affected by the translation. Because of this remarkable property, the cumulants are also called "*semi-invariants*".

The cumulants of the sum of independent random variables. The sum s of two *independent* random variables, x and y, is investigated first; i.e. $s = x+y$. With the definition (B.15b), the cumulant-generating function $\psi_s(t)$ follows

$$\psi_s(t) = \ln\Big(\mathrm{E}\left\{\mathrm{e}^{\imath t(x+y)}\right\}\Big) = \ln\left(\mathrm{E}\left\{\mathrm{e}^{\imath tx}\mathrm{e}^{\imath ty}\right\}\right) = \ln\left(\mathrm{E}\left\{\mathrm{e}^{\imath tx}\right\}\,\mathrm{E}\left\{\mathrm{e}^{\imath ty}\right\}\right)$$
$$\psi_s(t) = \ln\mathrm{E}\left\{\mathrm{e}^{\imath tx}\right\} + \ln\mathrm{E}\left\{\mathrm{e}^{\imath ty}\right\} = \psi_x(t) + \psi_y(t). \tag{B.26}$$

Hence, one finds with (B.19)

$$\kappa_k\{x+y\} = \kappa_k\{x\} + \kappa_k\{y\}, \text{ if } x, y \text{ are statistically independent.} \tag{B.27}$$

The generalization to a sum of N *statistically independent* random variables is trivial. As an instance, the sum s of N *identically* distributed random variables x_i is considered

$$s = \sum_{i=1}^{N} x_i \qquad \Rightarrow \qquad \kappa_k\{s\} = N\,\kappa_k\{x\}. \tag{B.28}$$

Let us denote the expected value of each random variable x_i as $\overline{x}$, and the standard deviation as $\mathrm{D}\{x\}$. One finds, using (B.21) and (B.22) for the cumulants κ_1 and κ_2 of s,

$$\kappa_1\{s\} = N\,\overline{x} = \mathrm{E}\{s\} \qquad \text{and} \qquad \kappa_2\{s\} = N\,\mathrm{D}^2\{x\} = \mathrm{D}^2\{s\}. \tag{B.29}$$

If the sum s is *standardized*, i.e. $z := (s - \mathrm{E}\{s\})\,/\mathrm{D}\{s\}$, one obtains, considering (B.21, B.22) and (B.25), because $a = 1/\sqrt{N}\mathrm{D}\{x\}$,

$$z := \frac{\sum_1^N x_i - \mathrm{E}\Big\{\sum_1^N x_i\Big\}}{\mathrm{D}\Big\{\sum_{i=1}^N x_i\Big\}} \quad \Rightarrow \quad \begin{cases} \kappa_1\{z\} = 0, \\ \kappa_2\{z\} = 1, \\ \kappa_k\{z\} = N^{1-\frac{1}{2}k}\,\dfrac{\kappa_k\{x\}}{(\mathrm{D}\{x\})^k}\,, \quad k \geq 3. \end{cases} \tag{B.30}$$

As one can see, the cumulants of the standardized sum of N independent, identically distributed random variables vanishes with order $\mathcal{O}(N^{-(k/2-1)})$.

Cumulants and central moments. For some questions, it is advantageous to express the cumulants $\kappa_k\{y\}$ of a random variable y using the central moments $\mu_k\{y\}$

$$\mu_k\{y\} := \overline{(y - \overline{y})^k}. \tag{B.31}$$

The Taylor expansion (B.18) of the random variable $y - \overline{y}$ reads

$$\psi_{y-\overline{y}}(t) = \sum_{k=0}^{\infty} \kappa_k\{y - \overline{y}\}\,\frac{(\imath t)^k}{k!} = 1 + \sum_{k=2}^{\infty} \kappa_k\{y\}\,\frac{(\imath t)^k}{k!}. \tag{B.32}$$

At this point, $\kappa_0\{y - \overline{y}\} = 1$ and $\kappa_1\{y - \overline{y}\} = 0$, Eqs. (B.20, B.21), as well as (B.25), have been taken into account. Furthermore, one obtains using (B.15b) and (B.17)

$$\begin{aligned} \psi_{y-\overline{y}}(t) &= \ln\left(\sum_{k=0}^{\infty} \overline{(y - \overline{y})^k}\,\frac{(\imath t)^k}{k!}\right) = \ln\left(1 + \sum_{k=2}^{\infty} \mu_k\{y\}\,\frac{(\imath t)^k}{k!}\right) \\ &= 1 + \sum_{k=2}^{\infty} \mu_k\{y\}\,\frac{(\imath t)^k}{k!} - \frac{1}{2}\left(\sum_{k=2}^{\infty} \mu_k\{y\}\,\frac{(\imath t)^k}{k!}\right)^2 \\ &\quad + \frac{1}{3}\left(\sum_{k=2}^{\infty} \mu_k\{y\}\,\frac{(\imath t)^k}{k!}\right)^3 - \dots . \end{aligned} \tag{B.33}$$

The second line is obtained using the Taylor expansion (B.3). Comparing term by term, after reordering by the exponents with respect to $(\imath t)^k$ the right sides of (B.33) and (B.32), one finds

$$\begin{aligned} &\kappa_2 = \mu_2, && \kappa_3 = \mu_3, \\ &\kappa_4 = \mu_4 - 3\mu_2^2, && \kappa_5 = \mu_5 - 10\mu_2\mu_3, \\ &\kappa_6 = \mu_6 - 15\mu_2\mu_4 - 10\mu_3^2 + 30\mu_2^3\,. \end{aligned} \tag{B.34}$$

Of course, the cumulants can also be expressed by the moments $\overline{y^k}$. The relation between the central moments μ_k and $\overline{y^k}$ should be established first for this purpose. This is possible by immediate calculation of (B.31). One obtains

$$
\begin{aligned}
\mu_2\{y\} &= \overline{y^2} - \overline{y}^2, \\
\mu_3\{y\} &= \overline{y^3} - 3\,\overline{y^2}\,\overline{y} + 2\,\overline{y}^3, \\
\mu_4\{y\} &= \overline{y^4} - 4\,\overline{y^3}\,\overline{y} + 6\,\overline{y^2}\,\overline{y}^2 - 3\,\overline{y}^4, \\
\mu_5\{y\} &= \overline{y^5} - 5\,\overline{y^4}\,\overline{y} + 10\,\overline{y^3}\,\overline{y}^2 - 10\,\overline{y^2}\,\overline{y}^3 + 4\,\overline{y}^5, \\
\mu_6\{y\} &= \overline{y^6} - 6\,\overline{y^5}\,\overline{y} + 15\,\overline{y^4}\,\overline{y}^2 - 20\,\overline{y^3}\,\overline{y}^3 + 15\,\overline{y^2}\,\overline{y}^4 - 5\,\overline{y}^6.
\end{aligned}
\tag{B.35}
$$

After inserting this in (B.34), the desired relations follow

$$
\begin{aligned}
\kappa_2 &= \overline{y^2} - \overline{y}^2, \\
\kappa_3 &= \overline{y^3} - 3\,\overline{y^2}\,\overline{y} + 2\,\overline{y}^3, \\
\kappa_4 &= \overline{y^4} - 4\,\overline{y^3}\,\overline{y} - 3\,\overline{y^2}^2 + 12\,\overline{y^2}\,\overline{y}^2 - 6\,\overline{y}^4, \\
\kappa_5 &= \overline{y^5} - 5\,\overline{y^4}\,\overline{y} - 10\,\overline{y^3}\,\overline{y^2} + 20\,\overline{y^3}\,\overline{y}^2 + 30\,\overline{y^2}^2\,\overline{y} - 60\,\overline{y^2}\,\overline{y}^3 + 24\,\overline{y}^5.
\end{aligned}
\tag{B.36}
$$

B.3.2 The Relation of the Weight Coefficients for the Density Approximation of a Standardized Random Variable

Assume that a density function of a standardized random variable z

$$
z := \frac{y - \mathrm{E}\{y\}}{\mathrm{D}\{y\}}
\tag{B.37}
$$

is approximated by a series of Hermite polynomials (see Appendix B.2), according to (B.13)

$$
p(z) = \frac{\mathrm{e}^{-\frac{1}{2}z^2}}{\sqrt{2\pi}} \sum_{m=0}^{\infty} \frac{a_m}{m!}\,\mathrm{He}_m(z).
\tag{B.38}
$$

The weight coefficients appearing in (B.16) can be calculated using (B.14). However, they are expressed here alternatively using cumulants.

The characteristic function $\varphi_z(t)$ for the density (B.38), according to the definition (B.15a), will be calculated first

$$
\varphi_z(t) = \int_{-\infty}^{\infty} \mathrm{e}^{\imath t z} p(z)\,\mathrm{d}z = \sum_{m=0}^{\infty} \frac{a_m}{m!} \frac{1}{\sqrt{2\pi}} \int_{-\infty}^{\infty} \mathrm{e}^{\imath t z} \mathrm{e}^{-\frac{1}{2}z^2} \mathrm{He}_m(z)\,\mathrm{d}z.
\tag{B.39}
$$

After inserting (B.7) for $\mathrm{He}_m(z)$ in (B.39), one obtains

$$
\varphi_z(t) = \sum_{m=0}^{\infty} (-1)^m \frac{a_m}{m!} \frac{1}{\sqrt{2\pi}} \int_{-\infty}^{\infty} \mathrm{e}^{\imath t z} \frac{\mathrm{d}^m}{\mathrm{d}z^m} \left(\mathrm{e}^{-\frac{1}{2}z^2}\right) \mathrm{d}z.
\tag{B.40}
$$

The integral can be solved by successively applying partial integration

$$
\begin{aligned}
\int_{-\infty}^{\infty} & \mathrm{e}^{\imath t z} \frac{\mathrm{d}^m}{\mathrm{d}z^m} \mathrm{e}^{-\frac{1}{2}z^2} \mathrm{d}z \\
&= \left[\mathrm{e}^{\imath t z} \frac{\mathrm{d}^{m-1}}{\mathrm{d}z^{m-1}} \mathrm{e}^{-\frac{1}{2}z^2} \right]_{-\infty}^{\infty} - (\imath t) \int_{-\infty}^{\infty} \mathrm{e}^{\imath t z} \frac{\mathrm{d}^{m-1}}{\mathrm{d}z^{m-1}} \mathrm{e}^{-\frac{1}{2}z^2} \mathrm{d}z \\
&= (-\imath t) \int_{-\infty}^{\infty} \mathrm{e}^{\imath t z} \frac{\mathrm{d}^{m-1}}{\mathrm{d}z^{m-1}} \mathrm{e}^{-\frac{1}{2}z^2} \mathrm{d}z \\
&\ \ \vdots \\
&= (-\imath t)^m \int_{-\infty}^{\infty} \mathrm{e}^{\imath t z} \mathrm{e}^{-\frac{1}{2}z^2} \mathrm{d}z = (-1)^m (\imath t)^m \int_{-\infty}^{\infty} \mathrm{e}^{-\frac{1}{2}(z-\imath t)^2} \mathrm{e}^{-\frac{1}{2}t^2} \mathrm{d}z \\
&= (-1)^m (\imath t)^m \, \mathrm{e}^{-\frac{1}{2}t^2} \sqrt{2\pi}.
\end{aligned}
\tag{B.41}
$$

Note, in the line before the last, the equality $2\imath t z - z^2 = -(z - \imath t)^2 - t^2$ was used. After inserting the result of (B.41) into (B.40), one gets for the characteristic function

$$
\varphi_z(t) = \mathrm{e}^{-\frac{1}{2}t^2} \sum_{m=0}^{\infty} \frac{a_m}{m!} (\imath t)^m. \tag{B.42}
$$

Secondly, the cumulant-generating function $\psi_z(t)$ will be written down. Note that $\kappa_0\{z\} = 1$ (see (B.20)); $\kappa_1\{z\} = 0$ and $\kappa_2\{z\} = 1$ because of (B.21) and (B.22), respectively. Therefore, one obtains for (B.18)

$$
\psi_z(t) = -\frac{t^2}{2} + \sum_{k=3}^{\infty} \kappa_k \frac{(\imath t)^k}{k!}. \tag{B.43}
$$

One has $\varphi_z(t) = \exp(\psi_z(t))$ because of (B.15b). After inserting (B.43), one can compare that with (B.42). This way one gets

$$
\sum_{m=0}^{\infty} \frac{a_m}{m!} (\imath t)^m \stackrel{!}{=} \exp\left(\sum_{k=3}^{\infty} \kappa_k \frac{(\imath t)^k}{k!} \right) = \prod_{k=3}^{\infty} \exp\left(\kappa_k \frac{(\imath t)^k}{k!} \right). \tag{B.44}
$$

The "exp"-term on the right-hand side of (B.44) is expanded into a Taylor series. The series obtained in this manner will be factorized out, and sorted with respect to the exponents of $(\imath t)^m$

$$
\begin{aligned}
\sum_{m=0}^{\infty} \frac{a_m}{m!} (\imath t)^m \stackrel{!}{=} & \left[1 + \kappa_3 \frac{(\imath t)^3}{3!} + \frac{1}{2} \left(\kappa_3 \frac{(\imath t)^3}{3!} \right)^2 + \ldots \right] \\
& \times \left[1 + \kappa_4 \frac{(\imath t)^4}{4!} + \ldots \right] \left[1 + \kappa_5 \frac{(\imath t)^5}{5!} + \ldots \right] \times \ldots \\
= 1 \ + & \ \frac{\kappa_3}{3!} (\imath t)^3 \ + \ \frac{\kappa_4}{4!} (\imath t)^4 \\
& + \frac{\kappa_5}{5!} (\imath t)^5 \ + \ \left(\frac{\kappa_6}{6!} + \frac{\kappa_3^2}{2 \cdot 3! \cdot 3!} \right) (\imath t)^6 \ + \ \ldots .
\end{aligned}
\tag{B.45}
$$

The coefficients before the $(\imath t)^m$-exponents can be compared now term-wise. The result reads

$$\begin{aligned} a_0 &= 1, & a_4 &= \kappa_4\{z\}, \\ a_1 &= 0, & a_5 &= \kappa_5\{z\}, \\ a_2 &= 0, & a_6 &= \kappa_6\{z\} + \frac{6!}{2\cdot 3!\cdot 3!}\left(\kappa_3\{z\}\right)^2, \\ a_3 &= \kappa_3\{z\}, & a_7 &= \kappa_7\{z\} + \frac{7!}{3!\cdot 4!}\,\kappa_3\{z\}\,\kappa_4\{z\}. \end{aligned} \tag{B.46}$$

Hence, the function series (B.38) for the approximation of $p(z)$ becomes

$$p(z) = \frac{\mathrm{e}^{-\frac{1}{2}z^2}}{\sqrt{2\pi}}\left[1 + \frac{\kappa_3}{3!}\,\mathrm{He}_3(z) + \frac{\kappa_4}{4!}\,\mathrm{He}_4(z) + \frac{\kappa_5}{5!}\,\mathrm{He}_5(z) + \left(\frac{\kappa_6}{6!} + \frac{\kappa_3^2}{2\cdot 3!\cdot 3!}\right)\mathrm{He}_6(z) + \left(\frac{\kappa_7}{7!} + \frac{\kappa_3\kappa_4}{3!\cdot 4!}\right)\mathrm{He}_7(z) + \ldots\right]. \tag{B.47}$$

Using this (regularly convergent) function series, it is possible to approximate density functions to any desired accuracy.

In the practical application of (B.47) as an approximation of a probability density, the question of the error made caused by the series cut emerges. It is difficult to make general and also practically useful statements: generally, one cannot assume that the magnitude of the $(k+1)$th series term is smaller than that of the kth term. The estimation of the error term requires concrete knowledge of the cumulants. The monograph of Gnedenko and Kolmogorov (1968) is recommended for further reading. Here, the convergence order of the density approximation of the standardized sum s, (B.30), composed of N identically distributed random variables x_i, with $\overline{x} = \mathrm{E}\{x_i\}$, $\mathrm{D}^2\{x\} = \mathrm{D}^2\{x_i\} = \mathrm{E}\left\{(x_i - \overline{x})^2\right\}$ and β_k, is considered

$$s := \sum_{i=1}^{N} x_i, \qquad z := \frac{s - N\,\overline{x}}{\sqrt{N}\,\mathrm{D}\{x\}}, \qquad \beta_k := \frac{\kappa_k\{x\}}{(\mathrm{D}\{x\})^k}. \tag{B.48}$$

The cumulants of (B.30), $\kappa_k\{z\} = \beta_k/(\sqrt{N})^{k-2}$, are inserted into (B.47), and sorted by the exponents of N [2]

$$p(z) = \frac{\mathrm{e}^{-\frac{1}{2}z^2}}{\sqrt{2\pi}}\left[1 + \frac{1}{\sqrt{N}}\left(\frac{\beta_3}{3!}\,\mathrm{He}_3(z)\right) + \frac{1}{N}\left(\frac{\beta_4}{4!}\,\mathrm{He}_4(z) + \frac{\beta_3^2}{2\cdot 3!\cdot 3!}\,\mathrm{He}_6(z)\right) + \frac{1}{\sqrt{N}\,N}\left(\frac{\beta_5}{5!}\,\mathrm{He}_5(z) + \frac{\beta_3\,\beta_4}{3!\cdot 4!}\,\mathrm{He}_7(z) + \frac{\beta_3^3}{(3!)^4}\,\mathrm{He}_9(z)\right) + \ldots\right]. \tag{B.49}$$

[2] Note that the He_9-term in (B.49) results from the a_9 coefficient in (B.45), which was not given explicitly.

Thus, we have obtained a generalization of the central limit theorem of statistics. For $N \to \infty$, one obtains the classical result. The case $N < \infty$ yields the correction terms vanishing at least with order $\mathcal{O}(1/\sqrt{N})$.

Considering that the density approximation (B.49) produces in the case of sum (B.48) a series of the form (B.49), one gets the reordering of the series terms in (B.47) according to the scheme in (B.49)[3]

$$p(z) = \frac{\mathrm{e}^{-\frac{1}{2}z^2}}{\sqrt{2\pi}} \left[1 + \frac{\kappa_3}{3!} \mathrm{He}_3(z) + \left(\frac{\kappa_4}{4!} \mathrm{He}_4(z) + \frac{\kappa_3^2}{2 \cdot 3! \cdot 3!} \mathrm{He}_6(z) \right) \right.$$
$$\left. + \left(\frac{\kappa_5}{5!} \mathrm{He}_5(z) + \frac{\kappa_3\, \kappa_4}{3! \cdot 4!} \mathrm{He}_7(z) + \frac{\kappa_3^3}{(3!)^4} \mathrm{He}_9(z) \right) + \ldots \right]. \tag{B.50}$$

This reordering is purely *formal*; it does not guarantee a certain convergence order of the approximation. Only in one case – the sum of identical random variables – does (B.50) ensures that the error terms vanish with order $\mathcal{O}(1/\sqrt{N}^k)$.

As an example for the density approximation, the density of the square of the standardized length of a mutation vector $\mathbf{h}$ with N $\mathcal{N}(0, \sigma^2)$ normally distributed random components h_i is considered. This can also be considered as a refinement of the normal approximation introduced in Sect. 3.4.2. Considering (B.48), we have

$$s = \sum_{i=1}^{N} h_i^2, \qquad \overline{x} = \overline{h_i^2} = \sigma^2, \tag{B.51}$$

and

$$\mathrm{D}\{x\} = \sqrt{\overline{h_i^4} - \overline{h_i^2}^2} = \sqrt{2}\sigma^2 \tag{B.52}$$

(see Table A.1, p. 329). Using (B.48), considering (B.25), one gets for β_k

$$\beta_k = \frac{\kappa_k\{(\mathcal{N}(0,1))^2\}}{(\sqrt{2})^k}. \tag{B.53}$$

With (B.36), κ_k can be expressed by the moments $(\mathcal{N}(0,1))^2$. These moments can be found in Table A.1. One gets for β_3

$$\beta_3 = \frac{\overline{\mathcal{N}^6} - 3\overline{\mathcal{N}^4}\,\overline{\mathcal{N}^2} + 3\,\overline{\mathcal{N}^2}^3}{(\sqrt{2})^3} = 2\sqrt{2} \tag{B.54}$$

as well as for β_4

$$\beta_4 = \frac{1}{4} \left(\overline{\mathcal{N}^8} - 4\,\overline{\mathcal{N}^6}\,\overline{\mathcal{N}^2} - 3\,\overline{\mathcal{N}^4}^2 + 12\,\overline{\mathcal{N}^4}\,\overline{\mathcal{N}^2}^2 - 6\,\overline{\mathcal{N}^2}^4 \right) = 12. \tag{B.55}$$

[3] This series is called an *Edgeworth* expansion, whereas (B.47) is sometimes called a *Gram-Charlier* series (Johnson & Kotz, 1970). However, according to Gnedenko and Kolmogorov (1968), the "originator" of these series is Chebyshev.

By inserting these results into (B.49), the square of the standardized length becomes

$$z = \frac{\sum_{i=1}^{N} h_i^2 - N\,\sigma^2}{\sqrt{2N}\,\sigma^2}, \tag{B.56}$$

and the density of the corresponding N-dimensional normally distributed vector $\mathbf{h}$ can be expressed as

$$p(z) = \frac{e^{-\frac{1}{2}z^2}}{\sqrt{2\pi}} \left[1 + \frac{1}{\sqrt{N}} \frac{\sqrt{2}}{3} \mathrm{He}_3(z) + \frac{1}{N} \left(\frac{1}{2} \mathrm{He}_4(z) + \frac{1}{9} \mathrm{He}_6(z) \right) + \ldots \right]. \tag{B.57}$$

Using the expansion (B.50), one can approximate the density function of a standardized distribution which is described by cumulants. The corresponding distribution function $P(z) = P(Z < z)$ follows by the integration, $P(z) = \int_{z'=-\infty}^{z'=z} p(z')\,dz'$. Considering (B.10), one has for the single terms in (B.50) the integral formula

$$\frac{1}{\sqrt{2\pi}} \int_{x=-\infty}^{x=z} e^{-\frac{1}{2}x^2} \mathrm{He}_k(x)\,dx = \begin{cases} \Phi(z), & k = 0 \\ -\dfrac{1}{\sqrt{2\pi}} e^{-\frac{1}{2}z^2} \mathrm{He}_{k-1}(z), & k \geq 1. \end{cases} \tag{B.58}$$

Hence, the cumulative distribution function reads

$$P(z) = \Phi(z) - \frac{e^{-\frac{1}{2}z^2}}{\sqrt{2\pi}} \left[\frac{\kappa_3}{3!} \mathrm{He}_2(z) + \left(\frac{\kappa_4}{4!} \mathrm{He}_3(z) + \frac{\kappa_3^2}{2 \cdot 3! \cdot 3!} \mathrm{He}_5(z) \right) + \left(\frac{\kappa_5}{5!} \mathrm{He}_4(z) + \frac{\kappa_3\,\kappa_4}{3! \cdot 4!} \mathrm{He}_6(z) + \frac{\kappa_3^3}{(3!)^4} \mathrm{He}_8(z) \right) + \ldots \right]. \tag{B.59}$$

B.4 Approximation of the Quantile Function

The distribution function $P(z)$ of a standardized random variable can be approximated by a series of Hermite polynomials, using the cumulants κ_k (cf. Appendix B.3). One can alternatively derive an approximation of the inverse function $P^{-1}(\cdot)$. The inverse function $P^{-1}(\cdot)$ is also called the quantile. A possible derivation will be sketched here below.[4]

The starting point is the idea that the expansion (B.59) for a $P(z) = f$,

$$f = \Phi(z) - \frac{\mathrm{e}^{-\frac{1}{2}z^2}}{\sqrt{2\pi}} \left[\frac{\kappa_3}{3!} \mathrm{He}_2(z) + \Big(\cdots \Big) + \ldots \right], \tag{B.60}$$

could be directly solved for z, provided that $\forall\, k \geq 3\colon\ \kappa_k = 0$ is fulfilled. As a result, $z = P^{-1}(f) = \Phi^{-1}(f)$ holds. Therefore, for $\kappa_k \neq 0$ it is reasonable to expand $P^{-1}(f)$ in a series of Hermite polynomials

$$P^{-1}(f) = \sum_{k=0}^{\infty} c_k \mathrm{He}_k \left(\Phi^{-1}(f) \right). \tag{B.61}$$

In the case of $P(x) \equiv \Phi(x)$, $c_1 = 1$ and $c_k = 0$ $(k \neq 1)$ follow. The deviation of the quantile from the normal distribution is expressed by the $c_k \neq 0$. If

$$y = \Phi^{-1}(f) \tag{B.62}$$

is inserted into (B.61), one obtains

$$P^{-1} \left(\Phi(y) \right) = \sum_{k=0}^{\infty} c_k \mathrm{He}_k(y). \tag{B.63}$$

The Fourier coefficients c_k result because of the orthogonality relation (B.12)

$$c_k = \frac{1}{k!} \frac{1}{\sqrt{2\pi}} \int_{-\infty}^{\infty} \mathrm{e}^{-\frac{1}{2}y^2} \mathrm{He}_k(y)\, P^{-1}(\Phi(y))\, \mathrm{d}y. \tag{B.64}$$

Substituting $z = P^{-1}\left(\Phi(y)\right)$ in (B.64) and considering (C.13), one finds

$$c_k = \frac{1}{k!} \int_{-\infty}^{\infty} \mathrm{He}_k \left(\Phi^{-1}(P(z)) \right) z\, p(z)\, \mathrm{d}z. \tag{B.65}$$

The density $p(z)$ and the distribution $P(z)$, which appear in (B.65), are given by the approximations (B.50) and (B.59). $\Phi^{-1}(P(z))$ can be expressed by a Taylor expansion at $P(z) = \Phi(z)$. Thus, one can integrate term-wise.

The calculations are simple but long, and therefore are not shown here in detail. Instead, the reader is referred to the original work of Cornish and Fisher (1937), which provides a derivation different to that shown here. The result can also be found in Abramowitz and Stegun (1984), p. 411. It is given

[4] The idea of the quantile approximation goes back to Cornish and Fisher (1937). An alternative derivation to that presented here is sketched in Johnson and Kotz (1970), pp. 33ff.

here directly. One obtains $P^{-1}(\Phi(y)) = z$ in terms of the expansion (B.63), collecting the terms analogous to the pattern in (B.50),

$$\begin{aligned} z = y &+ \frac{\kappa_3}{3!}\,\mathrm{He}_2(y) \\ &+ \frac{\kappa_4}{4!}\,\mathrm{He}_3(y) - \frac{\kappa_3^2}{36}\,(2\mathrm{He}_3(y) + \mathrm{He}_1(y)) \\ &+ \frac{\kappa_5}{5!}\,\mathrm{He}_4(y) - \frac{\kappa_3\kappa_4}{4!}\,(\mathrm{He}_4(y) + \mathrm{He}_2(y)) \\ &\qquad + \frac{\kappa_3^3}{324}\,(12\mathrm{He}_4(y) + 19\mathrm{He}_2(y)) \\ &+ \ldots . \end{aligned} \tag{B.66}$$

The ordering of the terms is similar to the pattern in (B.49). That is, if the sum of identically distributed random variables is standardized according to (B.48), the κ_3 term in the first line of (B.66) is of order $\mathcal{O}(1/\sqrt{N})$, the second line is of order $\mathcal{O}(1/N)$, and the third one of order $\mathcal{O}(1/N\sqrt{N})$.

Thereafter, (B.66) is ordered after exponents of y for the application in the $(1 \overset{+}{,} \lambda)$ quality gain analysis. One obtains

$$\begin{aligned} z = &\left[\left(-\frac{\kappa_3}{6}\right) + \left(\frac{17}{324}\kappa_3^3 - \frac{\kappa_3\kappa_4}{12} + \frac{\kappa_5}{40}\right) + \ldots\right] \\ &+ \left[1 + \left(\frac{5}{36}\kappa_3^2 - \frac{\kappa_4}{8}\right)\right. \\ &\qquad \left. + \left(-\frac{107}{648}\kappa_3^4 + \frac{107}{288}\kappa_3^2\kappa_4 - \frac{7}{60}\kappa_3\kappa_5 - \frac{29}{384}\kappa_4^2 + \frac{\kappa_6}{48}\right) + \ldots\right] y^1 \\ &+ \left[\left(\frac{\kappa_3}{6}\right) + \left(-\frac{53}{324}\kappa_3^3 + \frac{5}{24}\kappa_3\kappa_4 - \frac{\kappa_5}{20}\right) + \ldots\right] y^2 \\ &+ \left[\left(-\frac{\kappa_3^2}{18} + \frac{\kappa_4}{24}\right)\right. \\ &\qquad \left. + \left(\frac{211}{972}\kappa_3^4 - \frac{103}{288}\kappa_3^2\kappa_4 + \frac{17}{180}\kappa_3\kappa_5 + \frac{\kappa_4^2}{16} - \frac{\kappa_6}{72}\right) + \ldots\right] y^3 \\ &+ \left[\left(\frac{\kappa_3^3}{27} - \frac{\kappa_3\kappa_4}{24} + \frac{\kappa_5}{120}\right) + \ldots\right] y^4 \\ &+ \left[\left(-\frac{7}{216}\kappa_3^4 + \frac{7}{144}\kappa_3^2\kappa_4 - \frac{\kappa_3\kappa_5}{90} - \frac{\kappa_4^2}{128} + \frac{\kappa_6}{720}\right) + \ldots\right] y^5 \\ &+ \ldots . \end{aligned} \tag{B.67}$$

Again, the result is ordered after $\mathcal{O}((1/\sqrt{N})^k)$ terms. In this book, only the terms up to and including order $\mathcal{O}(1/N)$ are used. Hence, (B.67) reduces to

$$\begin{aligned} z = -\frac{\kappa_3}{6} &+ \left[1 + \left(\frac{5}{36}\kappa_3^2 - \frac{\kappa_4}{8}\right)\right] y + \left[\frac{\kappa_3}{6}\right] y^2 \\ &+ \left[\frac{\kappa_4}{24} - \frac{\kappa_3^2}{18}\right] y^3 + \ldots . \end{aligned} \tag{B.68}$$

C. The Normal Distribution

C.1 $\mathcal{N}(0,1)$ Distribution Function, Gaussian Integral, and Error Function

Besides $\Phi(x)$, the standard normal distribution function $\mathcal{N}(0,1)$,

$$\Phi(x) := P(t < x) = \frac{1}{\sqrt{2\pi}} \int_{t=-\infty}^{t=x} \mathrm{e}^{-\frac{1}{2}t^2} \, \mathrm{d}t, \tag{C.1}$$

other functions are also used in the literature. $\Phi(x)$ gives the cumulative distribution of the probability density function of the standard normal distribution $t = \mathcal{N}(0,1)$ with the density $p(t) := \phi(t)$

$$\phi(t) := \frac{\mathrm{d}}{\mathrm{d}t}\Phi(t) = \frac{1}{\sqrt{2\pi}} \mathrm{e}^{-\frac{1}{2}t^2}, \qquad \phi(t) = \phi(-t). \tag{C.2}$$

The Gaussian integral $\Phi_0(x)$ (Bronstein & Semendjajew, 1981, p. 71), and the error function $\mathrm{erf}(x)$ (Abramowitz & Stegun, 1984, p. 84)

$$\Phi_0(x) := \frac{1}{\sqrt{2\pi}} \int_{t=0}^{t=x} \mathrm{e}^{-\frac{1}{2}t^2} \, \mathrm{d}t, \qquad \mathrm{erf}(x) := \frac{2}{\sqrt{\pi}} \int_{t=0}^{t=x} \mathrm{e}^{-t^2} \, \mathrm{d}t, \tag{C.3}$$

are also closely related. The conversion between $\Phi_0(x)$ and $\mathrm{erf}(x)$ can be simply done by the variable substitution in the integrals of (C.3). One obtains

$$\Phi_0(x) = \frac{1}{2} \mathrm{erf}\left(\frac{x}{\sqrt{2}}\right), \qquad \mathrm{erf}(x) = 2\Phi_0\left(\sqrt{2}x\right). \tag{C.4}$$

Furthermore, by comparing (C.1) with (C.3), considering the symmetry in (C.2) with respect to the origin, one gets

$$\Phi(x) = \frac{1}{2} + \Phi_0(x) \qquad \text{and} \qquad \Phi(x) = \frac{1}{2}\left(1 + \mathrm{erf}\left(\frac{x}{\sqrt{2}}\right)\right). \tag{C.5}$$

The special values of Φ are

$$\Phi(-\infty) = 0, \qquad \Phi(0) = \frac{1}{2}, \qquad \Phi(\infty) = 1. \tag{C.6}$$

Because of the point-symmetry of $\phi(x)$ (alternatively, one can say that $\phi(x)$ is an *even* function of x), it follows from (C.3) that

$$\Phi_0(x) = -\Phi_0(-x) \qquad \text{and} \qquad \mathrm{erf}(x) = -\mathrm{erf}(-x). \tag{C.7}$$

After adding 1/2 to the first equation in (C.7), and comparing with (C.5), it follows from the symmetry relation of $\Phi(x)$

$$\Phi(x) = 1 - \Phi(-x). \tag{C.8}$$

The inverse function $\Phi^{-1}(x)$ of the Gaussian distribution (C.1) is also called the xth order quantile in probability theory. It plays a certain role in the approximations of the ES theory. Hence, its relation to the inverse functions $\mathrm{erf}^{-1}(x)$ and $\Phi_0^{-1}(x)$ will be established next. Since

$$x \equiv \Phi(\Phi^{-1}(x)) = \frac{1}{2} + \Phi_0(\Phi^{-1}(x)),$$

it follows that

$$\begin{aligned} \Phi^{-1}(x) &= \Phi_0^{-1}\left(x - \frac{1}{2}\right), & 0 \le x \le 1 \\ \Phi_0^{-1}(y) &= \Phi^{-1}\left(y + \frac{1}{2}\right), & -\frac{1}{2} \le y \le \frac{1}{2} \end{aligned} \tag{C.9}$$

and because $x \equiv \Phi(\Phi^{-1}(x)) = \frac{1}{2}\left(1 + \mathrm{erf}\left(\Phi^{-1}(x)/\sqrt{2}\right)\right)$,

$$\begin{aligned} \Phi^{-1}(x) &= \sqrt{2}\,\mathrm{erf}^{-1}(2x-1), & 0 \le x \le 1 \\ \mathrm{erf}^{-1}(z) &= \frac{1}{\sqrt{2}}\,\Phi^{-1}\left(\frac{z}{2} + \frac{1}{2}\right), & -1 \le z \le 1. \end{aligned} \tag{C.10}$$

Analogously, one finds

$$\begin{aligned} \Phi_0^{-1}(y) &= \sqrt{2}\,\mathrm{erf}^{-1}(2y), & -\frac{1}{2} \le y \le \frac{1}{2} \\ \mathrm{erf}^{-1}(z) &= \frac{1}{\sqrt{2}}\,\Phi_0^{-1}\left(\frac{z}{2}\right), & -1 \le z \le 1. \end{aligned} \tag{C.11}$$

The derivatives of $\Phi^{-1}(x)$ are required for some Taylor expansions. By differentiating the identity

$$x \equiv \Phi(\underbrace{\Phi^{-1}(x)}_{:=a}) \tag{C.12}$$

one gets

$$\frac{\mathrm{d}}{\mathrm{d}x}x \equiv 1 = \frac{\mathrm{d}\Phi(a)}{\mathrm{d}a}\frac{\mathrm{d}\Phi^{-1}(x)}{\mathrm{d}x}, \qquad \frac{\mathrm{d}\Phi(a)}{\mathrm{d}a} = \frac{1}{\sqrt{2\pi}}\,\mathrm{e}^{-\frac{1}{2}a^2}$$

$$\frac{\mathrm{d}}{\mathrm{d}x}\Phi^{-1}(x) = \frac{1}{\frac{\mathrm{d}\Phi(a)}{\mathrm{d}a}} = \sqrt{2\pi}\,\mathrm{e}^{\frac{1}{2}a^2}$$

and, consequently,

$$\frac{\mathrm{d}}{\mathrm{d}x}\Phi^{-1}(x) = \sqrt{2\pi}\exp\left[\frac{1}{2}\left(\Phi^{-1}(x)\right)^2\right]. \tag{C.13}$$

The second derivative follows by applying $\frac{\mathrm{d}}{\mathrm{d}x}$ to (C.13)

$$\frac{\mathrm{d}^2}{\mathrm{d}x^2}\,\Phi^{-1}(x)=\sqrt{2\pi}\exp\left[\frac{1}{2}\left(\Phi^{-1}(x)\right)^2\right]\Phi^{-1}(x)\frac{\mathrm{d}}{\mathrm{d}x}\,\Phi^{-1}(x)$$

$$\frac{\mathrm{d}^2}{\mathrm{d}x^2}\,\Phi^{-1}(x)=\left\{\sqrt{2\pi}\exp\left[\frac{1}{2}\left(\Phi^{-1}(x)\right)^2\right]\right\}^2\Phi^{-1}(x) \tag{C.14}$$

and for the next three derivatives, one gets

$$\frac{\mathrm{d}^3}{\mathrm{d}x^3}\,\Phi^{-1}(x)=\left\{\sqrt{2\pi}\exp\left[\frac{1}{2}\left(\Phi^{-1}(x)\right)^2\right]\right\}^3\left[2\left(\Phi^{-1}(x)\right)^2+1\right],$$

$$\frac{\mathrm{d}^4}{\mathrm{d}x^4}\,\Phi^{-1}(x)=\left\{\sqrt{2\pi}\exp\left[\frac{1}{2}\left(\Phi^{-1}(x)\right)^2\right]\right\}^4\left[6\left(\Phi^{-1}(x)\right)^3+7\Phi^{-1}(x)\right],$$

$$\frac{\mathrm{d}^5}{\mathrm{d}x^5}\,\Phi^{-1}(x)=\left\{\sqrt{2\pi}\exp\left[\frac{1}{2}\left(\Phi^{-1}(x)\right)^2\right]\right\}^5 \times\left[24\left(\Phi^{-1}(x)\right)^4+46\left(\Phi^{-1}(x)\right)^2+7\right].$$

C.2 Asymptotic Order of the Moments of $\frac{x}{R}$

The N-order of the random quantity x/R will be estimated in this appendix. The magnitude of the expected value

$$\mathrm{E}\left\{\frac{x^k}{R}\right\} = \frac{1}{R}\int_{-\infty}^{\infty} x^k P_\mathrm{a}(x)\, p_x(x)\,\mathrm{d}x \tag{C.15}$$

is investigated for this purpose. $P_\mathrm{a}(x)$ stands for any arbitrary probability function with $0 \le P_\mathrm{a}(x) \le 1$, and $p_x(x)$ for the density (3.12) of the coordinate x of the mutation vector $\mathbf{z}$. The magnitude is estimated as follows

$$\begin{aligned}\left|\mathrm{E}\left\{\frac{x^k}{R}\right\}\right| &\le \frac{1}{R}\frac{1}{\sqrt{2\pi}\,\sigma}\int_{-\infty}^{\infty} \left|x^k\right| P_\mathrm{a}(x)\,\mathrm{e}^{-\frac{1}{2}\left(\frac{x}{\sigma}\right)^2}\mathrm{d}x \\ &\le \frac{1}{R}\frac{1}{\sqrt{2\pi}\,\sigma}\int_{-\infty}^{\infty} \left|x^k\right|\,\mathrm{e}^{-\frac{1}{2}\left(\frac{x}{\sigma}\right)^2}\mathrm{d}x \\ &= \frac{2}{R}\frac{1}{\sqrt{2\pi}\,\sigma}\int_{0}^{\infty} x^k \mathrm{e}^{-\frac{1}{2}\left(\frac{x}{\sigma}\right)^2}\mathrm{d}x.\end{aligned} \tag{C.16}$$

After substitution using $t = \frac{1}{2}\left(\frac{x}{\sigma}\right)^2$, considering (3.49) and (2.22), one gets

$$\begin{aligned}\left|\mathrm{E}\left\{\frac{x^k}{R}\right\}\right| &\le \frac{\sigma}{R}\,\frac{(\sqrt{2})^k}{\sqrt{\pi}}\,\sigma^{k-1}\int_{0}^{\infty} t^{\frac{k+1}{2}-1}\,\mathrm{e}^{-t}\mathrm{d}t \\ &= \frac{\sigma^*}{N}\left(\frac{R}{N}\right)^{k-1}\sigma^{*k-1}\frac{(\sqrt{2})^k}{\sqrt{\pi}}\,\Gamma\left(\frac{k+1}{2}\right).\end{aligned} \tag{C.17}$$

Obviously, $\overline{x/R} = \mathcal{O}(1/N)$ and the moments for $k > 1$ give quantities of order $\mathcal{O}(1/N^k)$. Hence, the random variable x/R is of the order $1/N$

$$\frac{x}{R} = \mathcal{O}\left(\frac{1}{N}\right). \tag{C.18}$$

C.3 Product Moments of Correlated Gaussian Mutations

C.3.1 Fundamental Relations

Definition. The product moments of an N-dimensional distribution of a random vector $\mathbf{x} = (x_1, x_2, \ldots, x_N)^{\mathrm{T}}$ with density $p(\mathbf{x})$ are defined by the integral

$$\overline{x_{k_1}^{r_1} x_{k_2}^{r_2} \cdots x_{k_N}^{r_N}} := \int\!\!\int \cdots \int x_{k_1}^{r_1} x_{k_2}^{r_2} \cdots x_{k_N}^{r_N} \, p\left(x_1, x_2, \ldots, x_N\right) \mathrm{d}^N x. \quad \text{(C.19)}$$

The symbol $\mathrm{d}^N x = \mathrm{d}x_1 \mathrm{d}x_2 \cdots \mathrm{d}x_N$ stands for the N-dimensional volume element.

In this appendix, the moments of the correlated normal distribution

$$p(\mathbf{x}) = \frac{1}{\left(\sqrt{2\pi}\right)^N} \frac{1}{\sqrt{\det[\mathbf{C}]}} \exp\left(-\frac{1}{2} \mathbf{x}^{\mathrm{T}} \mathbf{C}^{-1} \mathbf{x}\right) \quad \text{(C.20)}$$

with the correlation matrix $\mathbf{C} = \mathbf{C}^{\mathrm{T}}$, also referred to as the covariance matrix, will be calculated. The approach uses the following characteristic function, which is defined as a generalization of definition (B.15a):

$$\varphi(\mathbf{t}) := \mathrm{E}\left\{ \mathrm{e}^{\imath \mathbf{t}^{\mathrm{T}} \mathbf{x}} \right\} = \int_{-\infty}^{\infty} \int_{-\infty}^{\infty} \cdots \int_{-\infty}^{\infty} \mathrm{e}^{\imath \mathbf{t}^{\mathrm{T}} \mathbf{x}} p(\mathbf{x}) \, \mathrm{d}^N x, \quad \text{(C.21)}$$

where $\mathbf{t}$ is an N-dimensional t vector, and $\mathbf{t}^{\mathrm{T}} \mathbf{x}$ is the usual scalar product in vector calculus.

Generalizing (B.16), one easily obtains the product moments by the partial differentiation of $\varphi(\mathbf{t})$ at $\mathbf{t} = \mathbf{0}$

$$\overline{x_{k_1}^{r_1} x_{k_2}^{r_2} \cdots x_{k_N}^{r_N}} = \frac{1}{\imath^{(r_1 + r_2 + \ldots r_N)}} \frac{\partial^{(r_1 + r_2 + \ldots r_N)}}{\partial t_{k_1}^{r_1} \partial t_{k_2}^{r_2} \cdots \partial t_{k_N}^{r_N}} \varphi(\mathbf{t}) \Bigg|_{\mathbf{t} = \mathbf{0}} . \quad \text{(C.22)}$$

The characteristic function of the correlated Gaussian mutations. Using the definition (C.21), the density (C.20) gives the integral

$$\varphi(\mathbf{t}) = \frac{1}{\left(\sqrt{2\pi}\right)^N} \frac{1}{\sqrt{\det[\mathbf{C}]}} \int \cdots \int \exp\left(-\frac{1}{2} \mathbf{x}^{\mathrm{T}} \mathbf{C}^{-1} \mathbf{x} + \imath \mathbf{t}^{\mathrm{T}} \mathbf{x}\right) \mathrm{d}^N x. \quad \text{(C.23)}$$

The exponent in (C.23) can be rewritten as

$$\mathbf{x}^{\mathrm{T}} \mathbf{C}^{-1} \mathbf{x} - 2\imath \mathbf{t}^{\mathrm{T}} \mathbf{x} = (\mathbf{x} - \imath \mathbf{C} \mathbf{t})^{\mathrm{T}} \mathbf{C}^{-1} (\mathbf{x} - \imath \mathbf{C} \mathbf{t}) \ + \ \mathbf{t}^{\mathrm{T}} \mathbf{C} \mathbf{t} \quad \text{(C.24)}$$

using quadratic completion. The integration thereafter yields the desired characteristic function

$$\varphi(\mathbf{t}) = \exp\left(-\frac{1}{2} \mathbf{t}^{\mathrm{T}} \mathbf{C} \mathbf{t}\right) = \exp\left[-\frac{1}{2}\left(\sum_{a,b} t_a t_b C_{ab}\right)\right], \quad \text{(C.25)}$$

where $C_{ab} := (\mathbf{C})_{ab}$ is the component expression of the matrix $\mathbf{C}$. All indices run between 1 and N, and the symmetry $C_{ab} = C_{ba}$ is valid for all components.

C.3.2 Derivation of the Product Moments

Applying (C.22) to the characteristic function (C.25), one can calculate step by step the product moments of the correlated Gaussian mutations (see, e.g., Miller, 1964).

For the first moments, using

$$\frac{\partial \varphi(\mathbf{t})}{\partial t_i} = \varphi(\mathbf{t}) \left(-\frac{1}{2}\right) \sum_{a,b} (\delta_{ai} t_b C_{ab} + \delta_{bi} t_a C_{ab}) = -\varphi(\mathbf{t})\, c_i(\mathbf{t}) \tag{C.26}$$

with

$$c_i(\mathbf{t}) := \sum_a C_{ia} t_a, \tag{C.27}$$

one obtains the result

$$\overline{x_i} = \frac{1}{\imath} \left.\frac{\partial \varphi}{\partial t_i}\right|_{\mathbf{t}=\mathbf{0}} = 0. \tag{C.28}$$

The symmetry of C_{ab} was used here. The result is immediately clear, since a $\mathcal{N}(\mathbf{0}, \mathbf{C})$ normal distribution is of concern. The application of the $\partial/\partial t_j$ operator to (C.26), taking

$$\frac{\partial c_i(\mathbf{t})}{\partial t_j} = \sum_a C_{ia} \delta_{aj} = C_{ij} \tag{C.29}$$

into account, yields

$$\frac{\partial^2 \varphi}{\partial t_i\, \partial t_j} = \varphi(\mathbf{t})\, (c_i c_j - C_{ij}). \tag{C.30}$$

Consequently, from (C.22) we get for the product moments

$$\overline{x_i x_j} = - \left.\frac{\partial^2 \varphi}{\partial t_i\, \partial t_j}\right|_{\mathbf{t}=\mathbf{0}} = C_{ij}. \tag{C.31}$$

The application of $\partial/\partial t_k$ to (C.30) gives

$$\frac{\partial^3 \varphi}{\partial t_i\, \partial t_j\, \partial t_k} = -\varphi(\mathbf{t})\, (c_i c_j c_k - c_i C_{jk} - c_j C_{ik} - c_k C_{ij}), \tag{C.32}$$

and, as a result,

$$\overline{x_i x_j x_k} = \imath \left.\frac{\partial^3 \varphi}{\partial t_i\, \partial t_j\, \partial t_k}\right|_{\mathbf{t}=\mathbf{0}} = 0. \tag{C.33}$$

Applying $\partial/\partial t_l$ to (C.32), one obtains

$$\begin{aligned} \frac{\partial^4 \varphi}{\partial t_i\, \partial t_j\, \partial t_k\, \partial t_l} = \varphi(\mathbf{t})\, \big(& c_i c_j c_k c_l - c_i c_l C_{jk} - c_j c_l C_{ik} - c_k c_l C_{ij} - c_j c_k C_{il} \\ & - c_i c_k C_{jl} - c_i c_j C_{kl} + C_{il} C_{jk} + C_{jl} C_{ik} + C_{kl} C_{ij}\big), \end{aligned} \tag{C.34}$$

and the product moments are obtained

$$\overline{x_i x_j x_k x_l} = \left.\frac{\partial^4 \varphi}{\partial t_i\, \partial t_j\, \partial t_k\, \partial t_l}\right|_{\mathbf{t}=\mathbf{0}} = C_{il}C_{jk} + C_{jl}C_{ik} + C_{kl}C_{ij}. \tag{C.35}$$

The product moments of higher orders are obtained in a completely analogous manner. Only the results will be given here due to the enormous effort of writing the derivation:

$$\begin{aligned}
\overline{x_i x_j x_k x_l x_m} &= 0, \\
\overline{x_i x_j x_k x_l x_m x_n} &= \quad C_{in}\,(C_{jk}C_{lm} + C_{jl}C_{km} + C_{jm}C_{kl}) \\
&\quad + C_{jn}\,(C_{ik}C_{lm} + C_{il}C_{km} + C_{im}C_{kl}) \\
&\quad + C_{kn}\,(C_{ij}C_{lm} + C_{il}C_{jm} + C_{im}C_{jl}) \\
&\quad + C_{ln}\,(C_{ij}C_{km} + C_{ik}C_{jm} + C_{im}C_{jk}) \\
&\quad + C_{mn}(C_{ij}C_{kl} + C_{ik}C_{jl} + C_{il}C_{jk}), \\
\overline{x_i x_j x_k x_l x_m x_n x_o} &= 0.
\end{aligned} \tag{C.36}$$

Here, the law of composing higher product moments,

$$\overline{x_{k_1} x_{k_2} \cdots x_{k_\Gamma}} = \overline{\prod_{\gamma=1}^{\Gamma} x_{k_\gamma}}, \tag{C.37}$$

should be clear: for an odd number Γ of factors, the moments are always zero. The moments with an even number of factors yield always a combination of C_{ab} matrices. Comparing (C.35) with (C.36), one recognizes the general building law

$$\overline{x_{k_1} x_{k_2} \cdots x_{k_\Gamma}} = \overline{\prod_{\gamma=1}^{\Gamma} x_{k_\gamma}} = \begin{cases} 0, & \text{if } \Gamma \text{ odd} \\ \displaystyle\sum_{\gamma=1}^{\Gamma-1} C_{k_\gamma k_\Gamma} \overline{\prod_{\substack{\alpha\neq\gamma \\ \alpha\neq\Gamma}} x_{k_\alpha}}, & \text{if } \Gamma \text{ even.} \end{cases} \tag{C.38}$$

D. $(1,\lambda)$-Progress Coefficients

D.1 Asymptotics of the Progress Coefficients $d_{1,\lambda}^{(k)}$

D.1.1 An Asymptotic Expansion for the $d_{1,\lambda}^{(k)}$ Coefficients

An expansion method for asymptotic series expressions of the progress coefficients $d_{1,\lambda}^{(k)}$ is demonstrated using the progress coefficients $c_{1,\lambda}$ and $d_{1,\lambda}^{(2)}$ as examples.

For the derivation of the series expansion, the $d_{1,\lambda}^{(k)}$ definition in (4.41) is transformed by the substitution $z = \Phi(t)$ to the integral

$$d_{1,\lambda}^{(k)} = \lambda \int_0^1 \left[\Phi^{-1}(z)\right]^k z^{\lambda-1} \mathrm{d}z. \tag{D.1}$$

This integral can be interpreted as an expected value integral of the function $f(z) = [\Phi^{-1}(z)]^k$ of a random variable z obeying the probability density $p(z) = \lambda z^{\lambda-1}$:

$$\overline{f} = \mathrm{E}\{f(z)\} = \int_0^1 f(z)\, p(z)\, \mathrm{d}z \qquad \text{and} \qquad \overline{z} = 1 - \frac{1}{\lambda+1}. \tag{D.2}$$

It is self-evident that we can expand $f(z)$ at $z = \mathrm{E}\{z\} = \overline{z}$

$$\overline{f} = \sum_{m=0}^{\infty} \frac{1}{m!} \left.\frac{d^m f}{dz^m}\right|_{z=\overline{z}} I_m^{(1)}(\lambda) \tag{D.3}$$

with

$$I_m^{(1)}(\lambda) := \lambda \int_0^1 \left(z - \frac{\lambda}{\lambda+1}\right)^m z^{\lambda-1} \mathrm{d}z. \tag{D.4}$$

The integrals I_m occurring in (D.4) can be calculated easily for small m; one obtains for the first four

$$\begin{aligned} I_0^{(1)}(\lambda) &= 1, & I_1^{(2)}(\lambda) &= 0, \\ I_2^{(1)}(\lambda) &= \frac{\lambda}{(\lambda+1)^2(\lambda+2)}, & I_3^{(1)}(\lambda) &= -\frac{2\lambda\,(\lambda-1)}{(\lambda+1)^3(\lambda+2)(\lambda+3)}. \end{aligned} \tag{D.5}$$

One can calculate the expected value $\overline{f}$ using the derivatives of $f(z)$.

For $c_{1,\lambda}$, $f(z) = \Phi^{-1}(z)$ holds because of Eq. (4.42). Consequently, using the expansion (D.3) up to and including the term with $m = 2$, and considering (C.14) one obtains

$$c_{1,\lambda} = \Phi^{-1}\left(1 - \frac{1}{\lambda+1}\right)\left[1 + \frac{1}{2}\frac{\lambda}{\lambda+2} \times \left\{\frac{\sqrt{2\pi}}{\lambda+1}\exp\left[\frac{1}{2}\left(\Phi^{-1}\left(1 - \frac{1}{\lambda+1}\right)\right)^2\right]\right\}^2\right] + \ldots \tag{D.6}$$

For $d^{(2)}_{1,\lambda}$, one has $f(z) = [\Phi^{-1}(z)]^2$. The second derivative of this function is obtained using (C.13) and (C.14)

$$\frac{\mathrm{d}^2 f}{\mathrm{d}z^2} = 2\left\{\sqrt{2\pi}\exp\left[\frac{1}{2}\left(\Phi^{-1}(z)\right)^2\right]\right\}^2\left[\left(\Phi^{-1}(z)\right)^2 + 1\right]. \tag{D.7}$$

Therefore, using the terms up to and including $m = 2$, it follows for $d^{(2)}_{1,\lambda}$ that

$$d^{(2)}_{1,\lambda} = \left[\Phi^{-1}\left(1 - \frac{1}{\lambda+1}\right)\right]^2\left[1 + \frac{\lambda}{\lambda+2} \times \left\{\frac{\sqrt{2\pi}}{\lambda+1}\exp\left[\frac{1}{2}\left(\Phi^{-1}\left(1 - \frac{1}{\lambda+1}\right)\right)^2\right]\right\}^2\right] + \frac{\lambda}{\lambda+2}\left\{\frac{\sqrt{2\pi}}{\lambda+1}\exp\left[\frac{1}{2}\left(\Phi^{-1}\left(1 - \frac{1}{\lambda+1}\right)\right)^2\right]\right\}^2 + \ldots. \tag{D.8}$$

D.1.2 The Asymptotic $c_{1,\lambda}$ and $d^{(2)}_{1,\lambda}$ Formulae

The expansion (D.6) and (D.8) yield for $\lambda \to \infty$ the asymptotically *exact* formulae

$$c_{1,\lambda} \simeq \Phi^{-1}\left(1 - \frac{1}{\lambda+1}\right) \quad \text{and} \quad d^{(2)}_{1,\lambda} \simeq \left[\Phi^{-1}\left(1 - \frac{1}{\lambda+1}\right)\right]^2. \tag{D.9}$$

Therefore, it follows that

$$d^{(2)}_{1,\lambda} \simeq c^2_{1,\lambda} \qquad \text{for} \qquad \lambda \to \infty. \tag{D.10}$$

Generally, one finds for $d^{(k)}_{1,\lambda}$

$$d^{(k)}_{1,\lambda} \simeq \left[\Phi^{-1}\left(1 - \frac{1}{\lambda+1}\right)\right]^k. \tag{D.11}$$

In order to prove this fact, one should note that for $\lambda \to \infty$ all further terms in the expansions (D.6), (D.8), etc. vanish. Therefore, it is to be shown that

$$h := \lim_{\lambda \to \infty} \left\{ \frac{\sqrt{2\pi}}{\lambda + 1} \exp\left[\frac{1}{2} \left(\Phi^{-1}\left(1 - \frac{1}{\lambda+1}\right)\right)^2 \right] \left(\Phi^{-1}\left(1 - \frac{1}{\lambda+1}\right)\right)^\alpha \right\}$$
$$= 0 \tag{D.12}$$

is always fulfilled. The expansion terms of higher orders are already considered here; they yield terms with $\alpha \neq 0$: therefore, for α, the inequality $0 \le \alpha < 1$ is valid.

In order to prove (D.12), one substitutes $\Phi^{-1}\left(1 - \frac{1}{\lambda+1}\right) =: z$ and then L' Hospital's rule is applied; it follows that

$$\begin{aligned} h &= \sqrt{2\pi} \lim_{z \to \infty} (1 - \Phi(z))\, e^{\frac{1}{2}z^2} z^\alpha \\ &= \sqrt{2\pi} \lim_{z \to \infty} \frac{\frac{d}{dz}\left[(1 - \Phi(z))\, z^\alpha\right]}{\frac{d}{dz}\left[e^{-\frac{1}{2}z^2}\right]} \\ &= \sqrt{2\pi} \lim_{z \to \infty} \frac{-\frac{z^\alpha}{\sqrt{2\pi}} e^{-\frac{1}{2}z^2} + (1 - \Phi(z))\; \alpha\, z^{\alpha-1}}{-z\, e^{-\frac{1}{2}z^2}} \\ &= \lim_{z \to \infty} \frac{z^\alpha}{z} = 0. \end{aligned} \tag{D.13}$$

D.1.3 An Alternative Derivation for $c_{1,\lambda}$

Instead of (D.6), one can obtain an equivalent integral expression for $c_{1,\lambda}$. To this end Definition (3.100) is transformed into

$$\begin{aligned} c_{1,\lambda} &= \frac{\lambda}{\sqrt{2\pi}} \int_{-\infty}^{\infty} \left(-\frac{d}{dt} e^{-\frac{1}{2}t^2} \right) [\Phi(t)]^{\lambda-1}\, dt \\ &= \frac{\lambda\,(\lambda-1)}{2\pi} \int_{-\infty}^{\infty} e^{-t^2} [\Phi(t)]^{\lambda-2}\, dt \end{aligned} \tag{D.14}$$

and the substitution $\Phi(t) = z$ is applied

$$c_{1,\lambda} = \frac{\lambda\,(\lambda-1)}{\sqrt{2\pi}} \int_0^1 \underbrace{\exp\left[-\frac{1}{2} \left(\Phi^{-1}(z)\right)^2 \right]}_{:= f(z)} z^{\lambda-2} dz. \tag{D.15}$$

Using the derivation method of Sect. D.1.1, one obtains

$$c_{1,\lambda} = \sum_{m=0}^{\infty} \frac{1}{m!} \left. \frac{d^m f}{dz^m} \right|_{z=\overline{z}} \frac{I_m^{(2)}(\lambda)}{\sqrt{2\pi}}, \tag{D.16}$$

$$I_m^{(2)}(\lambda) := \lambda\,(\lambda-1) \int_0^1 (z - \overline{z})^m z^{\lambda-2} dz, \quad \text{with} \quad \overline{z} = 1 - \frac{1}{\lambda}. \tag{D.17}$$

One finds for the first four $I_m^{(2)}(\lambda)$

$$\begin{aligned} I_0^{(2)}(\lambda) &= \lambda, & I_1^{(2)}(\lambda) &= 0, \\ I_2^{(2)}(\lambda) &= \frac{\lambda - 1}{\lambda(\lambda + 1)}, & I_3^{(2)}(\lambda) &= -\frac{2(\lambda - 1)(\lambda - 2)}{\lambda^2(\lambda + 1)(\lambda + 2)}. \end{aligned} \tag{D.18}$$

Considering (C.13), the first derivative of $f(z)$ yields $\frac{\mathrm{d}f}{\mathrm{d}z} = -\sqrt{2\pi}\Phi^{-1}(z)$. Therefore, one finds for the second derivative $\frac{\mathrm{d}^2 f}{\mathrm{d}z^2} = -2\pi \exp\left[\frac{1}{2}(\Phi^{-1}(z))^2\right]$. Using these intermediate results, one gets for (D.16)

$$\begin{aligned} c_{1,\lambda} = {} & \frac{\lambda}{\sqrt{2\pi}} \exp\left[-\frac{1}{2}\left(\Phi^{-1}\left(1 - \frac{1}{\lambda}\right)\right)^2\right] \\ & - \frac{1}{2}\frac{\lambda - 1}{\lambda + 1}\frac{\sqrt{2\pi}}{\lambda} \exp\left[\frac{1}{2}\left(\Phi^{-1}\left(1 - \frac{1}{\lambda}\right)\right)^2\right] + \ldots . \end{aligned} \tag{D.19}$$

If one considers $(\lambda + 1) \sim \lambda$ for the asymptotic $\lambda \to \infty$ case, the second term as well as all further terms in the expansion (D.19) are again of the form (D.12). Hence, one gets the asymptotically exact expression

$$c_{1,\lambda} \simeq \frac{\lambda}{\sqrt{2\pi}} \exp\left[-\frac{1}{2}\left(\Phi^{-1}\left(1 - \frac{1}{\lambda}\right)\right)^2\right]. \tag{D.20}$$

This expression is more complicated than that in (D.9); however, it is used in Chap. 6 in the investigation of the $c_{\mu/\mu,\lambda}$ progress rate coefficients.

D.2 Table of Progress Coefficients of the $(1, \lambda)$-ES

Table D.1. Progress coefficients of the $(1, \lambda)$-ES and related quantities. For the corresponding definitions, see Eqs. (3.100), p. 72, and (4.41), p. 119, as well as the derivations related to Eq. (7.168) on p. 301.

λ	$c_{1,\lambda}$	$d^{(2)}_{1,\lambda}$	$d^{(3)}_{1,\lambda}$	$d^{(4)}_{1,\lambda}$	$d^{(5)}_{1,\lambda}$	$d^{(6)}_{1,\lambda}$	$\sqrt{d^{(2)}_{1,\lambda} - c^2_{1,\lambda}}$	$\varsigma^*_{\psi 0}$
1	0.0000	1.0000	0.0000	3.0000	0.0000	15.000	1.0000	–
2	0.5642	1.0000	1.4105	3.0000	6.0650	15.000	0.8256	0.8862
3	0.8463	1.2757	2.1157	4.1945	9.0976	21.800	0.7480	0.9165
4	1.0294	1.5513	2.7004	5.3891	11.881	28.599	0.7012	1.0213
5	1.1630	1.8000	3.2249	6.5234	14.539	35.227	0.6690	1.1179
6	1.2672	2.0217	3.7053	7.5974	17.095	41.681	0.6449	1.2009
7	1.3522	2.2203	4.1497	8.6170	19.558	47.974	0.6260	1.2722
8	1.4236	2.3995	4.5636	9.5879	21.939	54.117	0.6106	1.3343
9	1.4850	2.5626	4.9512	10.515	24.243	60.119	0.5978	1.3890
10	1.5388	2.7121	5.3158	11.403	26.477	65.990	0.5868	1.4376
12	1.6292	2.9780	5.9866	13.075	30.754	77.370	0.5689	1.5210
14	1.7034	3.2092	6.5928	14.628	34.807	88.310	0.5547	1.5905
16	1.7660	3.4137	7.1464	16.080	38.664	98.858	0.5432	1.6499
18	1.8200	3.5970	7.6565	17.446	42.347	109.05	0.5334	1.7016
20	1.8675	3.7632	8.1298	18.736	45.876	118.92	0.5251	1.7474
25	1.9653	4.1210	9.1843	21.688	54.120	142.36	0.5084	1.8424
30	2.0428	4.4187	10.097	24.326	61.677	164.28	0.4958	1.9183
40	2.1608	4.8969	11.629	28.915	75.215	204.52	0.4775	2.0349
50	2.2491	5.2740	12.892	32.841	87.166	240.97	0.4644	2.1227
60	2.3193	5.5856	13.970	36.292	97.929	274.48	0.4545	2.1928
70	2.3774	5.8512	14.914	39.382	107.76	305.61	0.4465	2.2509
80	2.4268	6.0827	15.755	42.187	116.84	334.77	0.4399	2.3005
90	2.4697	6.2880	16.514	44.762	125.28	362.26	0.4343	2.3436
100	2.5076	6.4724	17.207	47.145	133.20	388.32	0.4294	2.3817
150	2.6492	7.1883	19.991	57.028	166.99	502.33	0.4121	2.5246
200	2.7460	7.7015	22.077	64.733	194.31	597.55	0.4009	2.6225
300	2.8778	8.4310	25.164	76.580	237.80	754.07	0.3865	2.7559
400	2.9682	8.9524	27.457	85.693	272.37	882.31	0.3772	2.8477
500	3.0367	9.3587	29.291	93.169	301.40	992.36	0.3704	2.9172
600	3.0917	9.6919	30.826	99.544	326.59	1089.5	0.3651	2.9731
700	3.1375	9.9744	32.148	105.12	348.96	1176.9	0.3608	3.0197
800	3.1768	10.220	33.311	110.09	369.13	1256.6	0.3572	3.0596
900	3.2111	10.436	34.351	114.58	387.55	1330.1	0.3541	3.0944
1000	3.2414	10.630	35.292	118.68	404.53	1398.5	0.3514	3.1253
2000	3.4353	11.914	41.729	147.69	528.49	1913.1	0.3349	3.3224
3000	3.5444	12.669	45.687	166.30	611.26	2270.1	0.3263	3.4333
4000	3.6199	13.207	48.578	180.23	674.78	2550.5	0.3206	3.5102
5000	3.6776	13.625	50.868	191.47	726.87	2784.3	0.3164	3.5688
6000	3.7241	13.967	52.768	200.92	771.28	2986.3	0.3130	3.6161
8000	3.7964	14.507	55.820	216.34	844.83	3325.6	0.3079	3.6896
10000	3.8516	14.927	58.232	228.72	904.85	3606.9	0.3042	3.7458

References

Aarts, E., & Korst, J. (1989). *Simulated Annealing and Boltzmann Machines.* Chichester: Wiley.

Abramowitz, M., & Stegun, I. A. (1984). *Pocketbook of Mathematical Functions.* Thun: Harri Deutsch.

Ahlbehrendt, N., & Kempe, V. (1984). *Analyse stochastischer Systeme.* Berlin: Akademie-Verlag.

Arnold, B. C., Balakrishnan, N., & Nagaraja, H. N. (1992). *A First Course in Order Statistics.* New York: Wiley.

Arnold, D. V., & Beyer, H.-G. (2001). Local Performace of the $(\mu/\mu_I, \lambda)$-ES in a Noisy Environment. In W. Martin & W. Spears (Eds.), *Foundations of Genetic Algorithms, 6.* San Mateo, CA: Morgan Kaufmann.

Asoh, H., & Mühlenbein, H. (1994). On the Mean Convergence Time of Evolutionary Algorithms without Selection and Mutation. In Davidor et al., 1994, pp. 88–97.

Bäck, T. (1992). The Interaction of Mutation Rate, Selection, and Self-Adaptation Within a Genetic Algorithm. In Männer & Manderick, 1992, pp. 85–94.

Bäck, T., Hammel, U., & Schwefel, H.-P. (1997). Evolutionary Computation: Comments on the History and Current State. *IEEE Trans. Evolutionary Computation, 1*(1), 3–17.

Bäck, T., & Schwefel, H.-P. (1993). An Overview of Evolutionary Algorithms for Parameter Optimization. *Evolutionary Computation, 1*(1), 1–23.

Banzhaf, W., Daida, J., Eiben, A., Garzon, M., Honavar, V., Jakiela, M., & Smith, R. (Eds.). (1999). *GECCO-99: Proc. Genetic and Evolutionary Computation Conf.* San Francisco, CA: Morgan Kaufmann.

Berg, L. (1968). *Asymptotische Darstellungen und Entwicklungen.* Berlin: VEB Deutscher Verlag der Wissenschaften.

Bernstein, H., Hopf, F. A., & Michod, R. E. (1987). The Molecular Basis of the Evolution of Sex. *Advances in Genetics, 24*, 323–370.

Beyer, H.-G. (1988). Ein Evolutionsverfahren zur mathematischen Modellierung stationärer dynamischer Systeme. In 10. ibausil-*Tagungsbericht, Sektion 4, Glas* (pp. 175–180). Weimar: Hochschule für Architektur und Bauwesen.

Beyer, H.-G. (1989a). *Ein Evolutionsverfahren zur mathematischen Modellierung stationärer Zustände in dynamischen Systemen.* Dissertation, Hochschule für Architektur und Bauwesen, Weimar. (Reihe: HAB-Dissertationen, No. 16)

Beyer, H.-G. (1989b). Evolutionsverfahren – Nutzung des Darwinschen Paradigmas zur Feldberechnung. *Wissenschaftliche Zeitschrift der Hochschule für Verkehrswesen Dresden, 51*(Sonderheft), 17–40. (Dresden, Germany)

Beyer, H.-G. (1990a). On a General Evolution Strategy for Dissipative Systems. In H.-M. Voigt, H. Mühlenbein, & H.-P. Schwefel (Eds.), *Evolution and Optimization '89* (pp. 69–78). Berlin: Akademie-Verlag.

Beyer, H.-G. (1990b). Simulation of Steady States in Dissipative Systems by Darwin's Paradigm of Evolution. *J. Non-Equilib. Thermodyn., 15*, 45–58.

Beyer, H.-G. (1992a). Some Aspects of the 'Evolution Strategy' for Solving TSP-like Optimization Problems. In Männer & Manderick, 1992, pp. 361–370.

Beyer, H.-G. (1992b). *Towards a Theory of 'Evolution Strategies'. Some Asymptotical Results from the* $(1 \overset{+}{,} \lambda)$*-Theory* (Tech. Rep. No. SYS-5/92). University of Dortmund: Department of Computer Science.

Beyer, H.-G. (1993). Toward a Theory of Evolution Strategies: Some Asymptotical Results from the $(1 \overset{+}{,} \lambda)$-Theory. *Evolutionary Computation, 1*(2), 165–188.

Beyer, H.-G. (1994a). Towards a Theory of 'Evolution Strategies': Progress Rates and Quality Gain for $(1 \overset{+}{,} \lambda)$-Strategies on (Nearly) Arbitrary Fitness Functions. In Davidor et al., 1994, pp. 58–67.

Beyer, H.-G. (1994b). *Towards a Theory of 'Evolution Strategies': Results from the* N*-dependent* (μ, λ) *and the Multi-Recombinant* $(\mu/\mu, \lambda)$ *Theory* (Tech. Rep. No. SYS-5/94). University of Dortmund: Department of Computer Science.

Beyer, H.-G. (1995a). *How GAs do* NOT *Work – Understanding GAs without Schemata and Building Blocks* (Tech. Rep. No. SYS-2/95). University of Dortmund: Department of Computer Science.

Beyer, H.-G. (1995b). Toward a Theory of Evolution Strategies: On the Benefit of Sex – the $(\mu/\mu, \lambda)$-Theory. *Evolutionary Computation, 3*(1), 81–111.

Beyer, H.-G. (1995c). Toward a Theory of Evolution Strategies: The (μ, λ)-Theory. *Evolutionary Computation, 2*(4), 381–407.

Beyer, H.-G. (1995d). *Towards a Theory of 'Evolution Strategies': The* $(1, \lambda)$*-Self-Adaptation* (Tech. Rep. No. SYS-1/95). University of Dortmund: Department of Computer Science.

Beyer, H.-G. (1996). Toward a Theory of Evolution Strategies: Self-Adaptation. *Evolutionary Computation, 3*(3), 311–347.

Beyer, H.-G. (1997). An Alternative Explanation for the Manner in which Genetic Algorithms Operate. *BioSystems, 41*, 1–15.

Beyer, H.-G. (1998). On the "Explorative Power" of ES/EP-like Algorithms. In V. Porto, N. Saravanan, D. Waagen, & A. Eiben (Eds.), *Evolutionary Programming VII: Proc. 7th Annual Conf. on Evolutionary Programming* (pp. 323–334). Heidelberg: Springer.

Beyer, H.-G. (1999). On the Dynamics of EAs without Selection. In W. Banzhaf & C. Reeves (Eds.), *Foundations of Genetic Algorithms, 5* (pp. 5–26). San Mateo, CA: Morgan Kaufmann.

Beyer, H.-G., & Arnold, D. (1999). Fitness Noise and Localization Errors of the Optimum in General Quadratic Fitness Models. In Banzhaf et al., 1999, pp 817–824.

Birge, E. A. (1994). *Bacterial and Bacteriophage Genetics.* New York: Springer.

Bischof, N. (1985). *Das Rätsel Ödipus.* München: Piper.

Blumer, M. (1980). *The Mathematical Theory of Quantitative Genetics.* Oxford: Clarendon Press.

Born, J. (1978). *Evolutionsstrategien zur numerischen Lösung von Adaptationsaufgaben.* Dissertation, Humboldt-Universität, Berlin.

Bremermann, H. (1958). *The Evolution of Intelligence* (ONR Tech. Rep. No. 1, Contract No. 477(17)). University of Washington, Seattle.

Bremermann, H., Rogson, M., & Salaff, S. (1966). Global Properties of Evolution Processes. In H. Pattec, E. Edelsack, L. Fein, & A. Callahan (Eds.), *Natural Automata and Useful Simulations* (pp. 3–41). Washington: Spartan Books.

Bronstein, I., & Semendjajew, K. (1981). *Taschenbuch der Mathematik.* Leipzig: BSB B.G. Teubner.

Christiansen, F., & Feldman, M. (1986). *Population Genetics.* Palo Alto, CA: Blackwell Scientific Publications.

Cornish, E., & Fisher, R. (1937). Moments and Cumulants in the Specification of Distributions. *Review of the International Statistical Institute, 5*, 307–320.

David, H. (1970). *Order Statistics.* New York: Wiley.

Davidor, Y., Männer, R., & Schwefel, H.-P. (Eds.). (1994). *Parallel Problem Solving from Nature, PPSN 3* (Lecture Notes in Computer Science, Vol. 866). Heidelberg: Springer.

Deb, K., & Beyer, H.-G. (1999). Self-Adaptation in Real-Parameter Genetic Algorithms with Simulated Binary Crossover. In Banzhaf et al., 1999, pp. 172–179.

DeJong, K. (1975). *An Analysis of the Behavior of a Class of Genetic Adaptive Systems.* Ph.D. Thesis, University of Michigan.

DeJong, K. (1992). Are Genetic Algorithms Function Optimizers? In Männer & Manderick, 1992, pp. 3–13.

Droste, S., Jansen, T., & Wegener, I. (1998). A Rigorous Complexity Analysis of the (1 + 1) Evolutionary Algorithm for Separable Functions with Boolean Inputs. *Evolutionary Computation*, *6*(2), 185–196.

Eiben, A., Raué, P.-E., & Ruttkay, Z. (1994). Genetic Algorithms with Multi-Parent Recombination. In Davidor et al., 1994, pp. 78–87.

Fisz, M. (1971). *Wahrscheinlichkeitsrechnung und mathematische Statistik.* Berlin: VEB Deutscher Verlag der Wissenschaften.

Fogel, D. (1992). *Evolving Artificial Intelligence.* Ph.D. Thesis, University of California, San Diego.

Fogel, D., & Ghozeil, A. (1996). Using Fitness Distributions to Design More Efficient Evolutionary Computations. In *Proc. 1996 IEEE Int. Conf. on Evolutionary Computation (ICEC'96)* (pp. 11–19). New York: IEEE Press.

Fogel, D., & Ghozeil, A. (1997). A Note on Representations and Variation Operators. *IEEE Trans. Evolutionary Computation*, *1*(2), 159–161.

Fogel, D., & Stayton, L. (1994). On the Effectiveness of Crossover in Simulated Evolutionary Optimization. *BioSystems*, *32*, 171–182.

Fogel, D. B., & Beyer, H.-G. (1996). A Note on the Empirical Evaluation of Intermediate Recombination. *Evolutionary Computation*, *3*(4), 491–495.

Fogel, L., Owens, A., & Walsh, M. (1966). *Artificial Intelligence through Simulated Evolution.* New York: Wiley.

Fraser, A. (1957). Simulation of Genetic Systems by Automatic Digital Computers. I. Introduction. *Australian J. Biological Sciences*, *10*, 484–491.

Gnedenko, B., & Kolmogorov, A. (1968). *Limit Distributions for Sums of Independent Random Variables.* Reading, MA: Addison-Wesley.

Goldberg, D. (1989). *Genetic Algorithms in Search, Optimization, and Machine Learning.* Reading, MA: Addison-Wesley.

Goldberg, D., & Segrest, P. (1987). Finite Markov Chain Analysis of Genetic Algorithms. In J. Grefenstette (Ed.), *Genetic Algorithms and Their Applications: Proc. 2nd Int. Conf. on Genetic Algorithms* (pp. 1–8). Hillsdale, NJ: Lawrence Erlbaum Associates.

Hansen, N., & Ostermeier, A. (1996). Adapting Arbitrary Normal Mutation Distributions in Evolution Strategies: The Covariance Matrix Adaptation. In *Proc. 1996 IEEE Int. Conf. on Evolutionary Computation (ICEC'96)* (pp. 312–317). New York: IEEE Press.

Hansen, N., & Ostermeier, A. (1997). Convergence Properties of Evolution Strategies with the Derandomized Covariance Matrix Adaptation: The $(\mu/\mu_I, \lambda)$-CMA-ES. In H.-J. Zimmermann (Ed.), *5th European Congress on Intelligent Techniques and Soft Computing (EUFIT'97)* (pp. 650–654). Aachen, Germany: Verlag Mainz.

Hansen, N., Ostermeier, A., & Gawelczyk, A. (1995). On the Adaptation of Arbitrary Normal Mutation Distributions in Evolution Strategies:

The Generating Set Adaptation. In L. Eshelman (Ed.), *Proc. 6th Int. Conf. on Genetic Algorithms* (pp. 57–64). San Francisco, CA: Morgan Kaufmann.

Herdy, M. (1991). Application of the 'Evolutionsstrategie' to Discrete Optimization Problems. In Schwefel & Männer, 1991, pp. 188–192.

Herdy, M. (1992). Reproductive Isolation as Strategy Parameter in Hierarchically Organized Evolution Strategies. In Männer & Manderick, 1992, pp. 207–217.

Holland, J. (1975). *Adaptation in Natural and Artificial Systems.* Ann Arbor: The University of Michigan Press.

Jansen, T., & Wegener, I. (1999). On the Analysis of Evolutionary Algorithms – A Proof that Crossover Really Can Help. In J. Nesetril (Ed.), *Proc. 7th Annual European Symposium on Algorithms (ESA'99)* (Vol. 1643, pp. 184–193). Berlin: Springer.

Jaynes, E. (1979). Where Do We Stand on Maximum Entropy? In R. Levine & M. Tribus (Eds.), *The Maximum Entropy Formalism* (pp. 15–118). Cambridge, MA: MIT Press

Johnson, N., & Kotz, S. (1970). *Continuous Univariate Distributions-1.* Boston: Houghton Mifflin.

Kirkpatrick, S., Gelatt Jr., C., & Vecchi, M. (1983). Optimization by Simulated Annealing. *Science, 220*, 671–680.

Kreyszig, E. (1968). *Differentialgeometrie.* Leipzig: AVG Geest & Portig.

Laarhoven, P. van, & Aarts, E. (1987). *Simulated Annealing: Theory and Applications.* Dordrecht, The Netherlands: D. Reidel.

Lenk, R., & Gellert, W. (Eds.). (1972). *Brockhaus ABC Physik.* Leipzig: VEB F.A. Brockhaus.

Lin, S., & Kernighan, B. (1973). An Effective Heuristic Algorithm for the Traveling Salesman Problem. *Oper. Res., 21*, 498–516.

Männer, R., & Manderick, B. (Eds.). (1992). *Parallel Problem Solving from Nature, PPSN 2.* Amsterdam: North Holland.

Margulis, L., & Sagan, D. (1988). The Cannibalistic Legacy of Primordial Androgynes. In R. Belling & G. Stevens (Eds.), *Nobel Conf. XXIII (1987): The Evolution of Sex* (pp. 23–40). San Francisco: Harper and Row.

Miller, B., & Goldberg, D. (1997). Genetic Algorithms, Selection Schemes, and the Varying Effects of Noise. *Evolutionary Computation, 4(2)*, 113–131.

Miller, K. (1964). *Multidimensional Gaussian Distributions.* New York: Wiley.

Mühlenbein, H. (1992). How Genetic Algorithms Really Work I: Mutation and Hillclimbing. In Männer & Manderick, 1992, pp. 15–25.

Mühlenbein, H., & Schlierkamp-Voosen, D. (1993). Predictive Models for the Breeder Genetic Algorithm. *Evolutionary Computation, 1*(1), 25–49.

Mühlenbein, H., & Schlierkamp-Voosen, D. (1994). The Science of Breeding and its Application to the Breeder Genetic Algorithm BGA. *Evolutionary Computation*, *1*(4), 335–360.

Muth, C. (1982). Einführung in die Evolutionsstrategie. *Regelungstechnik*, *30*(9), 297–303.

Nürnberg, H.-T., & Beyer, H.-G. (1997). The Dynamics of Evolution Strategies in the Optimization of Traveling Salesman Problems. In P. Angeline, R. Reynolds, J. McDonnell, & R. Eberhart (Eds.), *Evolutionary Programming VI: Proc. 6th Annual Conf. on Evolutionary Programming* (pp. 349–359). Heidelberg: Springer.

Ostermeier, A., Gawelczyk, A., & Hansen, N. (1994). Step-Size Adaptation Based on Non-Local Use of Selection Information. In Davidor et al., 1994, pp. 189–198.

Ostermeier, A., Gawelczyk, A., & Hansen, N. (1995). A Derandomized Approach to Self-Adaptation of Evolution Strategies. *Evolutionary Computation*, *2*(4), 369–380.

Ott, K. (1993). *Einfluß stochastischer Störungen auf das Konvergenzverhalten von Evolutionsstrategien.* Diploma thesis, University of Dortmund, Department of Computer Science.

Oyman, A. (1999). *Convergence Behavior of Evolution Strategies on Ridge Functions.* Ph.D. Thesis, University of Dortmund, Department of Computer Science.

Oyman, A. I., & Beyer, H.-G. (2000). Analysis of the $(\mu/\mu, \lambda)$-ES on the Parabolic Ridge. *Evolutionary Computation*, *8*(3), 267–289.

Oyman, A. I., Beyer, H.-G., & Schwefel, H.-P. (1998). Where Elitists Start Limping: Evolution Strategies at Ridge Functions. In A. E. Eiben, T. Bäck, M. Schoenauer, & H.-P. Schwefel (Eds.), *Parallel Problem Solving from Nature, PPSN 5* (pp. 34–43). Heidelberg: Springer.

Oyman, A. I., Beyer, H.-G., & Schwefel, H.-P. (2000). Analysis of a Simple ES on the "Parabolic Ridge". *Evolutionary Computation*, *8*(3), 249–265.

Prügel-Bennett, A., & Shapiro, J. (1994). An Analysis of Genetic Algorithms using Statistical Mechanics. *Phys. Rev. Lett.*, *72*(9), 1305.

Prügel-Bennett, A., & Shapiro, J. (1997). The Dynamics of a Genetic Algorithm for Simple Random Ising Systems. *Physica D*, *104*, 75–114.

Rappl, G. (1984). *Konvergenzraten von Random Search Verfahren zur globalen Optimierung.* Dissertation, HSBw München, Germany.

Rappl, G. (1989). On Linear Convergence of a Class of Random Search Algorithms. *Zeitschrift f. angew. Math. Mech. (ZAMM)*, *69*(1), 37–45.

Rattray, M., & Shapiro, J. (1997). Noisy Fitness Evaluation in Genetic Algorithms and the Dynamics of Learning. In R. Belew & M. Vose (Eds.), *Foundations of Genetic Algorithms, 4.* San Mateo, CA: Morgan Kaufmann.

Rechenberg, I. (1965). Cybernetic Solution Path of an Experimental Problem. *Royal Aircraft Establishment, Farnborough*, Library Translation 1122.

Rechenberg, I. (1973). *Evolutionsstrategie: Optimierung technischer Systeme nach Prinzipien der biologischen Evolution.* Stuttgart: Frommann-Holzboog.

Rechenberg, I. (1978). Evolutionsstrategien. In B. Schneider & U. Ranft (Eds.), *Simulationsmethoden in der Medizin und Biologie* (pp. 83–114). Berlin: Springer.

Rechenberg, I. (1984). The Evolution Strategy. A Mathematical Model of Darwinian Evolution. In E. Frehland (Ed.), *Synergetics - From Microscopic to Macroscopic Order* (pp. 122–132). Berlin: Springer.

Rechenberg, I. (1994). *Evolutionsstrategie '94.* Stuttgart: Frommann-Holzboog.

Reed, J., Toombs, R., & Barricelli, N. (1967). Simulation of Biological Evolution and Machine Learning. I. Selection of Self-reproducing Numeric Patterns by Data Processing Machines, Effects of Hereditary Control, Mutation Type and Crossing. *J. Theoretical Biology, 17*, 319–342.

Rosenberg, R. (1967). *Simulation of Genetic Populations with Biochemical Properties.* Ph.D. Thesis, Univ. Michigan, Ann Arbor, MI.

Rudolph, G. (1994). An Evolutionary Algorithm for Integer Programming. In Davidor et al., 1994, pp. 139–148.

Rudolph, G. (1997). *Convergence Properties of Evolutionary Algorithms.* Hamburg: Verlag Dr. Kovač. (Ph.D. Thesis)

Scheel, A. (1985). *Beitrag zur Theorie der Evolutionsstrategie.* Dissertation, Technical University of Berlin, Berlin.

Schwefel, H.-P. (1974). *Adaptive Mechanismen in der biologischen Evolution und ihr Einfluß auf die Evolutionsgeschwindigkeit* (Tech. Rep.). Technical University of Berlin. (Abschlußbericht zum DFG-Vorhaben Re 215/2)

Schwefel, H.-P. (1975). *Evolutionsstrategie und numerische Optimierung.* Dissertation, Technische Universität, Berlin.

Schwefel, H.-P. (1977). *Numerische Optimierung von Computer-Modellen mittels der Evolutionsstrategie.* Basel: Birkhäuser.

Schwefel, H.-P. (1981). *Numerical Optimization of Computer Models.* Chichester: Wiley.

Schwefel, H.-P. (1987). Collective Phenomena in Evolutionary Systems. In P. Checkland & I. Kiss (Eds.), *Problems of Constancy and Change – the Complementarity of Systems Approaches to Complexity* (Vol. 2, pp. 1025–1033). Budapest: Int. Soc. for General System Research.

Schwefel, H.-P. (1995). *Evolution and Optimum Seeking.* New York: Wiley.

Schwefel, H.-P., & Rudolph, G. (1995). Contemporary Evolution Strategies. In F. Morana, A. Moreno, J. Merelo, & P. Chacon (Eds.), *Advances in Artificial Life. 3rd ECAL Proc.* (pp. 893–907). Berlin: Springer.

Shapiro, J., Prügel-Bennett, A., & Rattray, L. (1994). A Statistical Mechanical Formulation of the Dynamics of Genetic Algorithms. *Lecture Notes in Computer Science, Vol. 865* (pp. 17–27). Heidelberg: Springer.

Sorkin, G. (1991). Efficient Simulated Annealing on Fractal Energy Landscapes. *Algorithmica*(6), 367–418.

Strickberger, M. (1988). *Genetik.* München: Carl Hanser.

Syswerda, G. (1989). Uniform Crossover in Genetic Algorithms. In J. Schaffer (Ed.), *Proc. 3rd Int. Conf. on Genetic Algorithms* (pp. 2–9). San Mateo, CA: Morgan Kaufmann.

Thierens, D., & Goldberg, D. (1994). Convergence Models of Genetic Algorithm Selection Schemes. In Davidor et al., 1994, pp. 119–129.

Voigt, H.-M., Born, J., & Treptow, J. (1991). *The Evolution Machine. Manual* (Tech. Rep.). Institute for Informatics and Computing Techniques, Berlin, Rudower Chaussee 5.

Index

$(.)_{m:l}$, *see* order statistics
$(.)_{m;l}$, *see* order statistics
$(\tilde{1} \overset{+}{,} \tilde{\lambda})$-ES, 80
– with repeating measurements, 92
– with scaled inheritance, 102
1/5th rule
– heuristics, 100
– idea, 99
– limits, 101
D_∞, 315
$F(\mathbf{y})$ fitness function, *see* fitness
$Q_\mathbf{y}(\mathbf{x})$ quality function, local fitness, 27, 35
T, 317
$\mathfrak{E}$, 5
$\mathrm{He}_k(x)$, *see* Hermite polynomials
$\Phi(x)$
– asymptotic, 65
– definition, 65, 351
Tr, *see* trace, definition
α angle, 194, 197, 199, 229, 230
β angle, 200, 231
$\mathcal{A}$, 5
$\mathcal{F}$, 4
$\mathcal{S}$, 5
$\mathcal{Y}$, 4
χ distribution, 62
χ^2 distribution, 54
η, *see* fitness efficiency
$\overline{Q}_{1+\lambda}$ formula, 120, 121
$\overline{Q}_{1,\lambda}$ formula, 119
$\overline{\varphi^*(\varsigma^{*(g)})}$, 314
ψ function, *see* SAR-function
ψ_∞, 283, 308
σ control
– 1/5th rule, 98
σ-SA, *see* self-adaptation
σ-self-adaptation
– foundations, 259
$\sigma^* \leftrightarrow p_m$ correspondence, 129, 130
$\sigma^*(g)$, in $(1,\lambda)$-σSA-ES, 314
σ^*_∞, 309
$\sqrt{g}$ rule, 196
τ-learning parameter, 262
– $1/\sqrt{N}$ rule, 303
φ, φ^*, *see* progress rate
φ^*_∞, 314, 322
$\varsigma^*_{\psi_0}$, 281, 324, 363
$\varsigma^*_{\varphi_0}$, 300
$\varsigma^*_{ss} \to \sigma^*_\infty$ shift, 321
ς^*_{ss}, 300
ϑ, *see* selection strength
$c_{1,\lambda}$ coefficient, 72, 74
– alternative integral expression, 361
– approximation, 75, 359
– asymptotes, 76
– expression as a series, 360
– relation to $d^{(2)}_{1,\lambda}$, 360
– table, 75, 363
$c_{\mu,\lambda}$ coefficient
– definition, 188
– table, 189
$c_{\mu/\mu,\lambda}$ coefficient
– expression using double sum, 247
– table, 216
$c_{\mu/\mu,\lambda}$ coefficient
– approximation, 248
– asymptotes, 249
– asymptotic order, 250
– definition, 216
– integral expression, 247
– series expression, 248
$d^{(k)}_{1,\lambda}$ coefficient
– definition, 119
– table, 363
$d^{(k)}_{1,\lambda}(x)$ progress function, 120
$e^{\alpha,\beta}_{\mu,\lambda}$ coefficient, 171
g_A adaptation time, 305, 325
$p(\varsigma^*_\infty)$ Ansatz, 319
$p_\infty(\varsigma^*)$, $p(\varsigma^*_\infty)$, 317
$p_{\mathrm{1n}} \leftrightarrow p_{\mathrm{II}}$ correspondence, 298

Q-matrix, 35

Aarts, E., 10, 17
acceptance probability
– $(\tilde{1} + \tilde{\lambda})$-ES, 83
– $(\tilde{1} \overset{+}{,} \tilde{\lambda})$-ES, 81
– $(\tilde{1} + \tilde{1})$-ES, 58
– $(\tilde{1}, \tilde{\lambda})$-ES, 82
adaptation time
– of mutation strength in $(1, \lambda)$-σSA-ES, 300
algorithms, *see* ES
Arnold, B. C., 70
Arnold, D. V., 47, 256
Asoh, H., 242

Bäck, T., 1, 131
BBH, *see* building block hypothesis
Berg, L., 61
Bernstein, H., 24
beta function
– complete, 247
– incomplete, regularized, 147
Beyer, H.-G., 1, 8, 9, 16, 18, 23, 24, 34, 47, 53, 57, 61, 78, 92, 105, 142, 204, 218, 223, 245–247, 256, 257, 263, 267, 324
Bienert, P., 4
Birge, E. A., 203
Blumer, M. G., 27
Born, J., 15, 19
Bremermann, H. J., 14, 36, 131, 246
building block hypothesis, 12, 21, 218, 222, 245

causality, 19
center of mass, *see* centroid
central limit theorem, 105, 116
– generalization, 347
centroid, 29, 204
Chapman-Kolmogorov equation, 25, 265, 309
characteristic function, 355
– definition, 341
Christiansen, F. B., 242
CMA, *see* covariance matrix adaptation
comparison
– $(\mu/\mu_D, \lambda)$-ES vs. $(\mu/\mu_I, \lambda)$-ES, 238
– $(\mu/\mu_I, \lambda)$-ES vs. (μ, λ)-ES, 218
convergence, 14
– global, 50
– linear order, 16, 255, 306, 314
– – multimodal fitness, 50
– local, *see also* evolution criterion
– sublinear order, 16
– Thales criterion, 90
convexity
– concept, 223
– local, 223
coordinate system, spherical, 194
Cornish, E. A., 349
correlation matrix, 10, 141, 355
correspondence principle, *see* normalization
corridor model, 34
covariance matrix, *see* correlation matrix
– isotropic mutations, 125
covariance matrix adaptation, 258
crossing-over, 218
CSA, 258, *see* cumulative step-size adaptation
cumulants, 118, 341
– definition, 341
– translation invariance, 167
cumulative path-length control, 258
cumulative step-size adaptation, 258

David, H. A., 8, 74, 146
Deb, K., 257
DeJong, K., 2, 8, 14
democracy principle vs. elitist principle, 93
design rules, *see* genetic operators
differential-geometrical model, 38, 41, 46
– applicability, 133
– in multimodal fitness landscapes, 50
discus model, 34
distribution
– excess, 149
– skewness, 149
distribution of the offspring of the (μ, λ)-ES, 144
Droste, S., 36
dynamics, 16, 47
– (μ, λ)-ES experiment, 199
– $(\mu/\mu_I, \lambda)$-ES experiment, 230
– $R(g)$, 48
– $\sigma^*(g)$, 304
– ~, *see also* evolution equation
– as stochastic mapping, 264
– differential equation
– – $(\tilde{1}, \tilde{\lambda})$-ES, 90
– recombination-mutation-~, 237

EA

– basic principles, 20, 240
– time complexity, 50
efficiency, *see* fitness efficiency
EIBEN, A. E., 246
elitist principle, 93
elitist selection, 95, 139
EP, 2, 324
EPP, 20, 138
– $(1+1)$-ES, 68
– $(1,\lambda)$-ES, 73
– (μ,λ)-ES, 189
– $(\mu/\mu_I,\lambda)$-ES, 219
– $(\mu/\rho_I,\lambda)$-ES, 224
– $(\tilde{1},\tilde{\lambda})$-ES, 88, 102
– and surrogate mutations, 234
– exploration behavior, 200
– `OneMax` quality gain, 130
– quadratic fitness model, 132
equation of evolution, *see* evolution equation
ES, 2
– algorithm
– – $(1+1)$-minimization, 52
– – $(1,\lambda)$-σ-self-adaptation, 261
– – (μ,λ)-ES, 144
– – $(\mu/\mu_D,\lambda)$-ES, 232
– – $(\mu/\mu_I,\lambda)$-ES, 205
– – general, 6
– divergence in (,) strategies, 72
– hierarchically organized, 5, 141
– Meta, 5
– nested, 5, 141
– noisy, *see* $(\tilde{1} \overset{+}{,} \tilde{\lambda})$-ES
– – convergence improvement, 92
– – improving convergence, 103
– – repeating measurements, 92
– parallelization, 74, 78, 225, 229, 256
ES dynamics, 50, 90, *see* dynamics, evolution equation
ES experiments
– $(1,\lambda)$-σSA
– – adaptation time, 305
– – progress rate, 315
– – stationary σ^* density, 318
– – stationary performance, 320
– $\varphi^*_{\mu,\lambda}$ from (μ,λ) simulations, 186
– $\varphi^*_{\mu,\lambda}$ from (μ,λ)-ES runs, 185
– $\varphi^*_{\mu/\mu_D,\lambda}$ from $(\mu/\mu_D,\lambda)$-ES runs, 239
– exploration behavior of the $(\mu/\mu_I,\lambda)$-ES, 230
– exploration behavior of the (μ,λ)-ES, 199
– one-generation experiments, 110, 135, 216, 282
ES hardness, 19
ES time complexity
– $(1,\lambda)$-ES, 77
– $(\mu/\mu,\lambda)$-ES, 255
evolution
– macroscopic level, 14, 299
– microscopic level, 14, 17, 141
evolution criterion, 90
– $(1,\lambda)$-σSA-ES
– – necessary, 300
– $(\tilde{1}+\tilde{\lambda})$-ES
– – necessary, 96, 103
– $(\tilde{1},\tilde{\lambda})$-ES
– – necessary, 90, 103
– – sufficient, 90
evolution equation
– R dynamics, 48
– $\sigma^*(g)$ differential equation, 304
– $\sim$, *see also* dynamics
– ς dynamics, 266
– ς^* dynamics, 268
– r dynamics, 265
– r dynamics, normalized, 266
– differential equations of σSA in deterministic approximation, 303
evolution strategy, *see* ES
evolution window, 17, 69
evolvability, 18, 20
excess of distribution, 149
exploration behavior, 201
– (μ,λ)-ES, 199
– $(\mu/\mu_I,\lambda)$-ES, 229
– β angle, 200, 231
– GA, 199

fading memory, 237
FELDMAN, M. B., 242
FISHER, R. A., 3, 349
FISZ, M., 25, 54, 105, 341
fitness
– bilinear form, 35
– biquadratic, 126
– function, 4, 84
– – power of R, 59
– – relative measurement error, 59
– multimodal, 199
– number of necessary evaluations, 256
fitness efficiency
– $(1+\lambda)$-ES, 78
– $(1,\lambda)$-ES, 73
– – order, 77

– (μ, λ)-ES, 190
– $(\mu/\mu, \lambda)$-ES
– – asymptotic, 252
– $(\mu/\mu_I, \lambda)$-ES, 227
– μ-optimal, 227
fitness landscape
– OneMax, 36
– and evolvability, 19
– causality, 19
– corridor, 34, 99
– discus, 34
– flat, 22
– hyperplane, 33
– multimodal
– – selection behavior, 142
– noisy, 36
– parabolic ridge, 141
– quadratic and higher, 35
– sharp ridge, 101
– sphere, 32, 99
fitness measurement
– measurement error
– – absolute, 37, 57, 60
– – relative, 59
– repeating measurements, 92
FOGEL, D. B., 3, 12, 15, 101, 218, 234, 262
FOGEL, L. J., 3
Fokker-Planck equation, 313
FRASER, A. S., 14
fundamental form
– first, 39
– second, 40

GA, 2
– "recombination dogma", 218
– alternative functioning principle, 245
– building block hypothesis, 12, 21, 218, 222, 245
– genetic drift, 242
– myth, 199
– real-coded, 257
gamma function
– complete, 62
– incomplete, 56
Gaussian diffusion law, 196
gene convergence, 14
genetic drift, 14, 200, 241
– fixation time, 242
genetic operators, 6
– design rules, 7–11, 13, 324
– wild growth, 7
genetic repair, 21
– and inbreeding, 224
– hypothesis, 12, 222, 244
– in dominant recombination, 242
– principle and mechanism, 220
GHOZEIL, A., 101, 234
GOLDBERG, D. E., 3, 4, 12, 18, 21, 199, 218, 222, 242, 245
GR, *see* genetic repair
gradient diffusion, 201, 229

HANSEN, N., 141, 258
HERDY, M., 5, 9, 141, 258
Hermite polynomials, 117
– definition, 339
HOLLAND, J. H., 2, 3, 18, 21, 26, 218
hyper-geometric function, 248
hyperplane, 33, 192
– restrictions, 34

inbreeding, 224
– in mating selection, 11
individual, 4
induced order statistics, *see* order statistics, $m;\lambda$ induced
inversion, 9
isolation time, 142

JANSEN, T., 12
JAYNES, E. T., 9

KERNIGHAN, B. W., 9
KIRKPATRICK, S., 3
KORST, J., 10
KREYSZIG, E., 37, 42

LAARHOVEN, P. J. M. VAN, 15, 17
Laplace operator, 47
learning parameter, *see* self-adaptation, learning parameter
learning rate, 281
learning strength, 281
LENK, R., 19
LIN, S., 9
Lin-2-Opt, 9
log-normal distribution
– σ mutation distribution, 262
– as $p(\varsigma^*_\infty)$ candidate, 319
– density $p^*_{\text{ln}}(\varsigma^*)$, 270
– moments, 290

MARGULIS, L., 6
Markovian process, 264
– inhomogeneous, 25
mating selection

– with replacement, 11
maximal performance
– scaling behavior
– – $(\mu/\mu, \lambda)$-ES, 251
mean curvature, 41
mean radius, 43
mean value dynamics
– symbols, 264
Meta-ES, 5, 141
– concept, 258
metric tensor, 39, 43
MILLER, B. L., 4
mimicry experiment, 11
minimum attractor, 19
MISR, 22, 241
– hypothesis, 242
Monte Carlo methods, 18
MÜHLENBEIN, H., 21, 27, 36, 131, 242
multiprocessor systems, *see* ES, parallelization
multirecombination, 13, 203
mutation rate, 11, 128
– optimal for `OneMax`, 130
mutation vector
– decomposition
– – additive, 52
– – multiplicative, 134
– density function, 10
– pdf, 11
mutations, 9
– σ-mutation operator
– – design condition, 301
– asymptotic length, 63
– correlated, 10
– design criteria, 9
– equality of the effects of mutation operators for σ, 298
– isotropic, 10, 324
– log-normal operator, 262
– meta-EP operator, 262
– multiplicative, 259
– normalization, 32
– physical vs. surrogate, 12, 235
– rescaled, 102
– two-point operator, 263

nabla operator, 31
neighborhood, 17
nested ES, 5
noise strength
– absolute, 37
– relative, 92
normal distribution
– as $p(r)$ approximation, 111
– distribution function $\Phi(x)$, 65
– moments, 329
normal progress φ_R, 31, 132, 134
normalization, 32
– correlated mutations, 133
– correspondence principle, 131
– definition for sphere model, 32
– for $\varphi_{\mu,\lambda}$, 145
– for biquadratic fitness, 134
– for noise strength, 59
– for the biquadratic fitness, 134
– in σSA theory, 266, 272
– with isotropic Gaussian mutations, 132
normalized noise strength, 59
NÜRNBERG, H.-T., 9, 16

object parameter, 4
offspring distribution
– (μ, λ)-ES, 146
– – approximation, 147, 149
– – stationary state, 147, 167
open problems, 7, 9, 15, 28, 48, 50, 74, 103, 125, 131, 134, 141, 142, 157, 168, 185, 189, 190, 201, 204, 220, 222, 233, 234, 240, 242, 245, 246, 249, 255, 256, 263, 276, 296, 298, 319, 320, 324–326
operator
– for mutation of σ, 261–263
– mutation, 9
– recombination, 12
– reproduction, 11
– selection, 8
optimization
– combinatorial, 8
– fitness maximization, 113
– global, 199
order statistics
– $1;\lambda$ induced, 271
– λth order statistic, 70, 114
– – derivation, 70
– $k\,l:\lambda$ joint density
– – derivation, 160
– $m:\gamma$-nomenclature, 8
– $m:\lambda$ order statistic
– – derivation, 146
– $m;\lambda$ induced, 210
– $x_{m;\lambda}$ order statistic, 211
– first-order statistic, 69, 105
– induced, 210, 271
OSTERMEIER, A., 141, 258, 325
OTT, K., 36

OYMAN, A. I., 142

parabolic ridge, 28, 48, 141
parallel computers, *see* ES, parallelization
path integrals, 313
performance, 14
performance comparison
– fair, 218
performance measures
– progress rate, 28
– comparison, 137, 141
– normal progress, 31
– quality gain, 26
population, 4
probabilistic algorithms, 15
probability density
– χ distribution, 62
– χ^2 distribution, 54
– log-normal distribution, 262
– normal distribution, 53
product moments, 355
progress coefficient, *see* $c_{1,\lambda}$ coefficient
progress coefficients, 72, 172, 359–361, *see also* $c_{1,\lambda}$
– $c_{\mu,\lambda}$, 188
– $c_{\mu/\mu,\lambda}$, 216, 247
– $d^{(k)}_{1,\lambda}$, *see also* $c_{1,\lambda}$
– analytical formulae, 121
– asymptotic expansion of $d^{(k)}_{1,\lambda}$, 359
– definition, 119
– generalized
– – definition, 171
progress function, 78
– first-order, 78, 79
– kth order, 120
progress rate, 4, 17
– $\varphi^*_{1 \overset{+}{,} \lambda}$
– – integral expression, 71
– φ^*_{1+1}
– – asymptotic, 67
– – integral form, 55
– $\varphi^*_{1+\lambda}$
– – asymptotic, 77
– $\varphi^*_{1,\lambda}$
– – N-dependent, 109, 172
– – asymptotic, 72
– $\varphi_{\mu \overset{+}{,} \lambda}$, 28
– $\varphi^*_{\mu,\lambda}$
– – N-dependent, 184
– – approximations, 188
– $\varphi^*_{\mu/\mu_D,\lambda}$
– – N-dependent, 238
– – asymptotic, 240
– $\varphi^*_{\mu/\mu_I,\lambda}$
– – N-dependent, 215
– – asymptotic, 217
– $\varphi_{\mu/\rho \overset{+}{,} \lambda}$, 29
– $\varphi^*_{\tilde{1}+\tilde{1}}$
– – asymptotic, 97
– $\varphi^*_{\tilde{1}+\tilde{\lambda}}$
– – asymptotic, 94
– $\varphi^*_{\tilde{1},\tilde{\lambda}}$
– – asymptotic, 88
– average, of the $(1,\lambda)$-σSA-ES, 314
– higher-order
– – definition, $\varphi^{(k)}$, 266
– normal progress φ_R, 31, 132
PRÜGEL-BENNETT, A., 4

quadratic “complexity”, 133
quality function
– local $Q_{\mathbf{y}}(\mathbf{x})$, 27, 35
quality gain
– selection response, 27
– advantage of ~, 126
– application spectrum, 122, 126, 131, 142
– isotropic mutations
– – statistical parameters, 125
– correlated mutations
– – statistical parameters, 122
– definition, 26
– $\overline{Q}_{1+\lambda}$, 120
– – approximation formula, 121
– $\overline{Q}_{1,\lambda}$, 119
– – approximation formula, 120
quantile, 115, 349
– approximation, 118

random walk, 241
RAPPL, G., 16
RATTRAY, L. M., 4
RAUÉ, P.-E., 246
RECHENBERG, I., 3–5, 9, 11, 15, 17, 19, 31, 34, 49, 67, 69, 72, 75, 99, 100, 102, 133, 137, 141, 189, 199, 203, 217, 229, 240, 247, 258, 259, 263
recombination
– μ-optimal, 219
– and convexity, 223
– benefit, 12, 222, 241, 244
– dominant, 13
– – (μ/μ_D), 232
– – (μ/ρ_D), 204

-- fixation time, 242
-- transient time, 239
- global discrete, *see* ∼ dominant
- intermediate, 13
-- (μ/μ_I), 204
-- (μ/ρ_I), 204, 225
- kinds of, 13
- recombination operator
-- design criteria, 13
- relation to the $(1+1)$-ES, 252
- similarity extraction, 12, 222, 224, 243, 244, 246
- system conditions, 222
- types, 203
- uniform crossover, 14, 246
REED, J., 14
resampling, 93, 100
residual distance, 91
- $(1, \lambda)$-ES, 91
- $(\tilde{1}, \tilde{\lambda})$-ES, 92
ridge functions, 28, 48, 101, 141
ROGSON, M., 131, 246
ROSENBERG, R. S., 3
RUDOLPH, G., 9, 36, 259
RUTTKAY, Z., 246

SAGAN, D., 6
SALAFF, S., 131, 246
SAR function
- analytic ψ formula, 276
- approximations, 289
-- ψ formula, log-normal operator, 295
-- ψ formula, two-point operator, 298
-- $\psi^{(k)}$ formula, general, 293
- definition, 267
- integral expression, 273
- numerical examples, 275, 280
SCHEEL, A., 264, 302
schema theorem, 218
SCHLIERKAMP-VOOSEN, D., 27, 36
SCHWEFEL, H.-P., 1, 3–5, 8, 9, 35, 100, 101, 134, 203, 258, 259, 262, 263, 303, 326
search behavior
- (μ, λ)-ES, 145
- $(\mu/\mu_I, \lambda)$-ES, 229
SEGREST, P., 242
selection
- $(\overset{+}{,})$, 6
- Boltzmann, 28
- deterministic, 8
selection pressure, 187, 190, 226
- for the (μ, λ)-ES optimal, 191
selection response, 27
selection sharpness, 89
selection strength, 248
- asymptotically optimal for the $(\mu/\mu, \lambda)$-ES, 252
self-adaptation, 258
- $(1, \lambda)$-σSA-algorithm, 261
- σ fluctuations, 325
-- influence of the ς^* variance, 315
-- reduction techniques, 325
- σ^* dynamics, 304
- adaptation time, 305
-- scaling behavior, 305, 317
- and (multi-)recombination, 325
- fluctuations in SAR, 295
- in ES, 3, 5
- in GAs, 3
- learning parameter
-- α, β, 262
-- τ, 262
-- τ scaling rule, 303
-- generation-dependent, 307
-- influence on φ^*, 276
- moving σ-average, 325
- nonoptimality for $\lambda < 4$, 303
- rudimentary by recombination, 13
- stationary $p(\varsigma^*)$ density
-- Ansatz, 319
-- integral equation, 317
self-adaptation response, 26, *see* SAR function
self-consistency condition, 167, 183
semi-causality, 19
semi-invariants, *see* cumulants
SHAPIRO, J. L., 4
sharp ridge, 101
simulated annealing, 3, 19
skewness, 149
SORKIN, G. B., 19
species, 241
sphere model, 32, 48
- deformed, 37
stability card
- $(\tilde{1}, \tilde{\lambda})$-ES, 90
state space $\mathcal{A}$, 5, 16
stationary state, 90, 166, 167, 241, 300
STAYTON, L. C., 3, 12, 218
steady-state, *see* stationary state
step-size factor, 134
strategy parameters
- adaptation of mutation strength, 257
-- 1/5th rule, 98
-- deterministic, 258

- endogenous, 5
- exogenous, 262
- individual life span, 259
- isolation time, 142
- optimal μ choice
-- in the $(\mu/\mu_I, \lambda)$-ES, 225
- optimal choice
-- of λ in $(1, \lambda)$-ES, 73
-- of λ in (μ, λ)-ES, 190
-- of ρ in the $(\mu/\rho_I, \lambda)$-ES, 225
STRICKBERGER, M. W., 241
strong causality, 19
success domain
- local, 39, 46, 140
success probability
- $(1+1)$-ES
-- asymptotic, 68
-- integral form, 55
- $(1+\lambda)$-ES
-- asymptotic, 79
- $(\tilde{1}+\tilde{1})$-ES
-- asymptotic, 97
- $(\tilde{1}+\tilde{\lambda})$-ES
-- asymptotic, 94, 95
- and recombination, 223
- and the optimal selection strength in the $(\mu/\mu, \lambda)$-ES, 254
- in the quality gain analysis, 115
- optimal, 99
summation convention, 38
surrogate mutations
- asymptotic model, 234
- concept, 12, 234
- isotropic density, 235
SYSWERDA, G., 12, 246

tensor
- $b_{\alpha\beta}$, 41, 44
- $g_{\alpha\beta}$, 39, 43, 197
THIERENS, D., 4
time complexity, 16, 50, 77, 255
trace, definition, 46
transient time
- for μ/μ_D recombination, 239
- for σ-self-adaptation, 305
translation invariance, 342, *see also* cumulants
traveling salesman problem, 9, 16, 101
truncation selection, 8
truncation threshold, 27, *see* selection strength

uniformity condition, 134

VOIGT, H.-M., 19

wall-clock-time, 78
WEGENER, I., 12
Wiener process, 196
wild-type, 241
working principles, 18

Natural Computing Series

W.M. Spears: Evolutionary Algorithms. The Role of Mutation and Recombination. XIV, 222 pages, 55 figs., 23 tables. 2000

H.-G. Beyer: The Theory of Evolution Strategies. XIX, 380 pages, 52 figs., 9 tables. 2001

L. Kallel, B. Naudts, A. Rogers (Eds.): Theoretical Aspects of Evolutionary Computing. X, 497 pages. 2001

M. Hirvensalo: Quantum Computing. XI, 189 pages. 2001

M. Amos: Theoretical and Experimental DNA Computation. Approx. 200 pages. 2001

L.F. Landweber, E. Winfree (Eds.): Evolution as Computation. DIMACS Workshop, Princeton, January 1999. Approx. 300 pages. 2001